Preface

The gestation period for space astronomical observatories is very long. The Space Infrared Telescope Facility, SIRTF, has been under development in various forms since the late 1970s. The concept was outlined in the 1980 Decadal Survey (the "Field Report" – *Astronomy and Astrophysics for the 1980s, Volume 1: Report of the Astronomy Survey Committee* 1982). SIRTF was ranked as the highest priority program in the 1990 Astronomy Decadal Survey (the "Bahcall Report" – *The Decade of Discovery in Astronomy and Astrophysics* 1991). SIRTF will launch in 2003.

The Next Generation Space Telescope NGST was the subject of extensive workshops and studies in the late 1980s and early 1990s (see, e.g., *The Next Generation Space Telescope* conference in 1989, and the *Technologies for Large Filled Aperture Telescopes in Space* conference in 1992). It received a major boost from a study in 1995 (the *HST & Beyond* Committee Report), and has now passed a major milestone with the selection of a prime contractor TRW/Ball. The earliest that NGST will be launched is late 2010.

The Hubble Space Telescope has been the most productive scientific space mission of all time. Its unique high resolution imaging and UV capabilities will disappear around 2008–2010 as it ages following its last servicing mission in 2005. As the above examples suggest, we are already late in taking the ground-breaking steps on a successor to HST that could be launched during the first half of the next decade.

The success of HST would suggest that a successor is an obvious choice. However, the very large investment in time, effort, and funding needed to bring about a mission of the scale of HST, SIRTF, or NGST requires us to make the scientific case for a follow-up mission. This scientific case must be compelling, since it will be evaluated with a very critical eye by skeptical scientific committees, NASA HQ senior scientists and managers, budget agencies, and congressional committees.

Two questions clearly play a central role in discussions of an HST successor: What are the major science issues of the next decade? What role would an optical-UV space telescope play in the resolution of those questions?

After considerable discussion with senior scientists at NASA and the Space Telescope Science Institute, the Space Telescope Institute Council recommended to AURA that they sponsor a workshop as the next step in addressing these questions. This workshop would build on the pioneering steps taken at the first workshop on this topic. This workshop, *Ultraviolet-Optical Space Astronomy Beyond HST*, was held in 1998. It led to a recommendation for a Space Ultraviolet/Optical Observatory (SUVO), a 4–8 m UV-optical space telescope. This recommendation is discussed in more detail in the UVOWG committee's report "The Emergence of the Modern Universe: Tracing the Cosmic Web."

The second workshop, *Hubble's Science Legacy: Future Optical-UV Astronomy from Space*, ultimately came to be co-sponsored by AURA, NASA, and ESA and was generously hosted by the University of Chicago. It was held in the Ida Noyes Hall at the University of Chicago from April 2–5, 2002. About 130 people attended. The breadth and quality of the discussion can be seen from the excellent talks that are reproduced in this volume.

The workshop's goal was to look beyond HST at the scientific challenges of the next decade, and then at the types of missions and the technologies needed to reach those goals.

Invited speakers included experts in the science areas along with experts on the requisite technologies and those with experience in missions and projects with overlapping scientific objectives. A special effort was made to include a diverse representation among the speakers, ranging from senior scientists with a broad experience and perspective to young scientists who will be the primary users of these future facilities.

The framework under which the meeting was developed was given in the announcement, which is reproduced here:

Workshop Announcement

In 2010 the Hubble Space Telescope is expected to have completed its two-decade mission. Its enormous scientific legacy will be furthered in part by other missions, including the Next Generation Space Telescope, but the lack of optical-UV capability in space after 2010 will likely have a significant scientific impact. Progress on many key astrophysical problems will need Hubble-like capability – namely diffraction-limited imaging at optical and near-ultraviolet wavelengths and high-throughput UV spectroscopy.

Recognizing the impact of the science produced by HST, and the scientific potential of a larger-aperture successor, AURA, NASA, and ESA are co-sponsoring a workshop to consider the opportunities presented by optical and UV astronomy in space in the next decade. We invite the international astronomy community to consider scientific goals and accomplishments achievable with a large new optical-UV telescope in space, with potentially multiples of Hubble's resolving power and even greater multiples of its collecting area. We encourage new initiatives and challenging questions in scientific areas not capable of being addressed by HST that will be amenable to discovery and study by such a powerful optical-UV telescope.

The most important objective of the workshop is to identify the compelling science questions that remain unanswered, and the corresponding scientific breakthroughs that could be achieved with optical-UV space telescopes. It will also address the likely performance requirements needed to achieve these scientific goals, and identify the technical challenges and hurdles that need to be overcome in the next decade if optical-UV successors to HST are to become a reality in the 2010–2020 time frame.

The scope of the scientific discussion will be very broad, encompassing problems in cosmology, extragalactic astronomy, stellar populations, the ISM/IGM, star and planet formation, stellar astrophysics, and Solar System astronomy, all considered in the context of other ground and space missions producing discoveries in the 2010–2020 time frame. Optical-UV capability was recognized as likely to play a key role for progress in most of these areas. The workshop is not intended to discuss or promote specific telescope or mission concepts, but instead to focus on the scientific promise and rationale for a large optical-UV space telescope, and the range of technical capabilities that might be needed to meet these scientific objectives.

Thanks

The organizers would like to thank all of the participants for their thoughtful contributions, excellent talks, and the extensive discussion. The organizers particularly appreciated the generous support from NASA that allowed us to bring a considerable number of young scientists to the workshop, and the support of ESA and ESO which enabled us to bring many international visitors. A workshop of this breadth and scale needs enthusiastic dedicated people and a lot of effort to make it a success. We particularly appreciated the time and effort spent by the Local Organizing Committee and others in Chicago — Chris Blades (STScI), Colleen Duke (Adler Planetarium), Marylu Evans (AURA), Lynne Jaicomo (Malone Jaicomo Caterers), Richard Kron (U. Chicago), Ryan Richmond (AURA), Ethan Schreier (STScI & AUI), Jennifer Smith (U. Chicago), and Mike Turner (U. Chicago). We reserve special thanks to our co-editors Ken Sembach and Chris Blades, who have labored long and hard to produce these outstanding proceedings.

Garth Illingworth
Rob Kennicutt
Co-chairs, Scientific Organizing Committee

Scientific Organizing Committee

Co-Chairs
G. Illingworth (UC Santa Cruz)
R. Kennicutt, Jr (Arizona)

S. Beckwith (STScI)
C. Cesarsky (ESO)
J. Crocker (BATC/JHU)
A. Dressler (Carnegie Obs)
A. Dupree (CfA)
S. Faber (UC Santa Cruz)
W. Freedman (Carnegie Obs)
A. Giménez (ESA)
H. Hasan (NASA HQ)
R. Kron (Chicago)
S. Lilly (NRC)
J. Lunine (LPL)
F.D. Macchetto (ESA/STScI)
J. Mould (NOAO)
W. Oegerle (NASA/GSFC)
E. Schreier (STScI & AUI)
J.M. Shull (Colorado)
L. Simmons (JPL)
E. Smith (NASA HQ)
D. Spergel (Princeton)
C.M. Urry (Yale)

Local Organizing Committee

J.C. Blades (STScI)
C. Duke (Adler Planetarium)
M. Evans (AURA)
R. Kron (Chicago)
E. Schreier (STScI & AUI)

Conference Participants

Kurt Adelberger	Harvard-Smithsonian Center for Astrophysics
Tom Ajluni	Swales Aerospace
B.-G. Andersson	The Johns Hopkins University
Roger Angel	University of Arizona, Steward Observatory
Thomas Ayres	University of Colorado
Joanne C. Baker	University of Oxford
Jill E. Bechtold	University of Arizona Steward Observatory
Steven Beckwith	Space Telescope Science Institute
Charles A. Beichman	Jet Propulsion Laboratory
David P. Bennett	University of Notre Dame
Chris Blades	Space Telescope Science Institute
William P. Blair	The Johns Hopkins University
Guiseppe Bono	Rome Astronomical Observatory
Tereasa G. Brainerd	Boston University
Thomas M. Brown	Space Telescope Science Institute
Kenneth G. Carpenter	NASA/Goddard Space Flight Center
Jane C. Charlton	Pennsylvania State University
Carol Christian	Space Telescope Science Institute
Chris W. Churchill	Pennsylvania State University
Mark Clampin	Space Telescope Science Institute
John T. Clarke	Boston University
Timothy A. Cook	Boston University
Sigismondi Costantino	University of Rome
Phillippe Crane	NASA Headquarters
James H. Crocker	Ball Aerospace & Technologies Corp.
Tim P. de Zeeuw	Leiden Observatory
Jean-Michael Deharveng	Laboratoire d'Astrophysique de Marseille
Jie Ding	Pennsylvania State University
Alan Dressler	Carnegie Observatories
Douglas K. Duncan	University of Chicago
Andrea K. Dupree	Harvard-Smithsonian Center for Astrophysics
Erica Ellingson	University of Colorado
Marylu Evans	AURA
John E. Everett	University of Chicago
Paul D. Feldman	The Johns Hopkins University
Henry C. Ferguson	Space Telescope Science Institute
Laura Ferrarese	Rutgers University
Francesco R. Ferraro	Observatorio Astronomico Bologna
Priscilla C. Frisch	University of Chicago
Rajib Ganguly	Pennsylvania State University

Jonathan P. Gardner	NASA/Goddard Space Flight Center
Peter M. Garnavich	University of Notre Dame
Donald R. Garnett	University of Arizona, Steward Observatory
Robert Gehrz	University of Minnesota
Alvaro Giménez	European Space Agency–ESTEC
Ana I. Gomez de Castro	Instituto de Astronomia y Geodesia
Ph. Gondoin	European Space Agency
Carol A. Grady	NASA/Goddard Space Flight Center
Robert Grange	Laboratoire Astronomie Spatiale CNRS
Eva K. Grebel	MPI for Astronomy
James C. Green	University of Colorado
Theodore R. Gull	NASA/Goddard Space Flight Center
Walter M. Harris	University of Wisconsin–Madison
Lee M. Hartmann	Harvard-Smithsonian Center for Astrophysics
Martin Harwit	Cornell University
Hashima Hasan	NASA Headquarters
Michael Hauser	Space Telescope Science Institute
Sara R. Heap	NASA/Goddard Space Flight Center
Jay B. Holberg	University of Arizona
J. Christopher Howk	The Johns Hopkins University
John B. Hutchings	Dominion Astrophysical Observatory
Garth Illingworth	University of California–Santa Cruz
Myungshin Im	California Institute of Technology
Edward Jenkins	Princeton University Observatory
Michael S. Kaplan	Boeing-SVS Inc.
Robert C. Kennicutt, Jr.	University of Arizona, Steward Observatory
Charles D. Keyes	Space Telescope Science Institute
Steven D. Kilston	Ball Aerospace & Technology Corp.
Chris Kochanek	Harvard-Smithsonian Center for Astrophysics
Richard Kron	University of Chicago
Tod R. Lauer	Kitt Peak National Observatory
James T. Lauroesch	Northwestern University
David S. Leckrone	NASA/Goddard Space Flight Center
Bruce Martin Levine	NASA/Jet Propulsion Laboratory
Charles F. Lillie	TRW Inc.
Jeffrey L. Linsky	University of Colorado
Jonathan I. Lunine	University of Arizona
F. Duccio Macchetto	Space Telescope Science Institute
John W. MacKenty	Space Telescope Science Institute
Maria L. Malagnini	University of Trieste
Bruce H. Margon	Space Telescope Science Institute
Christopher Martin	California Institute of Technology
Benjamin A. Mazin	California Institute of Technology

Stephan R. McCandliss	The Johns Hopkins University
Melissa A. McGrath	Space Telescope Science Institute
Michael T. Menzel	Lockheed Martin
Michael R. Meyer	University of Arizona, Steward Observatory
Edwin J. Mierkiewicz	University of Wisconsin–Madison
Bruno Millard	Laboratoire Astronomie Spatiale
Warren Moos	The Johns Hopkins University
Jon A. Morse	University of Colorado
Jeremy Mould	National Optical Astronomy Observatory
William R. Oegerle	NASA/Goddard Space Flight Center
Anthony Peacock	European Space Agency–ESTEC
Bradley M. Petersen	Ohio State University
Jason X. Prochaska	Carnegie Observatories
Terrence Rettig	University of Notre Dame
Sun Hong Rhie	University of Notre Dame
Jason Rhodes	NASA/Goddard Space Flight Center
Michael Rich	University of California–Los Angeles
Fred L. Roesler	University of Wisconsin–Madison
Roger W. Romani	Stanford University
Chuck E. Rudiger	Lockheed Martin
Alex R. Ruelas	Instituto de Astronomia
Elena Sabbi	University of Bologna
Meena Sahu	NASA/Goddard Space Flight Center
Ravi Sankrit	The Johns Hopkins University
Blair D. Savage	University of Wisconsin–Madison
Ethan J. Schreier	Space Telescope Science Institute
Kenneth R. Sembach	Space Telescope Science Institute
J. Michael Shull	University of Colorado
Oswald H. Siegmund	University of California–Berkeley
Larry L. Simmons	Jet Propulsion Laboratory
William Smith	AURA, Inc. Headquarters
Rachel Somerville	University of Michigan
George Sonneborn	NASA/Goddard Space Flight Center
John R. Spencer	Lowell Observatory
David N. Spergel	Princeton University Observatory
Phillip Stahl	NASA/Marshall Space Flight Center
Matthias Steinmetz	University of Arizona, Steward Observatory
Mark SubboRao	University of Chicago
Domenick Tenerelli	Lockheed Martin
Scott Texter	TRW Inc.
Harley Thronson	NASA Headquarters
Scott Trager	Carnegie Observatories
Wesley A. Traub	Harvard-Smithsonian Center for Astrophysics

John T. Trauger	Jet Propulsion Laboratory
Todd M. Tripp	Princeton University Observatory
Michael Turner	University of Chicago
Melville P. Ulmer	Northwestern University
Christian Veillet	Canada-France-Hawaii Telescope Corp.
Bart P. Wakker	University of Wisconsin – Madison
Willem Wamsteker	European Space Agency – VILSPA
Lifan Wang	Lawrence Berkeley Laboratory
Daniel E. Welty	University of Chicago
Arthur L. Whipple	University of Texas, McDonald Observatory
Erik Wilkinson	University of Colorado
Bruce E. Woodgate	NASA/Goddard Space Flight Center
Robert A. Woodruff	Boeing SVS Inc.
Donald York	University of Chicago
Idit Zehavi	University of Chicago

Acronyms

Space Observatories/Missions/Instruments (existing, planned, or concept)		
ASCA	–	Advanced Satellite for Cosmology and Astrophysics
BOOMERANG	–	Balloon Observations of Millimeter Extragalactic Radiation and Geophysics
CBI	–	Cosmic Background Imager
CHIPS	–	Cosmic Hot Interstellar Plasma Spectrometer
COBE	–	Cosmic Background Explorer
CODEX	–	Coronagraph Optimized for Detection and Exploration
COROT	–	Convection, Rotation, and Planetary Transits (mission)
DASI	–	Degree-Angular Scale Interferometer
DEEP	–	Deep Extragalactic Evolutionary Probe
ESPI	–	Extra-Solar Planetary Imager
FAIR/SAFIR	–	Filled-Aperture Infrared / Single-Aperture Far Infrared (telescope)
FUSE	–	Far Ultraviolet Spectroscopic Explorer
GALEX	–	Galaxy Evolution Explorer
HST	–	Hubble Space Telescope
HST++	–	successor to HST (also HST-II, HST2, NHST)
HST/ACS	–	HST Advanced Camera for Surveys
HST/COS	–	HST Cosmic Origins Spectrograph
HST/FOC	–	HST Faint Object Camera
HST/GHRS	–	HST Goddard High Resolution Spectrograph
HST/NICMOS	–	HST Near Infrared Camera and Multi-Object Spectrometer
HST/STIS	–	HST Space Telescope Imaging Spectrograph
HST/WFPC2	–	HST Wide Field & Planetary Camera 2
HUT	–	Hopkins Ultraviolet Telescope
IMAPS	–	Interstellar Medium Absorption Profile Spectrograph
INTEGRAL	–	International Gamma-Ray Astrophysics Laboratory
IRSI/Darwin	–	Infrared Space Interferometer / Darwin (mission)
ISO	–	Infrared Space Observatory
JPF	–	Jovian Planet Finder
JWST	–	James Webb Space Telescope (formerly NGST)
IUE	–	International Ultraviolet Explorer
LISA	–	Laser Interferometer Space Antenna
MAHRSI	–	Middle Atmosphere High Resolution Spectrograph Investigation
MAP	–	Microwave Anisotropy Probe
MAXIMA	–	Millimeter Anisotropy Experiment Imaging Array
MIRI	–	Mid-Infrared Instrument (NGST)
NGST	–	Next Generation Space Telescope (see also JWST)
NHST	–	Next Hubble Space Telescope (also HST++, HST-II, HST2)
NIRSpec	–	Near Infrared Spectrograph (NGST)
PI	–	Planet Imager

SHIMMER	–	Spatial Heterodyne Imager for Mesospheric Radicals
SI	–	Stellar Imager
SIM	–	Space Interferometry Mission
SIRTF	–	Space Infrared Telescope Facility
SMART	–	Small Missions for Advanced Research in Technology
SNAP	–	Supernova Acceleration Probe
SOFIA	–	Stratospheric Observatory for Infrared Astronomy
SOHO	–	Solar and Heliospheric Observatory
STS	–	Shuttle Transport System (Space Shuttle)
SUVO	–	Space Ultraviolet-Visible Observatory
TPF	–	Terrestrial Planet Finder
XEUS	–	X-ray Evolving Universe Spectroscpy (mission)
XMM-Newton	–	X-ray Multi-Mirror Newton (mission)

Ground-Based Observatories/Instruments/Surveys (existing, planned, or concept)

2dF	–	Two Degree Field (galaxy redshift survey)
ALMA	–	Atacama Large Millimeter Array
AMBER	–	Astronomical Multiple Beam Recombiner
CELT	–	California Extremely Large Telescope
CFHT	–	Canda-France-Hawaii Telescope
CLASS	–	Cosmic Lens All Sky Survey
EVLA	–	Expanded Very Large Array
FALCON	–	Fiber-spectrograph with Adaptative-optics on Large-fields to Correct at Optical and Near-infrared (wavelengths)
FIRST	–	Faint Images of the Radio Sky at Twenty Centimeters
GENIE	–	Ground-based Nulling Interferometer Experiment
GEST	–	Galactic Exoplanet Survey Telescope
GSMT	–	Giant Segmented Mirror Telescope
GTC	–	Gran Telescopio Canarias
HET	–	Hobby-Eberly Telescope
JVA	–	Jodrell Bank VLA Astrometric (survey)
LBT	–	Large Binocular Telescope
LSST	–	Large Synoptic Survey Telescope
MDS	–	Medium Deep Survey (HST)
MUSE	–	Multi Unit Spectrographic Explorer (VLT)
NIFS	–	Near Infrared Integral-Field Spectrograph (Gemini)
NOAO	–	National Optical Astronomy Observatories
OASIS	–	Optically Adaptive System for Imaging Spectroscopy (CFHT)
OWL	–	OverWhelmingly Large Telescope
PMN	–	Parkes-MIT-NRAO (survey)
SCUBA	–	Submillimeter Common-User Bolometer Array
SAURON	–	Spectrographic Areal Unit for Research on Optical Nebulae
SDSS	–	Sloan Digital Sky Survey

SINFONI	–	Singgle Far Object Near-IR Investigation (VLT)
SNO	–	Sudbury Neutrino Observatory
TIGER	–	Traitement Integral des Galaxies par l'Etude de leurs Raies
VLA	–	Very Large Array
VLBI	–	Very Long Baseline Interferometer
VLT	–	Very Large Telescope
VLTI	–	Very Large Telescope Interferometer
VVDS	–	VIRMOS-VLT Deep Survey
WHAM	–	Wisconsin H-alpha Mapper (also WHαM)
WHT	–	William Herschel Telescope
WIYN	–	Wisconsin-Indiana-Yale-NOAO (Observatory)
VIRMOS	–	Visible and Infrared Multi-Object Spectrograph (VLT)

Institutions and Organizations
(Astronomical, Corporate, Military, Government)

ATP	–	Astrophysics Theory Program (NASA)
AURA	–	Associated Universities for Research in Astronomy
CfA	–	Center for Astrophysics (Cambridge, MA)
ESTEC	–	European Space Research and Technology Center
DoD	–	Department of Defense
GSFC	–	Goddard Space Flight Center
JPL	–	Jet Propulsion Laboratory
LMATC	–	Lockheed Martin Advanced Technologies Center
NASA	–	National Aeronautics and Space Administration
NIST	–	National Institute of Standards
NRO	–	National Reconaissance Organization
NSF	–	National Science Foundation
OS	–	Origins Subcommittee (NASA)
OSS	–	Office of Space Science (NASA)
SEU	–	Structure and Evolution of the Universe (NASA theme)
SEUS	–	Structure and Evolution of the Universe Subcommittee (NASA)
SScAC	–	Space Science Advisory Committee (NASA)
STScI	–	Space Telescope Science Institute
NSF	–	National Science Foundation
ST-ECF	–	Space Telescope European Coordinating Facility
ESA	–	European Space Agency
UVOWG	–	Ultraviolet-Optical Working Group (NASA)

Astronomical Terms

AGB	–	asymptotic giant branch
AGN	–	active galactic nucleus
BAL	–	broad absorption line
BH	–	black hole
BBN	–	big bang nucleosynthesis
BLR	–	broad line region (AGN)

BSC	–	blue spheroid(al) candidate
CDM	–	cold dark matter
CMB	–	cosmic microwave background
CMD	–	color-magnitude diagram
CNM	–	cold neutral (interstellar) medium
CV	–	cataclysmic variable
DLA	–	damped Lyman-α (system[s])
EGP	–	exo-solar giant planet
FOV	–	field of view
FUV	–	far-ultraviolet
GC	–	globular cluster
GMC	–	giant molecular cloud
GRB	–	gamma ray burst(er)
HB	–	horizontal branch
HDF	–	Hubble Deep Field (-N north, -S south)
HH	–	Herbig-Harro (object)
HR	–	Hertzsprung-Russell (diagram)
HVC	–	high velocity cloud
IBH	–	intermediate-mass black hole
IR	–	infrared
ICM	–	intracluster medium
IGM	–	intergalactic medium
IMF	–	initial mass function
ISM	–	interstellar medium
L2	–	Langrangian point L2
LIC	–	Local Interstellar cloud
LINER	–	low-ionization nuclear emission-line region
LMC	–	Large Magellanic Cloud
LOS	–	line of sight
LSB	–	low surface brightness
LTE	–	local thermodynamic equilibrium
MACHO	–	massive compact halo object
MC	–	Magellanic Clouds
MSTO	–	main sequence turn-off
NIR	–	near-infrared
NLR	–	narrow line region (AGN)
NUV	–	near-ultraviolet
PMS	–	pre-main sequence
QAL	–	quasar absorption line
QSO	–	quasi-stellar object
RGB	–	red-giant branch
SBH	–	supermassive black hole (also SMBH)
SED	–	spectral energy distribution
SFR	–	star formation rate
SMBH	–	supermassive black hole (also SBH)
SMC	–	Small Magellanic Cloud
SN	–	supernova
SNR	–	supernova remnant

S/N	–	signal-to-noise ratio
TGRB	–	tip of the red giant branch
UV	–	ultraviolet
UVOIR	–	ultraviolet-optical-infrared
WIM	–	warm ionized (interstellar) medium
WNM	–	warm neutral (interstellar) medium

Technical/Instrumentation/Other

ADR	–	adiabatic demagnetization refrigerator
AMSD	–	Advanced Mirror System Demonstrator
AO	–	adaptive optics
CCD	–	charge coupled device
CMOS	–	complimentary metal oxide silicon
CPU	–	central processing unit
DC-SQUID	–	direct current superconducting quantum interference device
DM	–	deformable mirror
EBCCD	–	electron beam charge coupled device
ETC	–	exposure time calculator
F-P	–	Fabry-Perot
FTS	–	Fourier transform spectroscopy
FWHM	–	full width (at) half maximum (intensity)
GPS	–	Global Positioning System
GTO	–	guaranteed time observer
HWHM	–	half width (at) half maximum (intensity)
MAMA	–	multi-anode microchannel array
MCAO	–	multi-conjugate adaptive optics
MEMS	–	micro-machined mechanical system
MCP	–	microchannel plate
OPD	–	optical path difference
PI	–	principal investigator
PSF	–	point spread function
RC	–	Ritchey-Chrétien (also R-C)
QDE	–	quantum detection efficiency
QE	–	quantum efficiency
SHS	–	spatial heterodyne spectroscopy
SQUID	–	superconducting quantum interface device
STJ	–	superconducting tunnel junction
TES	–	transition edge sensor
TRL	–	technology readiness level
ULE	–	ultra-low expansion

Hubble's Science Legacy: Future Optical-Ultraviolet Astronomy from Space
ASP Conference Series, Vol. 291, 2003
K.R. Sembach, J.C. Blades, G.D. Illingworth, R.C. Kennicutt, Jr. (eds.)

The UVOIR Astronomy Landscape in 2015

Alan Dressler

Carnegie Observatories, 813 Santa Barbara Street, Pasadena, CA 91101

Abstract. The likely time frame for a next-generation UV-optical space telescope — a successor to HST — is in the middle of the next decade. Here I review the landscape of ultraviolet-optical ground-based and space facilities in 2015 and the science topics that are likely to be active, identifying a few that I believe will be particularly exciting. These can be used as a basis for developing and promoting a 4 m class HST successor that will significantly advance goals of the Origins theme and also enable a wide variety of astronomical programs in the Astronomy & Physics Division of the NASA Office of Space Science.

1. Introduction

Without doubt, the Hubble Space Telescope (HST) has achieved a status that is special in the history of scientific instruments. Since its repair in December of 1993 HST has captured the attention and imagination of the public and its importance to our view of the Universe is comparable only to Galileo's first telescope and the Mount Wilson 100 inch Hooker telescope. Because of the servicing program carried out by NASA astronauts from the Space Shuttle, its capabilities have increased over time, and there is every reason to believe that HST will remain remarkably productive until the end of the decade, the planned end of mission.

In 1995 the HST & Beyond Committee that I chaired recommended the extension of HST to 2010, and it also placed top priority for the future on a Next Generation Space Telescope (NGST) with a $1-5\mu$m sensitivity that could take advantage of the huge gains available to cold space telescopes over ground-based telescopes. It was hoped at that time that a period of overlap of a few years might allow both HST and NGST to complement each other, but the realities of schedule and budget have made that ever more unlikely. Hence we might find ourselves with a many-year gap between the retirement of HST and its replacement with another space telescope with excellent performance in the space ultraviolet through visible wavelengths. If NGST were to launch in the 2010 time frame, as presently planned, the first available window for a Hubble successor would be in 2015, I believe, with years of slip possible if NGST should itself be delayed.

In considering the place of such a follow-on mission to HST, I would like to offer some abstracted guesses and opinions on four topics: (1) an inventory of likely ground and space based UV-optical-IR facilities in 2015; (2) the science priorities in 2015 as I imagine them; (3) a case for a HST successor; and (4) the relation of this mission to the themes in the Astronomy & Physics Division of NASA's Office of Space Science.

2. U.S. Ground- and Space-Based UVOIR Facilities in 2015

It seems certain that the ten or more mature U.S. 8 m class general purpose telescopes that are now in operation or soon to be will be the principal research tools on the ground. I believe there will also be, by 2015, a Large Synoptic Survey Telescope (LSST) as proposed in the McKee & Taylor 2000 Decadal Survey *Astronomy & Astrophysics in the New Millennium,* and probably many smaller synoptic survey instruments. In particular, I imagine at least one 4 m or larger rapid-acquisition ($t \sim 1$ min) photometric/spectroscopic "target of opportunity telescope".

It seems likely that at least one, and perhaps 2 telescopes of the 20-30 m class will have been built by the U.S. by 2015. These will be semi-specialized to take advantage of their apertures in D^2 or D^4 applications — this will include adaptive optics, multi-conjugate adaptive optics, near-IR spectroscopy, and highly-multiplexed seeing-limited optical spectroscopy. I also think that there will be, by 2015, some distributed aperture telescopes on the ground, both multiplexed (synoptic) and interferometric, up to 8 m class elements. Of course, there will still be of order ten 4 m class telescopes that will be primarily (but not entirely) dedicated to support work for other facilities (surveys, calibration, instrumentation development, etc.) Smaller telescopes than these are likely to be even more specialized.

Given the success of the Sloan Digital Sky Survey, one might guess that there will be a near-IR sky survey project; this will require continued rapid development in near-IR detectors to cover large field. (A deeper optical survey of a large portion of the sky will likely be accomplished by LSST.)

In space, we can look forward to several NASA flagship missions: NGST, a 4 to 6 m class cold telescope optimized for $1 - 5\mu$m imaging and moderate resolution spectroscopy, should have been in service for a few years. We hope that a $5 - 20\mu$m imaging and moderate resolution spectroscopic capability will survive the challenges of budget and schedule. SIM, with its unprecedented astrometric capability of several micro-arcsecond precision down to 20th magnitude, should be nearing the end of its principal mission. The airborne SOFIA, with its mid-IR moderate- to high-resolution spectroscopic capability, should be in its mature and most productive years.

Beyond this things become more speculative. My guess is that by 2015, a 2-element interferometer with ~ 1 m apertures, capable of high resolution, high dynamic range mid-IR imaging, will have been built or well on the way. This is, of course, a step toward the "book version" of TPF, but it also a testbed for future space interferometers. I think this will be important, for developing a high-resolution far-IR interferometer, for example, to study star and planet formation. Before that happens, however, I think there could be a 6 m class large-aperture mid-to-far-IR telescope — a follow-up to SIRTF — in the later stages of development. Alternatively, or maybe in addition if the coronagraphic approach to TPF looks especially promising, the development of a 4 m class UV-optical imager/coronagraph could be underway.

3. Hot Science Topics in 2015

Cosmology research may be still be a vital field of astrophysics in 2015, but my guess is that the UVOIR part of it will be played out and cosmic microwave background (CMB) observations will dominate. By this time there may have already been a considerable effort to chart the Hubble expansion using supernovae, perhaps something like the SNAP mission that Saul Perlmutter and colleagues are proposing. Likewise, I think that mapping of large scale structure will be largely complete out to fairly high redshift, but there may be a continuing, important effort to map the galaxy distribution at very high redshift ($z > 3$) in order to connect the formation and early evolution of galaxies with the gas and dark matter distributions at those epochs, and to study directly the merging history of galaxies in the hierarchical clustering picture.

Galaxy birth should be a mature but still active field thanks to NGST and giant ground-based telescopes. We expect to see the first stars and their assembly into galactic fragments, the growth of the first black holes (AGN), the first supernovae (heavy element enrichment), and the role of feedback from star formation in forming the structures and morphological types we recognize today. The redshift interval $1 < z < 2.5$ is difficult to study from the ground (although 20 m and larger telescopes will help) but apparently crucial, since it is within this epoch that the galaxies we see today seem to have been assembled. Specific studies will include the evolution of the intergalactic medium and interstellar medium, the history of star formation rates, of stellar populations, and of heavy element production, as well as structure (morphology and stellar dynamics).

I'm optimistic about the progress that will be made around 2015 to answer the last remaining questions about how galaxies like our own were born. Two concerns I have is that theoretical astrophysicists will lose interest in galaxy evolution once it becomes a question of "weather," that is to say, all of the basic physical law is known but the emergent complexity dominates the process. I also worry that if NGST becomes as small as a 4 m, it will be hard to satisfy the expectations that have developed in the last 7 years about observing first starlight, and that with this mission we will be limited to the brightest, and not typical systems at the epoch of galaxy birth.

It's my guess that the study of AGN engines will rely less and less on UVOIR observations by 2015. There may be ongoing adaptive-optics or interferometric observations of the relation of gas, dust, and star formation to the feeding of the black holes, but the vital work in this field will, I think, be in the X-ray band. Similarly, I expect stellar dynamics in nearby galaxies, including the study of complex cores, anisotropies, bars, and the search for massive black holes to be cleaned up by 2015 by 8 m class telescopes with integral field spectroscopy fed by laser-guide-star adaptive optics.

On the other hand, there will be a long way to go in studying the stellar populations of nearby galaxies. One would like to extend through the Local Supercluster the color-magnitude diagrams of galactic stellar populations, below the main-sequence turnoff in Local Group galaxies (which will be important in understanding mergers and the star formation history of bulges and spheroids), and down to the horizontal branch in Virgo Cluster galaxies. The problems of crowding, field size, and high Strehl ratios suggest that space telescopes will be crucial to the work. Both space and ground-based cameras, particularly in the near-IR, will be important to understanding the initial mass function of stars, for example, in different environments and at different heavy element abundances — this may be better understood in 2015 but still require a vigorous research program. An intensive program of spectroscopy of giant and supergiant stars will provide complementary information relating the history of heavy element enrichment, from the general, to the specific r- and s-process enrichment patterns. The required high spectral resolutions are best achieved with large ground-based telescopes.

The interstellar medium in our Galaxy, especially the evolution of molecular clouds and the chemistry and dynamics of molecular cloud cores, will be a central program in 2015, accomplished with moderate spatial and high spectral resolution observations. Although SOFIA will be a major contributor to this effort, and a contribution from a successor to SIRTF may be important, the principal results should come mostly from millimeter and sub-millimeter ground-based observations. Directly connected, and probably one of the most active research areas in 2015, will be the study of the star and planet formation environment. Central issues are grain and ice chemistry, the life history of organics, dynamics in debris disks and the growth of large-bodies. Such studies will require very high resolution (1 AU $\Rightarrow < 0.1''$) mid-IR imaging of accretion and post-accretion disks in nearby star forming regions and high-resolution mid-IR spectroscopy from SOFIA and imaging/spectroscopy with next-generation ground-based telescopes. Achieving very high spatial resolution in the mid-IR may require an interferometric space telescope of considerable dimensions.

Finally, I consider briefly the detection and characterization of planetary systems, sure to be a major, if not *the* major component of Origins research in 2015. I think it would be a mistake to overlook the importance of maintaining, or increasing, our capabilities to observe weather on the planets in our own Solar System and to study cloud chemistry in their atmospheres. The composition and distribution of comets and asteroids, especially Kuiper Belt objects, will be key evidence in understanding the assembly history of the Solar System. Without HST, some important capabilities in the visible wavelengths are likely to be absent in 2015.

I think we all expect that there will be an extensive catalog of giant planets around nearby stars by 2015, and that for these there will be vigorous program of characterization — studies of the atmospheres of these giant planets found by ground-based interferometers such as the Keck and the Large Binocular Telescope, and with NGST. With the exception of some sort of "giant planet finder," the tools required for such studies are already in place or well along in the planning stage. It is my opinion that, in 2015, we could extend the detection of planets down to "super-Earth" sizes for stars within, say 10 parsecs, by techniques that are now under consideration for TPF — near-IR interferometry and visible coronagraphy. Coupled with information about giant planets, this should begin to provide a substantial data base for studying the demographics of planetary systems. However, given our rate of progress toward NGST and SIM, and the probability that missions to find true terrestrial-sized planets will be at least as challenging, I regard it as overly optimistic to think that the TPF we have been dreaming of — with an ability to detect an Earth twin at 10-15 pc and to provide rudimentary characterization — will be in orbit in 2015. One hopes it might be in construction by this time, or at least well along in the planning stage.

4. The Case for a "Hubble Legacy" Space Telescope

With apologies to all my colleagues who might feel their science to be undervalued in my personal and no doubt biased perspective, the preceding are my guesses of what the "hot science topics" in UVOIR Origins related research will be in 2015. Among these are several that would benefit greatly from a UV-optical telescope that would follow the amazingly successful path of HST. Indeed, some of these topics may not substantially advance without one. I have chosen three that are also key Origins theme goals which I think could be combined into a mission for such a space telescope that is neither hopelessly ambitious nor overly complex. Of course, such a facility would have a significant impact on dozens of other important programs.

Beginning with high-redshift galaxy evolution, I identify the evolution of the gaseous component of the Universe at $z < 2.5$ as a critical component in understanding the origins and evolution of galaxies which will require such a facility. What is needed are damped Lyα and Lyα forest spectral observations — moderate-to-high resolution in the vacuum UV — with background quasars and perhaps gamma-ray-bursts as the probes. Even the much more sensitive COS on HST will increase the sample of such measurements only to a couple of dozen, an insufficient number to represent fairly the range of massive halos that can be intercepted. To increase this to several hundred — a decent though by no means extravagant sample — will require a ten-fold gain over HST/COS, probably achievable with a combination of a somewhat larger aperture and much more quantum-efficient UV detectors than those which are flight-ready today. A continuing investment in UV detector technology is, then, crucial in order to fill in this missing chapter in the history of galaxy evolution. Jason Prochaska, Jill Bechtold, Jane Charleton, and Ken Sembach spoke on this topic at this workshop.

Second, extending the resolution of stellar populations out to galaxies as far as the Virgo Cluster is a job for which such a UV-optical space telescope is uniquely suited. Diffraction-limited, high-Strehl ratio imaging in the UV-optical would allow us to reach the main sequence turnoff in all local galaxies and down to the giant branch in hundreds of major galaxies. Again, only a modest field (a few arcminutes) is needed, so a relatively compact imager with a manageable number of pixels is required. These kinds of observations will be crucial in applying what we learn from observations of star formation in high redshift galaxies to our own Galaxy and its neighbors. Note that I am not hauling out the old saw that we will be unable to interpret high-redshift observations without full stellar population histories of nearby galaxies — I believe that the star formation history seen "as it happened" may be in fact be *easier* to interpret than any integrated population, even one resolved star by star. My point is that a prime Origins goal is to understand what happened in our corner of the Universe, in particular the history of the Milky Way galaxy that we inhabit, and stellar population studies are the way to move from the general patterns we see at early epochs to what actually happened *here*. Michael Rich, Eva Grebel, and Harry Ferguson addressed these issues at this workshop.

The third opportunity for a more-powerful UV-optical telescope is in searching for planets around nearby stars. The key here is for very high-dynamic range imaging (factors of $> 10^6$) within $0.1''$ of bright stars. As is reviewed at this conference by Chas Beichman, coronagraphs and/or shaped pupil masks on a modest-size (4 m) UV-optical telescope could be a powerful device for detecting Jovian and sub-Jovian size planets out to tens of parsecs, with the possibility of detecting even Earth-like worlds out to about 5 parsecs. Although a full-up TPF capable of detecting Earth-like worlds at 10-15 pc is likely to be a much more formidable challenge (see the 4 x 10 m telescope proposed by David Spergel and referred to in Chas Beichman's presentation), a more-modest version would be an excellent stepping stone both technically and scientifically. The challenges are great, but the potential to test this approach for TPF make this an exciting prospect, where the contrast ratio exceeds 10^9.

I have chosen to highlight these three scientific programs not only because they are among the most exciting but also because one can imagine combining them into a 4 m Hubble Legacy telescope that is not the proverbial "Christmas tree." That is, the technology would be driven mainly by the planet-finding capability, with a small field camera and an even smaller field (single object) high-resolution spectrograph going for the ride. Of course, even these three capabilities would support a broad range of programs, but the idea would be to keep the focus on high-dynamic range imaging with a couple of straightforward additions.

Let me close this section by giving an example of an important program which requires capabilities that would add substantial complexity to what I've just described. Another unique contribution to galaxy evolution could be made in studying the star formation history over the assembly epoch, $1 < z < 3$, information that is complementary to the evolution of the gas component and required for understanding how the present suite of galaxies came to be. NGST will provide crucial contributions to this effort by detecting, with far greater sensitivity than is possible with ground-based telescopes, the $\sim 2\mu$m rest frame light that records the buildup of aging stellar populations, and the Hα emission from star forming regions. This needs to be compared to star formation rates measured from rest-frame UV light and also recorded in type II supernovae in order to understand the production of dust and its distribution in developing galaxies, and to look for possible evolution of the initial mass function.

This work can be carried on with HST through this decade, and the new generation of ground-based 8 m class telescopes will be important contributors, but neither can provide a spatially resolved spectroscopic capability at high enough resolution ($D < 500$ pc) to adequately follow the evolution of star formation in massive galaxies and relate it to structural evolution and the merger history. A multi-object spectrograph with programmable slits, as is under development for NGST, on a new UV-optical telescope with a moderate field ($\sim 10'$) could be operated in a stepped mode to map large numbers of galaxies with the required spatial resolution and sensitivity. Deployable integral-field units are another possible approach. Either way, this is the kind of addition that would substantially increase the cost and extend the schedule of a next-generation UV-optical space telescope, so this is an example of the kind of tough choice that the community may have to make if an HST successor is to become a reality within the next decade. No matter how compelling, not all science will be able to be accommodated in a moderate-sized mission.

5. Hubble Legacy Telescope for the Origins Theme?

At the present time it is especially helpful to think about proposing missions within the theme structure that has developed at NASA's Office of Space Science, and in the Astronomy & Physics Division in particular. A next generation UV-optical space telescope fits naturally into the Origins theme, particularly with its planet finding potential and the priority given to this area in Origins. Although, like HST, such a telescope would share many science goals with the Structure and Evolution of the Universe (SEU) theme, the major missions highlighted in that theme will likely stress other priorities from the 2000 Decadal Report, specifically Constellation-X and LISA.

A UV-optical space telescope of 4 m or larger would almost certainly fall under the category of a "flagship mission," in NASA parlance. The line of Discovery-class missions that is under consideration by the SEU theme would, I think, be unable to accommodate even a 4 m version with multiple instruments; perhaps a very focused mission — for example, a 4 m UV-spectroscopic "light-bucket" to study the intergalactic medium — could be investigated in this context. A cost of more than $500M (the dividing line for "large missions" in the Decadal report) for phase CD implementation of a multi-purpose telescope seems realistic, although present Astronomy & Physics Division projects suggest that even this might be a substantial underestimate. Although there is nothing in the organizational structure, to my knowledge, that prevents the development of such a flagship mission exclusive of the Origins and SEU themes, it appears that the road is paved and well marked for the theme approach while the "independent" path has become more ill-defined and perhaps rocky.

Of course, considering the time scale under discussion, this begs the question of whether the theme structure will in fact survive a decade or more. This attempt to organize missions principally around science themes with only secondary division by wavelength or target (for example, the Outer Planets vs. the Inner Planets distinction that has existed in the Solar System Exploration theme) will only succeed if scientists see themselves as part of larger intellectual enterprises and if they commit to working for a larger program as opposed to one or another mission that serves their scientific specialty. At the present time many in our community still view the theme structure as a marketing tool devised to better explain our programs to Congress, the Executive Branch, and the public. Organizing by themes may indeed accomplish this better than the previous system, but it is not the heart of the idea, in my opinion — the themes are not just a marketing tool. I'm cautiously optimistic that our colleagues will give the theme structure a try and to support these broader visions of their research in space science, but only time will tell.

If the Origins theme continues with its present funding level, the opportunity to incorporate a UV-optical space telescope — a worthy successor to HST — will present itself in the next decade. The Origins Roadmapping process, feeding into the NASA Strategic Plan for the Office of Space Science, is the appropriate way to develop and promote this project in the near term, with of course the 2010 Decadal Study presenting the key hurdle that must be cleared to achieve a high priority for an HST successor. We may wish we could reach this destination sooner, but the most important thing is to plan wisely and develop a scientifically compelling motivation, so that we may insure that we will in time reach our goal.

Hubble's Science Legacy: Future Optical-Ultraviolet Astronomy from Space
ASP Conference Series, Vol. 291, 2003
K.R. Sembach, J.C. Blades, G.D. Illingworth, R.C. Kennicutt, Jr. (eds.)

Thoughts on a Powerful New Optical/Ultraviolet Facility

Martin Harwit

511 H St., SW, Washington, DC 20002

Cornell University, Ithaca, NY 14853

Abstract. I attempt to examine the potential of a large ultraviolet space telescope and its prospective instrumental components in the context of the astronomical community's continuing search through the Universe.

1. Introduction

I begin with a summary of the evolution of astronomical instrumentation in the past few decades to show how technical advances have shaped our world view. I emphasize this perspective because of the prevailing, countervailing belief that it is our scientific world view that should shape our request for new facilities in space. That is the theme of this workshop assembled to define the major scientific questions that an optical/ultraviolet facility might permit us to answer. But in considering a new facility we should look not only at how to best provide observational answers to longstanding theoretical questions. Rather, we should also ponder the choice of instruments that will provide novel kinds of observations that could raise new incisive questions. Defining the right questions in astrophysics is often much harder than looking for answers, and a thoughtful choice of new instruments, together with inspired judgment on how to put them to best use, would be a valuable contribution from this workshop. With this view in mind, I suggest some new opportunities we might consider as we plan the next optical/ultraviolet mission in space. This choice is not meant to be unique, but just one way of examining the past and seeing where earlier explorations left gaps in the ways we have viewed the Universe.

2. The Evolution of Astronomical Instrumentation

The evolution of astronomical instrumentation and the resulting astronomical advances can provide a useful guide for the future. Instrumental capabilities across the entire electromagnetic band have steadily improved in recent decades, as shown in Figures 1 to 3. Each figure exhibits the available instrumental resolving power in different years plotted as a function of wavelength. Advances are plotted at twenty year intervals. The status in 1939, shown in the darkest shading, reveals the capabilities we had just before World War II. By 1959 and the birth of NASA, capabilities had advanced, as shown in a somewhat lighter shading, mainly in radio astronomy and with a start in the ultraviolet regime. Then, by 1979, about the time that astronomy was turning from little science with rockets and small satellites to big science in space with the Great Observatories, huge strides had been made in infrared, X-ray and gamma-ray astronomy, as shown in a lighter shading still. Finally, the status in 1999, just before the turn of the millennium, is shown in the lightest shading.

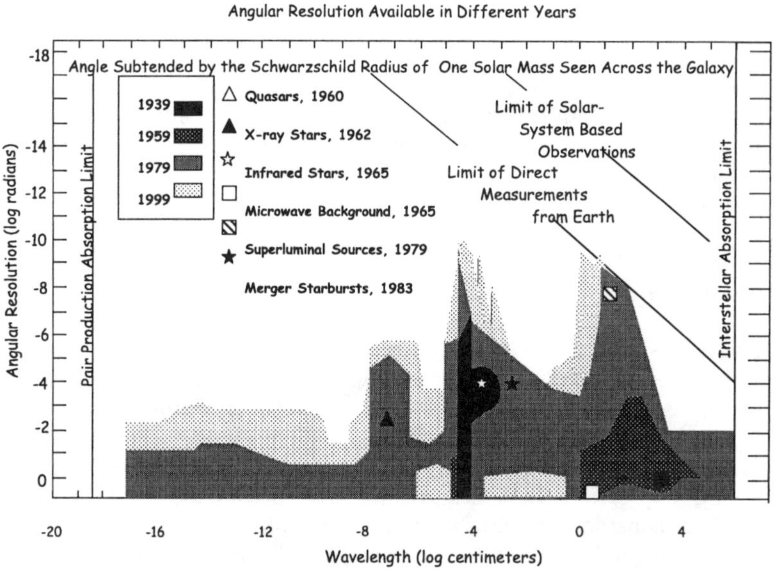

Figure 1. Angular resolving power available in different years. Symbols indicate the resolving power that was required to bring about the listed discoveries. Many of these findings followed quickly once the necessary instrumentation was available.

Figure 1 shows the advances attained in angular resolution – image quality. The shortest wavelength at which observations can be contemplated is the limit at which electromagnetic waves rapidly collide with microwave background photons to produce electron-positron pairs. This prevents ultra-high energy radiation from crossing distances exceeding a few kiloparsecs. At the longest wavelengths, electromagnetic frequencies drop below the interstellar plasma frequency and are readily absorbed, again preventing traversal of appreciable distances. The two slanted lines on the right side of the plot give the limiting angular resolving power for very long baseline interferometry at each wavelength. The lower line gives the resolving power for a baseline equaling Earth's diameter. The upper line shows how much more might be achieved with a baseline stretching across much of the Solar System.

The darkest shading in the figure shows the angular resolution available in 1939, mainly in the optical regime, with some excursions into the infrared through the use of thermopiles. And there is a small domain of radio astronomy representing Karl Jansky's detection in 1932 of radio noise from the Galaxy's center.

Historically we had of course known about planets, comets, stars, and galaxies for a long time, but now we became aware of phenomena that had escaped all notice. In 1962, two discoveries hit the scene with tremendous impact. The first was the finding that there are astronomical sources that copiously emit X-rays (Giacconi et al. 1962). This was followed the same year by the discovery of quasi stellar radio sources that appeared point-like at even the highest angular resolution (Hazard, Mackey, & Shimmins, 1963). Then, in 1965, a few stars were found to emit solely in the infrared without discernible emission at optical wavelengths (Neugebauer, Martz, & Leighton, 1965). That same year, the discovery of the microwave background radiation came as a surprise (Penzias & Wilson 1965). In 1971, superluminal sources, unbelievably energetic outbursts that were detectable only at the highest angular resolutions, obtained through very long baseline radio interferometry, again astounded the community (Knight, et al. 1971). This was followed, thirteen years later, by the discovery of ultraluminous infrared galaxies rivaling the output of quasars but clearly the result of collisional mergers on enormous scales (Soifer et al. 1984).

The prevalence of mergers changed our views on star formation in a galaxy and the origins of the initial mass function of stars. The initial mass function was quite evidently not ordained by some intrinsic property of interstellar gas clouds. Rather, it was strongly influenced by the collisions a galaxy had undergone. The concept of galaxy formation at some well defined primeval epoch – each galaxy independently evolving in isolation thereafter – also had to be discarded.

None of this had been dreamed of. Over the decades, the instrumental advances that brought with them increasing clarity of view were raising new questions at a breathtaking rate, breaking up cherished views of the timeless, imperturbable cosmos that James Jeans had so eloquently described in the early decades of the 20th century (Jeans 1931). By the year 2000, our views could not have been more different. They pictured a roiling, active cosmos, flaring up here and there, in violence on unimaginable scales.

Similar advances were taking place in time resolution. Both high speed and long-term observations made a decisive impact on scientific views. In 1939, John C. Duncan of Mount Wilson observatory compared plates of the Crab Nebula taken over a thirty year period from 1909 to 1939 and found a systematic expansion of 1/5 of a second of arc per year (Duncan 1939). This measurement involved both high angular resolution and long duration observations. Three years later Duyvendak (1942), and Oort & Mayall (1942) made the connection between the expanding Crab Nebula and the supernova explosion of 1054 A.D.

Ever faster instrumental response yielded the 1949 discovery of giant flaring of stars on scales of one second (Luyten 1949); the 1968 discovery of radio pulsars keeping time on scales of tens of milliseconds (Bell Burnell 1978); the 1973 announcement of gamma ray bursts (Klebesadel et al. 1973); and, as recently as 1996, the finding of quasiperiodic X-ray oscillations (QPOs) currently thought to emanate from the accretion disks or jets associated with black holes, and the enigmatic rapid X-ray repeaters (Miller 1996; Kouveliotou 1996). We will have to wait to assess the impact of these last two findings on astronomical thought.

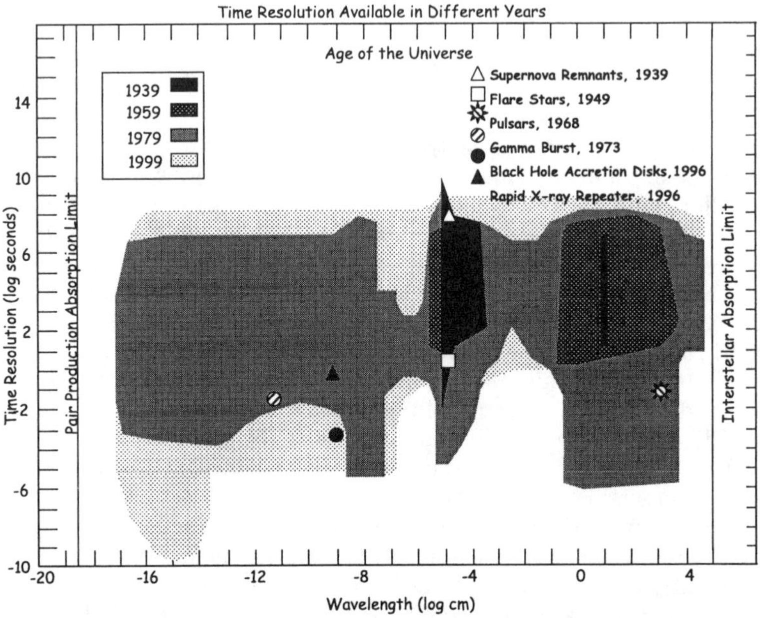

Figure 2. Time resolution available in different years. The observational capabilities needed to make several discoveries are indicated.

Finally, a quick review of the rise in spectral resolving power and how this has shaped our thinking. In 1946, Horace Babcock detected magnetic fields of a baffling 30,000 to 40,000 gauss in the star 78 Virginis by resolving the small Zeeman shift of spectral lines (Babcock 1947). Similarly-high spectral resolution led to the detection of cosmic masers (Weaver et al. 1965). But plain broadband photometry also was necessary to discover X-ray galaxies that same year. This same need for both low and high spectral resolution capabilities is, respectively, illustrated more recently also by the discovery of TeV blazars (Punch et al. 1992) and exoplanets (Mayor & Queloz 1995).

Two points are worth noting. The first is the vast unexplored domain shown by the blank areas in Figures 1 to 3. The second is the broad range of observational capabilities required to reveal the novel phenomena that raised fundamentally new questions. These capabilities usually differed from each other by many orders of magnitude, generally factors of a thousand, either in wavelength or in instrumental resolving power.

Progress in polarization capabilities have had similar impact, particularly on the understanding of magnetic fields on all scales from stellar, to interstellar and, most recently, intergalactic. Both plane and circular polarization capabilities have played a crucial role. Radio and infrared astronomers have capitalized on these. But, for reasons that are not obvious, the optical/ultraviolet community has not made much use of polarization studies.

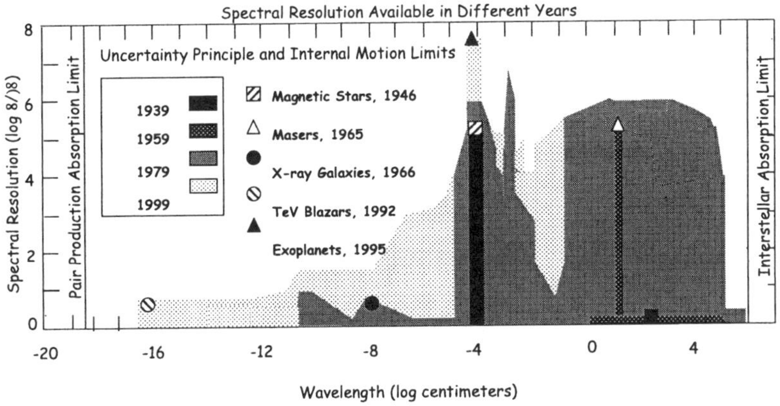

Figure 3. Spectral resolution available in different years. Observational capabilities required to make a number of discoveries are indicated.

Solely one of the major discoveries I have cited was theoretically predicted, the microwave background radiation foreseen by Ralph Alpher and Robert Herman (1949). However, their work had been long forgotten by 1965 and did not play a role in the discovery of the microwave background.

So, where do we stand? Do observations drive our thinking, or does our thinking drive the observations we undertake? Clearly it is both, and we need to carefully consider the choice of instruments we ask for to strike the right balance between pursuing current questions and looking for new phenomena that will inspire novel thinking. New phenomena are likely to be found primarily by systematic searches through the empty spaces shown in white in the figures. This is where the *terrae incognitae* reside, the unknown worlds that will sooner or later emerge to raise tantalizing new questions as we further hone our tools. We all realize this, of course. But as the history of our discipline has made clear, we would never have made most discoveries – or at least not made them as quickly – if the astronomical community had prevailed. Astronomers would have wanted the leading scientific questions of the day to drive requests for new instrumentation, and we would have missed out on the great discoveries of our time.

This seems hard to believe, because our successes loom so large. But a closer look reveals that radio astronomy was started not because the astronomical community became interested, but because radar techniques had undergone tremendous advances during World War II, and this equipment was beginning to detect radio noise from the Sun, from meteor showers, and from the first-discovered extragalactic source, Cygnus A. James Hey, a young troubleshooter for British radar in World War II, initially brought these to attention in 1946 and continued to use wartime equipment to conduct further observations (Harwit 1980). Most technical advances at radio wavelengths since then have been further driven by military needs. The same holds true for virtually every infrared device used to date. Even had we wanted to, astronomers would not have had the funding on the scales of billions of dollars that the DoD has spent on infrared detection techniques to discern and intercept ballistic missiles in space (Jamieson 1995). CCDs for the optical regime were all developed by the military at enormous expense. The gamma ray equipment that led to the discovery of gamma-ray bursts was first developed for detecting surreptitious Soviet nuclear tests. The very discovery of the gamma-ray bursts was kept classified for four years before Ray Klebesadel and his colleagues were permitted to publish their findings (Klebesadel et al. 1973). Moreover, most of the early discoveries were made not by astronomers but by physicists and engineers who had pioneered the new instrumental techniques, had assembled simple instrumentation to look at the sky, and made the first breakthroughs. Astronomers only joined in later in order not to be left out.

These may seem like distant events, but current efforts seem to follow along the same traditional lines. The search for neutrinos from the Sun and from supernova explosions is conducted entirely by high energy physicists. Efforts to detect gravitational waves are similarly led by physicists. Virtually all of the theoretical work on particle physics in the early universe is carried out by physicists. Work on the physical nature of dark matter appears primarily in journals published by physical, rather than astronomical societies. Traditionally trained astronomers obtaining their doctoral degrees in astronomy departments are not equipped to carry out these projects. Decadal plans carried out by the astronomical community do not include them on their wish lists, whereas comparable plans produced by the physics community do. Of course, there is common agreement that these projects are important, and this tends to be reflected in the mutual endorsement of long-term planning documents by both communities. In addition, the two disciplines have tended to merge through the creation of university departments of "Physics and Astronomy." But the two communities, nevertheless, do ply quite different trades, and view the world from rather different perspectives.

Physicists are more interested in fundamental processes that would permit us to see astrophysical phenomena with entirely new eyes – to raise novel questions that normal astronomical investigations would be unlikely to broach. In contrast, astronomers are more content to examine what they already know, though more closely, to see whether previously blurred details might be examined more clearly to reveal insights that a coarser look did not provide.

Even if we recognize the existence of such differences between astronomers and outsiders who come into our field with entirely new tools, we astronomers do need to recognize that we should not be content merely to ask, "What questions can we answer with our next big mission?" Rather, we should be asking "What novel questions will be able to raise, given new observational opportunities? What may we have overlooked in the past that we have a chance to search for in our next major mission?"

One such example may be a search for faint sources in the immediate vicinity of very bright ones. A large telescope lends itself to observing fine spatial features. The place where looking for these may be most important is in the immediate vicinity of whatever compact source may dominate a region. However, this is precisely where our high resolution instruments tend to fail. Some years ago, John Bahcall and colleagues tried to look at host galaxies of quasars in the immediate vicinity of the bright central source (Bahcall et al. 1997). But the surface brightness of the diffracted light within one arc second of the quasars was as high as one part in 10^4 of the central source. Diffraction has also limited our study of winds emanating from O stars, observations of close binaries and nova ejecta, and exoplanet searches.

Such studies should now become possible as a consequence of the recent work of Nisenson and Papaliolios at Harvard, who have investigated a powerful optical apodization method. This permits detection of a faint object within a few tens of milli-arcseconds from a source more than a billion times brighter (Nisenson & Papaliolios 2001). Spergel (2001) has come out with a similar theoretical proposal. Nisenson has demonstrated his method in the laboratory by detecting a faint source separated by only a few Airy fringe widths, in this case 200 milli-arcseconds from a laser source 3×10^8 times brighter. And this was done with off-the-shelf equipment. The main optical requirement for success is extremely high surface quality mirrors. The Terrestrial Planet Finder community has been energetically pursuing this lead. So, we may expect that there will be a concerted push for industry to refine its mirror polishing techniques. To provide significant benefits, industry does not need to go far beyond what they have already achieved in polishing the mirrors for Chandra. With such efforts going on, the time appears ripe to produce an optical/ultraviolet telescope with extremely good surface quality.

In summary, if we do get to place an optical/ultraviolet telescope in space, we should look hard for means of investigating the Universe in ways that we have systematically ignored in the past. This is an opportunity to correct the neglect. I have mentioned searches for faint sources very close to bright ones. These may or may not be the most important areas we have neglected. There may be more pressing neglects elsewhere, possible a systematic search for rapidly varying sources of the kind common in radio, X-ray, and gamma-ray astronomy, but not yet fully investigated in the optical/ultraviolet regime.

If there is one lesson the history of our discipline seems to be crying out to tell us, as we begin to plan the next big optical/ultraviolet mission, it is: "Let us come to terms with what we have so far failed to observe; let us not rely entirely on what we think we understand today. And let us put this attitude to work right from the start as we plan a new observatory, rather than wait until it is launched. By then it may be too late to seek observational innovations."

References

Alpher, R.A., & Herman, R. 1949, Phys. Rev. 75, 1091

Babcock, H.W. 1947, ApJ 105, 105

Bahcall, J.N., Kirhakos, S., Saxe, D.H., & Schneider, D.P. 1997, ApJ, 479, 642

Bell Burnell, S.J. 1978, Sky & Telescope, 55, March issue, 218

Duncan, J.C. 1939, ApJ 89, 482

Duyvendak, J.J.L. 1942, PASP 54, 91

Giacconi, R., Gursky, H., Paolini, F.R., & Rossi, B.B. 1962, Phys. Rev. Lett. 9, 439

Harwit, M. 1980, Cosmic Discovery – The Search, Scope, & Heritage of Astronomy, (New York: Basic Books)

Hazard, C., Mackey, M. B., & Shimmins, A.J. 1963, Nature 197, 1037

Jeans, J. 1931, The Universe Around Us, (London: Cambridge University Press)

Jamieson, J.A. 1995, Proc. Infrared Instrumentation Society (IRIS) 40, No1., 13

Klebesadel, R.W., Strong, I.B., & Olson, R.A. 1973, ApJ 182, L85

Knight, C.A., et al. 1971, Science, 172, 52

Kouveliotou, C., et al. 1996, Nature 379, 799

Luyten, W.J. 1949, ApJ 109, 532

Mayall, N.U., & Oort, J.H. 1942, PASP 54, 95

Mayor, M., & Queloz, 1995, Nature,378, 355

Miller, G. 1996, Nature, 381, 472

Neugebauer, G., Martz, D.E., & Leighton, R.B. 1965, ApJ 142, 399

Nisenson, P., & Papaliolios, C. 2001, ApJ, 548, L 201

Penzias, A.A., & Wilson, R. W. 1965, ApJ 142, 419

Percival, J.W., et al. 1993, ApJ 407, 276

Punch, M., et al. 1992, Nature 358, 477

Soifer, B.T., et al. 1984, ApJ 283, L1

Spergel, D. 2001, astro-ph/0101142

Weaver, H., Williams, D.R.W., Dieter, N.H., & Lum, W.T. 1965, Nature, 208, 29

Hubble's Science Legacy: Future Optical-Ultraviolet Astronomy from Space
ASP Conference Series, Vol. 291, 2003
K.R. Sembach, J.C. Blades, G.D. Illingworth, R.C. Kennicutt, Jr. (eds.)

Highlights of the SUVO Study

J. Michael Shull

CASA, Dept. of Astrophysical & Planetary Sciences, University of Colorado, Boulder, CO 80309

Abstract. This paper summarizes the scientific and technological goals of the Space Ultraviolet-Visible Observatory (SUVO) mission study, commissioned by NASA in 1998-99. The full report (Shull et al. 1999) may be obtained as the SUVO white paper, available at astro-ph/9907101 and on the website: http://origins.Colorado.EDU/uvconf/UVOWG.html. The UVOWG was asked to study the scientific rationale for new missions in ultraviolet/optical space astronomy. A large-aperture telescope in space would provide a major facility for understanding the "Emergence of the Modern Universe". Key science applications include: (1) mapping dark baryons, metals, large-scale structure, and dark-energy equation of state; (2) detecting unseen matter through weak gravitational lensing; (3) studying the feedback from star formation; and (4) measuring the evolution of stars, planets, and elements from redshifts $z = 3 - 4$ down to the current epoch. In the last several years, it appears possible that the SUVO concept might also address exciting new goals in the study of extra-solar planets and their atmospheres, through UV/optical coronagraphy. The UVOWG studied both 4 m and 8 m telescopes, and made recommendations about the critical technology development for high-quantum-efficiency ultraviolet detectors, gratings, mirrors, spectrographs, and wide-field optical/UV imagers (at least 16 K $\times$ 16 K). In the current epoch of realism in mission costs and schedules, I believe that a 4 m UV/optical telescope would provide a powerful scientific mission for the next decade.

1. Introduction

This article provides a focussed summary of the direct connections of the proposed Space Ultraviolet-Visible Observatory (SUVO) mission to the NASA/OSS and the Origins/SEU strategic plans. The full scientific and technological aspects are contained in the report of the Ultraviolet-Optical Working Group (UVOWG). The UVOWG was commissioned by NASA to study the scientific rationale for new missions in ultraviolet/optical space astronomy. A large-aperture UV/optical telescope in space will provide a major facility in the 21st century for solving scientific problems in the NASA/OSS Strategic Plan and roadmaps.

Since the SUVO report (Shull et al. 1999, astro-ph/9907171), many of the scientific highlights have gained greater attention in the astrophysical community. The HST has just been upgraded with two exciting new imagers (ACS and

NICMOS), and it will achieve further gains in capability after Servicing Mission-4 (March 2004) with the Cosmic Origins Spectrograph (COS) and Wide Field Camera 3 (WFC3). Nevertheless, unless NASA can find the additional funds to augment the Hubble Project in a 5th servicing mission, the UV/optical community will need to accept the eventual mortality of HST.

The SUVO science goals are listed below. Some of these goals are complementary to those of interest to the infrared and X-ray communities. However, several important scientific goals can only be achieved using the images and sensitive spectral diagnostics of the UV/optical wavelength bands. In particular, images of weak gravitational lensing and large-scale surveys of Lyα and metal absorption lines (O VI, C IV) may be the only way to find the dominant components of the "invisible universe". SUVO will allow astronomers to map out the large-scale distribution of dark matter and the nearly invisible warm photoionized and hot shocked components of the "cosmic web" of matter that may dominate the missing baryons. It will also study the assembly of galaxies, and map out the distribution of the nucleosynthetic products of star formation in these high-redshift objects.

1.1. Major SUVO Scientific Goals

- Map dark baryons, metals, large-scale structure, and the dark-energy equation of state
- Detect unseen matter through weak gravitational lensing
- Study the feedback from star formation
- Determine the evolution of stars, planets, and the elements

SUVO will extend the scientific heritage of HST and NGST. No space astrophysics mission has demonstrated wider public appeal than the Hubble Space Telescope. Discoveries from the Hubble program provide an effective combination of captivating images and compelling scientific questions. SUVO will continue this tradition, through its wide-field images of galaxies, nebulae, clusters, stars, and planets. Its spectroscopic studies will produce answers to fundamental astrophysical and cosmological questions about the origin and evolution of matter in the Universe. To inform and engage the public, these answers must be cast in a framework of large questions about galaxy formation, the evolution of large-scale structure, and the transport of heavy elements.

The UVOWG studied the science of both 4 m and 8 m UV/optical telescopes; AURA is currently studying a large NHST (Next-Generation Hubble Space Telescope) to be launched in the decade after $\sim$ 2010. I believe that realistic estimates of mission cost and schedule dictate that we should proceed first with a 4 m option. This mission would focus on UV spectroscopy, wide-field imaging, and planet-finding (coronagraphy). However, no high-throughput UV/optical mission will be possible without significant investments in technology, including UV detectors, gratings, mirrors, spectrographs, and imagers.

2. Key Science Goals of SUVO

The NASA strategic plan goals are directly related to key science identified for SUVO, focussed on three themes: (1) Measuring the dark matter and baryons; (2) Understanding the chemical evolution of the elements; (3) Observing the major construction phase of galaxies and quasars. Our overall theme is entitled "The Emergence of the Modern Universe", defined as the period from redshifts $z \approx 3 - 4$ down to the present epoch. This period occupies over 80% of cosmological time, and captures the evolution of nearly every major astronomical structure, from galaxies and clusters to quasars and the intergalactic medium (IGM). During this time (8–10 Gyr), large-scale structure developed in both dark matter and baryons. Much of the intergalactic gas collapsed onto galaxies, over 90% of the heavy elements were formed, and the energy sources of radiation, hot gas, and dynamic outflows acted back on the surrounding galaxies and gas clouds. This feedback from galaxy formation is the least understood aspect of galaxy evolution and assembly. The SUVO mission will provide definitive measurements, with unprecedented accuracy, to fill the evolutionary gap of galaxies, IGM, and large-scale structures from infancy to maturity. SUVO will also provide accurate (nearby objects, low-confusion) low-z templates required to understand the high-z phenomena. Complementing the distant frontiers, SUVO will thoroughly characterize nearby bright objects, with exquisite imaging and spectra.

2.1. Major Scientific Questions for SUVO

- Where is the rest of the unseen universe?
- What is the interplay of the dark and luminous universe?
- How did the IGM collapse to form the galaxies and clusters?
- What is the role of star-formation feedback (radiation, metals, energy)?
- When were galaxies, clusters, and stellar populations assembled?
- Are massive black holes a natural part of most galaxies?
- What is the history of star formation and galactic chemical evolution?
- How do extra-solar planetary systems evolve, and what do they look like?

SUVO will connect the high-redshift universe observable by NGST to the low-redshift universe, which SUVO can study in detail by wide-field imaging and quantitative spectroscopy. SUVO will also relate infrared and sub-mm observations of distant galaxies (NGST, FAIR/SAFIR) to optical and ultraviolet observations of the local universe. NGST observes the rest-frame ultraviolet light of the first stars, galaxies, and quasars, redshifted to 1–5 μm by cosmological expansion. For cosmological studies of the distant universe, space astronomy must adopt a multi-spectral view, in which NGST and SUVO complement one another, both in wavelength and across cosmological time.

3. Specific Connections to the NASA/OSS Strategic Plan

Many scientific programs can be accomplished with wide-field imaging and high-throughput spectroscopy on a 4–8 m telescope in space (SUVO or NHST). Among these, several projects are unique to SUVO and worth special emphasis:

3.1. Mapping Dark Baryons and Large-Scale Structure

From recent galaxy redshift surveys, astronomers have detected the existence of an organized large-scale structure in the galaxy distribution, which takes the form of large filamentary walls and "empty" voids. With H I, He II, and O VI absorption, we are starting to detect the variety of warm photoionized and hot shocked baryons and test the predictions of Λ-CDM cosmological simulations of structure formation and feedback of galaxy formation. By 2010, galaxy surveys will have outlined the distribution of luminous matter in fine detail, but the dark, gaseous universe (the IGM) will remain largely unexplored at $z < 1.6$. SUVO will measure the distribution of dark baryonic matter and heavy elements in these filaments and voids, tracing the cosmic web using UV spectra of H I, C IV, O VI, and other ions toward QSOs. The goal is to conduct an IGM baryon and metal survey on sub-degree angular scales, comparable to that of the MAP explorer and to the structure seen in galaxy surveys. In doing so, we will connect the high-redshift seeds of galaxies and clusters with the distributions of galaxies and metals in the modern epoch, at $z < 1$. This project is impossible without improvements in UV detectors, gratings, and spectrographs. To achieve a frequency of one QSO every $10' - 20'$ on the sky requires using QSOs at $m_B = 18 - 20$ as background targets. In the next several years, the GALEX mission is expected to identify 10^5 QSOs with $18 < m_B < 20$, and the Sloan survey will provide redshifts for $\sim 10^5$ QSOs brighter than $i = 19$.

3.2. Detecting Unseen Matter in the Modern Universe

SUVO can use large-scale weak gravitational lensing to probe the underlying matter in galaxy clusters and superclusters, extending over $5 - 20h^{-1}$ Mpc. A $10' - 15'$ field of view (much larger than HST or NGST instruments) provides a good match to expected correlation lengths of cosmological structures at $z < 1$. Through its wide-field imaging, SUVO will use weak gravitational lensing to map out the distribution of dark matter in clusters of galaxies and perhaps on the supercluster scale. The inversion of the lensed arclets is best done in space, with the stable point spread function over a wide field in the optical band. One can also characterize dark energy equations of state.

3.3. Studying the Feedback from Star Formation

SUVO will follow the evolution of the Universe from the "dark ages" down to the "renaissance" of star formation, supermassive black holes (quasars) and metal production. Measuring the evolution of the Lyα baryon content is vital if we are to understand the mass evolution of galaxies, black-hole co-evolution, the rate at which gas in the IGM is incorporated into galaxies, and the rise of metallicity over cosmic time. We can study the interplay between the formation of galaxy structures and the IGM and study mass exchange: the depletion of the reservoir of intergalactic gas into galaxies, and the flow of mass from galaxies back to the IGM through galactic winds and tidal stripping.

3.4. Studying Galactic Star formation

With its wide-field imagers, SUVO will be able to map, in a single pointing, entire Galactic H II regions in diagnostic emission lines and broad-band colors. Within the Milky Way, but also in nearby external galaxies, SUVO will study global star formation and the interactions between young hot clusters and the interstellar medium at scale lengths necessary to measure important physical processes (winds, shock waves, ablation of disks). The dynamic range of these images will allow studies of the stellar initial mass function from the most massive stars down to brown dwarfs.

4. Required Key Technologies

As discussed in great depth in the UVOWG White Paper, the SUVO Mission will require significant NASA technology investments. Throughput is the single most important technology driver for the future of UV/optical space astronomy, both for spectroscopy and for wide-field imaging. The key areas are: detectors, large light-weight precision mirrors, optical materials and coatings, and precise optical elements (gratings and micro-mirrors). The following short list highlights the most critical needs identified by the UVOWG:

- Develop more sensitive UV detectors, with low background noise (including cosmic-ray exclusion), high quantum efficiency (QE), large dynamic range, and large formats. Increasing detector QE in the UV can be as important as building larger mirrors.
- Space-qualify large mosaics of low-noise, high-QE CCD detectors (at least 16K $\times$ 16K) for wide-field imaging cameras in the optical and near-UV.
- Develop micro-mirror arrays for use in multi-object and integral-field spectrographs in the UV and visible.
- Develop large, lightweight precision mirrors for use in the UV/visible. Although SUVO could be done with a 4 m monolith, the extension to an 8 m aperture will likely require segmented deployable optics.
- On a very long time scale, develop "3D" energy-resolving detectors such as photon-counting superconducting tunnel junction (STJ) or transition-edge sensor (TES) devices. These cryogenic detectors have the potential to revolutionize UV/optical astrophysics. This cryogenic technology would be shared with other missions operating at longer (e.g., far-IR and sub-mm) wavelengths.

Hubble's Science Legacy: Future Optical-Ultraviolet Astronomy from Space
ASP Conference Series, Vol. 291, 2003
K.R. Sembach, J.C. Blades, G.D. Illingworth, R.C. Kennicutt, Jr. (eds.)

The Future of Optical Astronomy in ESA's Science Program

Alvaro Giménez

Research and Scientific Support Department, ESA-ESTEC, Noordwijk, The Netherlands

Abstract.
A summary of the present status of ESA's Science Program is given together with a more detailed discussion of the main activities in optical space astronomy. ESA missions are defined in response to specific and relevant scientific questions rather than to provide general-purpose facilities. Within this context, GAIA is designed to deliver the necessary tools to understand the Galaxy and its content by surveying around one billion stars astrometrically, photometrically, and spectroscopically. The Eddington mission, on the other hand, is aimed at the study of stellar structure through the measurement of oscillations and the search for extrasolar planets within the habitable zone by means of the transit technique. In addition, technology developments are being pursued to ensure the availability of focal plane arrays of detectors with the required performances and to test space interferometry techniques.

1. Introduction

The Science Program of the European Space Agency (ESA) is aimed at providing the scientific community with the required support in space infrastructure for their research activities. The areas covered by the science program include the exploration of the Solar System (including the Sun), astronomy, and fundamental physics. ESA's activities may be optional or mandatory. The scientific program is mandatory and accounts for between 10% and 12% of the total budget. The contribution of each member state is made according to its gross national product. The current annual budget for science is around 360 million euros.

The selection of individual missions is based on community-wide peer review involving ESA's advisory bodies and, finally, the delegate body of the Science Program Committee. Each project is then implemented in close collaboration with institutes in the member nations. ESA's budget is basically used for the spacecraft development, launch, and operations, while national institutes contribute to scientific payloads through a competitive process and their own funding. The mandatory nature of the science program allows for the definition of a long-term plan which helps in the preparation of national efforts as well as the establishment of international co-operation. Optical astronomy from space is, of course, part of ESA's planning, given the exceptional conditions that can be achieved in space for global surveys, accurate photometry, and stability.

Several missions are now on their way to the launch pad. Within half a year, four missions will be launched. Three are in the area of Solar System exploration: Rosetta, to explore a comet; SMART-1, to test electric propulsion technologies and revisit the Moon; and Mars Express, to investigate the red planet by means of an orbiter and a lander. The other mission to be launched shortly is INTEGRAL. This is a large high-energy observatory that will provide images and spectra of astrophysical sources in the gamma ray domain with unprecedented sensitivity and resolution. INTEGRAL uses a slightly modified version of the XMM-Newton bus and is to be launched with a Proton vehicle in October 2002. In the mean time, XMM-Newton, launched in 1999, is providing a new view of the Universe at X-ray energies thanks to its large photon-collecting capability and throughput. After a phase of commissioning, verification and calibration, XMM-Newton entered a successful phase of routine scientific operations as an observatory.

Also, work is continuing with previous missions in their post-operational phase aimed at obtaining a homogeneous recalibration of the observations and the delivery of a final archive to the scientific community. The International Ultraviolet Explorer (IUE) final archive was already handed over to national organizations, and we are now within the active archive phase for the Infrared Space Observatory (ISO), increasing the quality and accessibility of the data by improving the previous global pipeline analyses that populated the existing archive.

For missions under development, the main activity now concentrates on preparing for two astronomy missions: Planck, to investigate the fine spatial variation of the cosmic background radiation, and Herschel, a powerful far-infrared and sub-millimeter observatory. They will be launched together in 2007 and delivered to the L2 Lagrangian point on the Earth-Sun line for their operational phase.

With respect to optical and ultraviolet (UV) astronomy, mention has to be made of the successful scientific outcome of the co-operation program with NASA that led to the Hubble Space Telescope (HST) mission. HST continues to generate exciting results even in areas that were unimagined during its design (e.g., the study of extrasolar planets). The recent servicing mission to HST brought back to Earth the European Faint Object Camera (HST/FOC) after it was replaced by the Advanced Camera for Surveys (HST/ACS).

2. Reworking the Long-Term Plan

The possibility of maintaining a comprehensive space science program in ESA, preserving the flexibility to incorporate new missions in a short time frame under the pressure of reduced overall budgets has been a recent task of the Science Directorate. The long-term planning of the program has proved to be robust and adaptable to new challenges and uncertainties. In 2001, a proposal was made to increase the number of missions, their scope, and the rate of launches. Unfortunately, the Ministerial Conference held in Edinburgh by the end of the year did not support the expected level of funding, and an entire re-planning exercise was needed. During the last months, the way of working that was in place for the implementation of the previous long-term plan of ESA, called

"Horizons 2000", has been fully revised. A fixed end date for the planning of individual missions up to 2012 was decided upon to avoid the temptation of stretching schedules without really modifying the way of working. Anything beyond the timeframe of the coming decade should be expressed in terms of guidelines, or scientific ideas, rather than specific missions.

The challenge was to maintain as many of the missions demanded by the European scientists while keeping to a realistic budget within the approved boundaries. Avoiding the initial possibility of concentrating all efforts on the so-called cornerstone missions, the aim was to keep an average launch rate close to one mission per year. Mission extensions were, of course, favored to maximize the scientific return per investment. The reuse of standard platforms was necessary, as was the need to procure smarter, more system-integrated spacecraft. International cooperation was also maintained and fostered whenever possible. Finally, the concept of "production groups" was introduced; the platforms are reused, and the engineering teams, budget, and schedule are tailored for optimum use of all available resources for a given group of missions. Thus, the Eddington mission is now grouped with the already-paired Herschel and Planck missions.

Given the changes introduced into the long-term plan of ESA's Scientific Program – both the missions and the way of working – the former "Horizons 2000" has been replaced by "Cosmic Vision".

Within the new planning, some missions are now under their study or definition phase. Some are planned for launches between 2010 and 2012, such as the European contribution to the Next Generation Space Telescope, the Laser Interferometer Space Antenna (LISA) gravitational wave observatory, and GAIA, which will produce an unprecedented three-dimensional map of the stars in the Galaxy. Beyond that, missions such as Darwin, to characterize extrasolar planets, and XEUS, a very large X-ray observatory deployed using the International Space Station, are under study. There have been some additions during recent updates of the program for missions with launches before 2010. An example in astronomy is the Eddington mission, which will measure stellar oscillations and search for Earth-like extrasolar planets.

3. The Future after HST

HST has represented the main source of information for European scientists in the optical domain from space. In the ultraviolet it provided the most relevant tool after the community interest grew considerably with the accumulated experience over the years using the long-lived IUE satellite. But we should now question ourselves about the future. For this purpose, it is important to point out the European effort in ground-based optical astronomy to provide a large infrastructure such as the Very Large Telescope (VLT) in Chile within the European Southern Observatory. It is clear that ground-based facilities should be fully exploited before attempting new steps in space. This implies scientific objectives as well as testing of new techniques and hardware such as adaptive optics, interferometry, and CCD-based focal-plane detectors. Moreover, given the current pressure on smaller smarter space missions, new projects should probably be based on specific science objectives rather than attempting to provide large general-purpose facilities. Opening possibilities for flexibility or new ideas through relatively small missions is the aim, as is increased international cooperation. Within this latter context, we should of course remain open to a possible European contribution to a large observatory.

In the current program, optical astronomy is considered through specific missions as well as the developing of enabling technologies for future missions. The relevant science goals in the optical domain from space are: a) understanding the structure, content, and evolution of the Galaxy, b) understanding the formation of planetary systems, like our own, and searching for Earth-like extrasolar planets. Both objectives require accurate global astrometry and high-precision photometry that are only available from space. Two projects have been designed to cope with this science using space-based optical measurements: GAIA and Eddington.

4. The GAIA Project

GAIA is an astrometry mission to be launched no later than 2012. When compared with Hipparcos, many differences are immediately identified. GAIA will have a limiting magnitude around 20–21, while Hipparcos was limited to $V = 12$. As a consequence, the number of stars surveyed will be around one billion for GAIA, while it was just $\sim 120,000$ for Hipparcos. In terms of astrometric precision GAIA will reach micro-arcsecond values, while Hipparcos was limited to a few milli-arcseconds. Instead of only observing in two broad-band filters, GAIA will provide photometry for all stars in 4 broad-band and 11 intermediate-band filters. GAIA will observe a large number of quasars and galaxies for reference and includes a global spectroscopic survey providing radial velocities with precisions of a few km s^{-1} at 16th magnitude. A very important difference is the unbiased, not pre-selected, GAIA survey, which will allow an unprecedented multi-dimensional study of the Galaxy with a large, statistically significant sample of its members.

The key objectives of GAIA are the structure and kinematics of the Galaxy, the study of stellar populations, and the test of galaxy formation models. In addition, GAIA will provide a major tool for stellar astrophysics by delivering comprehensive luminosity calibration and stellar properties. Moreover, the mission will allow the study of important problems of the Solar System through a deep and uniform detection of all moving objects, will measure many extra-galactic objects. It will provide essential tests of some problems in fundamental physics.

The astrometric focal plane will consist of more than 100 large-format CCDs and will be divided into a main section for astrometric measurements and two side regions for the identification of objects and broad-band photometry. An independent telescope and focal-plane assembly will provide the necessary radial velocity measurements as well as intermediate-band photometry. Photometric measurements of around one billion stars in an unbiased sample requires a careful analysis of the best possible color indices and wavelength bands. In addition to the classical problem of interstellar extinction determinations, a large number of low mass stars will be observed, as will rare objects needing classification tools. In order to be able to extract the necessary information from the huge database that is anticipated, the identification and proper grouping of stars are needed.

It is worth mentioning that GAIA will provide orbits for more than 100,000 resolved binaries with separations larger than 20 milli-arcseconds. Masses for more than 10,000 objects with precisions of the order of 1% will be obtained throughout the Hertzsprung-Russell diagram. The astrometric measurements will even permit the study of interacting binaries, systems with brown dwarf components, and some planetary companions. Millions of binary systems will be identified through their motions or variable photometry.

5. The Eddington Project

Ground-based studies using accurate radial velocity measurements have allowed the discovery of extrasolar planets but the uncertainty about the orbital inclination does not allow measurements of the masses of the companion objects. After a survey of $\sim$ 3000 G and K stars in the solar neighborhood, less than 100 of them have been found to have companions with masses that could be compatible with giant planets. These discoveries have led to even more problems since many Jupiter-sized planets have been found with short periods ($\sim$ days), and even eccentric orbits, contrary to our Solar System experience and what we expect from available planetary system formation models. Nevertheless, observations indicate a significant increase of planets for lower masses despite the observational bias favoring larger masses. The real mass distribution, the existence of planets or satellites around giant planets, like our Earth, or their possible location in the habitable zone are among the most challenging questions of modern astronomy. GAIA can observe stars with planetary companions but cannot detect Earth-like planets.

In order to answer some of these questions, Eddington was designed and proposed as an ESA mission to be launched in the timeframe between 2007 and 2008. Eddington has two main science goals: a reliable, tested theory of stellar evolution to be used in astrophysics, and the detection of habitable Earth-like planets, their frequency, and their properties. The former science goal requires studies of the ages of stars, stellar structure, chemical evolution, and nucleosynthesis, and is achieved through the technique of astroseismology (i.e., the measurement of stellar oscillations). In this research activity, the Solar and Heliospheric Observatory (SOHO) has already demonstrated the validity and efficiency of this method in studying the Sun. The second science goal involves the detection and study of planets, with a wide range of characteristics, and the properties of planetary systems through the method of transits. In both cases, long duration, high-precision photometry from space is required.

Studies of stellar structure and evolution with Eddington will provide essential information about the physical parameters of stars, such as pressure, density, and rotation, as well as key tests of the physics of convective overshoot, mixing, and diffusion. Furthermore, detailed information about the chemical evolution of stars will be available, and the ages of individual stars and stellar systems will provide an accurate dating method.

In the case of extrasolar planets, the detection of thousands of planets is expected, with hundreds located in the habitable zone and a significant subsample of Earth-like objects. Due to the unbiased selection of targets to check for planetary transits, the first reliable statistic on the abundance of planets will be acquired. Earth-like planets will be discovered around stars as faint as V = 15, and the habitable zone will be covered for stars with G, K, and M spectral types. In addition, the long term monitoring of transits will allow the detection of massive satellites around giant planets, and hot giant planets may be detected through reflected light.

As a precursor to Eddington, the French-led mission COROT is aimed at astroseismology of a limited number of bright stars and the search for relatively massive extrasolar planets. ESA contributes to this project a 30 cm off-axis telescope in low polar orbit that will be launched before the end of 2005.

6. Other Activities

Some fundamental questions will be raised after the results of the previously mentioned missions on extrasolar planets. One of them is whether conditions on newly discovered planets can sustain life as we know it. In order to find and characterize Earth-like planets and detect biospheres, the Darwin project is now being discussed. The initial requirements for the study of star-planet separations and luminosity ratios indicate the need of nulling interferometry in the mid-infrared range. Six telescopes in hexagonal configuration with a beam combiner are being considered as a possible design scenario.

In order to provide the necessary technology available for this future mission, a Ground-based Nulling Interferometer Experiment (GENIE), is currently being developed. GENIE is an instrument for the VLT in Chile that will see first light in 2006. Furthermore, to demonstrate precision formation flying and to verify the need for optical delay-line control systems in space, a technology mission is being studied. This could be either an upgrade of SMART-2, initially aimed at tests required for LISA, or a dedicated SMART-3. Parallel activities are also taking place to ensure the availability of the best possible detectors to be used in space. This applies to ultraviolet as well as optical astronomy and includes the use of superconducting tunnel junction detectors for time-resolved spectrophotometry or deep field imaging combined with spectrometry. Further information about these technologies, and also GENIE, is provided in separate contributions.

Hubble's Science Legacy: Future Optical-Ultraviolet Astronomy from Space
ASP Conference Series, Vol. 291, 2003
K.R. Sembach, J.C. Blades, G.D. Illingworth, R.C. Kennicutt, Jr. (eds.)

Ground-Based Astronomy in 2010-2020: ESO's View

Roberto Gilmozzi

European Southern Observatory, Garching bei München, Germany

Abstract. This paper is an extended summary of my talk at the Workshop, with emphasis on the European Southern Observatory's plans for second generation instrumentation at the Very Large Telescope (VLT), the plans for VLT Interferometer (VLTI) and the Atacama Large Millimeter Array (ALMA), and the concept study of the 100 m OverWhelmingly Large (OWL) Telescope. The scientific complementarity between these facilities and a possible optical/ultraviolet space mission are briefly discussed.

1. Third Decade, Second Millennium A.D.

The third decade of the second millennium A.D. will see the maturity of the current generation of telescopes [VLT, Keck, Gemini, Subaru, Gran Telescopio Canarias (GTC), the Hobby-Eberly Telescope (HET), etc.] equipped with a second generation of instruments often performing at the diffraction limit through advanced adaptive optics (AO) systems. Interferometry will have come out of its infancy to operate in the faint object regime (K$\sim$ 20) and to produce astrometric results in the micro-arcsecond range. ALMA will provide mm and sub-mm astronomers with a facility equivalent to optical ones (both in terms of service offered to the community and in resolution and sensitivity). A new generation of ground based optical/near-infrared 30 to 100 m telescopes now on the drawing board [the California Extremely Large Telescope (CELT), OWL, the Giant Segmented Mirror Telescope (GSMT)] may start producing unprecedented results, with a resolution of $\sim$ 1 milli-arcsecond and sensitivity hundreds or even thousands of times beyond what is available today.

2. Evolution of Existing Facilities

2.1. Adaptive Optics

AO, now in its early stages, will soon outgrow the current limitations — single natural or laser guide star (GS) systems, limited field of view (FOV), small sky coverage — through the development of multi-conjugate adaptive optics (MCAO). MCAO uses multiple natural/laser GS systems to provide a wider corrected FOV, and is now being developed at several existing observatories. For example, Gemini is building an MCAO system for its instrumentation. ESO is building the MCAO Demonstrator (MAD) to see first light at the VLT in early 2004 as an enabling experiment for the new VLT instruments and for

OWL. AO is considered one of the most critical developments for astronomical instrumentation and a "go/no-go" milestone for future giant telescopes.

2.2. Second Generation Instruments

Among the second generation instrumentation considered by ESO are a multi-micro-mirror classical AO system instrument (FALCON) to study many individual objects in the telescope FOV at the same time, AO-fed planet finders using nulling interferometry coronagraphs, near-infrared multi-object wide-field spectral imagers, image-slicer-based multi-integral-field spectrographs, and very wide wavelength coverage "fast shooters" able to do simultaneous spectroscopy from 0.3 to 3 μm. The underlying philosophy is one of sampling the instrumentation parameter space (wavelength, resolution, FOV, image quality, multiplexing ability, complementarity with other space or ground facilities, etc.) based on clear science requirements.

2.3. Interferometry

VLTI has achieved fringes in 2001, both with test siderostats and with the VLT 8 m telescopes, and is currently in continuous operation performing technical commissioning as well as scientific observations for the community. It will soon evolve towards imaging, both with the present generation of instruments (AMBER – the Astronomical Multiple BEam Recombiner – using 3-telescope measurements of phase closure) and with PRIMA (Phase Referenced IMAging, a dual feed facility providing stabilization of the fringes of a faint object by tracking the fringes of a bright reference star within one arcminute). The series of instruments will image planetary systems, the inner regions of active galactic nuclei, and objects as faint as $K \sim 20$. VLTI will also provide astrometric measurements down to a few micro-arcseconds thus enabling the possibility of direct detection of extrasolar planets and determination of their orbits.

3. ALMA

The Atacama Large Millimeter Array, an example of big project collaboration between Europe and the United States, is a variable configuration array of 64 12 m antennae working in the 0.3 to 10 mm wavelength range to be located at an altitude of 5000 m in Chajnantor in the Atacama Desert. ALMA is a 50/50 partnership, with ESO managing the European side. It will have high angular resolution ($< 0.01''$ with a baseline > 10 km) and high sensitivity (area $\sim 7,200\,\mathrm{m}^2$). ALMA will be able to study galaxy formation in the very early Universe, resolve the far-infrared background, study star formation deep in dark clouds, search for protostars, analyze star and planet formation processes, study the bodies of the Solar System. The project is entering into Phase 2 (construction in 2003-2010) and will start interim operations with a reduced number of antennae as early as 2006.

4. The Next Generation Telescopes: OWL

Since 1997 ESO is studying a concept of a ground optical/near-infrared 100 m telescope dubbed OWL for the eponymous bird's keen night vision and for "Over-Whelmingly Large". The study is in Phase A. OWL's design relies on proven technology and on a mass-production approach to contain the cost to the maximum possible extent. The baseline design (alternatives are also being studied) has a spherical segmented F/1.42 primary ($\sim 2,000$ identical segments), a segmented flat secondary (to relax mechanical tolerances), and a corrector based on two 8 m monolithic mirrors with a size chosen to take advantage of VLT developments. Two further smaller (adaptive) mirrors are used for focusing and field stabilization to deliver a final F/6 beam. The design provides a $3.0'$ diffraction-limited FOV in the visible, and a $10'$ field with an image quality of $\sim 0.07''$. OWL is conceived as an adaptive telescope delivering diffraction-limited performance. The mechanical structure is based on a "lego" approach (mass produced elements), with low cost, distributed kinematics (friction drives) and built-in maintenance. The structure is self-supporting (no scaffolding during construction). OWL will operate in open air, with a sliding hangar (similar to, but half the size of one built for dirigibles near Berlin) for weather protection. Wind studies confirm that the stiffness of the lightweight steel structure ($\sim 8,000$ tons, comparable to the weight of today's 100 m radio telescopes) allows laminar wind flow operations under moderate wind speeds ($\geq 10\,\mathrm{m\,s^{-1}}$).

OWL's design has several advantages compared to traditional ones: relaxed mechanical constraints (cm instead of μm), "built-in" adaptive optics and field stabilization, economy of scale, mass production, proven technology. It also has diffraction-limited performance and a better wavefront root-mean-square residual (3 times better than an equivalent Ritchey-Chrétien). The drawback is worse throughput (more surfaces), which could be offset by an appropriate choice of coating (e.g., Ag instead of Al).

With milli-arcsecond-resolution and limiting magnitude V$\sim$ 38, OWL will be capable of imaging Solar System objects at resolutions comparable to those offered by space probes, over much longer time scales. It will unveil the intricate processes underlying the formation of stellar and planetary systems. It will resolve solar-type stars at the distance of Virgo. It will be able to image exoplanets and exo-Earths, determine their atmospheric composition, and possibly reveal the existence of biospheres. It will peer into the deepest reaches of the Universe and witness the birth of the very first stars and galaxies. It will perform spectroscopy of supernovae out to a redshift of 5. It will study the evolution of the cosmological parameters. Indeed, it may, eventually, revolutionize our perception of the Universe as much as Galileo's telescope did.

The primary objectives of the OWL study are to verify the feasibility of a 100 m class optical telescope, explore potential science cases, define a baseline conceptual design and operations scheme, and establish reliable cost, schedule, and performance estimates. This study would eventually lead to a proposal for the detailed design and construction of the OWL telescope. The current cost estimate is 900 MEuro (backed by industrial studies). The timeline from start of Phase B to completion is 12 years, but science could be started earlier with a partially filled primary (the "grow a telescope" concept). It is likely that a project of this scale will need an international collaboration similar to ALMA.

5. Conclusions

The second decade of the third millennium will witness the maturity of the present generation of telescopes with new instruments and new capabilities afforded by advances in adaptive optics. It will also see imaging interferometry on faint sources and the advent of ALMA. In parallel to this, a new generation of giant telescopes delivering milli-arcsecond-resolution at magnitudes 35-38 (if AO lives up to expectations) will possibly also see first light in that decade. I have presented ESO's plans and projects in these areas.

There are many synergies among the facilities foreseen in space and on the ground. ALMA, NGST, and OWL (or another extremely large telescope) will play complementary roles in studying the Universe, in detecting planets, and in improving our view of many astronomical phenomena and the cosmos.

Some questions related to this stem from the consideration that a 70 m+ near-infrared telescope on the ground could detect exo-Earths and possibly observe their spectra. Is this argument strong enough to try to have one as soon as possible, and should it be a collaborative unified effort? Should it be an alternative to or complementary to space missions with similar science goals? Could it be used to accelerate follow up missions like the Terrestrial Planet Imager? The answers to these questions may have a profound impact on the future of projects like OWL.

A possible optical/ultraviolet successor to HST was the topic of the meeting, and many interesting science cases have been presented here to support its construction. The UV is of course a domain unique to space. However, in view of the possibilities opened up by adaptive optics, the optical capabilities of a 4-8 m telescope in space may not be competitive with those of 30-100 m telescopes on the ground. It is not inconceivable that 10 years from now it may make more sense to go to space only for those wavelengths for which the advantage is overwhelming (X-ray, ultraviolet, thermal infrared, etc.), leaving the optical and near-infrared to adaptive ground-based telescopes that for similar costs could provide much higher angular resolution and sensitivity. I think it is not premature to consider such a possibility, even acknowledging how much it depends on very demanding developments in adaptive optics.

Hubble's Science Legacy: Future Optical-Ultraviolet Astronomy from Space
ASP Conference Series, Vol. 291, 2003
K.R. Sembach, J.C. Blades, G.D. Illingworth, R.C. Kennicutt, Jr. (eds.)

Stellar Astrophysics: Accretion Binaries and Early-Type Stars

J.B. Hutchings

Herzberg Institute of Astrophysics, NRC of Canada, 5071 West Saanich Rd., Victoria, B.C. V9E 2E7, Canada

Abstract. This contribution discusses the problems in these topics that are expected to be of interest ten years from now. Some observational programs to address them are described, and the new ultraviolet/optical space telescope facility that would be needed to perform them. A few other stellar astrophysics topics on hot stars are also mentioned.

1. Introduction: Unique Capabilities

In discussing the future problems for space-based telescopes in the ultraviolet (UV) and optical spectral regions, it is important to note the performance space where they have a compelling advantage over ground-based or other space-based facilities, in the timeframe of the next decade.

We expect that adaptive optics will be routinely used on 8 m-class ground-based telescopes, and that this will enable high resolution in small fields anywhere in the sky, with best performance in the near-infrared. Multi-conjugate adaptive optics (MCAO) developments may well be extending these fields to a few arcminutes. New 20–30 m class ground-based telescopes, also with some narrow-field AO, will be coming into use. The Next Generation Space Telescope (NGST) will be in operation, producing diffraction-limited performance in the 10–30 micron wavelength range, but dropping rapidly in sensitivity and performance down to 0.6 microns, with no throughout below this wavelength.

The unique strengths of a new space-based telescope lie in wide-field high strehl imaging at optical and UV wavelengths. The demise of the Hubble Space Telescope (HST) will end some 30 years of spectroscopic access to the 120–300 nm UV range, and can only be restored by a new space telescope for this region. Such a telescope offers the highest spatial resolution per aperture by going to these short wavelengths. The UV sky is dark, and solar-blind detectors will enable work on hot stars in regions that are very crowded at optical or infrared wavelengths.

Space-based telescopes also have the advantage of being able to monitor objects without the diurnal sampling imposed from the ground, and even offer extended continuous viewing zone observations.

One potential advantage is that of a stable and simple point spread function, compared with AO on the ground. However, this can only be achieved with attention to design, and by learning lessons from the HST point spread function problems.

2. Accretion Binaries

This general class of objects refers to systems where one compact star accretes, almost always through a disk, from its companion. There are many parameters in play in these systems: mass ratio, rate and type of mass-flow, strength and geometry of magnetic field, ionizing radiation from the disk or either star, and mass-loss from the whole system. While this represents a large amount of parameter space, there are many systems to study, so that we can separate the parameters and study accretion under a wide range of conditions. Because the binary periods are short (hours to days), we are also able to use our azimuthally varying sight line – including eclipsing systems – as a tool in the studies.

As accretion disks are found in other situations of current interest, these binaries offer a way to learn about the processes and hence the other topics that are not nearly as easy to observe – such as accretion processes in active galactic nuclei (AGN), in star-formation, and in planet-formation. These are topics that will be of considerable interest ten years from now. The more exotic phenomena that occur in accretion binaries are:

- The channeling of the inner disk (or the entire mass-exchange) by strong magnetic fields, which we can observe in great detail in AM Her and DQ Her type systems.
- Quasi-periodic oscillations that arise from the spiraling in of inhomogeneities in disks, thus revealing the viscous and dispersion phenomena and resonances that occur in different types of disks.
- Jets are found in a few binary systems with very bright disks, which in principle enable detailed investigation into the jet-formation that accompanies much AGN activity and gives rise to the large extragalactic radio sources. In addition, there are winds that blow off disks that can be studied, also related to the formation of the broad-line profiles in AGN, and the disk ionization by a high-energy source at the disk center.
- Accretion binaries are uniquely important in that they provide the most direct information on neutron stars, white dwarfs, and stellar black holes, as well as many aspects of stellar evolution.

A space UV-optical telescope will be uniquely able to observe the flux and spectral features from the inner parts of the disks, where the main energy source is. I list below some investigations that are not currently possible with the HST, but which will be topical and possible with a larger future space telescope. The recent results from the FUSE observatory, with its unique access to the rich 90 to 120 nm wavelength region, makes future access to this part of the spectrum very desirable.

3. Accretion Binary Programs

All of these programs will require an orbiting UV-optical telescope of 4 m aperture with sensitivity to 100 nm.

1. Central disk and interface with compact star. The principal tool is the O VI doublet at 103.5 nm, which has been shown to arise in the inner disk (and some other shock-ionized regions) in the few systems accessible to FUSE. These show a very broad profile and velocity amplitude not seen in the lower ionization lines such as C IV. A larger facility is needed to extend this sample and cover all the variables. Polarization analysis of strong UV lines (including the lower ionization features in the HST range) is also a valuable tool.

2. Direct observation of jets and spectroscopy of the UV resonance lines in jet systems. We see spectroscopic signatures in the visible of collimated outflows out to several thousand km s^{-1}. These are faint in the Galaxy due to UV extinction and faint in the Magellanic Clouds because of their distance.

3. A related issue is discovering more systems like the currently unique SS433, found in our galaxy. This system shows precessing twin jets seen spectroscopically with velocities of $0.26\,c$. A few other systems are seen in the Galaxy, but only at radio wavelengths and with relativistic expansion velocities (see Mirabel & Rodriguez 1999). These systems may provide key evidence that can never be resolved in the much more distant AGN. These systems will be found as UV-bright stars in Local Group galaxies.

4. There is a class of X-ray sources found in galaxies that have luminosities of $\sim 10^{40}$ erg s^{-1} cm^{-2}, suggesting an accreting mass of 100 to 10^4 solar masses. These are not in galactic nuclei and are more massive than any stellar remnant that is currently understood. They may be resolved into close groups of normal stellar sources, or they may represent a stage in forming massive black holes, or are possibly remnants of the earliest star-formation. We need high spatial resolution in the UV, and UV spectra of these faint objects to understand them.

5. X-ray binaries are a fossil record of star-formation in galaxies, but need to be identified and classified spectroscopically to separate the parent populations. We know there are large differences between these populations in the Galaxy, LMC, and SMC, and need to use this tool to include the M31 system of galaxies, which is the main mass concentration in the Local Group.

4. Hot Star Winds

This is an old topic that has had some revival recently with HST and FUSE spectroscopy. It is likely to be of interest still, a decade from now, as there are unsolved problems that need a bigger UV facility, and the phenomenon has impact on various wider topics.

Massive stars drive stellar winds that affect the kinematics and composition of the ISM. More importantly, the process varies fundamentally with the elemental abundances in the host galaxy (the winds being radiatively driven by evolved element resonance absorptions), and this has implications for high redshift stellar populations and galaxy evolution. In the extreme, we know nothing about the first generation of stars, particularly their intital mass function (IMF), which may well be top-heavy and go to higher masses than in the present epoch (may be related to binary issue #4 above). Another wider implication is that models of winds in different environments are showing that the temperature scale for the hottest stars is not well understood, and this affects estimates of the ionizing radiation in the early universe (or anywhere in the universe) by significant amounts. Here too, we need a large UV-optical telescope to resolve and obtain spectra of such stars in external galaxy environments.

Astrophysical processes that occur in these stars that need further investigation are deviations from spherical symmetry; X-ray and shock ionization; magnetic fields in winds; anisotropies in the wind flows; and winds and mass exchange in massive binary systems. All of these play into the bigger picture and will need work from a new 4 m class facility in orbit. Finally, it is increasingly suggested that the broad emission line regions in AGN, are winds driven off the central accretion disks around massive black holes, and that much of the theory (and observation) of hot star winds may apply to this question too.

Some observational programs are listed below, that help define the capability required for this work, which is summarized in italics in each case.

1. High S/N UV spectra in Local Group galaxies. This will enable detailed wind models in several different host galaxy environments. [$R \sim 1000$, $m_V = 20$, wavelengths 100–160 nm]

2. Hot star UV spectra in galaxies out to Virgo. This will probe wider environment effects, with classification-quality data (like IUE low dispersion). [$R = 250$, $m_V = 25$, wavelengths 120–170 nm, high spatial resolution]

3. UV spectral monitoring with high S/N in several wind line troughs. This is required for the time-dependent and non-uniform understanding of the wind phenomena, for known single and binary stars in the Galaxy and Magellanic Clouds. [$R = 10000$, $m_V = 14$, wavelengths 95–160 nm]

4. Medium resolution UV spectra of starburst galaxies and regions. The understanding of distant young galaxies involves bridging the gap between individual stars and stellar populations. There are examples at redshifts out to a few tenths that require UV observations. *[$R = 2000$, $m_V = 18$, wavelengths 120–160 nm, high spatial resolution]*

5. Some Other Hot Star Programs

The NGST Design Reference Mission (DRM) contains a number of programs that relied on high resolution at wavelengths through the blue-visible. These will no longer be possible with the NGST instrumentation and telescope performance, but are pressing problems. Some are dealt with in more detail by other presentations, but I list here a couple of others, plus some problems not in the DRM.

1. The stellar population that is responsible for the UV flux from the central regions of galaxies is not understood. A high-resolution UV telescope will

make considerable progress in identifying this population by its spatial distribution, number counts, and spectra that reveal the properties of the stars.

2. It will be possible to do microlensing observations to the Virgo cluster, to investigate the unseen population of objects in that environment. This will require monitoring at optical wavelengths with resolution $0.03''$, and imaging S/N of about 200 at I = 25.

3. Imaging of Local Group and Virgo galaxies with high resolution and wide field is needed to determine the lower IMF and the globular cluster populations. This needs to reach $m_V = 32$. This will provide new information on stellar populations and early star formation that is unlikely to be done any other way.

4. Other pressing problems such as the white dwarf cooling for oldest stellar populations, and supernova studies at high redshift are likely to be done within the decade, but there may be new issues that emerge in these topics as they evolve.

6. Summary

While the topics outlined here may not on their own be the compelling reasons to build the new UV-optical space telescope, they are of wide significance and a new space facility will be used uniquely and effectively in addressing them.

The programs outlined all call for a 4 m class telescope with diffraction limited performance down to wavelengths of 400 nm (less is better), and UV operation down to 100 nm. A combination of well-sampled imaging and low to medium resolution spectroscopy is needed for instrumention. There will be a noticeable gap in many capabilites that have high demand at the present, once the HST is no longer operating, and no ground-based or anticipated space telescope will provide these capabilities.

In addition to the science-driven needs, the capability for high strehl imaging at optical-UV wavelengths maintains unique discovery space, and will be the source of continued imaging data of the kind that have made HST enormously popular with the public and educational facilities.

References

Mirabel, I.F., & Rodrguez, L.F. 1999, ARA&A, 37, 409

Hubble's Science Legacy: Future Optical-Ultraviolet Astronomy from Space
ASP Conference Series, Vol. 291, 2003
K.R. Sembach, J.C. Blades, G.D. Illingworth, R.C. Kennicutt, Jr. (eds.)

Cool Stars and the Future in Space

A.K. Dupree

Harvard Smithsonian Center for Astrophysics, 60 Garden Street, Cambridge, MA 02138

Abstract. Recent results selected from ultraviolet spectroscopy and imaging of cool stars reveal challenging questions for future observations ranging from imaging of stellar surfaces and environments, to pursuit of faint objects populating the depths of globular cluster cores. Spectroscopy of fiducial Galactic stars and star clusters and study of abundances, atmospheric structures, winds and mass loss drive future opportunities in studies of stars and clusters. A 6-8 m class diffraction-limited ultraviolet telescope (perhaps with outriggers) will make discoveries and address fundamental problems associated with cool stars.

1. Introduction: General Considerations

A major new space facility must offer unique and compelling science and capabilities to justify the use of substantial resources. 'Unique' becomes a significant adjective in this context, for preparations of a decade or more should not lead to modest advancements over previous results. Such an obvious requirement underscores the advantages of the ultraviolet (UV) spectral region ($\lambda\lambda$ 912–3000). Optical astronomy from the ground is flourishing with rapid advances in technology of large mirrors and active optics. UV space astronomy, with results from the International Ultraviolet Explorer (IUE), Far Ultraviolet Spectroscopic Explorer (FUSE), and Hubble Space Telescope (HST), is making significant contributions to astronomy and astrophysics and in many cases changing concepts of the Universe. 'Discovery' science in the UV is challenging, plentiful, and addressing fundamental physical questions. What happens on the face of a star? What exotic objects lurk in cluster cores? How do planets and their stars interact? Where and what are the strange hot dense structures on active stars? How does accretion work in young stars?

Basic physics determines that the UV spectral region will never be accessed from the ground. Another advantage derives from shorter wavelengths that cause the angular resolution of a telescope to be smaller in the ultraviolet than the same aperture operating in the optical. Spatial resolution will be optimized in the UV – an advantageous characteristic for both imaging and spectroscopy. Atoms, ions, and molecules in the region below 2000 Å are represented in astonishing variety before interstellar absorption of hydrogen begins at 912 Å. Molecular transitions of CO and H_2 mark the coolest gases found in the interstellar medium and cool stellar atmospheres; species of abundant elements, H, He, C, N, O are prevalent; and plasmas of several million degrees can be probed by Fe XII,

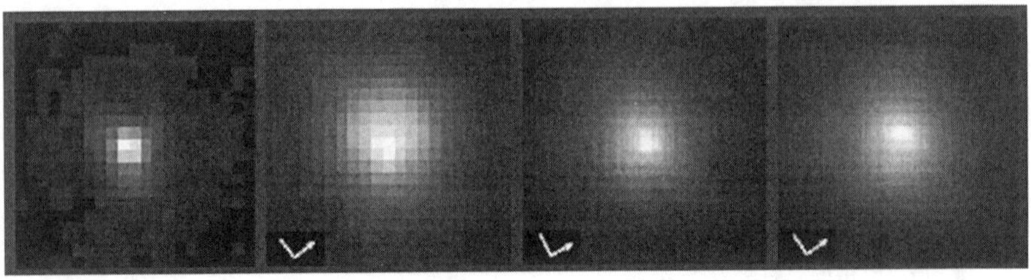

Figure 1. Sequential HST/FOC images of α Ori (Betelgeuse) in the λ2550 band with display scaled to the brightest pixel. A point source comparison star (HZ 4) is shown in the left image. Dates of observation are: 3/95, 10/96, and 9/97 (from Dupree & Lobel 1999). Betelgeuse is a nearby cool supergiant with an extended atmosphere that varies in appearance most likely resulting from the star's semi-regular pulsation.

Fe XVIII, and Fe XXI emission where fine structure transitions have already been detected by HST/STIS and FUSE.

This broad array of species accessible in the ultraviolet, spanning equilibrium temperatures from 100 K to 10 million K provides powerful diagnostics of temperature, densities, and abundances of a wide variety of objects and physical conditions. There are many ultraviolet targets: Solar System objects, stars (both single and multiple of all shapes and sizes), accretion disks, the interstellar medium, and galaxies. Practically every scientific problem can be addressed using the tools of the ultraviolet.

New classes of sources are revealed because UV imaging, by singling out hot objects, eliminates the confusion of optical sources. A profusion of faint new objects is likely also after the launch of the first UV-survey instrument, the Galaxy Evolution Explorer (GALEX), that will demand follow up spectroscopy of faint targets. Imaging and UV spectroscopy of stars have complementary but different goals and requirements.

2. UV Stellar Imaging

Direct imaging of stellar surfaces and their surroundings is unexplored territory with huge potential for addressing problems of surface convection, magnetic dynamos, differential rotation, and atmospheric structures. HST has made a start with the 38 milli-arcsecond resolution of the Faint Object Camera (FOC) followed by the Space Telescope Imaging Spectrograph (STIS), which has imaging capability. Concurrent with advances in imaging capability is the vastly increased power of computational astrophysics. Computational advances are remarkable in elucidating the total structure of stars from their central cores to the photosphere and extended atmospheres (cf., Freytag 2001), making direct confrontation of theory with observation possible. This represents a totally new and exciting area of astrophysics.

Indirect methods exist to image stellar surfaces. The well-known Doppler (or Zeeman-Doppler) imaging techniques which utilize high resolution spectra are confined to rapidly rotating stars (Vogt et al. 1999; Donati & Collier-Cameron 1997); photometric techniques can be used to reconstruct spot dis-

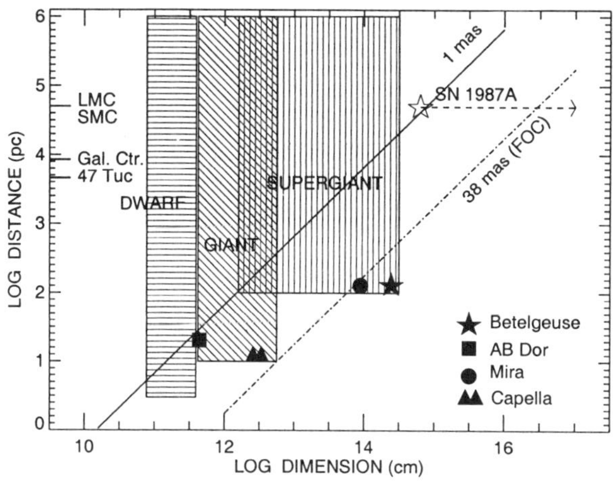

Figure 2. Sizes of various classes of objects with the locus of a HST/FOC UV (λ2700) point spread function (38 mas) and 1 mas marked. Chromospheres of giants and supergiants will appear larger in the ultraviolet than their photospheric size by a factor of 2 or more; the chromospheric sizes are shown here. SN 1987A is denoted by its diameter at 2 days; the next 4000 days brings the appearance of hot spots in the surrounding interstellar material at greater distances from the supernova.

tributions over the surface of active stars (cf., Budding 1996). Eclipse mapping particularly applied to the light curves of cataclysmic variables (cf., Rutten 1996) or to X-ray flux modulation (cf., Schmitt 1996) can pinpoint structures. Recently a microlensing event has been used to characterize the structure of the surface of a red giant (Castro et al. 2001). Plans are underway for direct optical and infrared imaging with unfilled apertures (cf., Pasquini et al. 2001). The ultraviolet region reveals the extremes of atmospheric structure.

Until recently, the Sun was the only star whose surface had been imaged directly. However, the capability of HST/FOC allowed direct images to be obtained of the M2 Ia supergiant Betelgeuse (Alpha Ori) (Gilliland & Dupree 1996), arguably the star with the largest apparent diameter in the sky. About 25 resolution elements could be obtained over the image of the star and its atmosphere at λ2500. Cool stars have chromospheres that are hotter and more extended than the stellar photospheres. Thus, they provide targets that are larger than indicated by optical or infrared interferometry which sample photospheric layers. For example, the optical diameter of Betelgeuse is $\sim$57 milliarcseconds; however, the UV continuum diameter is 2 times larger (Gilliland & Dupree 1996) and, if imaged in a resonance line, such as the Mg II multiplet at λ2800, the chromosphere extends much further (another factor of 2) in apparent size (Uitenbroeck et al. 1998).

The first image of Betelgeuse revealed a hot spot in the stellar chromosphere. Here, the ultraviolet spectral region is a real advantage because temperature fluctuations are more easily detectable. The hot ultraviolet spot showed at least a 200 K temperature increase and a contrast of $\sim$1.5 over the surrounding disk

(Gilliland & Dupree 1996); at longer wavelengths, for example at 8300 Å, this hot spot would cause only a 4% change in the flux, making it difficult to identify and measure. Subsequent ultraviolet images (see Figure 1) revealed changes in the appearance of the hot spots (Dupree & Lobel 1999), which seem to occur when the star is bright. These most probably represent the break-out of a chromospheric shock wave near the polar regions of Betelgeuse.

In addition to single stars (main sequence and giants), binaries offer the opportunity to image interactions directly. This has been accomplished with Mira and its companion (Karovska et al. 1997) where HST/FOC imaging revealed asymmetries in the atmosphere of Mira A (an M2–7 III giant) and an elongated structure between the giant and its companion (most probably a white dwarf), suggesting interaction between the two. Other stellar systems will become accessible with higher spatial resolution.

Novae and supernovae could be resolved during the early stages of their evolution. Milli- arcsecond resolution in the Large Magellanic Cloud could have resolved SN 1987A two days after outburst. The appearance of hot spots in rings surrounding SN 1987A as HST is now following, would be an easy target at even larger distances. A summary of typical sizes of targets for ultraviolet imaging is given in Figure 2.

Crowded stellar fields also demand high spatial resolution. A good example of the challenges can be found in studies of cores of globular clusters, where HST and the Chandra X-ray Observatory have revealed a number of 'weird' objects – faint blue stars and many variables in cluster cores (Edmonds et al. 1999; Grindlay et al. 2001). Identification of binaries is particularly important for the role they play in cluster evolution and dynamics. One rich cluster, 47 Tuc, has been well studied with HST to search for signs of planetary transits (Gilliland et al. 2000), and its core harbors quiescent low-mass X-ray binaries composed of neutron stars (Edmonds et al. 2002). Finding charts for these fields reveal variable sources with separations ~30 milli-arcseconds or less. Such images clearly show the need for higher resolution to identify individual objects in such crowded fields, and subsequently to follow up with spectroscopy to determine their nature.

3. UV Stellar Spectroscopy

The scientific goals of UV stellar spectroscopy focus on composition, structure, and evolution of young and old stars. Aspects of behavior of real stars go beyond the classic equations of stellar structure, and can usually be traced to rotation, magnetism, pulsation, and accretion. These are generally time-dependent processes that need to be studied with large apertures and high spectral resolution. Current UV instrumentation generally restricts spectroscopic targets to the brightest examples.

With increased effective area, ultraviolet grasp can be extended throughout the galaxy and to cluster objects leading to a comprehensive view of stellar activity and evolution. Cluster members are especially important because our current picture of atmospheric evolution including mass loss is based simply on a few of the brightest stars. Pre-main-sequence stars in star-forming regions can be studied beyond the handful of nearest objects. The puzzling interactions of cool

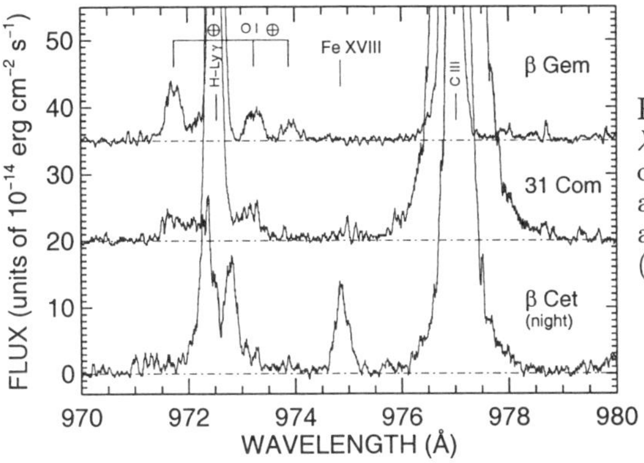

Figure 3. Emission from Fe XVIII $\lambda 974$ observed in the FUSE spectra of the rapidly rotating giant 31 Com and the active clump giant β Ceti, and not in the quiet giant β Gem (Dupree et al. 2002).

molecular gas with warm transition region gas discovered in FUSE spectra of AB Aur (Roberge et al. 2001) or the surprising absence of molecular hydrogen in Beta Pictoris (Lecavelier et al. 2001) can trace out the star formation history or evaluate the geometry of star formation. The ultraviolet region remains unique by allowing simultaneous measures of vastly different plasma conditions in pre-main-sequence objects.

Even the highest temperatures of 6×10^6 K and above can be identified in UV spectra of magnetically active systems (see Figure 3) to allow the dynamics of X-ray emitting gas to be determined. High spectral resolution confirmed that the hottest material in many active cool stars is confined to a stellar surface and not expanding in a wind from a corona (Redfield et al. 2002). Early ultraviolet observations with FUSE and HST reveal the changing strength of coronae in the bright binary Capella (cf., Johnson et al. 2002). Modeling of coronal structures in active stars is beginning (Jardine et al. 2002) driven by the ultraviolet results and measures of surface magnetic fields using techniques of Zeeman-Doppler imaging.

Rapidly rotating stars exhibit enormous magnetic activity, and the magnetic field configurations and location of hot spots on the surface of the stars themselves suggest gross discrepancies when compared to the Sun. It is known that magnetic flux emerging in the photosphere of a rapidly rotating star is swept to the polar regions by Coriolis and centrifugal forces (Schüssler et al. 1996). Observations of eclipsing stars suggest that the poles are special places with unusual non-solar like structures accompanied by high densities (Dupree 2001). Spectra from the Chandra X-ray Observatory have located these dense regions at high latitudes by Doppler imaging a binary with a 6 hour orbital period (Brickhouse et al. 2001). Ultraviolet spectra would be more fruitful in this application because of the wide range of plasma temperatures that can be sampled as well as stars with longer periods of rotation. Determination of coronal structures and their appearance, location, and differential rotation can allow the theoretical link to be made between solar activity and the greatly enhanced versions apparent in active stars.

A focus on cool stars with Earth-like planets will emerge from the Kepler Discovery mission scheduled for launch in 2006. Four years of photometry of about 100,000 non-active main sequence stars will identify those with Earth-size

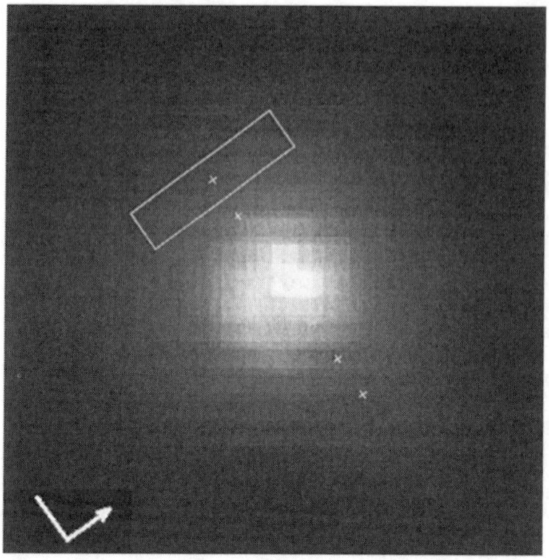

Figure 4. Co-aligned HST/FOC images in the λ2550 continuum band (where 1 pixel = 14.35 mas) of α Ori on 14 September 1998. Slit positions of the spatially resolved spectral raster scan with STIS are shown for the 100×25 mas aperture. Note the arc-like structure of hot emission (from Lobel & Dupree 2000).

planets in the habitable zone. The precise number that will be detected depends on many factors, but reasonable estimates indicate there will be several hundred (Remund et al. 2002). Giant planets would add to that number. Spectroscopic study of the host stars in the ultraviolet will be needed to determine their chromospheric activity to characterize the planetary environments and the stars themselves. The host stars are not active, and hence have weak chromospheric and coronal emissions. The targets monitored by Kepler are optically faint (between visual magnitude 7 and 14). Spectroscopy of such stars is challenging, if not impossible, for the current generation of ultraviolet telescopes.

A very powerful combination of high spatial resolution and spectroscopy in the ultraviolet is evidenced by the spectroscopy of the surface of Betelgeuse, in which the dynamics of the atmosphere can be identified. Individual spectra obtained at multiple positions (see Figure 4) across the extended atmosphere sample distinctly different parts of the stellar surface. Line profiles can be modeled in order to determine the mass motions present in the atmosphere (see Figure 5). A sequence of observations over a span of 14 months revealed that the predominant down flowing atmosphere turned around in the western hemisphere of the star and began to expand. The expanding region grew substantially in extent during the following 6 months suggesting the presence of non-radial chromospheric oscillations (Lobel & Dupree 2001).

4. Instrumental Needs

To address the problems of stellar structure, evolution and interaction, and to single out targets in complex regions such as cluster cores, enhancements over our current capabilities are needed. It is challenging to provide answers now to questions posed in the future!

4.1. Imaging

In the ultraviolet (~2700 Å), HST/FOC offered a point spread function of 38 milli-arcsecond (FWHM). To drive this spatial resolution down by an order of magnitude at shorter wavelengths, say 1200 Å, and achieve ~ 5 milli-arcsecond resolution, we need to contemplate a single diffraction-limited telescope with an

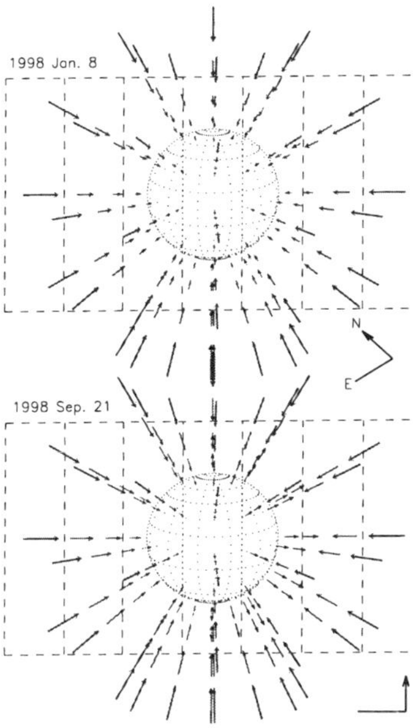

Figure 5. Schematic representation of the fluid movements in the extended chromosphere of α Ori based on spatially resolved semi-empirical modeling of the Si I 2516Å line. Note the change from inflow to outflow over the western hemisphere in the fall of 1998 that extends to the east in 1999 (from Lobel & Dupree 2001).

aperture on the order of 8 m. An increase in spatial resolution would be possible with outrigger telescopes that could be added subsequently to form an unfilled aperture, either on extended beams or with an associated free-flying swarm of smaller reflector telescopes. In addition to giving substantial improvement in spatial resolution, this would also provide a pathway to development of a larger interferometer system.

4.2. Spectroscopy

The technical capabilities required of UV stellar spectroscopy are twofold: fainter sources and time resolution of variable emission. Both must be achieved with spectral resolution sufficient to resolve thermally broadened line profiles. Advances in detector technology can introduce substantial increases in efficiencies of spectrographs which, when coupled to a larger aperture (6–8 m) than presently available, should give more than an order of magnitude improvement in sensitivity. Spectral resolution ($\lambda/\Delta\lambda$) needs to be ~40,000 to yield a velocity resolution of 7.5 km s^{-1} at 1200 Å to adequately sample line profiles, to trace line shapes, and to aid in separating spectral blends. Thermal broadening of Si II transitions at 20000 K is 4.6 km s^{-1} (FWHM); the important O VI resonance lines formed at 300,000 K have thermal widths of 24 km s^{-1} (FWHM). A reliable onboard wavelength fiducial is absolutely necessary to measure atmospheric radial velocities.

References

Brickhouse, N.S., Dupree, A.K., & Young, P.R. 2001, ApJ, 562, L75

Budding, E. 1996, in IAU Symp. 176, Stellar Surface Structure, ed. K. Strassmeier & J. Linsky, (Dordrecht:Kluwer), 95

Castro, S., Pogge, R.W., Rich, R.M., DePoy, D.L., & Gould, A. 2001, ApJ, 548, L197

Donati, J.-F., & Collier Cameron, A. 1997, MNRAS, 291, 1

Dupree, A.K. 2001, in ASP Conf. Ser. 223, 11th Cambridge Workshop on Cool Stars, Stellar Systems, and the Sun, eds. R.J.G. Lopez, R. Rebolo, & M.R. Zapaterio Osorio, (San Francisco, ASP), 333

Dupree, A.K., & Lobel, A. 1999, BAAS, 31, 929

Dupree, A.K., et al. 2002, ApJ, submitted

Edmonds, P.D., et al. 1999, ApJ, 516, 250

Edmonds, P.D., Heinke, C.O., Grindlay, J.E., & Gilliland, R.L. 2002, ApJ, 564, L17

Freytag, B. 2001, in ASP Conf. Ser. 223, 11th Cambridge Workshop on Cool Stars, Stellar Systems, and the Sun, eds. R.J.G. Lopez, R. Rebolo, & M.R. Zapaterio Osorio, (San Francisco: ASP), 785

Gilliland, R.L., & Dupree, A K. 1996, ApJ, 463, L29

Gilliland, R.L., et al. 2000, ApJ, 545, L47

Grindlay, J.E., Heinke, C.O., Edmonds, P.D., Murray, S.S., & Cool, A.M. 2001, ApJ, 563, L53

Jardine, M., Collier Cameron, A., & Donati, J.-F. 2002, MNRAS, 333, 339

Johnson, O., et al. 2002, ApJ, 565, L971

Karovska, M., Hack, W., Raymond, J., & Guinan, E. 1997, ApJ, 484, L175

Lecavelier des Etangs, A., et al. 2001, Nature, 412, 706

Lobel, A., & Dupree, A.K. 2000, ApJ, 545, 454

Lobel, A., & Dupree, A.K. 2001, ApJ, 558, 815

Pasquini, L., Hatzes, A., von der Lühe, O., & Wiedemann, G. 2001 in ASP Conf. Ser. 223, 11th Cambridge Workshop on Cool Stars, Stellar Systems, and the Sun, eds. R.J.G. Lopez, R. Rebolo, & M.R. Zapaterio Osorio, 423

Redfield, S., et al. 2002, ApJ, submitted

Remund, Q.P., Jordan, S.P., Updike, T.F., Jenkins, J.M., & Borucki, W.J. 2002, Proc. SPIE, 4495, 182 (see also http://www.kepler.arc.nasa.gov/)

Roberge, A. et al., 2001, ApJ, 551, L97

Rutten, R. 1996, in IAU Symp. 176, Stellar Surface Structure, eds. K. Strassmeier & J. Linsky, (Dordrecht: Kluwer), 69

Schmitt, J.H.M.M. 1996 in IAU Symp. 176, Stellar Surface Structure, eds. K. Strassmeier & J. Linsky, (Dordrecht: Kluwer), 85

Schüssler, M., et al. 1996, A&A, 314, 503

Uitenbroek, H., Dupree, A.K., & Gilliland, R.L. 1998, AJ, 116, 2501

Vogt, S.S., Hatzes, A.P., Misch, A.A., & Kürster, M. 1999, ApJS, 121, 547

Hubble's Science Legacy: Future Optical-Ultraviolet Astronomy from Space
ASP Conference Series, Vol. 291, 2003
K.R. Sembach, J.C. Blades, G.D. Illingworth, R.C. Kennicutt, Jr. (eds.)

Cosmic Distances: Current Odds and Future Perspectives

Giuseppe Bono

INAF - Rome Astronomical Observatory, Via Frascati 33, 00040 Monte Porzio Catone, Italy

Abstract. We discuss recent theoretical and empirical results concerning cosmic distances and discuss the problems affecting the Cepheid and the RR Lyrae distance scales. In particular, we outline the key role that first overtone Cepheids can play in improving the accuracy of distance determinations to nearby galaxies. We also address the positive features of the K-band period-luminosity-metallicity (PLZ_K) relation of RR Lyrae stars when compared to the M_V versus [Fe/H] relation. Moreover, we discuss the impact that accurate multi-band data for external galaxies can have on the evolutionary properties of intermediate and low-mass stars. Finally, we introduce possible avenues of future research in stellar astrophysics.

1. Introduction

Dating back to Baade (1944, 1958) the interplay between investigations on stellar populations and on radial variables as primary distance indicators is a well-established route to improve the knowledge on the stellar content of simple and complex stellar systems. During the last half century, paramount observational and theoretical efforts have been devoted to single out the evolutionary and pulsation properties of population I and population II stars. In particular, spectroscopic measurements disclosed that the metallicity was not the leading physical parameter to split Baade's populations. Actually, Galactic RR Lyrae stars present a metallicity distribution that ranges from $[Fe/H] \approx -2.2$ up to solar metallicities (Preston 1964; Lub 1977). At the same time, classical Cepheids have been discovered in dwarf galaxies, such as Phoenix and IC 1613, that are 1 dex more metal-poor ($[Fe/H] \approx -1.9$; $[Fe/H] \approx -1.3$) than the Large and the Small Magellanic Cloud (LMC, $[Fe/H] \approx -0.3$; SMC, $[Fe/H] \approx -0.7$). This is the reason why we subdivided the two most popular groups of variables located in the Cepheid instability strip into low and intermediate-mass radial variables.

The microlensing experiments (EROS, OGLE, MACHO) provided an unprecedented amount of photometric data. The number of variables for which pulsation parameters (periods, mean magnitude and colors, amplitudes) are available have doubled, and in some cases increased by one order of magnitude. In this context, the Hubble Space Telescope (HST) also played a crucial role, and indeed the two fundamental projects on stellar distances succeeded in the measurement of Cepheids in several spirals of the Virgo cluster (Freedman et al. 2001; Saha et al. 2001). This unique opportunity allowed a substantial improvement in the calibration of secondary distance indicators, and in the evaluation of the Hubble constant.

This is the coarse-grained scenario and no doubt the perspectives for the near future are quite promising. However, the fine-grained scenario still presents several unsettled problems. In this article, we discuss the impact that they might have on cosmic distance determinations and on stellar populations.

2. Classical Cepheids and Intermediate-mass Stars

Classical Cepheids are the most popular distance indicators since they are bright stars and they can be easily identified in the Galaxy, in Local Group (LG) galaxies, and in spiral galaxies of the Virgo and Fornax clusters. Pros and cons of the Cepheid distance scale have been discussed countless times in the recent literature (Sandage et al. 1999; Tanvir 1999; Feast 1999; Freedman et al. 2001). Physical arguments suggest that individual Cepheid distances can be obtained using the period-luminosity-color (PLC) relation. The use of the PLC relation is hampered by two problems: *i)* two accurate mean magnitudes are required; *ii)* the use of optical bands does require an accurate knowledge of the reddening. To overcome the first problem, the two H_0 projects decided to properly sample the V band and then to adopt a constant amplitude ratio to improve the accuracy of the mean I magnitude. This empirical trick is supported by theoretical predictions. Two different routes can be followed to overcome the latter problem: *i)* Near-infrared (NIR) magnitudes are marginally affected by reddening. However, we still lack accurate mean J and K magnitudes for Cepheids in external galaxies. Ground-based NIR observations with 8 m-class telescopes and/or H-band HST/NICMOS observations will become available in the near future, but certainly the Next Generation Space Telescope (NGST) will play a major role in improving both the accuracy and the sample size. *ii)* the Wesenheit magnitudes (W=V-2.45(V-I)) are reddening free. This approach mimics the use of a PLC relation in the V, V-I band, but the color coefficient attains slightly different values. The strengths and weaknesses mentioned before are the main reasons why Cepheid distances are estimated using the period-luminosity (PL) relation. The main problem in using the PL relation is its intrinsic width in effective temperature. This thorny problem is severe in the optical but negligible in the NIR bands (Madore et al. 1987).

On top of these problems, there is the long-standing debate concerning the universality of both PL and PLC relations (i.e., the fact that the zero-point and the slopes of these relations might depend on the metal content). During the last few years this crucial point has been addressed in several theoretical and empirical investigations (Sasselov et al. 1997; Kennicutt et al. 1998; Macri et al. 2001). These investigations do not agree on the magnitude of the effect, but they agree on the sign. In particular, they found that metal-rich Cepheids, at fixed period, are brighter than metal-poor ones. Theoretical models constructed by adopting different physical assumptions concerning the coupling between radial displacements and convection (Chiosi, Wood, & Capitanio 1993; Bono et al. 1999a,b; Alibert et al. 1999) also do agree on the sign among themselves but do not agree on the magnitude of the metallicity dependence. What is really puzzling is that both linear and nonlinear Cepheid models are at odds concerning

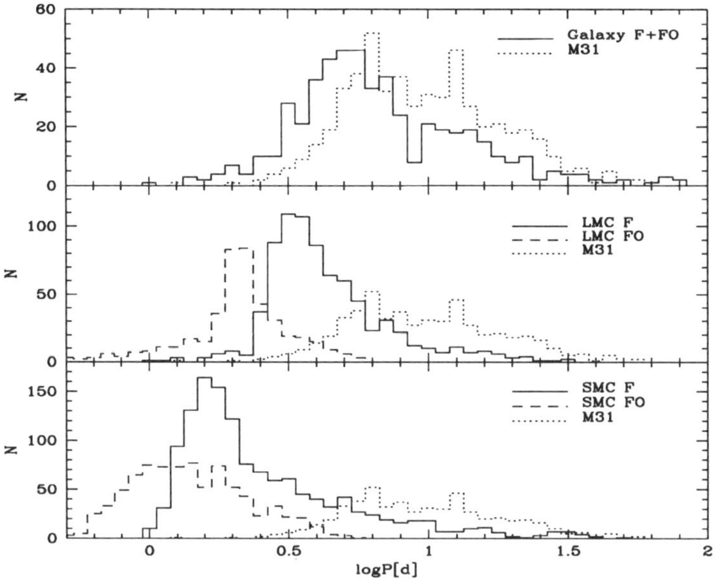

Figure 1. Comparison between the period distribution of Cepheids in M31 with Galactic (top) and Magellanic (middle and bottom) Cepheids. Solid and dashed lines show the period distribution of F and FO Cepheids, respectively. The period distribution of Galactic Cepheids accounts for both F and FO Cepheids. It is not clear whether the current sample of M31 Cepheids (Baade & Swope 1965; Mochejska et al. 2001) includes FOs (dotted line).

the sign, since they predict that metal-rich Cepheids, at fixed period, are fainter than metal-poor ones. Two pieces of evidence somehow support theoretical predictions: *i)* Theory and observations suggest that metal-rich Cepheids are redder than metal-poor ones; *ii)* Recent detailed analyses of V, I, and K-band data of a sizable sample of Magellanic Cloud (MC) Cepheids support the theoretical sign (Groenewegen & Oudmaijer 2000; Groenewegen 2000). This problem requires a firm solution, since it affects the calibration of both zero-point and slope. The approach adopted by Pietrzynski et al. (2002) to measure Cepheids in a substantial portion of NGC 300 is quite promising to settle this problem.

However, recent theoretical and empirical results do suggest that the metal content marginally affects the PL and the PLC relation of first overtone (FO) Cepheids. As a matter of fact, Bono et al. (2002) found that predicted and empirical PL_K relations (K band data from the Two Micron All Sky Survey) and Wesenheit functions (V,I band data from the Optical Gravitational Lensing Experiment) supply mean distances to the Magellanic Clouds that agree very well with each other. In particular they estimated a distance modulus for the LMC of 18.53 ± 0.08 (theory) and 18.48 ± 0.13 (obser.), as well as for the SMC of 19.04 ± 0.11 (theory) and 19.01 ± 0.13 (obser.). The reasons for the agreement are: *i)* K-band data and the Wesenheit function are marginally affected by uncertainties on reddening corrections; *ii)* The PL_K relation and the Wesenheit function have a mild dependence on metal content (Bono et al. 1999b). *iii)* The

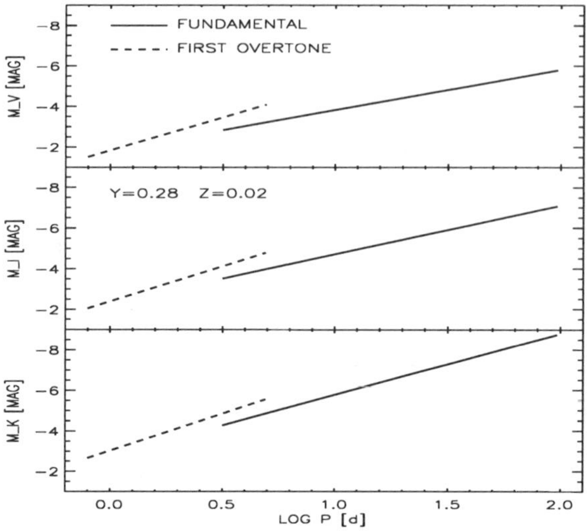

Figure 2. Theoretical PL relations in three different photometric bands: V (top), I (middle), and K (bottom), constructed by adopting a solar chemical composition and a canonical ML relation (Bono et al. 2002). Solid and dashed lines display the F and FO PL relations.

previous intrinsic features do apply to fundamental (F) Cepheids; however, they are magnified for FO Cepheids, since the width in temperature of the latter pulsators is systematically smaller than for the former ones. Therefore, distances based on these variables are less affected by the typical spread of PL relations.

The improvement in using FOs to constrain the accuracy of the Cepheid distance scale is not in doubt. The simultaneous detection of both F and FO Cepheids in nearby stellar systems should allow us to estimate on a quantitative basis the systematic errors affecting optical and NIR PL/PLC relations. Unfortunately, current photometric data for FO Cepheids are scanty, and indeed we still lack detailed information for FOs in nearby dwarfs such as Phoenix (Caldwell et al. 1988) and for spiral galaxies such as M 31. Figure. 1 shows the period distribution of F and FO Cepheids in the Galaxy, M31, and in the MCs. Data plotted in this figure clearly show that the mean metallicity affects the period distribution. In fact, the peak decreases from $\log P = 0.7 - 0.8$ in the Galaxy and in M 31 ($Z \approx 0.02$) to $\log P = 0.2$ in the SMC ($Z \approx 0.004$). The same outcome applies to FO periods, and indeed the period distribution peaks around $\log P = 0.3$ (LMC) and around $\log P = 0.0 - 0.1$ (SMC). The sample of Galactic FOs is too small to draw any firm conclusion. The main drawback that limits the use of FOs is that they present short periods (see Figure. 1), and, therefore, they are on average 1-2 mag fainter than F variables (see Figure. 2).

The detection of a sizable sample of FOs in nearby galaxies is important to improve the accuracy of Cepheid distances as well as to improve our knowledge of He-burning phases and the occurrence of blue loops. The lower limit in the period distribution of FOs depends on the minimum mass whose blue loop crosses the instability strip. Current evolutionary predictions (Bertelli, Bressan,

& Chiosi 1985; Castellani, Chieffi, & Straniero 1990; Stothers & Chin 1993; Bono et al. 2000; Limongi et al. 2000) suggest that the occurrence and the extent in temperature of blue loops strongly depend on the chemical compositions, on the efficiency of the mass loss as well as on the input physics, and on the physical assumptions adopted to handle mixing processes in the convective layers.

Up to now a detailed comparison between theory and observations has been performed for field stars in the MCs for a few LMC clusters, and in particular for NGC 1866. Although several theoretical and observational (HST) studies have been devoted to this cluster, we still lack firm quantitative constraints on the size of the convective core around the turn-off region (Testa et al. 1999; Barmina et al. 2002; Walker et al. 2002). This cluster might be crucial to address some problems concerning the main-sequence fitting and the Cepheid distance scale, since it hosts more than 20 Cepheids (Walker et al. 2001). However, the pulsation parameters are only available for half of them, and we still lack homogeneous photometric data for bright and faint cluster stars.

3. RR Lyrae Variables and Low-mass Stars

RR Lyrae stars are very good tracers of the old stellar component, and play a key role in the absolute age determination of globular clusters. Although, RR Lyrae stars present several advantages, the difference between distance estimates based on different calibrations (Baade-Wesselink method, horizontal branch [HB] models) of the M_V versus [Fe/H] relation is systematically larger than the empirical uncertainties, and cover a range of ≈ 0.3 mag on the distance modulus. This indicates that either the reddening accuracy is poor or RR Lyrae distances are still affected by systematic uncertainties. Moreover, recent theoretical and empirical studies confirm that the M_V vs [Fe/H] relation is not linear when moving from metal-poor to metal-rich RR Lyrae (Caputo et al. 2000).

Some of the problems affecting the RR Lyrae distance scale can be overcome using the K-band period-luminosity (PL_K) relation. In a seminal empirical investigation, Longmore et al. (1990) demonstrate that RR Lyrae do obey a PL_K relation. This finding was further strengthened by a recent theoretical investigation (Bono et al. 2001) suggesting that RR Lyrae obey to a very tight PLZ_K relation connecting the period, the luminosity, the K-band absolute magnitude, and the metallicity. Current models also disclosed that the PLZ_K relation is, in contrast with the M_V versus [Fe/H] relation, marginally affected by off-ZAHB evolution and by a spread in stellar masses (see Figure 3). The intrinsic accuracy of previous predictions was tested using the prototype RR Lyr. Recently, Benedict et al. (2002) on the basis of new astrometric data collected with HST/FGS3, provided an accurate estimate of the absolute trigonometric parallax of RR Lyr. The uncertainty of the new parallax ($\pi_{abs} = 3.82 \pm 0.20$ mas) is ≈ 3 times smaller than the uncertainty on the Hipparcos parallax ($\pi_{abs} = 4.38 \pm 0.59$ mas). Interestingly, Bono et al. (2002) using the predicted PLZ_K relation found a pulsation parallax $\pi_{puls} = 3.858 \pm 0.131$ mas, which agrees quite well with the HST parallax and presents a smaller formal error.

It is noteworthy that HB stars are at the crossroad of several astrophysical problems such as the second-parameter problem, the UV-upturn in ellipticals, as well as the dependence of $M_V(RR)$ on metallicity. Detailed extragalactic

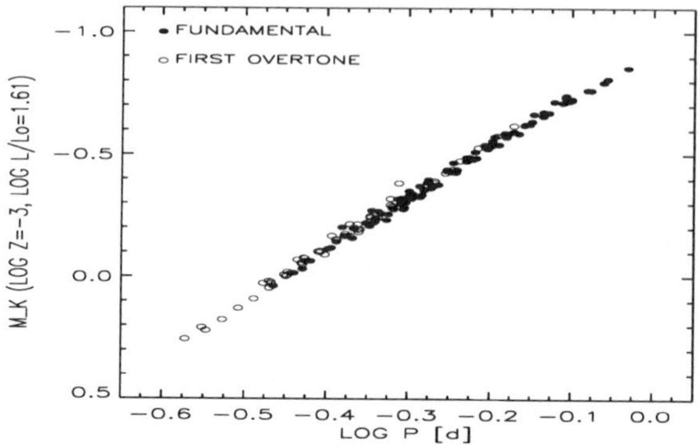

Figure 3. Projection onto a bidimensional plane of the predicted PLZ_K relation for F and FO pulsators. M_K magnitudes were scaled to the same metallicity ($\log Z = -3$) and stellar luminosity ($\log L/L_{\odot} = 1.61$). FO periods were fundamentalized ($\log P_F = \log P_{FO} + 0.127$). Pulsation models cover a wide range in metallicity ($0.0001 \leq Z \leq 0.02$), stellar mass ($0.53 \leq M/M_{\odot} \leq 0.75$), and luminosity.

samples of HB and RR Lyrae stars are only available for MCs and a few LG dwarfs (Mateo 1998; Bersier & Wood 2002). New homogeneous data of HB stars in stellar systems with different chemical and dynamical histories might play a crucial role in constraining their evolution properties.

4. Final Remarks

Some problems affecting the evolutionary and pulsation properties of classical Cepheids and RR Lyrae stars can be settled during the next ten years using the unprecedented sensitivity and spatial resolution of the CCD cameras that are available (HST/ACS) or will become available (HST/WFPC3). Sizable samples of FO Cepheids might be detected and measured across the disk of M 31. The number of orbits necessary to accomplish this experiment is rather modest, since this galaxy hosts regions rich of Cepheids and also because the apparent magnitudes around the minimum luminosity, assuming a distance modulus of 24.5 and a mean reddening of E(B-V)=0.08, range from $V \approx 22$ ($\log P = 0.6$) to $V \approx 24$ ($\log P = -0.1$). The same conclusions apply to RR Lyrae stars, and indeed their apparent magnitudes range from $V \approx 25$ to $V \approx 26$ (Clementini et al. 2001). Note that up to now we still lack complete sample of HB and RR Lyrae stars in elliptical galaxies such as M 32.

A substantial improvement concerning the intrinsic accuracy of the Cepheid distance scale will also be provided by NGST. The superior sensitivity in the NIR bands will allow us to provide accurate measurements of Cepheid mean K magnitudes out to the Virgo and the Fornax clusters. The new data will supply robust distance determinations based on both the PL_K and the $PLC(V, V-K)$ relations. At the same time, NGST will supply the unprecedented opportunity to estimate the distance of several galaxies out to the edge of the Local Group using NIR relations for RR Lyrae (PLZ_K) and classical Cepheids. Future astrometric missions, such as the SIM and GAIA, may also supply accurate empirical constraints concerning the accuracy of absolute (zero-point) and relative (slope) distance estimates as well as on their dependence on metallicity.

The previous improvements would supply tight constraints on the systematic uncertainties affecting primary distance indicators, and hopefully a determination of the Hubble constant with a global accuracy of the order of 5–7%. The next step that can allow us to nail down the systematic uncertainty affecting the evaluation of H_0 is to use only Cepheids and to by pass secondary distance indicators. Data available in the literature clearly show that distance modulus determinations for the Coma cluster based on different zero-points and secondary indicators range from 34.64 ± 0.25 (SBF in the K-band, Jensen et al. 1999) to 35.29 ± 0.11 (various methods, Tammann et al. 1999), while the H_0 values range from 85 ± 10 to 60 ± 6 km s^{-1} Mpc^{-1}.

The main advantage in using classical Cepheids is that we know the physical mechanisms that drive the pulsation instability in these objects. On the other hand, the experiment is challenging from an observational point of view, since it is necessary to detect and measure Cepheids in giant spirals of the Coma Cluster. This means that we should be able to perform accurate photometry down to $V \approx 31 - 32$ mag with a spatial resolution that is at least a factor of five better than current HST capabilities. We also note that this limit magnitude would supply the unique opportunity to estimate the age of old stellar component in Local Group galaxies (NGC 3109, Antlia), and in turn to supply a robust lower limit to the age of the Universe. Even though we expect that these two parameters are tightly correlated, we could agree with Baade's statement: *Although I am quite certain that I do not mistake pink elephants for pink mice an unprejudiced check is always reassuring* (as quoted by Feast 2000).

It is a pleasure to thank N. Panagia for a detailed reading of a draft of this paper. I am indebted to my collaborators for helpful discussions and suggestions.

References

Alibert, Y., Baraffe, I., Hauschildt, P., & Allard, F. 1999, A&A, 344, 551

Baade, W. 1944, ApJ, 100, 79

Baade, W. 1958, in Stellar Populations, ed. D.J.K. O'Connell (Amsterdam: North Holland), 3

Baade, W. & Swope, H.H. 1965, AJ, 70, 212

Barmina, R., Girardi, L., & Chiosi, C. 2002, A&A, 385, 847

Benedict, G.F., et al. 2002, AJ, 123, 473

Bersier, D., & Wood, P.R. 2002, AJ, 123, 840

Bertelli, G., Bressan, A., & Chiosi, C. 1985, A&A, 150, 33

Bono, G., Caputo, F., Castellani, V., & Marconi, M. 1999b, ApJ, 512, 711

Bono, G., Caputo, F., Castellani, V., Marconi, M., & Storm, J. 2001, MNRAS, 326, 1183

Bono, G., Caputo, F., Castellani, V., Marconi, M., & Storm, J. 2002, MNRAS, 332, L78

Bono, G., Castellani, V., Marconi, M., 2002, ApJ, 565, L83

Bono, G., Groenewegen, M.A.T., Marconi, M., Caputo, F. 2002, ApJ, 574, L33

Bono, G., et al. 2000, ApJ, 543, 955

Bono, G., Marconi, M., & Stellingwerf, R.F. 1999a, ApJS, 122, 167

Caldwell, N., & Schommer, R. 1988, in The Extragalactic Distance Scale, (San Francisco: ASP), 77

Caputo, F., Castellani, V., Marconi, M., & Ripepi, V. 2000, MNRAS, 316, 819

Castellani, V., Chieffi, A., & Straniero, O. 1990, ApJS, 74, 463

Chiosi, C., Wood, P.R., & Capitanio, N. 1993, ApJS, 86, 541

Clementini, G., Federici, L., Corsi, C., Cacciari, C., Bellazzini, M., & Smith, H.A. 2001, 2001, ApJ, 559, L109

Feast, M. 1999, PASP, 111, 775

Feast, M. 2000, JHA, 31, 29

Freedman, W.L., HST Key Project 2001, ApJ, 553, 47

Groenewegen, M.A.T. 2000, A&A, 363, 901

Groenewegen, M.A.T., & Oudmaijer, R.D. 2000, A&A, 356, 849

Jensen, J.B., Tonry, J.L., & Luppino, G.A. 1999, ApJ, 510, 71

Kennicutt, R.C. Jr., HST Key Project 1998, ApJ, 498, 181

Limongi, M., Straniero, O., & Chieffi, A. 2000, ApJS, 129, 625

Longmore, A.J., Dixon, R., Skillen, I., Jameson, R.F., Fernley, J.A., 1990, MNRAS, 247, 684

Lub, J. 1977, A&AS, 29, 345

Macri, L.M. et al. 2001, ApJ, 559, 243

Madore, B.F., Welch, D.L., McAlary, C.W., & McLaren, R.A. 1987, ApJ, 320, 26

Mateo, M. 1998, ARA&A, 36, 435

Mochejska, B.J., et al. 2001, AJ, 122, 2477

Pietrzynski, G., Gieren, W., Fouqué, P., & Pont, F. 2002, AJ, 123, 789

Preston, G.W. 1964, ARA&A, 2, 23

Saha, A., Sandage, A., Tammann, G.A., Dolphin, A.E., Christensen, J., Panagia, N., Macchetto, F.D. 2001, ApJ, 562, 314

Sandage, A., Bell, R.A., & Tripicco, M.J. 1999, ApJ, 522, 250

Sasselov, D.D., et al. 1997, A&A, 324, 471

Stothers, R.B., & Chin, C.W. 1993, ApJ, 412, 294

Tammann, G.A., Sandage, A., Reindl, B. 1999, in 19th Texas Symposium on Relativistic Astrophysics and Cosmology, astro-ph/9904360

Tanvir, N.R. 1999, in Post-Hipparcos Cosmic Candles, eds. A. Heck & F. Caputo, (Dordrecht: Kluwer), 17

Testa, V. et al. 1999, AJ, 118, 2839

Walker, A.R., et al. 2002, to be submitted

Walker, A.R., Raimondo, G., Di Carlo, E., Brocato, E., Castellani, V., & Hill, V. 2001, ApJ, 560, L139

Hubble's Science Legacy: Future Optical-Ultraviolet Astronomy from Space
ASP Conference Series, Vol. 291, 2003
K.R. Sembach, J.C. Blades, G.D. Illingworth, R.C. Kennicutt, Jr. (eds.)

Beyond the Cosmic Veil: Star and Planet Formation Research After SIRTF and NGST

Michael R. Meyer

Steward Observatory, The University of Arizona, 933 N. Cherry Avenue, Tucson, AZ 85721

Abstract. How do planets form from circumstellar disks of gas and dust? What physical processes are responsible for determining the final masses of forming stars and ultimately the initial mass function (IMF)? The Hubble Space Telescope (HST) has made major contributions in helping to address these fundamental questions. In the next decade, the Space Infrared Telescope Facility (SIRTF) and the Next Generation Space Telescope (NGST) will build on this heritage in the near- to far-infrared. However several crucial questions will remain. We review recent progress made in star and planet formation with HST, summarize key science objectives for SIRTF and NGST, and suggest problems that would be uniquely suited to a large aperture ultraviolet/optical space-based telescope. We focus on studies that take advantage of high spatial resolution and the unique wavelength range not accessible from the ground such as: 1) circumstellar disk structure and composition (resolved images of dust and ultraviolet spectroscopy of gas); and 2) extreme populations of young stars in the Local Group (ultraviolet imaging and spectroscopy of massive star-forming regions). An 8 m ultraviolet/optical space telescope operating from 1100–6000 Å over fields of view 4–10′ and with spectroscopic capabilities from R = 3,000–10,000 down to $\lambda < 1000$ Å would be a powerful tool for star and planet formation research.

1. Introduction

Star and planet formation will remain key themes in the NASA *Origins* Program through the next decade and beyond. During this time, the community will enjoy access to: 1) continued HST operation of STIS, NICMOS, ACS, COS, and WFC3 with wavelength coverage from 0.1150–2.5 μm; 2) SIRTF, to be launched in early 2003 (lifetime $\sim$ 5 years), covering the 3–160 μm wavelength range; and 3) NGST, scheduled for launch in 2010 with a projected 5 year mission, covering the 0.6–28 μm wavelength range. Because important processes in star and planet formation occur over a wide range of temperatures, multi-wavelength observations are required in order to make significant progress (see overview by Hartmann this volume).

Both SIRTF and NGST will execute programs aimed at understanding the emergence of planetary systems from circumstellar disks around young stars, and the origins of stellar masses. Approved SIRTF guaranteed time programs will focus on circumstellar disk evolution and the nature and frequency of brown dwarf objects. In addition, three of the six adopted Legacy Science Programs (*Galactic Legacy Infrared Mid-Plane Survey Extraordinaire*, *From Molecular Cores to Planet-Forming Disks*, and *Formation and Evolution of Planetary Systems*) will directly impact star and planet formation research (http://sirtf.caltech.edu/SSC/legacy/). Two of the five themes in the Design Reference Mission for the NGST (The Birth and Formation of Stars and The Origins and Evolution of Planetary Systems) deal with similar topics (http://ngst.gsfc.nasa.gov/science/drm.html).

Observations in the ultraviolet (UV) and optical wavelength ranges are also important to obtain a complete picture of star and planet formation. Key observations include: 1) sampling the Wien peak of massive star energy distributions; 2) measuring important gas diagnostics for infall/outflow; and 3) studying resolved images of dust disks. As in the infrared spectral regime, space offers unique capabilities for UV/optical astronomy, such as the accessible range from 912–3000 Å, and diffraction-limited imaging over a large field of view at $\lambda < 1\,\mu$m.

Here, we concentrate on two areas where a large aperture UV/optical space telescope could make substantial contributions. We begin by identifying key issues, review the legacy of HST in each area, summarize the promise of SIRTF for studies of circumstellar disks and NGST for investigations concerning the origin of the IMF, and suggest parameters required in order for a UV/optical space telescope to realize its potential for star and planet formation research.

2. Evolution of Circumstellar Disks

Why should we care about the evolution of circumstellar disks around young stars? Because their study holds the promise of connecting observations of disks observed in the Milky Way as a function of age with the origin and evolution of the Solar System. Disks are the mostly likely sites of planet formation giving rise to extra-solar planets such as those detected around Sun-like stars in the solar neighborhood. Ultimately we wish to know whether planetary systems are common or rare in the Universe. In addition, disk accretion is an important part of the star formation process. Does it help provide "feedback" into the surrounding interstellar medium and thus act as a regulating mechanism for the determination of stellar masses? What effects does it have on pre-main sequence (PMS) evolution?

Our current understanding of the evolution of circumstellar disks surrounding young stars is limited by available observations. High resolution spectroscopic observations have suggested a connection between the accretion of material through a disk and mass loss in powerful outflows (e.g., Edwards et al. 1994). Through ground-based near-infrared studies of young clusters, it appears that inner disk accretion terminates on time scales < 10 Myr for most stars (Haisch et al. 2001; Hillenbrand et al. 2002). Photometric monitoring and spectroscopic observations have suggested a link between stellar rotation and the presence or absence of a disk (e.g., Bouvier et al. 1993). Studies of outer disks (1–10 AU) require mid- to far-infrared observations. A number of groups have suggested that substantial evolution occurs in dust disks at these radii from 10–100 Myr

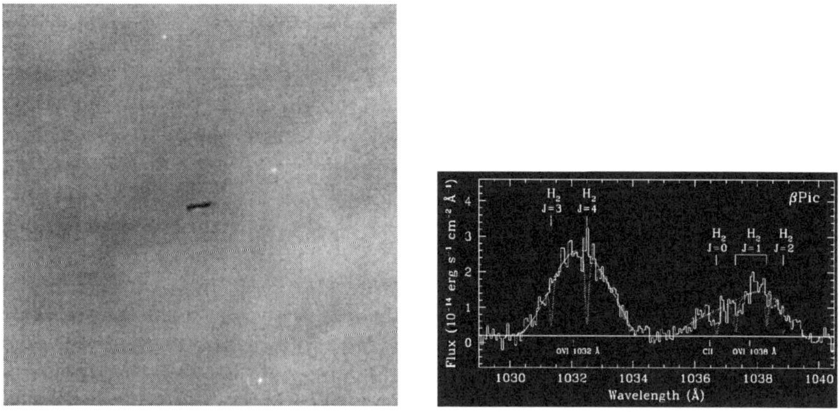

Figure 1. *Left*: HST/WFPC2 image of an edge-on disk seen in silhouette against the emission-line background of the Orion Nebula (Credit: NASA, J. Bally, H. Throop, and C.R. O'Dell). *Right*: FUSE observation of β Pictoris from Lecavelier des Etangs et al. (2001). The dotted line is a model for cold molecular hydrogen gas absorption seen against the continuum provided by O VI emission lines.

(see Meyer & Beckwith 2000 for a recent review). Millimeter wave observations probe evolution of the coolest dust found at the largest radii (e.g., Zuckerman & Becklin 1993).

2.1. Legacy of Hubble

The Hubble Space Telescope has provided a rich legacy of imaging and spectroscopic observations that have revolutionized the study of circumstellar disks around young stars. The first images of the dark disks seen in silhouette against the bright background of the Orion Nebula provided direct confirmation of the disk hypothesis that even the most ardent skeptics find difficult to refute (e.g., McCaughrean & O'Dell 1996; Figure 1). Subsequent studies have offered numerous examples of objects seen edge-on, where a favorable viewing geometry provides a wealth of detailed knowledge about individual objects (e.g., Burrows et al. 1996). Monitoring programs over several years have also enabled proper motion studies of accretion-powered jets that provide direct tests of theoretical models (Hartigan et al. 2001). Coronagraphic observations with STIS and NICMOS (Schneider & Silverstone, this volume) utilize high contrast imaging as a tool to probe dust structure and composition. Wide-field imaging studies have provided insight concerning large-scale mass-loss and interactions with the surrounding interstellar medium (Bally et al. 2002). Finally, UV spectroscopy is proving to be a powerful diagnostic tool to study accretion processes (Ardila et al. 2002) and the atomic and molecular gas content of disks through absorption-line observations along favorable lines of sight (e.g., Vidal-Madjar et al. 1994). Long-slit diffraction-limited imaging spectroscopy has also shed light on the complex relationships between star/disk systems and the surrounding interstellar medium (Grady et al., this volume).

2.2. The Promise of SIRTF

The Space Infrared Telescope, the last of NASA's "Great Observatories" and the first new mission of the NASA *Origins* program, will be launched in early 2003 and provide unprecedented mid- and far-infrared sensitivity for studies of circumstellar disks. Through a suite of guaranteed time programs and the Legacy Science Program, SIRTF will make major contributions to our understanding of the structure and evolution of circumstellar disks around young stars. One program, the Formation and Evolution of Planetary Systems (Meyer et al. 2002), is aimed directly at studying dust disks around solar-type stars and placing our Solar System in context [1]. Through construction of spectral energy distributions from 3–160 μm for 350 stars of spectral type F8–K3 we hope to: 1) characterize the transition from primordial dust disks to debris disks by tracing evolution in the amount, distribution, and composition of dust; and 2) examine the diversity of planetary systems through their dynamical effect on dust. Through spectroscopic observations of warm (50–200 K) molecular hydrogen gas we hope to constrain the time scale for gas disk dissipation and giant planet formation. These programs, which build directly on the heritage of the Infrared Astronomical Satellite (IRAS) and the Infrared Space Observatory (ISO), will leave a rich legacy for follow-up observations with NGST as well as a large aperture UV/optical space telescope.

2.3. Requirements for an Ultraviolet/Optical Space Telescope

What should we require from a UV/optical space telescope for studies of planet formation? Spatial resolution of order 0.3 AU for the nearest T Tauri stars (3 milli-arcseconds at 100 pc and 1100 Å) would enable us to: 1) possibly resolve the disk/jet interface providing a detailed understanding of the accretion process and insight into the star/disk interaction that may mediate stellar angular momentum evolution; and 2) detect large gaps in disks created through the dynamical interaction of giant planets with dust debris. In addition to spatial resolution, sensitivity to study gas disks at R = 10,000 is required in order to: 1) measure accretion rates $< 10^{-10}$ $M_\odot$/year (c.f., Johns-Krull et al. 2000) and 2) extend limits on H_2 and CO gas in disks to levels comparable to SIRTF (c.f., Figure 1). Finally, 4–10′ fields of view from 1100 to 6000 Å are required to trace the accretion-driven mass loss history of young stars by observing shock diagnostics over a range of velocities. Such studies can constrain: 1) the angular momentum loss of PMS star/disk systems through energetic winds; and 2) the deposition of kinetic energy into the turbulent interstellar medium providing a support mechanism for collapsing clouds. An 8 m UV/optical space telescope meets these requirements.

3. The Origins of Stellar Masses

How did the universal star formation rate evolve over cosmic time? How do galaxies, including our own Milky Way, assemble themselves? How are stars

[1] For additional information see http://feps.as.arizona.edu.

and planets formed? In order to quantitatively address all of these issues, one fundamental question must be answered: what physical processes determine the shape of the initial mass function of stars and sub-stellar objects? This question has plagued astronomers since the initial mass function was first introduced by Salpeter (1955). Its construction requires translation of an observed luminosity function into a mass function through adoption of an appropriate mass-luminosity (M-L) relation. Because high mass stars exist on the mass sequence for shorter lifetimes and because they are rare in comparison to lower mass stars, even construction of the field star IMF requires the assumption that: 1) the IMF is the same everywhere; and 2) it has remained the same for all time. These two assumptions have not been tested over the full range of available star-forming environments. Recent work by Massey et al. (1995), Scalo (1998), Reid et al. (1999), Kroupa (2001) and references therein have led to extensions of the field star IMF into new mass regimes as well as refinements in the characterization of its shape. Yet, even with the strong assumptions outlined above, uncertainties in the galactic birthrate complicate corrections for massive star evolution and adoption of time-dependent M-L relationships for sub-stellar objects in well-mixed stellar populations. Newly formed star clusters are ideal laboratories to investigate the IMF over the full range of stellar and sub-stellar masses with a minimum of assumptions.

3.1. Legacy of Hubble

Through observations that have complemented ground-based work, HST has provided intriguing hints concerning answers to the above questions. Studies of mass functions with HST/WFPC2 and HST/NICMOS in the field (Gould et al. 1997), the bulge (Holtzman et al. 1998), spheroid (Gould et al. 1998), and globular clusters (Paresce & De Marchi 2000) all point toward a stellar IMF that is similar in a variety of star-forming environments. Observations of young clusters with HST/NICMOS, where brown dwarfs are more easily detected due to their high "pre-main sequence" luminosities, suggest that the universality of the IMF extends into the sub-stellar range (Luhman et al. 2000; Najita, Teide, & Carr 2001). Massey & Hunter (1998) have utilized the HST/FOS to study the IMF in the low metallicity R 136 cluster in the Large Magellanic Cloud (Figure 2). The rapid post-main sequence evolution of massive stars creates degeneracies in the analysis of color-magnitude diagrams that require classification spectroscopy in order to accurately derive the IMF. Massey and collaborators find that the IMF above 10 $M_\odot$ does not vary over a range of 2.0 dex in metallicity. However, recent near-infrared photometric studies of massive star formation in the inner galaxy with NICMOS on HST have raised the possibility of an unusual power-law IMF above 5 $M_\odot$ in the heavily embedded Arches cluster (Figer et al. 1999).

3.2. The Promise of NGST

Does the IMF truncate somewhere below the hydrogen burning limit? Theoretical speculations abound. One conjecture is that the star-forming process is essentially self-regulating through some negative feedback mechanism. For example, accretion of material onto the central protostar drives a powerful bipolar outflow which could disrupt the remnant infalling envelope (Adams & Fatuzzo

1996). Balancing the mechanical energy between infall and outflow for typical molecular cloud cores suggests a characteristic mass of ~ 0.25 $M_\odot$. An opposing view holds that the distribution of star-forming units in a collapsing cloud core is determined by fragmentation processes which in turn depend on the initial conditions of the local interstellar medium (e.g., temperature, density, metallicity). For example, the mass scale for a gravitational perturbation to become unstable in a uniform density medium, the Jean's Mass, is roughly 0.7 $M_\odot$ when averaged over the observed range of densities in nearby molecular clouds. Indeed such a preferred scale is observed as a break in the power-law distributions that characterize the surface density of companions in the Taurus dark cloud (Larson 1995). As the fragments collapse, the local gas density increases. Provided the material remains at a constant temperature, the Jean's mass will decrease driving smaller and smaller regions unstable (hierarchical fragmentation). However, the minimum mass for fragmentation is fixed at the point where the condensing core becomes opaque to its own radiation, preventing further sub-fragmentation at a constant temperature (e.g., Spitzer 1978). Under conditions typical of nearby clouds this limit is $\sim 10 M_{\rm Jup}$. Large spectroscopic surveys will be needed to build robust estimates of the IMF and distinguish between these theories. With its combination of high angular resolution, accessible wavelength range, and sensitivity, NGST will be able to obtain NIR spectra for hundreds of 10^6 yr objects down to 1 $M_{\rm Jup}$ viewed through $A_V < 25^m$ of extinction within 500 pc of the Sun in a few hours of integration.

Nearby regions of massive star formation (R 136, NGC 3603, the Arches cluster) have commanded considerable attention as local analogs to "super star clusters" observed in interacting galaxies (e.g., O'Connell et al. 1994) that are proximate enough for detailed studies of the resolved stellar populations. With few exceptions, most studies show that the IMF of these regions of extreme star formation are consistent with the field star IMF that characterizes the solar neighborhood down to varying low mass limits > 1 $M_\odot$. Each of these regions contains dozens of stars > 10 $M_\odot$, which according to the field star IMF should be accompanied by $\sim 10^5$ stars down to the hydrogen burning limit. Yet very little is known concerning the IMF in regions of extreme star formation below 1.0 $M_\odot$. Theory suggests that lower metallicity regions might contain hotter gas due to a lack of molecular cooling and therefore exhibit higher Jean's mass on average (Nakamura & Umemura 2002). The starburst phenomenon is often associated with a top-heavy IMF as well as an upper mass cutoff suggested by models of their spectral evolution. In order to constrain the mass-to-light ratio in unresolved stellar populations over cosmic time, it is crucial to understand the shape of the IMF as a function of metallicity and environment in regions of extreme star formation. Our best hope is to study in detail star-forming events in the Milky Way and the Local Group that approach the activity of starbursts. NGST will be able to obtain spectra for objects below the hydrogen burning limit in clusters aged 1–3 Myr with $A_V = 0$–10 from the Galactic Center out to the distance of the LMC.

3.3. Requirements for an Ultraviolet/Optical Space Telescope

What capabilities are required for a UV/optical space telescope in order for it to make major advances in understanding the origin of the initial mass function?

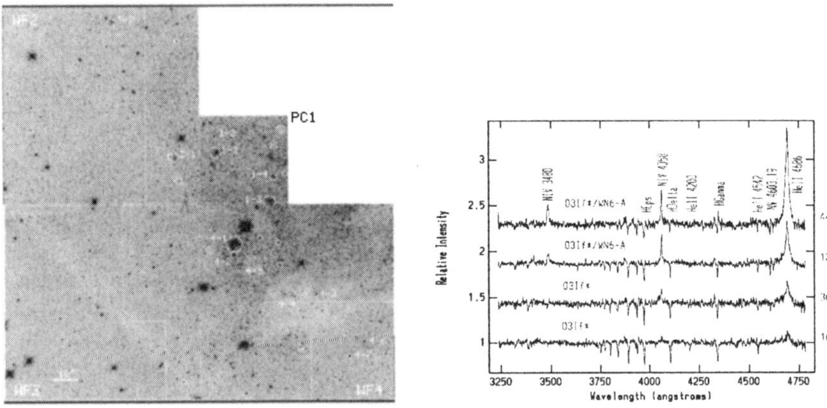

Figure 2. *Left*: HST/WFPC2 image of the nearby starburst IC 10 from Hunter (2001) with individual star clusters and W-R candidates marked. *Right*: HST/FOS spectra of massive stars in the R 136 cluster of the LMC from Massey & Hunter (1998). With an 8 m UV/optical space telescope, studies of the IMF in IC 10 would be comparable to extant studies of the LMC with HST.

Maintaining a wavelength range from 1100–6000 Å will enable a UV/optical space telescope to study the formation of high mass stars in regions of low extinction throughout the Local Group over a wide range of stellar temperatures (spectral types later than K0 to earlier than O5). Spatial resolutions of 200 AU (3 milli-arcseconds at the distance of the LMC and 1100 Å) will result in resolution of many wide binary systems and minimize the nebular background often associated with regions of massive star formation. Spectral resolution (and sensitivity) at $R > 3000$ is required for classification of the hottest stars, which is a necessary step in order to derive an accurate IMF in regions of extreme star formation. Fields of view of 4–10′ are required in order to separate clusters from field populations and assess local environments which could influence their dynamical evolution. An 8 m UV/optical space telescope would enable studies of nearby starbursts with resolution and sensitivity comparable to HST studies of massive star formation in the LMC (Figure 2).

Acknowledgments. I would like to thank the SOC for organizing such a stimulating meeting. Special thanks to C. Grady, P. Hartigan, P. Massey, and K. Stapelfeldt for discussions concerning priorities for a UV/optical space telescope in star and planet formation research, and M. McCaughrean for helpful comments on a draft of this manuscript. Thanks also to the the Max-Planck-Institut für Astronomie in Heidelberg and the Astrophysikalishes Institut Potsdam for the hospitality they provided during the preparation of this manuscript.

References

Adams, F.C., & Fatuzzo, M. 1996, ApJ, 464, 256

Ardila, D.R., et al. 2002, ApJ, 567, 1013

Bally, J., et al. 2002, AJ, 123, 2627
Bouvier, J., et al. 1993, A&A, 272, 176
Burrows, C.J., et al. 1996, ApJ, 473, 437
Edwards, S., Hartigan, P., Ghandour, L., & Andrulis, C. 1994, AJ, 108, 1056
Figer, D.F., et al. 1999, ApJ, 525, 750
Gould, A., Bahcall, J.N., & Flynn, C. 1997, ApJ, 482, 913
Gould, A., Flynn, C., & Bahcall, J.N. 1998, ApJ, 503, 798
Haisch, K.E., Lada, E.A., & Lada, C.J. 2001, ApJ, 553, L153
Hartigan, P., et al. 2001, ApJ, 559, L157
Hillenbrand, L.A., Meyer, M.R., & Carpenter, J.M. 2002, in preparation
Holtzman, J.A., et al. 1998, AJ, 115, 1946
Hunter, D.A. 2001, ApJ, 559, 225
Johns-Krull, C.M., Valenti, J.A., & Linsky, J.L. 2000, ApJ, 539, 815
Kroupa, P. 2001, MNRAS, 322, 231
Larson, R.B. 1995, MNRAS, 272, 213
Lecavelier des Etangs, A., et al. 2001, Nature, 412, 706
Luhman, K.L., et al. 2000, ApJ, 540, 1016
Massey, P., Johnson, K.E., & Degioia-Eastwood, K. 1995, ApJ, 454, 151
Massey, P., & Hunter, D.A. 1998, ApJ, 493, 180
McCaughrean, M.J., & O'dell, C.R. 1996, AJ, 111, 1977
Meyer, M.R., & Beckwith, S.V.W. 2000, ISO Survey of a Dusty Universe, eds. D. Lemke, M. Stickel, & K. Wilke, Lecture Notes in Physics, 548, 341
Meyer, M.R., et al. 2001, The Origins of Stars and Planets: The VLT View, eds. J. Alves & M. McCaughrean (astro-ph/0109038)
Najita, J.R., Tiede, G.P., & Carr, J.S. 2000, ApJ, 541, 977
Nakamura, F., & Umemura, M. 2002, ApJ, 569, 549
O'Connell, R.W., Gallagher, J.S., & Hunter, D.A. 1994, ApJ, 433, 65
Paresce, F., & De Marchi, G. 2000, ApJ, 534, 870
Reid, I.N., et al. 1999, ApJ, 521, 613
Salpeter, E.E. 1955, ApJ, 121, 161
Scalo, J. 1998, ASP Conf. Ser. 142, The Stellar Initial Mass Function (38th Herstmonceux Conference), eds. G. Gilmore & D. Howell, (San Francisco: ASP), 201
Spitzer, L. 1978, New York Wiley-Interscience, 290
Vidal-Madjar, A., et al. 1994, A&A, 290, 245
Zuckerman, B., & Becklin, E.E. 1993, ApJ, 414, 793

Hubble's Science Legacy: Future Optical-Ultraviolet Astronomy from Space
ASP Conference Series, Vol. 291, 2003
K.R. Sembach, J.C. Blades, G.D. Illingworth, R.C. Kennicutt, Jr. (eds.)

Problems in Star and Planet Formation

Lee Hartmann

Harvard-Smithsonian Center for Astrophysics, 60 Garden St., Cambridge, MA 02138

Abstract. In this contribution I outline some important issues in star and planet formation which could be addressed by a space-based, optical-ultraviolet telescope of large aperture: determining the stellar initial mass function, detecting molecular hydrogen in disks and envelopes, measuring disk accretion and possibly detecting disk evaporation, and investigating details of disk structure.

1. Why Space Optical-UV?

If the earliest phases of star formation occur in dusty opaque regions, why are optical and near-ultraviolet observations of these processes even possible, let alone important? For one thing, powerful bipolar outflows create cavities in the circumstellar medium which allow direct observation of embedded young stars if viewed along the outflow; T Tauri itself is an example of this (Calvet et al. 1994). In addition, clusters of at least moderate mass ($\gtrsim 10^3 M_\odot$) have high-mass stars, whose dissociating/ionizing fluxes and winds can remove molecular gas and dust on short time scales. A good example of this is the Orion Nebula Cluster (ONC); roughly half the stellar population can be observed directly at optical and ultraviolet (UV) wavelengths, despite the youth of the cluster ($\lesssim$ 1 Myr; Hillenbrand 1997) and the high extinction in the region ($A_V \sim 50 - 100$; Goldsmith, Bergin, & Lis 1997) because the O7 star θ_1 C Ori has evaporated the near side of the cloud (e.g., O'Dell 2001). Finally, whether or not a massive star is nearby, most star-forming clouds have dispersed by ages $\sim$ 3 Myr (Hartmann, Ballesteros-Paredes, & Bergin 2001), so that one can pursue optical and ultraviolet studies of circumstellar disks as they evolve and form planets (as long as the disk is not viewed edge-on).

The very nature of the complex physical processes involved in star and planet formation requires observations across a broad range of wavelengths. Although cold dusty disks and envelopes are best studied at mid-infrared to radio wavelengths, stars are best studied at optical to near-infrared wavelengths, and important accretion processes and diagnostics require ultraviolet studies. A space UV-optical telescope could provide complementary data essential to the interpretation of results from the new missions and projects operating at longer wavelengths [e.g., the Space Infrared Telescope Facility (SIRTF), the Stratospheric Observatory for Infrared Astronomy (SOFIA), the Next Generation Space Telescope (NGST), the Atacama Large Millimeter Array (ALMA), etc.].

2. The Initial Mass Function

There can hardly be a more fundamental question in the field of star formation than the origin of the stellar mass function, nor is there a problem for which there are more competing theories. Observationally, the fundamental problem is that the IMF spans a large range of masses and a huge range in luminosities, requiring detection over a very large dynamic range. In addition, there are other difficulties in trying to characterize the initial form of the mass function from cluster observations, especially accounting for the effects of mass segregation and dispersal of cluster members.

Whether there is statistically significant evidence for IMF variations as a function of environment continues to be debated (e.g., Scalo 1998; Eisenhauer 2001; Gilmore 2001). I think that the reason for this has less to do with the likelihood of intrinsic variations in the IMF than the difficulty with making the measurements. In Figure 1 I show our recent effort (Briceño et al. 2002) to compare the IMF in the low-density Taurus star forming region to that of the high-density ONC. The difference in the two IMFs is statistically significant, although this does not address the usual problem of selection effects. The point is simply that such a comparison of the low-mass IMF has only become possible in recent years, as sensitive wide-field surveys at optical and near-infrared wavelengths have been conducted in star-forming regions. We have a long way to go before environmentally-driven variations in the low-mass IMF can be confirmed or disproved.

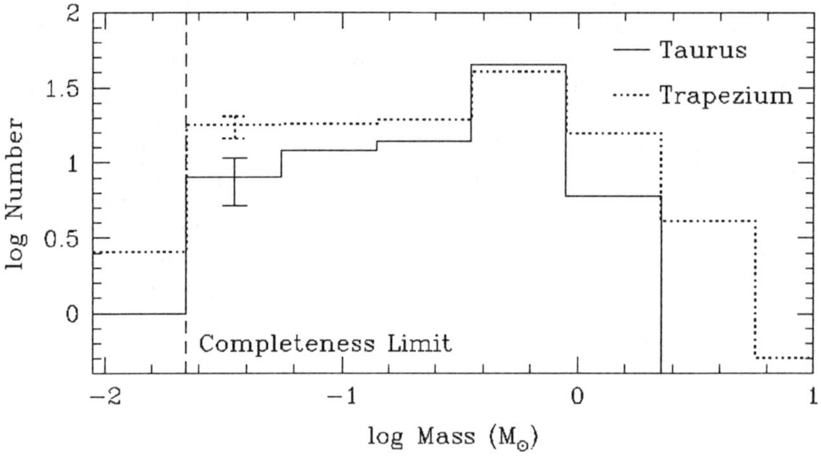

Figure 1. A recent detailed comparison of stellar IMFs in young star-forming regions, comparing the low-density Taurus complex and the central regions of the Orion Nebula Cluster (from Briceño et al. 2002).

The high-mass end of the IMF is perhaps of more importance for understanding effects on the interstellar medium, and more broadly, galaxy formation. Current estimates (see Massey & Hunter 1998, and references therein; Gilmore 2001) suggest that the slope of the high-mass IMF may be relatively invariant. However, I suspect that the high-mass IMF probably does depend upon environment, based on an observational and physical argument. I am unable to identify any gaseous region in Taurus or other low-density nearby molecular clouds which could possibly collapse to form a high-mass star; i.e., regions with large enough masses and high enough densities that they can collapse sufficiently rapidly to accrete a high mass star before it moves off the main sequence. Even though each individual low-density star-forming region has limited numbers of stars, so that the formation of a high-mass star would be unlikely on a statistical basis, I suspect that if one added up many such low-mass regions there still would be no high mass stars because the physical conditions are not right. Perhaps if conditions are right for forming high-mass stars, the Salpeter slope could still be universal, as suggested by Bonnell et al. (2001) in schemes of competitive accretion, but the mass of the most-massive star could be a function of environment.

The requirements for studying the IMF in a wide variety of star-forming environments are (1) large sample sizes and (2) wide range of environments. Both imply extragalactic studies. Blue-optical photometry and spectra are needed to assign masses of hot stars; high spatial resolution is essential. A good demonstration of what is required is given by the Hubble Space Telescope (HST) study of the remarkable R136 super-star cluster in the LMC by Massey & Hunter (1998). Here a space telescope with better spatial resolution and blue light-gathering power than HST is needed. Moreover, the results would be complementary to studies with NGST probing the low-mass end of the same star-forming regions using near-infrared imaging. Michael Meyer addresses the observational requirements for such projects in more detail in his contribution (this volume).

3. H_2

Perhaps the clearest case for the essential nature of ultraviolet observations can be made using one (two?) word(s): H_2. Molecular hydrogen is the dominant constituent of protostellar cloud cores, infalling envelopes, and protoplanetary disks, yet it has rarely been detected. Instead, we use trace constituents, such as molecular species or dust, as proxies for the mass. Because substantial evolution of dust from their interstellar state is predicted (e.g., Weidenschilling & Cuzzi 1993) and inferred observationally (e.g., D'Alessio et al. 2001), the dust opacity per unit gas mass is very uncertain. Molecular observations at millimeter and sub-mm wavelengths also have their difficulties; CO is frequently optically thick and freezes out in outer disk regions, and many other molecular species exist only on warm, ionized upper disk surfaces (Aikawa et al. 2002 and references therein).

Molecular hydrogen emission from young stellar objects has been observed at near-infrared (Weintraub, Kastner, & Bary 2000) and mid-to far-infrared (Thi et al. 2001) wavelengths. The problem with interpreting these transitions is that they require excitation temperatures of order 10^3 K and 10^2K, respectively, and so do not trace the colder gas. The Lyman and Werner bands lying between the Lyman limit and $\sim$ 1150 Å are electronic ground state transitions, and so can probe colder regions. Of course, dust strongly limits the column densities through which molecular hydrogen can be detected, and one may need to correct for foreground H_2 absorption. Nevertheless, the possibility of detecting cold H_2 is so important that the technical difficulties in obtaining good throughput at $\lambda < 1150$Å should not automatically rule out such a capability.

In the "Vega-like" or debris disk systems, the dust remaining in the disk is optically thin, permitting direct observations through the disk for H_2 absorption. To show the potential importance of such observations, one need only look to the recent study of the famous β Pic system by Lecavelier des Etangs et al. (2001). Using FUSE, these authors were unable to detect H_2 in the nearly edge-on disk against the strong O VI emission lines of the star. The upper limits obtained were in strong disagreement with Infrared Space Observatory (ISO) results for the mid-infrared emission lines of H_2 (Thi et al. 2001), probing hotter gas. Although the interpretation of the results is complicated by the possibility of strong spatial gradients, it is quite possible that the ISO observations overestimated the amount of gas present. In general, far-UV observations of H_2 absorption in debris disks (edge-on systems) could provide us with the best determinations of gas masses in these disks.

Another area where H_2 observations could be important is in understanding the dispersal of protoplanetary disk gas. Current theories suggest that giant planets open up gaps in disks (Lin & Papaloizou 1986), and then are carried inward along with the accreting disk material. While this disk-driven migration could account for the presence of "hot Jupiters", formed farther out in the disk but carried in near central stars, if the accretion mechanism is not shut off it could result in the accretion of most (all?) giant planets by the central star. Some theories suggest that the process stops because outer disk gas is rapidly lost due to some unknown process (e.g., Trilling et al. 1998), but the nature of the miracle is not clear.

Ultraviolet radiation from the central star (or neighboring stars) can heat the outer disk and thus evaporate the gas (e.g., Hollenbach et al. 2000). This mechanism is responsible for disk evaporation in the Orion Nebula Cluster, in which the powerful ultraviolet radiation field of the central O7 star can remove disks around low-mass on very short time scales, as dramatically revealed by HST observations (Johnstone, Hollenbach, & Bally 1998). It is far less clear whether the typical T Tauri star has sufficient ultraviolet emission to remove outer disk gas on a sufficiently short time scale (see following section).

One way of detecting material being ejected from the disk would be to observe it in absorption against the central star. Observations of outflow in H_2 could be used to detect the amount of cold gas being ejected, whatever the mechanism, and to determine whether this gas loss is significant in an evolutionary sense. Such disk mass loss might also be detected in other, near-ultraviolet lines such as many transitions of Fe. However, far-UV H_2 detection of many lines from differing rotational states would better determine the temperature of the gas and thus help localize where the (putative) mass is being ejected from; colder material would suggest larger disk radii.

Finally, H_2 can be detected in the 1150-1600 Å wavelength range due to resonance fluorescence with hydrogen Lyman α. An example of this is shown in Figure 2. This long-slit STIS spectrum of T Tauri shows spatially extended (over hundreds of AU) fluorescent molecular hydrogen emission. In this case, the H_2 must be at temperatures of order 1000-2000 K to have enough molecules

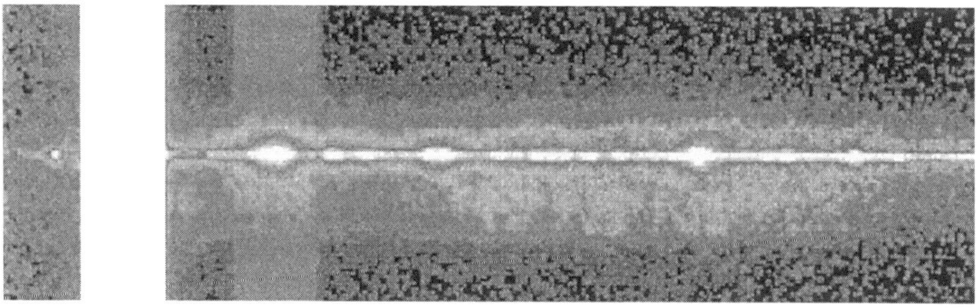

Figure 2. Long-slit Space Telescope Imaging Spectrograph (STIS) spectrum of T Tauri, showing the star (bright continuum streak) and extended diffuse emission which is dominated by fluorescent H_2 lines (from Calvet et al. 2002a).

in the excited states responsible for the fluorescence, and so in principle near-infrared emission lines would also be present. However, experience with T Tauri suggests that this near-IR emission can be difficult to detect against the strong background of the stellar and dust continuum emission. Conversely, the fluorescent emission, driven by Lyα, stands out strongly in contrast to the low background at the wavelengths of emission.

The detection of fluorescent H_2 at large distances from the star, partially aligned with the outflow cavity of T Tauri, suggest that we are observing slow shocks where the wind collides with the dusty envelope (Calvet et al. 2002). Such observations are of potentially great importance in providing us with evidence for slow, wide-angle winds surrounding the highly-collimated, fast jets of young stars, which might be essential to the jet collimation (Shu et al. 1995).

It is possible that an even more robust detection of disk evaporation could be made by detecting the Lyα-driven fluorescent H_2 emission, as the evaporating gas would have temperatures of order 1000 K (Hollenbach et al. 2000). While no one knows whether such evaporation is occurring, disk mass loss rates can't be much smaller than outflow loss rates to have evolutionary significance, and thus should be detectable.

The case for a larger telescope and better UV throughput is simple. HST observations probe only the brightest and nearest objects. The image shown in Figure 2 has very low signal-to-noise in the extended emission, and took about 1/2 hour. Other spectra taken of the brightest T Tauri stars showing fluorescent H_2 emission (Ardila et al. 2002) required $\sim 1-3$ hours of exposure time to obtain reasonable S/N. Sensitivity requirements for far-UV absorption measurements of H_2 are even more stringent, as resolving powers $\geq 1 \times 10^4$ are desirable.

4. Accretion

The rate of accretion through the disk helps to set the time scales of star and planet formation. In T Tauri stars, the accreting material passes from the disk through the stellar magnetosphere before landing on the central star. This material falls in at velocities of a few hundred $\mathrm{km\,s^{-1}}$; the kinetic energy is radiated away in the accretion shock, in both line and continuum emission; the accretion rate can then be derived from $\dot{M} \sim L_{acc}/(GM_*/R_*)$, where M_* and R_* are the stellar mass and radius.

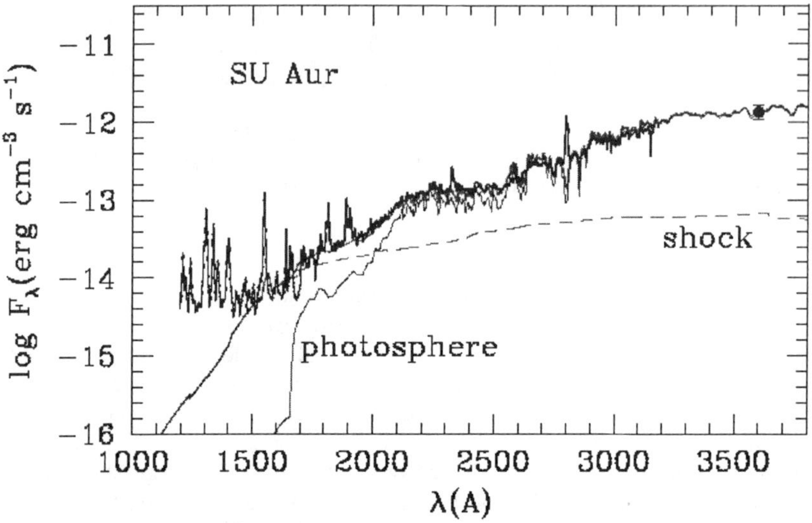

Figure 3. HST UV plus optical spectrum of SU Aur, showing the observed excess ultraviolet emission, along with a standard model for the accretion shock radiation (from Calvet et al. 2002b).

Much of the accretion luminosity is radiated away at ultraviolet wavelengths, making observations below the atmospheric cutoff crucial to measuring accretion rates. This is especially true for stars earlier than spectral type M, for which emission from the stellar photosphere tends to dominate the accretion-powered continuum in optical bands. Figure 3 illustrates this dramatically for the G-type T Tauri star SU Aur. It would be essentially impossible to measure the mass accretion rate without ultraviolet data. Currently, the sample of stars for which we have ultraviolet fluxes is very small. A follow-on to HST would make it possible to determine accretion in large, statistically-significant samples of young stars, providing a vastly clearer picture of the variation of disk accretion as a function of age, stellar mass, environment, and binary companions.

Perhaps the most intriguing and important feature of Figure 3 is the substantial emission far in excess of that predicted by a simple accretion shock model at wavelengths $\lambda < 1500$ Å. Much of the emission is in strong lines – C IV, Si IV, C II, O I – but there is also evidence for an underlying continuum as well whose origin is uncertain. Recall from the previous section that the electronic ground state transitions of H_2 lie in the wavelength range < 1150 Å. These observations suggest that the background fluxes against which molecular hydrogen absorption lines can be observed are much higher than one might expect (see previous section). Perhaps most importantly, we need direct observations of far-UV fluxes to constrain the importance of disk gas photo-evaporation by the central star, as discussed in the previous section.

5. Disk Structure

Observations of scattered light at short wavelengths can be used in principle to detect very small amounts of dust, especially if good coronagraphic capability exists to block out the central star. For this reason short-wavelength scattered light images can play an important complementary role to imaging thermal emission of dusty disks at mm and sub-mm wavelengths with ALMA. Scattered light can then detect structures involved in planet formation at much lower masses than with ALMA. In addition, scattered light is more sensitive to the distribution of small dust in upper disk layers, again complementary information to what ALMA will provide.

Glen Schneider discusses disk imaging in detail in his contribution (this volume), and I leave it to that paper and future considerations whether imaging of disks at shorter wavelengths than NGST will provide is an advantageous project for an optical-UV successor to HST. I simply wish here to stress the potential importance of these observations for understanding planet formation. In particular, I am vastly intrigued by the STIS imaging observations of the Herbig Ae/Be star AB Aur made by Grady et al. (1999). This imaging suggests the presence of large scale spiral-like structures in the disk. One wonders whether this structure is the signature of a gravitating body producing disk "wakes".

The disk of AB Aur is very optically thick, but because the disk is illuminated somewhat obliquely by the central star, ripples in the surface are highlighted. In principle, this means that the contrast in scattered light images could be much higher than that of corresponding thermal emission. Moreover, disks at radii $\lesssim$ 10 AU might well be optically thick at sub-mm wavelengths, making it almost impossible to detect small ripples in surface density. Thus, the speculative possibility is that one might be able to detect the first stages of planet formation from disk ripples by observing scattered light, before one can detect structure at sub-mm wavelengths. Much more theoretical investigation is needed to see whether this is a reasonable possibility, and just what are the advantages of imaging at wavelengths below those for which NGST is optimized.

Acknowledgments. I wish to thank Nuria Calvet for providing Figures 2 and 3. This work was supported in part by NASA grant NAG5-9670, Origins of Solar Systems.

References

Aikawa, Y., van Zadelhoff, G.J., van Dishoeck, E.F., & Herbst, E. 2002, A&A, 386, 622

Ardila, D.R., Basri, G., Walter, F.M., Valenti, J.A., & Johns-Krull, C.M. 2002, ApJ, 566, 1100

Bonnell, I.A., Clarke, C.J., Bate, M.R., & Pringle, J.E. 2001, MNRAS, 324, 573

Briceño, C., Luhman, K.L., Hartmann, L., Stauffer, J.R., & Kirkpatrick, J.D. 2002, ApJ, in press

Calvet, N., Hartmann, L., Kenyon, S.J., & Whitney, B.A. 1994, ApJ, 434, 330

Calvet, N., et al. 2002a,b, in preparation.

D'Alessio, P., Calvet, N., & Hartmann, L. 2001, ApJ, 553, 321

Eisenhauer, F. 2001, in Starbursts: Near and Far, eds. L.J. Tacconi and D. Lutz, (Heidelberg: Springer-Verlag), 24

Gilmore, G. 2001, in Starburst Galaxies: Near and Far (eds. L.J. Tacconi and D. Lutz), (Heidelberg: Springer-Verlag), 34

Goldsmith, P.F., Bergin, E.A., & Lis, D.C. 1997, ApJ, 491, 615

Grady, C.A., Woodgate, B., Bruhweiler, F.C., Boggess, A., Plait, P., Lindler, D.J., Clampin, M., & Kalas, P. 1999, ApJ, 523, L151

Hartmann, L., Ballesteros-Paredes, J., & Bergin, E.A. 2001, ApJ, 562, 852

Hillenbrand, L.A. 1997, AJ, 113, 1733

Hollenbach, D.J., Yorke, H.W., & Johnstone, D. 2000, Protostars and Planets IV, eds. V. Mannings, A.P. Boss, & S.S. Russell (Tucson: University of Arizona Press), 401

Johnstone, D., Hollenbach, D., & Bally, J. 1998, ApJ, 499, 758

Lecavelier des Etangs, A., et al. 2001, Nature, 412, 706

Lin, D.N.C., & Papaloizou, J. 1986, ApJ, 309, 846

Massey, P., & Hunter, D.A. 1998, ApJ, 493, 180

O'Dell, C.R. 2001, PASP, 113, 29

Scalo, J. 1998, in ASP Conf. Ser. 142, The Stellar Initial Mass Function (38th Herstmonceux Conference), eds. G. Gilmore & D. Howell, (San Francisco: ASP), 201

Shu, F.H., Najita, J., Ostriker, E.C., & Shang, H. 1995, ApJ, 455, L155

Thi, W.F., et al. 2001, ApJ, 561, 1074

Trilling, D.E., Benz, W., Guillot, T., Lunine, J.I., Hubbard, W.B., & Burrows, A. 1998, ApJ, 500, 428

Weidenschilling, S.J., & Cuzzi, J.N. 1993, in Protostars and Planets III, eds. E.H. Levy & J.I. Lunine (Tucson: University of Arizona Press), 1031

Weintraub, D.A., Kastner, J.H., & Bary, J.S. 2000, ApJ, 541, 767

Hubble's Science Legacy: Future Optical-Ultraviolet Astronomy from Space
ASP Conference Series, Vol. 291, 2003
K.R. Sembach, J.C. Blades, G.D. Illingworth, R.C. Kennicutt, Jr. (eds.)

High Contrast Imaging and the Disk/Planet Connection

Glenn Schneider and Murray Silverstone

Steward Observatory, University of Arizona, 933 N. Cherry Ave. Tucson, Arizona 85721

Abstract. The Hubble Space Telescope (HST) second generation instruments, NICMOS and STIS, have enabled the direct imaging of circumstellar debris disks and hot Jovian-mass exo-planets, which were previously beyond the reach of ground-based instrumental systems. Such contrast-challenged observations are extremely difficult from ground-based platforms, even with adaptive optics (AO) systems, because of residual uncompensated effects of atmospheric turbidity. The resulting temporal instabilities in ground-based point spread functions (PSFs) set systemic detectability limits which are often under-appreciated, and overlooked, when instrumental performance is expressed with presumptively static metrics such as Strehl ratios. These limitations are greatly relaxed with space-based imaging, as unequivocally demonstrated by HST. HST repeatably achieves panchromatic Strehl ratios in excess of 98%, and does so with intrinsic stability in PSF structures, routinely permitting removal of mid and high spatial frequency optical artifacts to a degree unchallenged from the ground. Coupled with coronagraphic optics to further reduce background instrumentally scattered and diffracted light, these capabilities have opened the door to new avenues of investigations requiring high-contrast imaging. As spectacular as the HST results have been, substantial improvements in high-contrast performance (perhaps three orders of magnitude in circumstellar light rejection) must be achieved in future space missions operating in the UV/optical domain. Such performance requirements must be baselined into mission/instrument concepts, design, and implementation to assure success.

1. Introduction

Imaging dust and debris that might agglomerate to form planets around young stars has been technologically challenging because of the very high intrinsic star-to-disk contrast ratios. Resolving structures on AU spatial scales is possible at ultraviolet (UV) to near-infrared (near-IR) wavelengths for nearby planet-forming systems if imaging instrumentation with sufficient dynamic range is employed. Until very recently the large and nearly edge-on disk around β Pictoris remained the only such disk imaged, and the direct imaging of planetary companions even to nearby stars remains elusive. Resolved imaging in the close circumstellar environments of young stars directly provides the spatial distribution of dust and debris from which planets and planetary systems may evolve and with which they may interact, influencing their growth and evolution. Radial and azimuthal asymmetries in the dust distributions in the forms of rings, central holes, gaps, clumps, arcs, arclets, and other spatially confined features may implicate the existence of low-mass perturbers (i.e., planets). Multi-wavelength observations help to elucidate the scattering and physical properties of the constituent grains. Planets that have recently formed in such environments, in association with their young central stars, may be imaged in the near-IR due to their self-luminosity, if the light from the star itself can be sufficiently suppressed. The enabling technologies and instrumentation for doing so currently exist, having been demonstrated on the HST. While imaging terrestrial-like planets and older gas-giants around stars is beyond the current scope of capabilities, HST and its instruments provide a pathway for the development of future systems with which such observations will be realized.

2. Breaking the Low Contrast Paradigm

Today, HST provides a unique venue for high contrast imaging, with its second-generation coronagraphic systems, offering diffraction limited imaging in optical and near-infrared wavelengths with Strehl ratios panchromatically exceeding 98%. Of equal or perhaps even greater import, HST provides a highly stable platform resulting in a very high degree of temporal invariability in the structure of the PSF sampled by the instrument detectors. The repeatability of the PSF, particularly at spatial frequencies comparable to the size of the PSF itself, routinely permits self-subtractions of the underlying stellar components in high-contrast fields with an efficacy greatly exceeding the capabilities of ground-based AO systems. This capability is further augmented by HST's pointing control system, which allows for highly precise intra-orbit field rotations about coronagraphically occulted targets, permitting ready discrimination between optical artifacts and features of celestial origin.

Differential (PSF-subtracted) coronagraphic imaging with HST/NICMOS consistently results in background rejections on the order of 10^6 per pixel at a distance of $1''$ from an occulted target in H-band, and 10^5 area-integrated over a $2''$-$3''$ radius annulus at 1.1μm. Similar contrast ratios are achievable at optical wavelengths with HST/STIS differential coronagraphy. With NICMOS, the non-destructive readouts of the HgCdTe detector in the coronagraphic channel results in a dynamic sampling range of 19.4 stellar magnitudes using the imaging strategy suggested by the NICMOS Instrument Definition Team (Schneider et al. 1998) and does not limit the high-contrast imaging range achievable by the optical system.

Scientific areas of inquiry enabled via PSF-subtracted coronagraphy include the search for and identification of young exo-solar planet and brown dwarf companions (with mass-equivalent detection limits of a few Jupiters at $1''$ for systems a few megayears old), and the direct spatially resolved imaging of circumstellar debris disks. Though the concern of this paper is in the arena of disk/planet imaging, it should be noted that the same capabilities can be applied to cosmological studies, such as in surveys for damped-Lyman alpha absorbers associated with host galaxies/quasars (e.g., Colbert & Malkin 2002).

3. HST Disk and Sub-Stellar Companion Detection

The detection of both debris disks and sub-stellar companions at small angular distances from their central stars are ultimately limited by the repeatability of the HST PSF. While atmospherics play no role in degrading the stability of the PSF, the varying near-Earth orbital thermal environment of HST drives very small changes in the metrology of the telescope assembly, resulting in fine-structure changes in the HST PSF (Schneider et al. 2001a), particularly on multi-orbit timescales and with large pointing excursions with respect to the Sun.

In practice, debris disk systems with fractional flux density excesses (stellar light scattered by the disk itself) exceeding a few times 10^{-4} at $1''$, or an order of magnitude lower surface brightness at or beyond $\sim 2''$, are readily detectable with both NICMOS and STIS. Stars with known large far-IR (e.g., IRAS) and sub-millimeterer flux density excesses are prime candidates for harboring disks within the regime of detectability by with HST. The high dynamic sampling range afforded by the NICMOS multi-accum readout strategy (important when target surface brightness are unknown) and small size of the circular region obscured by its coronagraph ($0\farcs3$ radius) taken together are ideal for conducting disk discovery/imaging surveys. Once the geometry and approximate scattering efficiencies of disks found by NICMOS have been established, the higher (wavelength-dependent) spatial resolving power of STIS may be used advantageously for subsequent follow-up observations, despite the larger size and greater obscuring area of its tapered coronagraphic wedge. Such a strategy was employed successfuly for a number of NICMOS-revealed dusty disks in HST Cycle 7. We illustrate this for HR 4796A (Figure 1), which owing to its small size ($b/a = 0\farcs27/1\farcs05$; Schneider et al. 1999), could easily have been obscured by the STIS wedge had it not first been imaged coronagraphically by NICMOS.

NICMOS detections of close companions are not limited by photon statistics but rather by the systematics imposed by small-scale structural variations in the HST+NICMOS PSF. These variations may be minimized by PSF self-subtractions using differentially oriented images spanning time intervals shorter than the thermal time constant of the telescope (in practice within single, ~ 50 minute, target visibility periods). On very moderate sub-orbit timescales (total integration time ~ 21 minutes) 5σ companion detection limits of $\Delta H = 10$ at $1''$, 12.5 at $2''$, and 14 at $3''$, are repeatably obtained from a single PSF-subtracted pair of observations (Figure 2). We illustrate this with an observation of a $\Delta H = 13.2$ object at $2\farcs5$ from TWA6 detected at $> 5\sigma$ significance, recently confirmed as a background object by a follow-up NICMOS Cycle 11 observation (Lowrance et al. 2002). These data, with a differential astrometric precision of ~ 10 mas and photometric accuracy of ~ 0.1 magnitudes at a contrast ratio of $\sim 1{:}2\text{x}10^5$, bespeak the performance levels routinely achieved by HST+NICMOS using this adopted two-orientation single-orbit observing strategy. For physical companions (i.e., those confirmed through common proper motions), expressing these luminosity ratios in terms of mass depends upon the age of the companion (which is presumed coeval with its primary). A hot young planet's bolometric luminosity will decline roughly as $L \sim t^{-1.2}$ (Burrows et al. 1997). At 10 Myr, sufficiently old for Jovian-like planets to form, these detection limits correspond to a 1 Jupiter-mass object at $2''$. Pushing into the sub-Jovian mass domain for

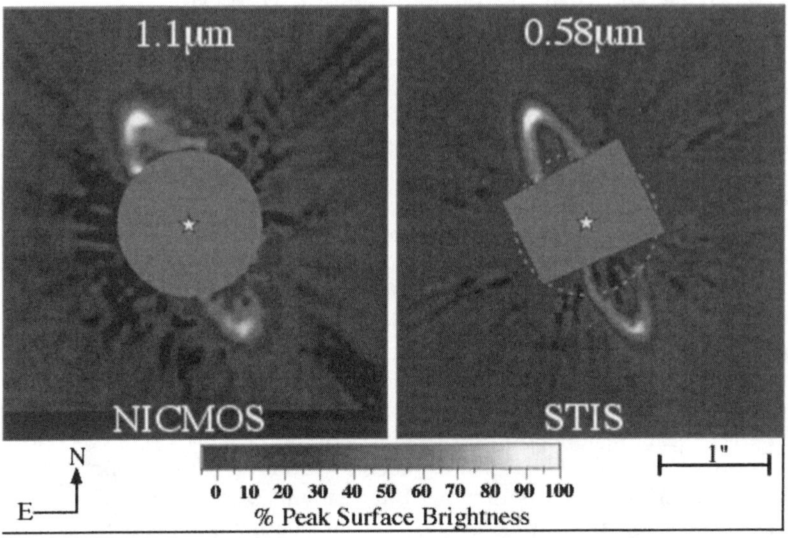

Figure 1. HST PSF-subtracted coronagraphic images of the ring-like debris disk circumscribing the ~ 8 Myr old A0V star HR 4796A.

hot gas giants is possible at ages of about 1 Myr, and suitable candidates may be found among the members of the newly identified nearby young associations in the southern hemisphere.

The level of stability and repeatability of the HST PSF is also a pathway to securing spectra of young exo-solar giant planets (EGPs), once suitable candidates have been identified. Strong, broad, near-IR molecular absorption features (notably H_2O, CH_4, CO and H_2-H_2) are predicted in atmospheric models of hot (young) EGP spectra and albedos (Sudarsky et al. 2000). NICMOS grism spectra of such objects at these ages (1 to 10 Myr with angular separations $> 1''$–$2''$) could be obtained by self-subtraction of the primary spectra with two-epoch observations acquired at diametrically opposed image orientation angles.

4. The Dust Disk/Planet Connection

Current theories of disk/planet evolution suggest an early epoch of planet-building via the formation and growth of planetary embryos followed by the accretion of gaseous atmospheres onto hot/giant planets (Boss 2001). Under this scenario, the circumstellar environments become increasingly dusty as the proto-stellar gas-to-dust ratios (presumed initially on the order of 100:1) in the circumstellar environments decline. In this critical phase of newly formed (or forming) exo-solar planetary systems, posited from a few megayears to a few tens of megayears, the circumstellar environments become dominated by a second-generation population of dust with size distributions skewed toward larger (evolved) particles arising from the collisional erosion of planetesimals. The spatial distribution of such dust debris may be influenced by gravitational

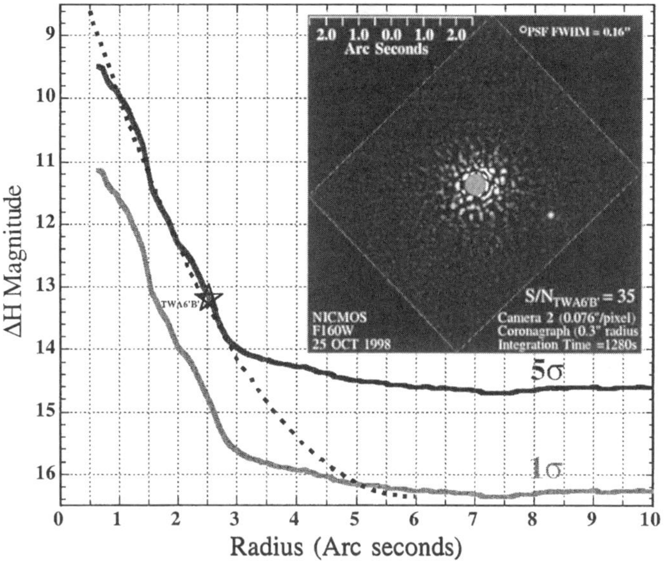

Figure 2. HST/NICMOS H-band azimuthally averaged detection limits for close companions to M & K stars with 2-roll coronagraphic PSF subtraction (21 minute total integration time). Dotted line: Median 5σ detection limit from fifty GTO program targets. Solid lines: 5 and 1 σ detection limits for TWA6 (H=6.9, read noise limited at $> 3''$) field. Inset image: TWA6 after high spatial frequency filtering with an H = 20.1 unresolved object at $2''\!.5$ (statistical S/N = 35).

interactions (sculpting, dynamical resonances, etc.) of co-orbital planetary-mass bodies. In Figure 3 we present two additional examples of dusty circumstellar disks detected by HST/NICMOS and later imaged at higher spatially resolution with HST/STIS. Their brightness anisotropies and complex morphologies are suggestive of such dynamical interactions and may implicate the existence of unseen planetary mass companions. High spatial resolution coupled with high levels of background light rejection enable intra-disk differential photometric mapping of the circumstellar environments, even in disks of intrinsically small angular sizes. We illustrate this for HR 4796A in Figure 4, where the STIS image of the disk (Figure 1, right) has been projected onto a "face-on" geometry (while conserving the flux density per unit area). Spatially resolved photometry of the ring indicates flux densities on the northeast side (left side in Figure 4) which are $\sim 20\%$ greater than at corresponding azimuths (reflected about the ring minor axis) on the souththwest side. Such a brightness anisotropy may result from an unequal spatial distribution of particles which may have been perturbed into (or out of) azimuthal sectors by resonances with one or more undetected planets. Photometry across the ring indicates a steeper inner truncation than an outer fall-off, which may also have implications for dynamical confinement mechanisms. Additionally, directionally preferential scattering, symmetric about the ring minor axis, is seen in an increase in efficiency of $\sim 36\%$

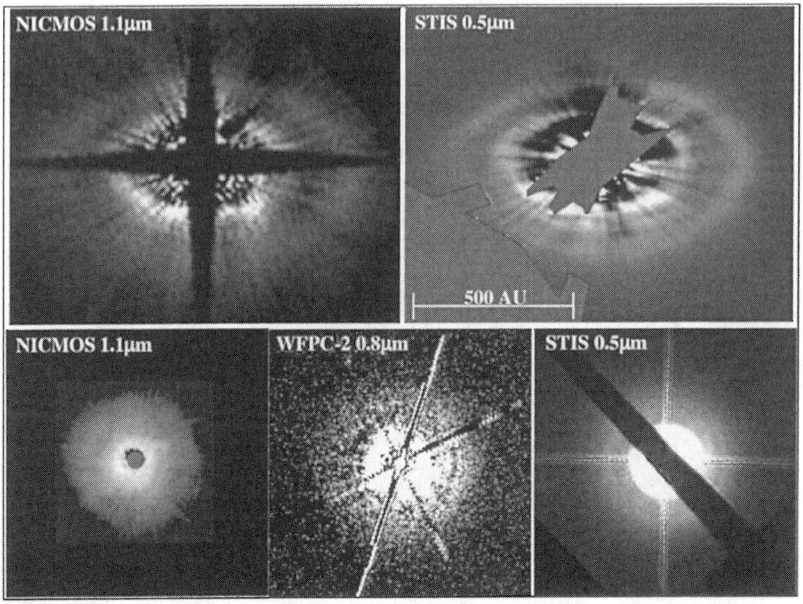

Figure 3. *Top*: HD 141569A (~ 5 Myr Herbig Ae/Be star) exhibits a 400 AU radius disk with a broad, partially filled, asymmetric gap (Weinberger et al. 1999; Mouillet et al. 2001) containing a spiral "arclet". *Bottom*: TW Hya (K7), an "old" classical T-Tauri (pre-main sequence) star of similar age has a pole-on circularly symmetric disk (Weinberger et al. 2001). A break in its surface brightness profile at 120 AU is also seen at lower S/N by HST/WFPC2 (Krist et al. 2000) and an arc-like depression is noted in both optical images (Schneider et al. 2001b). Both disks are shown at the same physical scale.

in the "forward" direction. The direction-dependent scattering properties of the disk grains, coupled with color information, can put strong constraints on the particle size distribution.

The panchromatic high-contrast imaging capabilities of NICMOS and STIS in concert can yield very useful photometric information on the disk particle properties. In the case of HR 4796A, though only very broad colors have been thus-far obtained (F160W and F110W with NICMOS, and 50CCD with STIS), it is quite apparent that the grains are intrinsically very red (see Figure 4, right panel). These colors are consistent with an evolved population of particles with characteristic sizes of at least a few microns, not primordial ISM grains. These colors are also consistent with laboratory irradiation experiments on a variety of organics (Andronico et al. 1987) to study reddening of D and P type asteroids (as a function of distance from the Sun). Indeed, it is of interest (though perhaps anecdotal) to note that the HR 4796A disk color indices are similar to the intrinsic colors of the brightest trans-Neptunian objects in our solar system (Barucci et al. 2001), when corrected for the color of the Sun (V-J = +1.07).

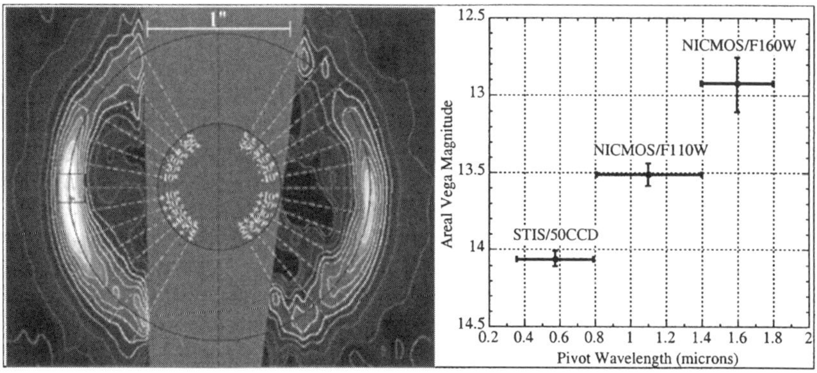

Figure 4. *Left*: Surface brightness isophotes of a "face-on" projection of the HR 4796A ring with a spatial resolution of $\sim$ 70 mas (4 AU at the 57 pc distance of HR 4796A). *Right*: HST/NICMOS and HST/STIS broad-band photometry of the HR 4796A ring indicate very red, and thus non-primordial, disk grains.

5. Moving Beyond HST into the Super-High Contrast Regime

While HST differential (PSF-subtracted) coronagraphy and spectroscopy (e.g., via grism imaging with NICMOS) permits the ready detection of young (few Myr) Jupiters at $> 1''$ from their primaries, and disks with scattering fractions of $\sim 10^{-4}$ at comparable distances, these capabilities are just a stepping stone to the "super" high contrast imaging which would be required for solar (few x 10^9 yr) age systems. For solar type (and age) stars, straylight/background rejections on the order of 10^9–10^{10} must be achieved to image zodiacal dust and planetary companions, and improvements in spatial resolution to a few mas (e.g., sub-AU at 200 pc) is highly desirable. Achieving this at least three-order of magnitude improvement in background light rejection over HST requires overcoming a number of significant technological challenges. Optical configurations which might lead to these levels of performance need not be limited to optimized coronagraphs. Polarimetric nulling, nulling interferometry, space-based systems with active wavefront correction, interferometric arrays, and station-keeping occulters warrant consideration.

Moving from the near-IR to the UV/short-wavelength optical is significantly advantageous simply from the purview of the smaller, and much better centrally contained, PSF as well as the commensurate improvement in spatial resolution with decreasing wavelength. The contrast advantage obtained in the near-IR for imaging young thermally emissive planets is not important for older systems whose planets (and dust debris) would be detected primarily by reflected light. The adoption of any such systems present a number of technology challenges which must be stringently controlled and addressed to enable success. These include micro-roughness of optical surfaces, particulate/contamination control, stray light management and control, pupil apodization (and shaping), metrological tolerancing and stability, wavefront/mirror sensing and control, and pointing control authority. Deficiencies in any of these areas would preclude achieving super-high contrast ratios of at least 10^9, critical for exo-planetary system imaging at evolved ages.

6. Summary

Optical imaging and spectroscopy of collisionally evolved circumstellar debris and co-orbital bodies will play a pivotal role in furthering our understanding of the formation, evolution, and maturation of exo-solar planetary systems. SIRTF, with its leap in mid and far-IR sensitivity over precursor survey missions, will undoubtedly discover a multitude of candidate systems to be subsequently imaged at higher spatial resolution and at other wavelengths. As a pathfinder, HST has shown how high contrast and high spatial resolution can elucidate the nature of such systems. With care and forethought, the legacy of Hubble in giving our first look at planetary systems in formation can be turned into a harvest to be reaped. To study physical processes acting over sub-AU spatial scales and timescales comparable to the age of our solar system will require a 3–4 order of magnitude improvement in instrumental stray light rejection over HST. Such an improvement is obtainable with a UV/optical 4–8 meter class space-based telescope with instrumentation optimized for high-contrast imaging, but requirements to achieve these levels of performance must be incorporated in the end-to-end design of the system at the outset.

Acknowlegements. This work is supported by NASA grants NAG5-3042 and NAG5-10843, and HST GO grants 8176 and 8624. Contributions from other members of the NICMOS and STIS IDTs and the aforementioned GO teams are gratefully acknowledged.

References

Andronico, G., et al. 1987, A&A, 184, 333
Barucci, M., et al. 2001, A&A, 371, 1150
Boss, A., 2002, Proc. IAU Symp. 202, in press
Burrows, A., et al. 1997, ApJ, 491, 856
Colbert, J., & Malkin, A., ApJ, 2002, 566
Krist, J.E., et al., 2000, ApJ, 538, 793
Lowrance, P., et al. 2002, ApJ, in prep.
Mouillet, D., et al. 2001, A&A, 372, L61
Schneider, G., et al. 1998, Proc. SPIE, 3356, 222
Schneider, G., et al. 1999, 513, L127
Schneider, G., et al. 2001a, AJ, 121, 525
Schneider, G., et al. 2001b, ASP Conf. Ser. 219, 499
Sudarsky, D., et al. 2000, ApJ, 538, 885
Weinberger, A., et al. 1999, ApJ, 522, L53
Weinberger, A., at al. 2001, AJ, 599, 409

Hubble's Science Legacy: Future Optical-Ultraviolet Astronomy from Space
ASP Conference Series, Vol. 291, 2003
K.R. Sembach, J.C. Blades, G.D. Illingworth, R.C. Kennicutt, Jr. (eds.)

Observations of Planetary Atmospheres with HST and Beyond

John T. Clarke

Center for Space Physics, Boston University, 725 Commonwealth Ave, Boston, MA 02215

Abstract. HST observations of the planets have been extremely successful in revealing new information about their atmospheres, both in the eyes of scientists and in the eyes of the public. This paper will present a short overview of some representative scientific highlights of HST observations of planetary atmospheres. A discussion will then be presented of the context and techniques for future progress in our knowledge of planetary atmospheres via a next-generation space telescope.

1. HST Observations of Planetary Atmospheres: Some Highlights

This discussion will begin with a short overview of some of the many exciting discoveries which have been made by Hubble Space Telescope (HST) observations of planetary atmospheres. Please see the review paper by James (1999) for a more detailed review of specific results from HST observations of the planets.

There have been no HST observations of Mercury, owing to the difficulty of avoiding sunlight entering the telescope forward light shield. This may eventually be attempted by observing Mercury for a short time before the Sun rises. Although Venus (Figure 1) gets as far as 47° from the Sun, it has only been observed on one occasion for the same reason. Near-ultraviolet (near-UV) images have shown the cloud patterns and wind motions in the middle atmosphere. Near-UV reflectance spectra have been used to track the O_2 composition, which may depend on the short-term history of volcanic eruptions at the surface. Future observations are planned to determine atomic abundances in the upper atmosphere, and potentially contribute to our understanding of the present rate of atmospheric escape. In the case of Mars (Figure 2), an extensive series of images has revealed the seasonal cycles in the polar caps and of the atmosphere, including dust storms which at times completely envelope the planet. HST images and spectra have revealed the abundances and spatial distributions of water and ozone in the Martian atmosphere. UV spectra have measured the atmospheric D/H ratio, with implications for the escape of H_2O from the Martian atmosphere over its lifetime.

In the case of Jupiter, UV images revealed the complex auroral structures and their time variations. UV spectra also measured the composition and temperature of Io's plasma torus (Figure 3). In dramatic fashion, just months after the installation of WFPC2, a campaign of images and spectra tracked the comet Shoemaker/Levy 9 impacts (Figure 4) and its effects on the atmosphere and

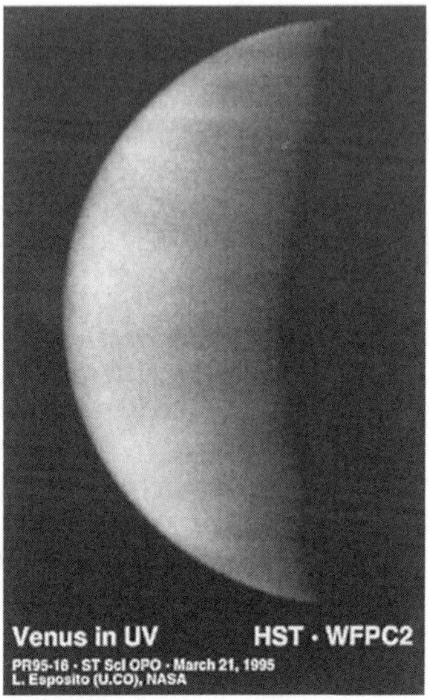

Figure 1. HST/WFPC2 near-UV image of Venus, showing the chevron-shaped upper cloud deck pattern (Na & Esposito 1995, STScI press release).

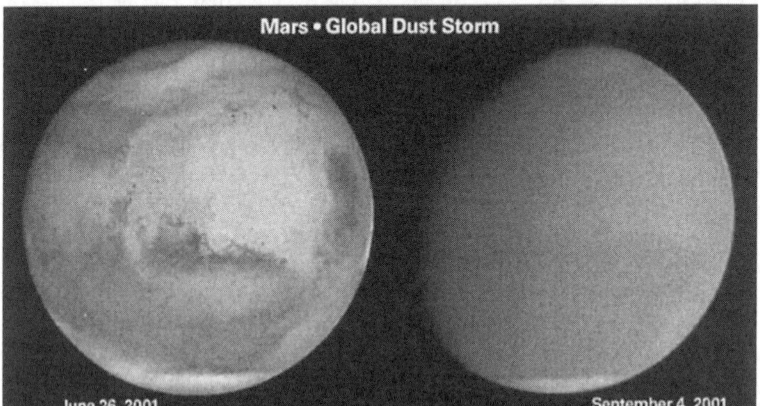

Figure 2. HST/WFPC2 visible enhanced-color images of the same face of Mars before and during the global dust storm which occurred over the summer of 2001 (Bell et al. 2001).

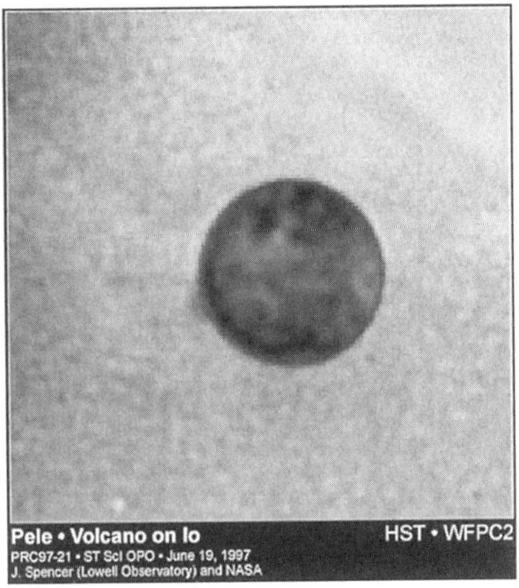

Figure 3. HST/WFPC2 near-UV image of Io in front of Jupiter, showing volcanic ejecta cloud above the Pele caldera (Spencer et al. 1997).

magnetosphere. The spectacular images of plumes rising above Jupiter's limb following the impacts are unprecedented, and have led to a major improvement in our understanding of the effects of large impacts in atmospheres in general. The atmospheres of the Jovian satellites have also been studied with spatially resolved spectra, mainly in the UV, including the discovery of oxygen on Europa. HST has the sensitivity needed to begin their study, but higher sensitivity would lead to great advances.

At Saturn, the series of images obtained during the ring plane crossing events of the mid-1990s revealed new satellites and new information about the structure of the rings, especially the E ring. UV images of Saturn's aurora have also shown that these emissions are not the rotationally symmetric emissions envisioned in the post-Voyager epoch. Visible images have tracked the development of a giant storm system in Saturn's atmosphere. Near-infrared (near-IR) images of Titan have revealed structure in the global haze layers and provided the first images of Titan's surface. In the cases of Uranus and Neptune (Figure 5), visible and IR images have provided tracking of cloud features and (from their motions) atmospheric winds. Seasonal cycles in their atmospheres have been revealed in series of images spanning several years. New planetary ring features and satellites have also been revealed in near-IR HST images. In the case of Pluto and Charon, HST images were the first to completely resolve these two bodies and image the surface of Pluto. A ground-based stellar occultation first revealed the altitude structure of Pluto's atmosphere, a result which could be greatly improved upon with a space-based telescope observing future occultations.

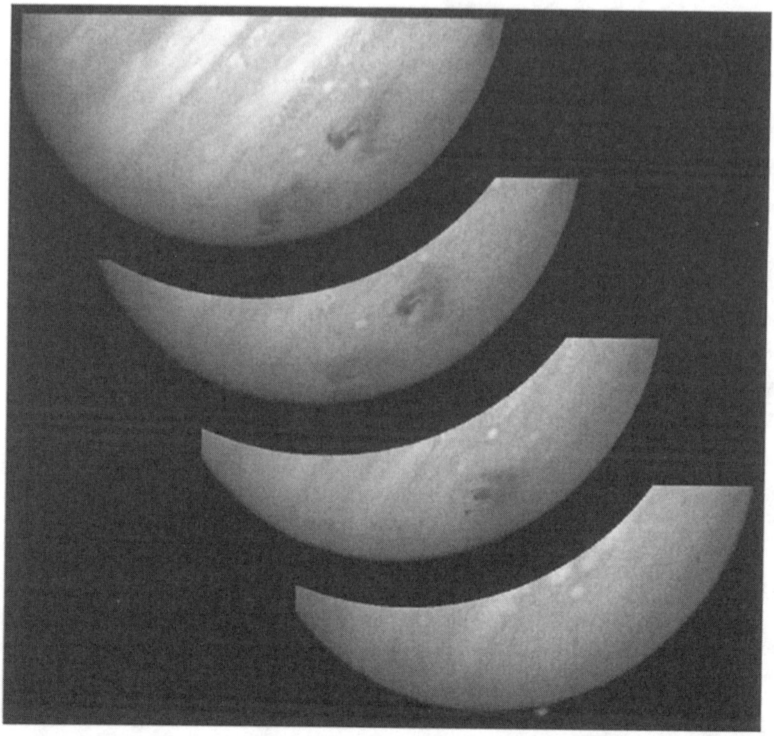

Figure 4. Evolution of G impact site and ejecta plume during the S/L-9 impacts on Jupiter, shown in HST/WFPC2 visible images (*Science* cover figure by J. Trauger and R. Evans (Hammel et al. 1995; Clarke et al. 1995).

2. Perspective: The Changing Nature of Planetary Science

The discovery of extrasolar planets is historic, opening new areas of research that overlap with classical astronomy, offering new targets for study, and changing the way planetary scientists think. At the same time, we need to first characterize the planets in the Solar System to be able to understand extrasolar planets. The intriguing question of the existence of life elsewhere than on the Earth is a central issue as we continue to study Mars, and extend our gaze to extrasolar planets. Some main questions have emerged from discussions among Solar System scientists, which are entirely consistent with the scientific goals of the NASA Strategic Plan:

1. How and why do solar systems form, and what do they look like when they do?

2. What makes planets evolve into habitable worlds?

3. What is the history of life in the Solar System?

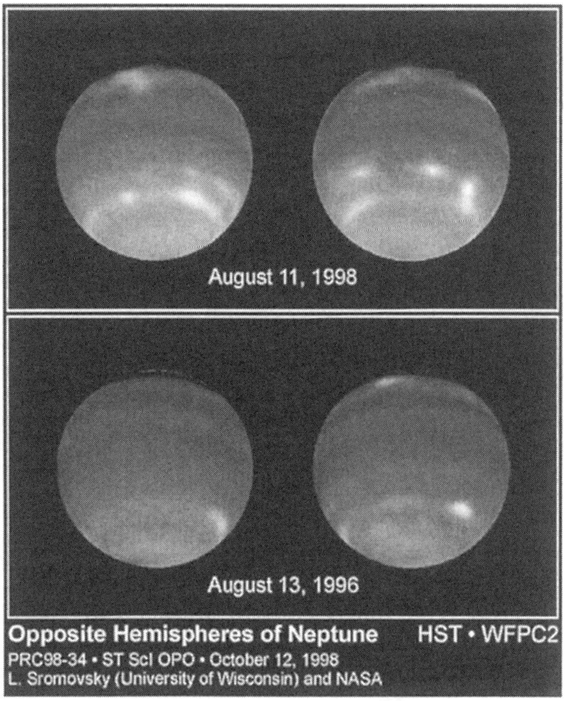

Figure 5. HST/WFPC2 visible images of Neptune showing the changing distributions of clouds in the atmosphere of Neptune (Sromovsky et al. 2001).

To understand the significance of future noisy measurements of spectral features in extrasolar planets, we will need first to understand the physical processes in planetary atmospheres in general for both terrestrial and giant planets. To address the issue of ancient life on the Earth, and extend this study to Mars and possibly Europa, we must understand the histories of their atmospheres and climatic conditions.

Today an increasing fraction of planetary science is being done by remote observations. Given the high cost and difficulty in sending missions to the more distant planets, we expect that the planets and satellites far from the Sun will continue to be studied mainly by remote observations. By virtue of their distance from the Sun, these objects are also the ones most likely to be detected around other nearby stars. In terms of the history of life, the questions of whether life ever began to evolve on Mars fascinates people from all walks of life, and the potential existence of liquid water beneath the surface of Europa is equally exciting. The scenario in which Mars has retained much of its original water presents the intriguing possibility that some form of life may exist today on Mars, and the possibility exists that Mars might be made habitable for humans in a reasonable period of time without an extreme cost.

The advent of a new mission with capabilities far superior to those of HST would make available important new measurements of distant planets and satellites in the Solar System, in addition to more detailed studies of Mars and the other planets and satellites. Outlined below are some of the capabilities that will be needed to determine these physical processes.

3. The Nature of Distant Planetary Atmospheres and Surfaces

Solar reflectivity spectra of planets and satellites give critical compositional information about their atmospheres and surfaces, and UV spectra at wavelengths below 2000 Å sample the most important and sensitive transitions of simple atoms and molecules. The present HST sensitivity limits spectra of distant and/or small objects (e.g., Galilean and more distant satellites, Neptune, Pluto) to near-UV wavelengths above roughly 2000 Å. A large increase in effective area could extend UV spectra to more distant and fainter objects, including Uranus and Neptune, asteroids, and Kuiper belt objects, and extend wavelengths down to the critical far-UV range where simple atoms and molecules have the strongest interactions.

4. Detailed Studies of Atmospheres by Stellar Occultations

Visible and UV stellar occultations provide measurements of the altitude profiles of the atmospheric composition of planetary and satellite atmospheres, with altitude resolution proportional to the time resolution and therefore signal to noise. UV occultations furthermore provide the highest sensitivity to small columns of planetary upper atmospheres and satellite atmospheres (e.g., Io, Ganymede, Triton, Pluto etc.). However, the present rate of suitable candidate events is one per several years with HST. A greatly increased effective area compared with HST will greatly increase the number of occultations available and the signal to noise of each event. UV observations are critical, especially when taken in conjunction with ground-based visible or IR occultation observations. Some expected benefits of UV occultations include the detection of haze or clouds, estimates of optical depth due to haze or clouds, and the determination of atmospheric constituents (e.g., CH_4, from its ionization cutoff). One example of a situation where UV observations would have been useful is the 1988 stellar occultation by Pluto. A kink in the occultation light curve at a Pluto radius of 1215 km may be due to (a) a haze layer, (b) a thermal inversion, or (c) some combination thereof. UV observations of this occultation would have resolved this basic question. A greatly increased effective area compared with HST would greatly increase the number of UV stellar occultations available and the signal to noise of each event.

5. Thermal Profiles and Escape Fluxes of Planetary Atmospheres

Another promising technique for studying planetary and satellite atmospheres is to make long-aperture spatially resolved emission scale height measurements, from a variety of atmospheres and for different emission lines. With sufficiently high angular resolution, one can measure atmospheric emission scale heights, yielding temperatures and the fraction of hot (superthermal) atoms produced by non-thermal processes. As examples of resolutions, 0.1″ observing Mars is roughly 40 km in altitude resolution, while 0.1″ at Jupiter is roughly 300 km resolution. Typical scale heights of planetary atmosphere are on the order of tens of km, and up to hundreds or thousands of km for the lightest atoms and highest temperatures. Higher angular resolutions and sensitivities will be needed to take advantage of this observing technique.

6. The Next Step: Advances in Planetary Sciences by Remote Observations

On the basis of what has been learned from observations by HST and other missions, the following techniques are suggested as promising methods for expanding the study of planetary systems and processes. To accomplish the scientific goals presented above, significant improvements in the capabilities of a space-based telescope will be needed. In planetary science, there are benefits to be derived from a range of different improvements.

- *Cause and effect measurements of variable phenomena (requires more observing time and more complete temporal coverage)*: Planetary emission processes are generally more variable than other astrophysical sources. To establish physical relationships between different variable phenomena, there is no substitute for temporal coverage. A real limitation with HST observations is that repeated Earth occultations interrupt the observability of all planetary objects lying in or near the ecliptic plane. Coupled with limited overall observing time, it is generally not possible to establish physical relationships between phenomena which vary on time scales of hours to months. A strong argument can be made for great scientific advances that would result from an observatory no more capable than HST, but in a high orbit with large continuous blocks of observing time for planetary targets.

- *High sensitivity UV spectroscopy: increase $A_{\rm eff}$ by 2–10 times*: A large increase in the effective area would directly extend the sensitivity for UV spectra of both small and distant planetary bodies to shorter wavelengths, permitting measurements of the key absorptions by simple atoms and molecules. This improvement is presently needed for objects as close as the Galilean satellites of Jupiter, and as distant as Pluto.

- *UV and visible stellar occultations: increase $A_{\rm eff}$ by 2–10 times*: A principle well known by astronomers is that the number of stars above a limiting magnitude increases at much more than a linear rate as one extends to fainter limits. A large increase in effective area would also greatly increase the number of stellar occultations that could be performed. This technique provides the highest altitude resolution presently possible in planetary atmospheres research.

- *Scale height measurements in planetary atmospheres: increase angular resolution by 2–10 times*: A large increase in angular resolution would make it possible to resolve altitude profiles of atmospheric emissions by the same factor. At present HST images can resolve the altitude extent of the auroral curtains on Jupiter. Such measurements could be extended to other planets by the direct ratio of angular resolution to that available today with HST.

References

Bell, J., et al. 2001, AGU fall meeting, abstract #P42A-0554

Clarke, J.T., et al. 1995, Science, 267, 1302

Hammel, H.B., et al. 1995, Science, 267, 1288

James, P.B. 1999, Ann. Rev. Earth & Plan. Sci., 27, 115

Spencer, J. Sartoretti, P., Ballester, G., McEwen, A., Clarke, J., & McGrath., M. 1997, Geophys. Res. Lett., 24, 2471

Sromovsky, L.A., P. Fry, T. Dowling, K. Baines, & S Limaye 2001, Icarus, 150, 244

Hubble's Science Legacy: Future Optical-Ultraviolet Astronomy from Space
ASP Conference Series, Vol. 291, 2003
K.R. Sembach, J.C. Blades, G.D. Illingworth, R.C. Kennicutt, Jr. (eds.)

Icy Bodies of the Outer Solar System: New Scientific Explorations in the Visible and Ultraviolet

Walter M. Harris

University of Wisconsin-Madison, 1150 University Ave., Madison, WI 53706

Abstract. The previous two decades have seen several milestones in our study of icy bodies in the solar system, ranging from the discovery and characterization of the Kuiper-Belt, to our first visits to cometary nuclei (1P/Halley; 19P/Borelly), to the new discoveries made from observations of recent active comets including C/1995O1 (Hale-Bopp) and C/1996B2 (Hyakutake). During and before this period many space-based remote sensing instruments have contributed to the advancement of our knowledge of these icy remnants of the formation of our planetary system. Here I review the characteristics of these bodics, the major questions that confront us, and the new capabilities that would most directly address them in the Hubble Science Legacy (HSL) telescope.

1. Introduction

It has become an accepted paradigm that comets and Kuiper Belt Objects (KBO) are the remnant "brick and mortar" of the outer Solar System. Both theoretical and observational studies have reached the basic conclusion that the cores of planets in the outer Solar System were formed from an aggregate of smaller bodies that slowly accumulated in the cold, dense regions of the remote solar nebula. As such, the remaining members of this object class are the oldest, least processed component of our planetary system. In them, we find a window to the era of the Solar System formation, where we may study the composition, density, and temperature of the solar nebula as a function of heliocentric distance. The modern spatial distribution and orbital characteristics of comets and KBOs contains an additional record of the orbital evolution of the planets and the subsequent interaction of these bodies with the planets and each other. It has even been suggested that the terrestrial biosphere owes its existence to icy planetesimals, which may have deposited much of the water (Meier & Owen 1999) and organic material currently tied up in living organisms.

In this paper I will address the role that a Hubble Science Legacy telescope could play in advancing our understanding of these small bodies in decades to come. The sections below provide a breakdown of comets and KBOs, their scientific utility, and the techniques used in their study. I then follow with a discussion of the technical and programmatic capabilities that would be of most use to this field from a new HSL instrument.

2. Comets

The observational study of comets dates back centuries. They have played a key role in the development of modern astronomy, including being the first time-variable objects demonstrated to be of extraterrestrial origin, and later as the best early examples of the generalized nature of orbit theory. Our modern understanding of these objects has evolved considerably. Comets are members of the original outer Solar System planetesimal population that are composed primarily of H_2O (up to 90%) and CO_2/CO (5-25%) ice, dust, and other minor chemical species. The relative amounts of these materials, their isotopic and chemical ratios, and the structure of the their ices combine with related information on dust grain characteristics to provide a detailed picture of the early Solar System. In addition to their scientific value as the primordial building blocks of planets, comets have been useful as probes of the solar wind and as astrophysical laboratories for the study of photo-chemistry, gas-dynamics, and plasma interactions. The coming two decades will provide yet more detailed information about these objects, first from the CONTOUR spacecraft, followed by Deep Impact in 2005, and Rosetta in 2012. Each of these missions are designed around the expectation of ground and space base remote sensing support.

The major caveat to our understanding is that the available comet sample consists of objects in the inner Solar System, where solar heating has melted their surface to form the coma and ion/dust tails by which they are characterized. This sample is thus biased toward the most numerous (and therefore available for interaction) end of the planetesimal size distribution, and, by virtue of solar heating and repeated visits, the least pristine members of the group (Figure 1a). The continuum of evolutionary states in comet nuclei complicates the task of extracting their underlying compositional characteristics and differences.

Comets fall into two orbital classes that also provide a general evolutionary (and possibly compositional) breakdown: long versus short period.

2.1. Long Period Comets (LPCs)

Long ($\gtrsim$200 yr) period comets exhibit very specific orbital characteristics that led Jan Oort (Oort 1950) to first suggest that they originated from a reservoir of objects (the Oort Cloud) located up to 50000 AU from the Sun. LPCs have highly eccentric, randomized orbits that suggest a spherical distribution of low energy orbits. They come from all directions and inclinations, with a nearly 50-50 percentile split of prograde versus retrograde motion with respect to the other planets. Models that produce an Oort cloud also lead to the ironic discovery that they were formed far closer to the Sun than the objects in the Kuiper-Belt. Their current circumstances appear to stem from interactions with the proto-Jovian planets, which gravitationally emptied the planetesimal population inside of Neptune, propelling a small, surviving fraction into a remote orbital distribution that has been shaped over time by internal interactions and/or passing stars. Modern evidence of this process is found from the existence of long period comets, which cannot survive close solar proximity perihelia over the life of the Solar System. At present we have very little direct information about the body size and orbit distribution of the undisturbed Oort Cloud.

2.2. Short Period Comets (SPCs)

The SPC (period $\lesssim$200 yr) population is thought to derive from within the Kuiper Belt, where they are presumed to have formed. The observed population again sample the most numerous (and presumably small) end of the sized distribution, although in this case we have supplemental data on the distribution of larger objects (see below). They are characterized by much shorter periods (2P/Encke has a period of 3.3 yr), reduced eccentricity relative to LPCs, prograde orbits, and a range of inclinations more or less confined to the ecliptic. They are also generally more evolved (Figure 1a) than long period comets, because their orbital periods are short and they encounter the Sun more frequently. It is unclear how this difference in relative surface evolution affects their properties in comparison with long period comets.

2.3. The Big Questions

The recent Solar System decadal survey report (Combi et al. 2002) has outlined several specific areas of importance in the study of comets

- *Are there compositional identifiers that can describe cometary nuclei on the basis of origin and/or evolutionary state?* [i.e., Can we identify a structural marker of LPCs versus SPCs? How do variations in gas/dust, CO/CO_2, and isotopic ratios relate to origin or evolution?]
- *Did comets contribute significantly to the water complement of the Earth?* [We have reliable data from D/H ratios on only a few objects that suggest the negative, but the sample is small.]
- *What is the relationship between comets and the Centaurs/KBOs?* [Our working model assumes they are from a single population.]
- *How do comets respond at the extremes of small and large heliocentric distances?* [Our sample is highly biased to objects well separated from the Sun or close enough for substantial outgassing to occur. Objects close to the Sun and very remote are poorly represented.]
- *Can we use comets to reliably probe the solar wind and as proxies for the interstellar medium (ISM)?* [As solar wind probes, comets represent an inexpensive means of sampling a range of heliocentric latitudes and distances. As presumed products of an unprocessed solar nebula, comets, particularly comet dust, are the only accessible example of the ISM.]

3. The Kuiper Belt

The discovery and initial characterization of the Kuiper-Belt is among the most significant modern developments in our understanding of the Solar System. Originally postulated by Kuiper and Edgeworth (Edgeworth 1943) as the outer rim of the planetary accretion disk, the first KBOs were discovered only 10 years ago. The Kuiper-Belt is believed to exist in a transition zone where enough mass was present to form planetesimals, but not larger bodies such as Neptune. The original distribution of debris has been modified considerably by the migration of major planets and major planet orbital resonances (Malhotra, Duncan, & Levison 2000), and it is believed that up to 99% of the original material has been lost. Since the first KBO discovery (1992QB1) (Jewitt, Luu, & Marsden 1992), more than 1000 bodies (Jewitt & Luu 2000) with diameters up to 900 km (Jewitt, Aussel, & Evans 2001) have been observed in the region at and beyond the orbit of Neptune. Statistical arguments suggest that more than 10^5 objects of this type may await discovery. The surveys conducted to date have found an intricate distributional pattern of resonant, nearly circular, and eccentric orbit groupings that give insight to the dynamical history of the Solar System. As with comets, the KBOs are statistically biased, although in this case to the largest (and therefore observable from Earth) bodies in the distribution. Surveys continue to search for KBOs, and we may an opportunity for direct study with the proposed Pluto-Kuiper Express (PKE) mission within 25 years.

The Kuiper Belt can be broken down into three primary groupings and one sub-group of less stable, transitional objects.

3.1. The Classical Belt

Classical Kuiper-Belt Objects (CKBO) make up the majority (about 70%) of the Belt (Trujillo, Jewitt, & Luu 2001). They exist in Neptune-avoiding, roughly circularized orbits between 41 and 46 AU. For the most part, they appear to be in the orbits in which they were formed, unmodified except for an unexplained inclination dispersion of up to 30°. The sharp drop in the population beyond 46 AU is not yet completely understood, and may result from a gravitational rejection or a downward shift in the maximum object size.

3.2. Plutinos

The Plutinos are a group of KBO objects that occupy orbital resonances with Neptune and are named for the largest of the objects in their class, Pluto. The orbits of the Plutinos are otherwise similar to the CKBOs and comprise roughly 20% of the total population of the belt. Of particular interest is the relatively large number of KBOs in the 3:2 (Pluto) resonance, compared to others (a few 2:1 and 4:3 resonant bodies have been found).

3.3. The Scattered Belt

Scattered KBOs (SKBO) have the smallest perihelia of the entire belt at about 35 AU. At this distance they are unable to avoid a weak gravitational interaction with Neptune and are scattered into highly eccentric orbits with aphelia as large as 200 AU (Trujillo, Jewitt, & Luu 2000). They make up a relatively small portion of the Kuiper-Belt ($\lesssim$10%), but a potentially important one, because it is thought that this population may be the primary source of both SPCs (Duncan & Levison 1997; Levison & Duncan 1997) and the Centaurs (see below).

3.4. Centaurs

The Centaurs are a small object class with chaotic orbits that carry them deeper into the Solar System than other KBOs (Millis et al. 2002). They are thought to be a transitory population of deflected objects that are dynamically biased to recent events in the same manner as SPCs. The largest known Centaur, Chiron, has a diameter of 175 km, but comet-sized ($\lesssim$10 km) members of this group are also observed. Due to their greater distance from the Sun, they are less processed by the solar heating, but nevertheless display some low level outgassing activity (Bockelée-Morvan et al. 2001). The Centaurs are thus extremely important as bridging bodies both between large and small KBOs and between the cold inactive occupants of the Kuiper-Belt and more evolved SPCs.

3.5. The Big Questions

Despite the discovery of more than 1000 objects, our knowledge of the Kuiper-Belt is rudimentary. Important questions awaiting complete answers include:

- *What is the size and orbital distribution of objects in the Kuiper belt?* [Does the above orbital fractionation hold? Are SPCs indicative of a large reservoir of small objects?]
- *What is the source of the high inclination scatter in the Classical belt?* [One suggestion is movement of Neptune's orbit during the late formation phase of the Solar System.]
- *Why does the Classical belt terminate at 49 AU?* [Does this represent a paucity of objects or only large bodies?]
- *Why is the 3:2 resonance with Neptune preferentially occupied?*
- *How do KBOs interact?* [Is the Pluto-Charon system unique or typical (Figure 1c)?]

4. Useful Capabilities for a HSL Instrument

There is an extensive history of observations of icy objects with ultraviolet/optical remote sensing instruments of various instrument types, bandpasses (extreme-ultraviolet to near-infrared), and field of view (FOV; arcseconds to degrees). Several examples of past and present instruments include IUE (e.g Feldman et al. 1987), Copernicus (e.g. Drake et al. 1976), HUT (Feldman et al. 1991), HST (e.g. Weaver et al. 1999), MSX (Kupperman et al. 1997), SOHO (Combi et al. 2000), FUSE (Feldman, Weaver, & Burgh 2001), and sub-orbital rockets (Harris et al. 1999; McPhate et al. 1999). Building from the strengths and weaknesses implied by these studies, the evolving capabilities of ground-based facilities, and our developing understanding of these objects, it is possible to develop a modern priority map for the HSL Observatory. In doing so here, I make no assumptions regarding the basic technology (aperture, instruments, bandpass, number of facilities) that will be incorporated into the HSL facility. This is driven partly by the current state of the design concept, but also by the fact that, for comets, KBOs, and Centaurs, the *most important new capability is time.* 100% of the time on an HST-class facility is generally preferable to $\lesssim$5% of the time on a telescope with 3 times the aperture.

Below, I break down the topics as in the above sections, between comets and the KBOs/Centaurs. This is as much topical as approach; each class of object has different instrumental requirements.

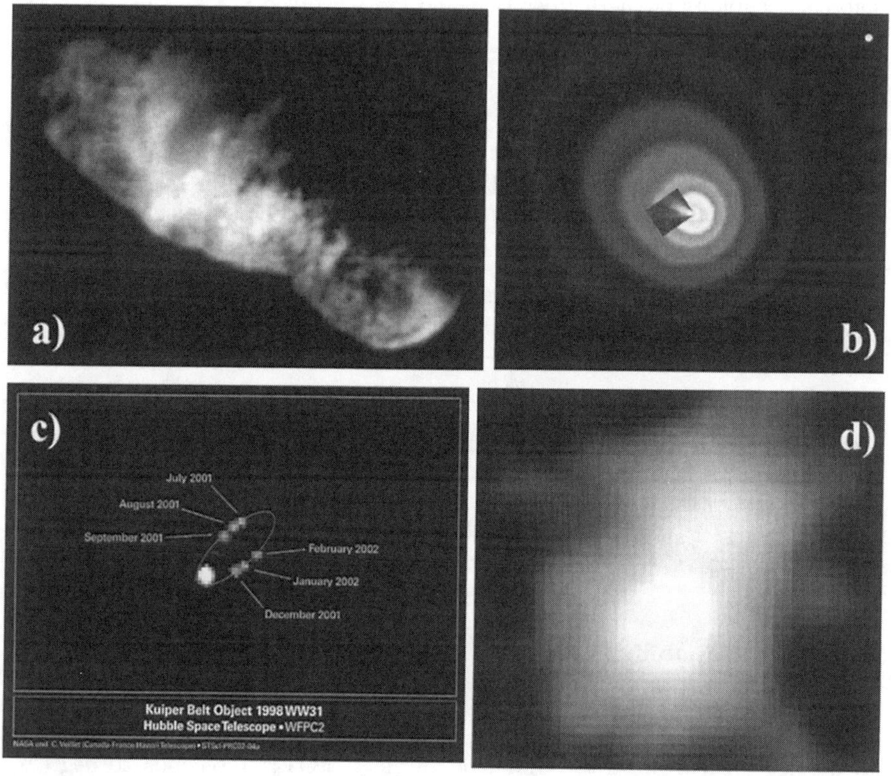

Figure 1. Two extremes of comet viewing perspective are shown. In a) the nucleus of 19P/Borrelly, as seen from the Deep Space 1 spacecraft, shows the non-spherical nucleus shape of the nucleus and the variations in surface structure. In b) we see the H coma of comet Hale-Bopp (Combi et al. 2000) with perspective images of the visible comet (at center) and a dot representation of the Sun's angular diameter (upper right). In the HST release in c) we see a series of images showing KBO 1998 WW31 and the motion of its companion. The largest separation between these two bodies is 0.56 arceseconds. In d) is a CFHT image from Veillet et al. (2002) of the same system. One advantage of a wide field imaging space telescope is the ability to obtain high resolution across the full field of view, as opposed to adaptive optics systems that correct less than an arcminute.

4.1. Comets

Comets are typically studied via emission lines (fluorescence and prompt) from neutral and ionized coma species, and by spatial maps of dust in the tail and inner coma. In the ultraviolet (UV) we gain access to element ground state emissions, reduced dust continuum, photochemical end state species, and higher energy interaction signatures. The nucleus is observed for rotational characteristics and the distribution of active sector. The major new requirements for comet remote sensing in the next several decades can be broken down into several areas;

- *Large ($\gtrsim 8'$) Field of View:* [Gas production and photochemistry play out in comets over nuclear distances up to 10^6 km (Figure 1b).]
- *Sensitivity into the EUV:* [EUV sensitivity will permit studies of noble gas emissions, He, and high ionization state solar wind charge exchange.]
- *Comet Specific Imaging Filters:* [In addition to the existing comet filter library (Farnham, Schleicher, & A'Hearn 2000), several atomic and molecular emissions may be studied in the UV with narrow band filters]
- *High Étendue Spectroscopy of H* I *Lyα:* [Hydrogen emissions (Figure 1b) are diagnostic of photochemical rates and branching ratios, and of the efficiency of collisional heating in the inner coma.]
- *Inner Solar System Targeting:* [The 45° solar exclusion angle typical of space observatories restricts study of comets when they are most active.]
- *Object Specific Campaigns:* [Comets are rarely followed over a full solar interaction. Campaigns following the development and decline of comet activity are one key to understanding the evolution of nuclei.]

4.2. Kuiper Belt Objects and Centaurs

KBOs and Centaurs are traditionally studied for their surface properties, rotation, and near surface outgassing. Such observations require techniques distinct from comets, a difference exaggerated by the need to survey for extremely faint objects over a wide FOV. The HSL observatory will not be competitive with ground-based telescopes for collecting area, but it has advantages from lower sky background and the ability to maintain diffraction limited imaging over its full FOV. In the UV, an HSL instrument can improve our characterization of KBO color (Jewitt & Luu 2001) and search for gas production signatures, particularly from Centaurs. These programs would benefit from several HSL capabilities.

- *4-6 m Class (or better) Collecting Area at the Diffraction Limit:* [An HSL telescope niche is the ability to survey for smaller, presumably more numerous, bodies over an intermediate FOV. Another is to perform pencil-beam searches for KBO satellites (e.g. Viellet et al. 2002) (Figure 1c).]
- *A UV Comet Filter Library:* [Focusing on high volatility species like CO.]

- *Broad-band Low Resolution (0.5-1 nm) Spectroscopy:* [To measure the surface reflectivity and ices of KBOs.]

- *Wide (10-15′) FOV:* [Large Kuiper-Belt surveys will extend our catalog of objects to smaller size, explore the 49 AU edge of the classical belt, and identify KBO companions (Figure 1c,d).]

References

Bockelé-Morvan, D., Lellouch, E., Biver, N., Paubert, G., Bauer, J., Colom, P., & Lis, D.C. 2001, A&A, 377, 343

Combi, M.R., Reinard, A.A., Bertaux, J.-L., Quemerais, E., & Mkinen, T. 2000, Icarus, 144, 191

Combi, M., et al. 2002, The Future of Solar System Exploration: 2003-2013, in press

Drake, J., Jenkins, E., Bertaux, J., Festou, M., & Keller, H. 1976, ApJ, 209, 302

Duncan, M.J., & Levison, H.F. 1997, Science, 276, 1670

Edgeworth, K.E. 1943, JBAA, 53, 181

Farnham, T.L., Schleicher, D.G., & A'Hearn, M.F. 2000, Icarus, 147, 180

Feldman, P.D., et al. 1991, LPI Contributions, 765, 60

—. 1987, A&A, 187, 325

Feldman, P., Weaver, H., & Burgh, E. 2001, AAS/DPS Meeting, 33

Harris, W., Nordsieck, K., Scherb, F., & Mierkiewicz, E. 1999, EM&P, 78, 161

Jewitt, D., Aussel, H., & Evans, A. 2001, Nature, 411, 446

Jewitt, D., Luu, J., & Marsden, B.G. 1992, IAUC, 5611, 1

Jewitt, D.C., & Luu, J.X. 2000, Protostars and Planets IV, 1201

—. 2001, AJ, 122, 2099

Kupperman, D., Paxton, L., Carbary, J., McEvaddy, P., Morrison, D., Meng, C.-I., Henry, R., Tedesco, E., & Price, S. 1997, AAS/DPS Meeting, 29

Levison, H. F., & Duncan, M. J. 1997, Icarus, 127, 13

Malhotra, R., Duncan, M., & Levison, H. 2000, Protostars and Planets IV, 1231

McPhate, J., Feldman, P., McCandliss, S., & Burgh, E. 1999, ApJ, 521, 920

Meier, R., & Owen, T.C. 1999, Space Science Reviews, 90, 33

Millis, R.L., Buie, M.W., Wasserman, L.H., Elliot, J.L., Kern, S.D., & Wagner, R.M. 2002, AJ, 123, 2083

Oort, J.H. 1950, Bull. Astr. Inst. Netherlands, 11, 91

Trujillo, C.A., Jewitt, D.C., & Luu, J.X. 2000, ApJ, 529, L103

—. 2001, AJ, 122, 457

Veillet, C., Parker, J., Griffin, I., Marsden, B., Doressoundiram, A., Buie, M., Tholen, D., Connelley, M., & Holman, M. 2002, Nature, 416, 711

Weaver, H.A., Feldman, P.D., A'Hearn, M.F., Arpigny, C., Brandt, J.C., & Stern, S.A. 1999, Icarus, 141, 1

Hubble's Science Legacy: Future Optical-Ultraviolet Astronomy from Space
ASP Conference Series, Vol. 291, 2003
K.R. Sembach, J.C. Blades, G.D. Illingworth, R.C. Kennicutt, Jr. (eds.)

Solar System: Satellites and Summary

Melissa A. McGrath
Space Telescope Science Institute, 3700 San Martin Drive, Baltimore, MD, 21218

Abstract. This paper presents a brief overview of the final of three topics chosen to represent Hubble's science legacy for the Solar System, namely Solar System satellites. This is followed by a short summary of the capabilities that are desirable for a future ultraviolet/optical space telescope to enable successful Solar System observations.

1. Hubble's Science Legacy for Solar System Satellites

One of the basic goals espoused in NASA's strategic plan for space science is to understand the formation and evolutionary processes that led to the diversity of Solar System bodies and the uniqueness of Earth. The importance of Solar System satellites in furthering this goal is eloquently articulated by McKinnon (1999) and illustrated in Figure 1: "Each giant planet, with its various rings and moons, is a miniature planetary system. The richness and complexity of these systems is what attracts our interest, for in their formation, structure, and evolution we glimpse some of the possibilities for planetary systems elsewhere in the galaxy. At the same time, the satellites themselves are natural laboratories for studying planetary evolution processes."

Within the context of the Hubble Space Telescope (HST) mission, Solar System science is unique because with only a few exceptions (Pluto, Kuiper Belt objects) every planet and major type of object in the Solar System has been explored in situ by spacecraft. Furthermore, most of the planets have witnessed not only an initial exploratory mission, but more detailed follow-on missions such as Magellan to Venus, Galileo to Jupiter, and the upcoming Cassini mission to Saturn. These missions typically produce spectacular high-resolution imaging (Galileo images of Io and Europa come readily to mind) far beyond the capability of HST on these same targets. Solar System targets for HST are therefore much more observationally mature than their counterparts in other astrophysics disciplines. Nonetheless, HST has made significant and widely-recognized contributions to the detailed understanding of our satellite neighbors. Many of its discoveries have not been possible with even the most proximate spacecraft because of various constraints, be they data-rate limitations, incomplete wavelength coverage, or the competition among numerous instruments for prime observing time near closest encounters. The study of satellites was recognized in the 10th anniversary symposium "Ten Years of HST Science" as being one of the four most significant contributions to Solar System science. HST's hundreds of satellite observations include some from every relevant planet in the Solar

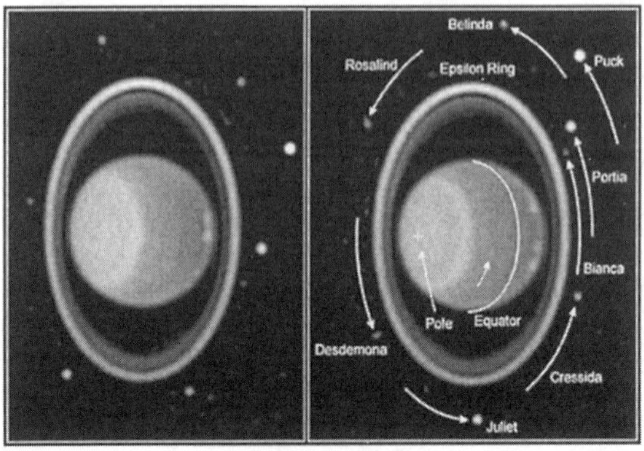

Figure 1. HST/NICMOS image of Uranus, its rings, and satellites. The image on the right was taken 6 hours after the one on the left, showing the rotation of Uranus' clouds and the orbital motion of several of its satellites. (PRC97-36a, November 1997, E. Karkoschka and NASA.)

System (Mercury and Venus have no satellites). We highlight a (subjective) few of the most significant results, many of which were unanticipated, to illustrate the potential discoveries that await the next steps in capability of future space telescopes.

2. Solar System Satellites: HST's "Greatest Hits"

Io

The most intensely observed satellite in the Solar System with HST has been the innermost Galilean satellite of Jupiter, the bizarre and enigmatic moon Io. Despite its small size (about the diameter of the Moon) and location buried deep within the gravitational well and magnetic cavity of the planet, Io has a unique and pervasive influence on the environment of its parent system. As an indirect result of its active volcanism Io supplies about one ton of material per second to the Jovian magnetosphere, material which has a number of profound impacts on the Jovian environment, including: implanting itself into the surfaces of the other Galilean satellites; precipitating along magnetic field lines to excite the Jovian aurorae; and escaping the system altogether to form a large interplanetary cloud of material observable to distances as large as 500 Jovian radii from the planet. Because the active volcanoes do not impart enough velocity to the plume material to allow it to escape, Io has a tenuous atmosphere which is stripped by plasma co-rotating with the Jovian magnetic field. The importance of Io's SO_2 atmosphere in the sequence of events triggered by Io's volcanoes has made it a subject of intense scrutiny since its discovery in 1979. Because it is notoriously difficult to detect via ground-based techniques, it was not detected again until 1990, and in the intervening 11 years interpretations of the few existing observations led to models discrepant by as much as 6 orders of magnitude in surface pressure. For these reasons, Io's surface, atmosphere, and volcanoes have been studied extensively with HST, work which has been resoundingly successful. The first direct remote sensing imaging and spectroscopic measurements of the Io's volcanic plumes have been accomplished with HST (Spencer et al. 1997b; Spencer et al. 2000), as well as numerous ultraviolet (UV) spectroscopic detections of the atmospheric SO_2 (Ballester et al. 1994; Trafton et al. 1996; McGrath et al. 2000; Feldman et al. 2000), which have resulted in reasonably good agreement regarding the abundance and distribution of this gas. Additionally, significant changes in the surface albedo patterns caused by active volcanism (Spencer et al. 1997a) and aurorae at Io (Roesler et al. 1999) have also been observed remotely for the first time with HST.

Icy Satellites (Jupiter, and Saturn)
In the Jovian system, plasma originating from Io is found to produce extensive radiolysis of the outer icy Galilean satellite surfaces. At Saturn, similar processes take place, with plasma originating from both rings and satellites. Solid state absorbers, the by-products of such radiolysis, have been discovered or mapped on many satellites using HST/FOS (Noll et al. 1995a,b, 1996, 1997; Calvin & Spencer 1997), including SO_2 in surface ice on Europa and Callisto, ozone (O_3) in surface ice on Ganymede, Rhea, and Dione, and O_2 in solid state form on Ganymede. Discrete spectral features are present in the spectra, including a broad feature at 280 nm on Europa's trailing hemisphere attributed to SO_2. The ratio of Ganymede's trailing to leading hemisphere spectrum results in a very broad absorption feature centered at ~260 nm. The position and width of the band is closely matched by the gas-phase Hartley band of ozone. Identification of a UV spectral feature in laboratory samples of Ar-irradiated H_2O ice as possibly due to O_3 strengthens the idea that particle irradiation of Ganymede's trailing hemisphere might form O_3 in radiation-damaged ice. The detection and mapping of spectral features attributed to O_2 in Ganymede's visible spectrum is also consistent with this hypothesis. Particle bombardment is also thought to produce the tenuous O_2 atmospheres first detected around Europa and Ganymede using HST/GHRS (Hall et al. 1995a; 1998). The oxygen UV emissions from Ganymede subsequently imaged with STIS confirmed the interpretation of the GHRS data as being due to polar aurorae on that satellite (Feldman et al. 2000).

Triton
Triton's occultation of a 10.6 magnitude star on 4 November 1997 was observed with HST's Fine Guidance Sensor. The star passed close enough to the center of Triton to record a central flash. From these data, Elliot et al. (1998) inferred a substantial increase in Triton's surface pressure since a previous occultation in 1995 and the Voyager encounter in 1989. Assuming that the predominantly N_2 atmosphere is in vapor-pressure equilibrium with surface frost, the increase in surface pressure implies a global warming of Triton's surface of 1–2 K over the 8-year period.

Titan
Titan's surface was first imaged at near-infrared wavelengths using HST/WFPC2 in 1994 (Smith et al. 1996). Because of the decreasing optical depth of Titan's dense N_2 atmosphere with increasing wavelength, filters centered at wavelengths of 850 and 1040 nm are able to image the surface of the satellite. Relative albedo maps of Titan's surface constructed from the images reveal surface features with

~10% maximum contrast. The surface albedo map is dominated by a large (4000×2500 km^2) feature centered at roughly 110° W, 10° S. Several explanations for the feature are possible, with that of an overall dark landscape broken by a bright feature suggesting highland washing by methane rainfall being considered most likely. These results have been used to help target the Cassini mission's Titan Huygens probe.

Saturn Ring Plane Crossing

The Earth and Sun crossed the Saturn ring plane four times from May 1995 to February 1996, providing a unique opportunity to search for faint satellites and observe the tenuous outer rings. These events are rare, occurring in groups of 2 or 4 every 14.5 years. Although Saturn was too close to the sun for HST to observe the last event on 11 February 1996, the 22 May, 10 August, and 19 November 1995 events were each observed. The observations yielded the recovery of several small satellites that had not been observed since Voyager, many far from their predicted positions based on their poorly known ephemeri. Several new satellites (S4, S5, and S6) were also discovered in the first two crossings, but the three new objects seem to have disappeared by 21 November 1995. Voyager found evidence of similar activity, but the new images marked the first time that observations from Earth's vicinity have documented the actual formation (in the 14 years since Voyager) and subsequent destruction of ring material in a matter of weeks. The discovery of new objects in this transitional phase is not totally unexpected since they orbit Saturn near the narrow F ring, which is a dynamic transition zone between the main rings and the larger satellites. Moonlets in this region can be easily disrupted by Saturn's tidal pull if they are fractured by an impact, forming a cloud of debris.

During the second ring plane crossing in August, Hall et al. (1995b) obtained UV spectra near the rings using HST/FOS that showed the presence of fluorescent OH $A^2\Sigma^+$ - $X^2\Pi(0,0)$ band emission. These observations showed that Saturn's rings are embedded in a tenuous gas of hydroxyl (OH) molecules with a scale height of 0.4 R_S. OH emission was also detected much further out in the Saturn magnetosphere near the orbit of Tethys (Shemansky et al. 1993), and subsequently mapped along many lines of sight in the magnetosphere (Richardson et al. 1998, Jurac et al. 2002). The OH is thought to come from H_2O extracted from the icy satellites, and previous models had assumed charged particle sputtering and micrometeorite bombardment as the primary sources. However, the source rate inferred from the observations is too large to be understood by either or both of these mechanisms, and is still unexplained.

3. Satellite Science with Future UV/Optical Telescopes

In considering the possibilities for satellite science with future UV/optical telescopes, it is important to bear in mind the scientific and mission context within which future telescopes will be operating. While future NASA Discovery, Small Explorer, and Medium Explorer missions to do Solar System science are unpredictable, there are several known or highly likely upcoming missions of relevance for satellite science in the time frame relevant for planning of the next UV/optical space telescope. The first is the Cassini mission to Saturn, with a nominal mission of 2004–2008, and potential mission extension beyond 2008. In the recently released decadal study for planetary science, "New Frontiers of Solar System Exploration", the three most highly ranked medium class missions are relevant to satellites: a Pluto/Kuiper Belt mission, with earliest possible arrival of 2015 at Pluto and 2025 at a Kuiper Belt object; the South Pole-Aitken lunar mission, which includes sample return from the Moon; and a Jupiter Polar Orbiter. Within the large mission class, a Europa orbiter mission (Europa Geophysical Explorer) is recommended.

Based both on the type of science results discussed in §1.1 and on the likely mission complement in the post-HST time frame, it is safe to predict that the Galilean satellites will continue to be primary targets, and the Saturn satellites will become increasingly important targets, especially in the UV, because studies of both are currently hampered by the relatively low signal-to-noise ratio obtainable on these targets. Enhanced capabilities of the next UV/optical telescope will allow detailed follow-up to Cassini with "Io-like" science for Titan and the mid-size Saturn satellites. Such enhanced capabilities will also open up the distant Solar System (asteroids, Uranus, Neptune, Pluto/Charon, the Kuiper Belt), which remains largely unexplored with HST. This will result in an increasing focus on nitrogen (N, N_2, NH_3) in the outer Solar System (Titan, Triton, Pluto, Charon atmospheres and surfaces). Cryo-volcanism may be important and observable on the numerous icy satellites in the outer Solar System. Finally, the Saturn ring plane crossings relevant in the post-HST time frame will take place in 2010 and 2024.

4. Solar System Science with Future UV/Optical Telescopes

The Solar System science highlighted in this conference proceedings amply demonstrates the success that HST's unique capabilities have achieved in addressing NASA's basic goals for Solar System exploration. An added advantage for Solar System observations is their popularity with the public, emphasized by a press release rate that is disproportionate to the actual amount of observing time ($\sim 5\%$ per cycle, averaged over all cycles). A further demonstration of this point was the extraordinary amount of publicity received by HST's observing campaign on the comet Shoemaker-Levy 9 impact with Jupiter in 1994. HST's Solar System observations have also provided an important legacy because of the many joint observing campaigns that have emphasized and achieved great synergy with other NASA missions including but not limited to Galileo, Cassini, Lunar Prospector, Mars Global Surveyor, Chandra, the Extreme Ultraviolet Explorer (EUVE), the International Ultraviolet Explorer (IUE), ASTRO-1 and -2, and the Far Ultraviolet Spectroscopic Explorer (FUSE).

5. Desired Capabilities

UV capability

UV capability is critical for Solar System science for several reasons. The solar-type spectra, combined with very low UV reflectivities, combine to give very low UV continuum compared to the visible, making energetic emission line processes such as aurorae and dayglow detectable. For example, the Jovian aurorae are not detectable in visible Hα from Earth, but are readily detectable via Lyα and H_2 emission in the UV from HST.

Ground-state resonance transitions of many atoms, ions and molecules common in Solar System objects occur in the UV: H_2, H, O, C, S, N, SO_2, N_2, CO, CO_2, to name a few. Io and its associated plasma torus have perhaps the richest S and O emission line spectrum known. Many UV absorbers important for planetary atmospheres also have strong absorption systems in the UV: hydrocarbons (CH_4, C_2H_2, C_2H_4, etc.), NH_3, SO_2, SO, S_2, etc.

Field of view and spatial resolution

It is highly desirable that the high resolution imaging channel of future space telescope cameras be large enough to accommodate Jupiter within the field of view (FOV). This requires a FOV $> 50''$, which is also large enough to accommodate Saturn and its rings in a single frame. This is particularly critical for imaging the north and south polar aurorae on Jupiter simultaneously, as is possible with HST/WFPC2 but is not possible with HST/STIS.

A large FOV (of order degrees) is desirable for Kuiper Belt surveys and comets. Regarding Kuiper Belt surveys, ~400 objects are now known, and many more are expected to be detected by 2010 via ground-based searches. Like HST and the Next Generation Space Telescope (NGST), the niche for the next generation UV/optical telescope will be the small, faint end of the distribution, which does not require a large FOV.

We are currently limited to using broad-band colors to deduce composition information for many of the smallest Solar System objects such as asteroids, Kuiper belt objects, and small/distant satellites. Increased spatial resolution would be highly desirable for advancing our understanding of these objects. Due to the very small apparent size of several important Solar System objects such as Pluto ($< 0.09''$) and Triton ($< 0.13''$) it is important to have as many resolution elements as possible in imaging these targets. With HST/FOC imaging, Pluto's diameter subtended ~7.5 pixels. The Pluto-Charon separation is $\sim 0.9''$. A fiducial is that an 8 m telescope can get ~ 15 resolution elements ($\lambda/D = 6$ milli-arcsec) across Pluto at 0.25 μm.

Sensitivity

For almost all the science discussed in the three Solar System contributions to this proceedings, we are still photon starved in the UV. Increased sensitivity is critical, both via the aperture size and detector technology improvements. As an example, the UV emission spectrum of Titan from atomic and molecular nitrogen has not been detected at all since the 1980 measurements by the Voyager ultraviolet spectrometer. HST/STIS does not have sufficient UV throughput to study the atmospheres of Titan (either the albedo or the atmospheric emissions), Triton, or make significant progress on either the albedos or UV auroral and dayglow emissions of Uranus or Neptune.

Spectral coverage, resolution, slits

Visible photon rejection (solar blind) detectors and filters are critical. The principal component of light from all Solar System objects is reflected sunlight, specifically that of a G2V star which peaks at ~5500Å and has 6 orders of magnitude more flux in the visible than the ultraviolet. If UV detectors are not solar blind, the detected photons are dominated by those from the visible spectral region. Three successive instruments on HST, the FOC, WFPC2, and STIS, have revolutionized the study of the Jovian aurorae, which has been one of the most extensively studied Solar System phenomena with HST. This success is a direct result of the increasingly higher UV throughput and visible photon rejection for the three successive instruments.

High Earth orbit is desireable. Lyα (1215.67Å) and OI 1304Å emissions are prevalent spectral components for many Solar System targets, which are difficult to detect and measure accurately from low Earth orbit.

Sensitivity below Lyα is important because there is virtually no intervening hydrogen absorption in the interplanetary medium. The near-UV spectral region is also important as many key minor species, such as hydrocarbons and NH_3, have strong absorption bands in this spectral region.

Spectral resolution of R = 10,000 is adequate for most problems, while 30,000 would yield Doppler shifts and allow studies of winds on Io.

Long, narrow slits like those available with STIS are important for Solar System observations because of the small angular size of many Solar System targets. Long slits are important to maximize throughput, and to obtain spatial resolution on extended targets. Narrow slits (e.g., slit width $\approx$ resolution element) are required to maintain adequate spectral resolution on extended targets. Spatial mapping with long slits provides much more efficient use of telescope time, instead of having to map large objects with small slits, as was done for Ganymede spectroscopy of O_2 (Calvin & Spencer 1997), and for the Jupiter atmosphere following the SL9 impacts (Yelle & McGrath 1996) using small HST/FOS slits.

Slitless spectroscopy such as that available with the wide slits on STIS is an important capability to maintain. This capability has produced many exciting results for the Galilean satellites, results which can be anticipated for Titan, Triton, Pluto, etc., on future space telescopes using this technique.

Temporal coverage (synoptic monitoring)

Time-tag/rapid read out (~0.1 s) is important for events such as occultations and the Jovian aurorae, which flare on very short time scales.

Low Earth orbit gives no CVZ for Solar System targets, and imposes an interuption to observing about every 60 minutes.

Many problems could be addressed with a relatively small (HST-class) aperture with continuous time coverage (planetary weather, climate, aurorae, satellite-magnetosphere interactions).

A small solar avoidance angle would allow observations of Venus and comets in the inner Solar System (the HST limit is 50°).

Other considerations

Do not preclude planetary observations in the early stages of planning. Then, at minimum, do the simple things that enable planetary observations. For example, many missions (IUE, HST, HUT) have a solar avoidance limit that barely precludes observations of Venus. Is this technically necessary, or historical?

The requirement for moving target tracking needs to be included from the beginning or it probably will not happen, or will be much more expensive if enabled late in the mission planning stages. Low-cost options that are relatively simple may be feasible (e.g., IUE).

Having more observing modes is not always the answer. For example, the planetary slits on STIS are not used frequently and have never been calibrated.

References

Calvin, W.M., & Spencer, J.R. 1997, Icarus, 130, 505

Elliot, J.L., et al. 1998, Nature, 393, 765

Feldman, P.D., McGrath, M.A., Strobel, D.F., Moos, H.W., Retherford, K.D., & Wolven, B.C. 2000, ApJ, 535, 1085

Feldman, P.D., et al. 2000, GRL, 27, 1787

Grundy, W.M., & Buie, M.W. 2002, Icarus, 157, 128

Hall, D.T., Strobel, D.F., Feldman, P.D., McGrath, M.A., & Weaver, H.A. 1995a, Nature, 373, 677

Hall, D.T., Feldman, P.D., Holberg, J.B., & McGrath, M.A. 1995b, Science, 272, 516

Hall, D.T., Feldman, P.D., McGrath, M.A., & Strobel, D.F. 1998, ApJ, 499, 475

Jurac, S., McGrath, M.A., Johnson, R.E., Richardson, J.D., Vasyliunas, & Eviatar, A. 2002, GRL, in press

McGrath, M.A., Belton, M.J.S., Spencer, J.R., & Sartoretti, P. 2000, Icarus, 146, 476

McKinnon, W.B. 1999, in The New Solar System, eds. J.K. Beatty, C.C. Petersen, & A. Chaikin, (Cambridge: Sky Publishing Corporation and Cambridge University Press), 297

Noll, K.S., et al. 1995a, Science, 267, 1307

Noll, K.S., Weaver, H.A., & Gonnella, A.M., 1995b, JGR, 100, 19057

Noll, K.S., Johnson, R.E., Lane, A. L., Domingue, D.L., & Weaver, H.A. 1996, Science, 273, 341

Noll, K.S., Johnson, R.E., McGrath, M.A., & Caldwell, J.J. 1997, GRL, 24, 1139

Richardson, J.D., Eviatar, A., McGrath, M.A., & Vasyliunas, V. 1998, JGR, 103, 20245

Roesler, F., et al., 1999, Science, 283, 353

Shemansky, D.E., et al., 1993, Nature, 363, 329

Smith, P.H., Lemmon, M.T., Lorenz, R.D., Sromovsky, L.A., & Caldwell, J.J. 1996, Icarus, 119, 336

Spencer, J.R., et al., 1997a, Icarus, 127, 221

Spencer, J.R., Sartoretti, P., Ballester, G.E., McEwen, A.S., Clarke, J.T., & McGrath, M.A. 1997b, GRL, 24, 2471

Spencer, J.R., Jessup, K.L., McGrath, M.A., Ballester, G.E., & Yelle, R. 2000, Science, 288, 1208

Yelle, R.V., & McGrath, M.A. 1996, Icarus, 119, 90

Hubble's Science Legacy: Future Optical-Ultraviolet Astronomy from Space
ASP Conference Series, Vol. 291, 2003
K.R. Sembach, J.C. Blades, G.D. Illingworth, R.C. Kennicutt, Jr. (eds.)

Recommended Architectures for the Terrestrial Planet Finder

Charles Beichman

Jet Propulsion Laboratory, California Institute of Technology, Pasadena, CA 91109

Abstract. The primary conclusion from an intensive, two year period of study is that with suitable technology investment, starting now, a mission to detect terrestrial planets around nearby stars could be launched within a decade. A visible light coronagraph using an 8-10 m telescope, or an infrared nulling interferometer, operated on either a $\sim$ 40 m structure or separated spacecraft, could survey over 150 stars, looking for habitable planets and signs of primitive life.

1. Introduction

In March 2000, the Terrestrial Planet Finder (TPF) project at JPL selected four university-industry teams to examine a broad range of instrument architectures capable of directly detecting radiation from terrestrial planets orbiting nearby stars, characterizing their surfaces and atmospheres, and searching for signs of life. Over the course of two years the four teams, incorporating more than 115 scientists from 50 institutions, worked with more than 20 aerospace and engineering firms. In the first year of study, the contractors and the TPF Science Working Group (TPF-SWG) examined over 60 different ideas for planet detection. Four main concepts, including a number of variants, were selected for more detailed study. Of these concepts, two broad architectural classes appear sufficiently realistic to the TPF-SWG, to an independent Technology Review Board, and to the TPF project that further technological development is warranted in support of a new start around 2010. *The primary conclusion from the effort of the past two years is that with suitable technology investment, starting now, a mission to detect terrestrial planets around nearby stars could be launched within a decade.*

The detection of Earth-like planets will not be easy. The targets are faint and located close to parent stars that are >1 million (in the infrared) to >1 billion times (in the visible) brighter than the planets. However, the detection problem is well defined and can be solved using technologies that can be developed within the next decade. We have identified two paths to the TPF goal of finding and characterizing planets around 150 stars out to distances of about 15 pc:

- At visible wavelengths, a large telescope (a 4×10 m elliptical aperture in one design and an 8×8 m square aperture in another) equipped with a selection of advanced optics to reject scattered and diffracted starlight (apodizing pupil masks, coronagraphic stops, and deformable mirrors) offers the prospect of directly detecting reflected light from Earth-like planets.

- At mid-infrared wavelengths, nulling interferometer designs utilizing from three to five 3–4 m telescopes located on either separated spacecraft or a large, 40 m boom can directly detect the thermal radiation emitted by Earth-like planets.

The TPF-SWG established that observations in either the optical/near-infrared or thermal infrared wavelength region would provide important information on the physical characteristics of any detected planets, including credible signposts of life. In fact, the two wavelengths provide complementary information so that in the long run, both would be desirable. The choice of wavelength regime for TPF will, in the estimation of the TPF-SWG, be driven by the technological readiness of a particular technique.

2. Science Goals for the Terrestrial Planet Finder

The TPF Science Working Group established a Design Reference Program to give broad guidelines for defining architectures for TPF. The goals for TPF were set out at the December, 2000, meeting of the TPF-SWG.

Primary Goal for the Terrestrial Planet Finder: TPF must detect radiation from any Earth-like planets located in the habitable zones surrounding 150 solar type (spectral types F, G, and K) stars. TPF must: 1) characterize the orbital and physical properties of all detected planets to assess their habitability; and 2) characterize the atmospheres and search for potential biomarkers among the brightest Earth-like candidates.

The Broader Scientific Context: Our understanding of the properties of terrestrial planets will be scientifically most valuable within a broader framework that includes the properties of all planetary system constituents, e.g. both gas giant and terrestrial planets, and debris disks. Some of this information, such as the properties of debris disks and the masses and orbital properties of gas giant planets, will become available with currently planned space or ground-based facilities. However, the spectral characterization of most giant planets will require observations with TPF. The ability of TPF to carry out a program of comparative planetology across a range of planetary masses and orbital locations in a large number of new solar systems is by itself an important scientific motivation for the mission.

Astrophysics with TPF: An observatory with the power to detect an Earth orbiting a nearby star will be able to collect important new data on many targets of general astrophysical interest. Architectural studies should address both the range of problems and the fundamental new insights that would be enabled with a particular design.

3. Biomarkers for TPF

Early TPF-SWG discussions made it apparent that observations in either the visible or mid-infrared portions of the spectrum were technically feasible and scientifically important. A sub-committee of the TPF-SWG was established under the leadership of Dr. Dave Des Marais to address the wavelength regimes for TPF. The conclusions of their report (Des Marais et al. 2002) can be summarized briefly as follows:

- Photometry and spectroscopy in either the visible or mid-infrared region would give compelling information on the physical properties of planets as well as on the presence and composition of an atmosphere.
- The presence of molecular oxygen (O_2) or its photolytic by-product ozone (O_3) are the most robust indicators of photosynthetic life on a planet. Even though H_2O is not a bio-indicator, its presence in liquid form on a planet's surface is considered essential to life and is thus a good signpost of habitability.
- Species such as H_2O, CO, CH_4, and O_2 may be present in visible-light spectra (0.7 to 1.0 μm minimum and 0.5 to 1.1 μm preferred) of Earth-like planets.
- Species such as H_2O, CO_2, CH_4, and O_3 may be present in mid-infrared spectra of Earth-like planets (8.5 to 20 μm minimum and 7 to 25 μm preferred).
- The influence of clouds, surface properties, and rotation can have profound effects on the photometric and spectroscopic appearance of planets and must be carefully addressed with theoretical studies in the coming years (e.g., Ford, Seager, and Turner 2001).

In conclusion, the TPF-SWG agreed that either wavelength region would provide important information on the nature of detected planets and that the choice between wavelengths should be driven by technical considerations.

4. TPF Architectural Studies

After an initial year during which the four study teams investigated more than 60 designs, the teams and JPL identified four architectural classes (with a number of variants) worthy of more intensive study (Table 1). High-level descriptions of these architectures are given below; more detailed information is available in the summary of the recent architecture studies (Beichman et al. 2002), the final reports from the teams, and, for the separated spacecraft interferometer, the TPF Book (1999).

4.1. Visible Light Coronagraphs

Two groups (Ball and Boeing-SVS) investigated the potential for a visible light coronagraph to satisfy the TPF goals. While there are differences between the designs, there are major similarities: 1) a large optical surface ($4 \times 10\,$m for Ball,

Table 1. Phase II Architecture Studies

Study Team	Architecture Class	Time to Survey 150 Stars (Days, One Epoch)
Ball Aerospace	Shaped Pupil Coronagraph (Vis.)	50
	Classical Coronagraph (Vis.)	15
Boeing-SVS	Non-Redundant Linear Array (IR)	55
	Apodized Square Aperture (Vis.)	262
Lockheed-Martin	Structurally Connected IR Interferometer	40
Lockheed-Martin	Separated Spacecraft IR Interferometer	106
TRW	28 m IR Coronagraph	N/A

8×8 m for Boeing-SVS); 2) a highly precise, lightweight primary mirror equipped with actuators for figure control with surface quality of order 1-5 nm depending on spatial frequency; and 3) a variety of pupil masks (square, Gaussian, or other (Spergel and Kasdin 2001) and/or Lyot stops) to suppress diffracted starlight. In the case of the Ball designs, a key component was a small deformable mirror with $\sim 100 \times 100 = 10^4$ elements capable of correcting residual mid-spatial frequency errors to $\lambda/3,000$ and stable to $\lambda/10,000$. In the Ball design, the combination of pupil masks and the deformable mirror reduces the ratio of starlight (scattered or diffracted) to planet light to approximately unity over an angular extent between $\sim 5\lambda/D$ and $100\lambda/D$.

With these features, the Ball systems are able to conduct a survey of 150 stars with images taken at 3 epochs for confirmation and orbital determination in less than half a year (Table 1). The Boeing-SVS system, as proposed, takes more time to complete such a survey because without a deformable mirror the ratio of starlight to planet light is about 100 times worse than in the Ball design; addition of a deformable mirror would result in comparable performance for the two telescopes. In under a day per star, the Ball system could detect ($S/N = 5$ at spectral resolution $R = \lambda/\Delta\lambda \sim 25 - 75$) various atmospheric tracers, including O_2, a critical signpost for the presence of photosynthetic life.

The study teams pointed out that the potential for ancillary science was particularly impressive for the visible systems, since it would be straight forward to add a complement of traditional astronomical instruments (e.g., ultraviolet (UV) and optical imagers and spectrographs). Operated on an 8-10 m telescope, such instruments would represent a giant advance over the present UV-optical performance of the Hubble Space Telescope (HST). Of particular interest would be the ability to make diffraction limited images at UV wavelengths with < 5 milli-arcsec resolution.

Future studies will have to assess whether the specialized requirements of a planet finding (e.g., an off-axis secondary) might compromise the general astrophysics potential of a visible system. Conversely, NASA will have to weigh whether specialized needs such as high UV throughput requiring special coatings and careful attention to contamination issues might significantly increase the cost of the observatory or compromise its planet-finding performance.

The greatest technical risk for the visible coronagraph is in the development, manufacturing, and implementation of a large primary mirror with ultra-low wavefront errors as well as components associated with starlight suppression. The coronagraphs themselves are functionally simple and, although the demands for system performance are challenging, none are thought to be insurmountable. However, the problem of fabricating and launching a large (8-10 m) mirror cannot be overemphasized. The TPF Project's independent Technology Review Board noted that there exists no capability to fabricate such a high precision (3-5 times better than the HST mirror, 5-10 times better than the NGST mirror), lightweight optical element for ground or space. But even if a 8-10 m system proves to be too difficult to implement on the TPF timetable, a 2-4 m class telescope could demonstrate high dynamic range coronagraphic imaging and carry out an exciting scientific program. Such a system could find Earths only around the closest dozen stars because of its degraded angular resolution, but it could find and characterize Jupiters around many more distant stars. A telescope of this scale might fit into budget of a Discovery mission.

4.2. Nulling Infrared Interferometers

Lockheed Martin and JPL examined two versions of the infrared nulling interferometer: structurally connected and separated spacecraft. The Lockheed Martin study concluded that a structurally connected infrared interferometer with four 3.5 m diameter telescopes on a fixed 40 m baseline comes close to achieving the TPF goals. The system uses four co-linear telescopes arranged as two interleaved Bracewell nulling interferometers to reject star light adequately so that stellar leakage does not compromise the overall system noise. The array would be rotated around the line of sight to the star over a 6-8 hour period. The telescopes can be combined in different pairs to achieve the short and long baselines needed to observe distant or nearby stars. The nulled outputs of the combined pairs are combined again to yield a θ^4 null or an effective θ^3 null with phase chopping.

The separated spacecraft version of the nulling interferometer was described in the 1999 TPF report (Beichman et al. 1999; also, see Woolf & Angel 1998). It uses a different arrangement of telescopes to produce a deeper, θ^6, null that can be tuned to resolve most effectively the habitable zone around each target. Because the stellar leakage is reduced in this design, the stability requirements are relaxed relative to the structurally connected interferometer. The θ^6 null is, however, less efficient in its use of baseline, requiring roughly 1.5-2 times longer baselines than the structurally connected system. Providing an 80-100 m baseline leads, in turn, to the likely requirement for a more complex separated spacecraft system. Thus, a near-term study must investigate whether a 40 m system can satisfy the TPF goals. If not, then NASA should pursue aggressively the development of a separated spacecraft nulling interferometer.

The infrared system is somewhat more susceptible to the effects of zodiacal emission than visible light coronagraphs. Figure 1 compares the relative time needed to detect an Earth-like planet as a function of the brightness of the zodiacal emission around the target star (in units of our own zodiacal cloud). At ten times the brightness of our zodiacal cloud, both the visible and infrared systems are adversely affected – the infrared system by a factor of 3-4, the visible system by a factor of 2-3.

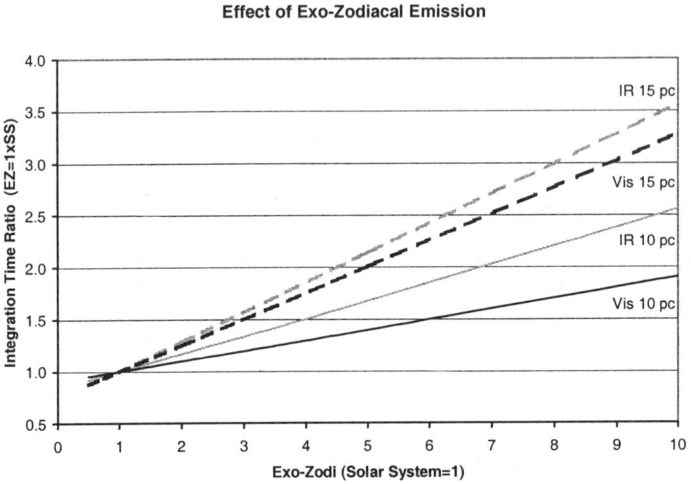

Figure 1. The effect of zodiacal emission around target stars at two distances on the integration time to detect an Earth-like planet for the Lockheed-Martin nulling interferometer and the Ball coronagraph.

It should be mentioned that the European Space Agency (ESA) has studied a two-dimensional, separated spacecraft array of infrared telescopes for its Darwin mission. An industrial study by Alcatel found that this version of a planet-finding mission was technically feasible.

The ancillary science possible with an interferometer is likely to be more specialized than for an 8-10 m visible telescope equipped with general purpose instruments. However, for a telescope with NGST-like sensitivity, but with 10 times better angular resolution, the prospect of imaging the cores of protostars, active galaxies, and high redshift quasars is an exciting one.

The largest area of technical risk for the infrared interferometers is not in the performance of the individual components but in the operation of the various elements as a complete system. Most of the required elements are either under development and making good progress, or are reasonable extensions of technology being developed for missions and ground observatories that will be in place well before TPF needs them. However, the system complexity of the separated spacecraft system cannot be overemphasized. Such a system would demand at least one precursor space mission: a formation flying interferometer, such as the Starlight project, to validate the complex control algorithms and beam transport needed for this version of TPF.

5. Other Concepts

TRW and Boeing-SVS investigated two architectures that in the opinion of the TPF project and the independent Technology Review Board were less promising in the near term than the infrared (IR) interferometer and the visible coronagraph.

- TRW studied a cooled 28 m telescope operating in the mid-IR. The major drawback of this system is its relatively poor angular resolution, $\lambda/D \sim 90$ mas at 10 μm, which would allow it to observe only the closest stars. This restriction becomes even more severe if one requires a typical inner working distance of $3 - 5 \times \lambda/D > 250$ mas. Interior to this separation, the ratio of planet light to diffracted starlight is very unfavorable ($< 10^{-3} - 10^{-4}$), demanding great stability for a large structure over many hours.

- Boeing-SVS studied a sparse aperture imaging system called the Non-Redundant Linear Array (NRLA). Working at mid-infrared wavelengths, the NRLA, or hyper-telescope, consists of an array of seven, passively cooled, 3 m meter telescopes distributed along a 100 m long structure. The system uses "densification" to rearrange the location of the apertures in the pupil to produce a compact, clean point spread function. A coronagraph consisting of a phase mask in the image plane rejects stellar light. The observatory, located at L2, rotates to collect data suitable for planet detection and characterization. The project and the Technology Review Board regarded the launch and on-orbit assembly of the required 100 m structure as potentially very expensive compared with some of the other systems.

6. Conclusions

The recently completed architectural studies have identified two plausible paths to TPF. With a substantial investment in technology over the next few years, the choice between a nulling infrared interferometer and a visible coronagraph will become clearer from the standpoint of technological maturity, cost, and risk. At the same time, a number of missions will have important scientific input into the design of TPF. By 2006, SIRTF, as well as the Keck, VLT, and LBT Interferometers will have measured zodiacal clouds around hundreds of TPF targets, giving us valuable information on the significance of this noise source. Thus by 2006, it should be possible to choose between the visible and infrared systems on both scientific and technical grounds. Another scientific milestone comes around 2010 when the Kepler mission (Borucki et al. 2001) will have measured the frequency of Earth-like planets around solar-type stars (Beichman 2000). This information will help to set the size of the TPF apertures and the required angular resolution by telling us whether we need to look out to 5, 8, or 15 pc to ensure a high probability of finding planets.

As the technology matures and the opportunity to start the mission approaches, NASA and the science community will have to reach a consensus on the scientific performance in the areas of planet finding and general astrophysics needed to justify the mission. Some of the observational capabilities of TPF will be affordable; others will have to be deferred to subsequent, and still more capable, missions. At the end of the TPF pre-formulation phase, NASA, together with its potential international partners, will be prepared to address the challenge of looking for habitable planets and seeking signs of life beyond the Solar System.

Acknowledgments. The research described in this paper was performed at the Jet Propulsion Laboratory, California Institute of Technology, under a contract with the National Aeronautics and Space Administration. This work was supported by the TPF project. The author acknowledges valuable contributions from TPF Science Working Group and the contractor teams, as well as the dedication and hard work of Dan Coulter and Chris Lindensmith at JPL.

References

Beichman, C.A. 2000, in Planetary Systems in the Universe, IAU Symposium 202, Manchester, England

Beichman, C.A., Coulter, D.R., Lindensmith, C.A., & Lawson, P.R. 2002, Summary Report on Architecture Studies for the Terrestrial Planet Finder, JPL Report, in press

Beichman, C.A., Woolf, N.J., & Lindensmith, C.A. 1999, The Terrestrial Planet Finder (TPF): A NASA Origins Program to Search for Terrestrial Planets, JPL Report 99-3

Borucki, W.J., Koch, D.G., Jenkins, J.M. 2001, BAAS, 199, 115.04

Des Marais, D., et al. 2002, Astrobiology, in press

Ford, E.B., Seager, S., Turner, E.L. 2001, Nature, 412, 885

Spergel, D., & Kasdin, J. 2001, BAAS, 199, 86.03

Woolf, N. & Angel, J.R. 1998, ARA&A, 36, 507

Hubble's Science Legacy: Future Optical-Ultraviolet Astronomy from Space
ASP Conference Series, Vol. 291, 2003
K.R. Sembach, J.C. Blades, G.D. Illingworth, R.C. Kennicutt, Jr. (eds.)

Strategies for Extra-Solar Planet Detection With UV/Visible Large Aperture Telescopes

Mark Clampin

Space Telescope Science Institute, 3700 San Martin Dr., Baltimore, MD 21212

Abstract. A large aperture ultraviolet/visible successor to the Hubble Space Telescope would be able to directly detect terrestrial planets and characterize them, by means of spectral biomarkers, to assess if there was evidence of life. We discuss the strategy proposed for terrestrial planet detection with a ultraviolet/visible telescope and summarize the technical approach and requirements for such an instrument.

1. Introduction

The next decade of astronomy will see a major expansion in the study of extra-solar planets. A key goal in this endeavor is the search for evidence of life. The major milestones on the path to this goal are the direct detection of terrestrial planets located in habitable zones (Kasting et al. 1993), followed by detection of biomarkers that indicate the presence of life or conditions appropriate for the emergence of life. In the last twelve years the Hubble Space Telescope (HST) has made several important contributions to the study of extrasolar planets and the formation of planetary systems, including the detection of an extrasolar planetary atmosphere (Charbonneau et al. 2002). A large aperture, UV/visible space telescope at the end of the decade would be capable of achieving these major milestones in the search for evidence of life.

The first step in this path is to determine the nature of planetary systems from studies of their frequency, sizes, distribution, and orbital properties. Major advances have already been made during the last seven years as the number of radial velocity detections has expanded the number of known planetary systems to ~90. Over the next decade, radial velocity surveys will start detecting longer period planetary systems, which are more likely to resemble Solar System analogs.

The Kepler mission will conduct a survey to detect stellar transits, and will provide the first detailed information on the frequency of terrestrial and larger planets in or near the habitable zone of a wide selection of stellar spectral types (Koch et al. 2000). Further data on the nature of planetary systems might be provided by the Galactic Exoplanet Survey Telescope (GEST) mission, which detects planets via microlensing and would complement Kepler with sensitivity to planets at separations from 0.7 AU to >30 AU. Both the Space Infrared Telescope Facility (SIRTF) and the Next Generation Space Telescope (NGST) will provide valuable insight into the nature of exo-zodiacal dust, and determine its likely impact on direct planet imaging programs. The Space Interferometry Mission (SIM) will also provide astrometric data on planetary systems associated with nearby stars. It will be able to conduct a census of gas giant planets and will be able to detect larger terrestrial planets ($\sim$3-10 $M_\oplus$). A number of mission concepts are also under consideration for the direct imaging detection of gas giant planets in visible light by coronagraphy. These missions include Jovian Planet Finder (JPF; Clampin et al. 2002), Eclipse (Trauger et al. 2002), and the Extra-Solar Planetary Imager (ESPI, Lyon et al. 2002). A direct imaging mission would survey nearby stars for gas giant planets with the goal of identifying Solar System analogs suitable for subsequent attempts at detecting terrestrial planets. Ground-based adaptive optics and interferometry programs are also working towards the goal of imaging younger extrasolar gas giant planets (Staplefeld 2002).

At the end of this decade, many of the scientific and technical elements required to mount a search for terrestrial planets will have been achieved. Concepts for the Terrestrial Planet Finder (TPF) mission have focused on two main architectures: nulling interferometry and coronagraphic imaging. A large aperture, UV/visible successor to HST would be an ideal instrument to mount a coronagraphic survey of nearby stars for terrestrial planets and conduct a low resolution spectroscopic characterization of these planets.

2. Coronagraphic Requirements for Planet Detection

Stellar coronagraphy is based on a technique first proposed by Lyot (1932) for observing the circumsolar environment. More recently this technique has been applied to ground-based observations of circumstellar environments. Stellar coronagraphs have been used to image circumstellar disks in reflected light such as the disk around β Pictoris (Smith & Terrile 1984; Golimowski et al. 1993), the brown dwarf Gliese 229B (Nakajima et al. 1995), and the stellar ejecta around HR Carinae (Clampin et al. 1995). Stellar coronagraphs have also been deployed on instruments installed in the HST including NICMOS (e.g., Weinberger et al. 1999) and STIS (e.g., Grady et al. 2001). The current state of the art for HST is the Advanced Camera for Surveys coronagraph (Krist et al. 2002), which has just been commissioned. Coronagraphy is already a well established technique in astronomy and has a strong record of scientific achievement.

In order to detect terrestrial planets, contrast ratios in the range 2×10^{-10} must be achieved (Brown et al. 2002). Thus, seen from 5 parsecs, the Earth is a 27.7 magnitude object located $0.2''$ from a 3.3 magnitude star. Clearly, the properties of individual planets will vary with albedo, mass, phase function, and central star properties. A planet will be detectable if the ratio of the central surface brightness of the planet divided by the surface brightness of the local background is $\geq$1. The factors that contribute to the contrast ratio achievable are the aperture size, the wavefront quality, and the suppression of diffraction.

Aperture size determines the size of the diffraction pattern of the parent star, which is proportional to λ/D. Wavefront quality is particularly important in the mid-spatial frequency regime where light is coherently diffracted by the residual polishing errors in the optics. In the HST, for example, there are roughly concentric polishing errors in the primary and secondary mirrors that produce

Figure 1. *Left and Middle*: Logarithmically scaled simulations of the HST V-band PSF, without and with the HST polishing errors included. *Right*: Map of the combined HST primary and secondary mid-frequency errors derived from phase retrieval measurements of on-orbit HST images (Krist & Burrows 1995). This map is displayed between ±30 nm surface amplitude.

an RMS wavefront error of 18 nm at spatial scales of between 5 to 100 cycles across the pupil (Krist & Burrows 1995). Such coherent errors (often called mid-frequency errors) at n cycles per aperture diffract light out to n Airy radii (e.g., errors of 10 cycles will diffract light into a zone of 0.16″ radius in an 8 meter telescope). In effect, the polishing errors act like a coarsely ruled diffraction grating. These features are not completely circular and concentric, nor are they smooth, so the diffraction pattern will contain numerous fixed secondary images of the point source (speckles), which smear out into streaks in broad-band images. Depending on the RMS wavefront error, the speckle amplitudes can be much greater than the peak pixel brightness of the planet. Figure 1 shows the halo in the HST point spread function (PSF) caused by diffraction from the coherent residual optical errors on the HST primary and secondary mirrors. Our requirement that $Q \geq 1$ requires that speckle amplitudes are at worst equal to expected amplitude of the planet.

The problem of mid-frequency spatial error can be addressed by fixed or active correction of the mid-frequency wavefront. In the case of smaller optics systems it is possible to build telescope optics that meet a specified wavefront quality goal in the mid-frequency regime. This was the adopted approach of the JPF mission (Clampin et al. 2002). Using this approach for modest sized telescopes (e.g., 1.5 m), it is possible to achieve wavefront quality of ~0.7 nm RMS. This approach requires active thermal control of the telescope, combined with stringent pointing requirements, and is limited to smaller aperture sizes.

However, to achieve the contrast ratio required for detection of terrestrial planets, the mid-frequency wavefront has to be controlled to between 0.2 nm and 0.02 nm, as is shown in Figure 2. Consequently, it is appropriate in this case to employ a deformable mirror to correct the mid-frequency wavefront errors. The required number of actuators is proportional to the square of the diameter of the corrected field of view. If we design an 8 m telescope that can search for and study planets over a 0.1 to 1.0 radius, the wavefront errors must be corrected between 6 and 63 cycles across the pupil. Assuming two actuators per cycle,

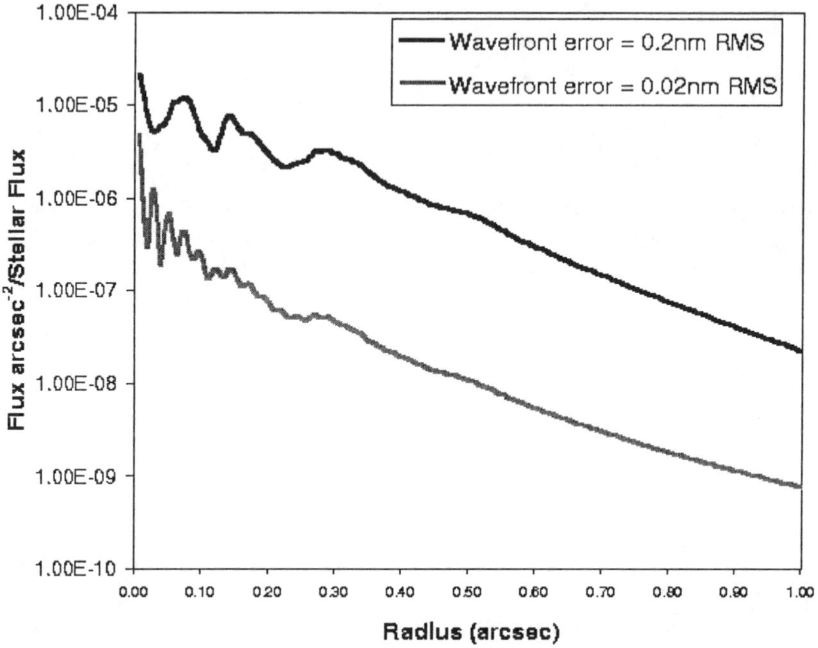

Figure 2. Simulations show a Lyot coronagraph with 0.075″ HWHM occulting spot and 70% diameter Lyot stop. imaging is with a V band filter and an 8 m unobscured telescope (Courtesy J. Krist)

~16,000 actuators are required. The requirements for such a deformable mirror differ from those of adaptive optics in that a high quality surface is required for the unactuated mirror and the actuator stroke is relatively small. There are two mirror technologies that can achieve the required performance by actuating a mirror faceplate, micro-machined mechanical system (MEMS) active optics, discussed by Olivier et al. (2002), and electrostrictive actuator mirrors (Brown at al. 2002). MEMS mirrors are a new technology with the potential for low cost scaling and thermally stable performance. A further technique that can be employed with a deformable mirror is the "dark hole" algorithm, where the mirror deformations are adjusted to create a region of higher contrast over half of the image (Malbet 1995).

Suppression of diffraction from the telescope optics is also a contributor to the final contrast. Traditional coronagraphs have a hard-edged occulting mask and employ a circular Lyot stop to achieve suppression of diffraction. Focal plane apodization using graded density occulting masks improves the suppression of diffraction by controlling diffraction from the edges of the occulting mask. More recently, proposals for shaped pupils (Spergel & Kasdin 2001; Nisenson and Papaliolios 2001) potentially obviate the need for occulting masks by optimally concentrating light in the focal plane. These apodizers relax pointing requirements, achieve higher levels of diffraction suppression, but also reduce through-

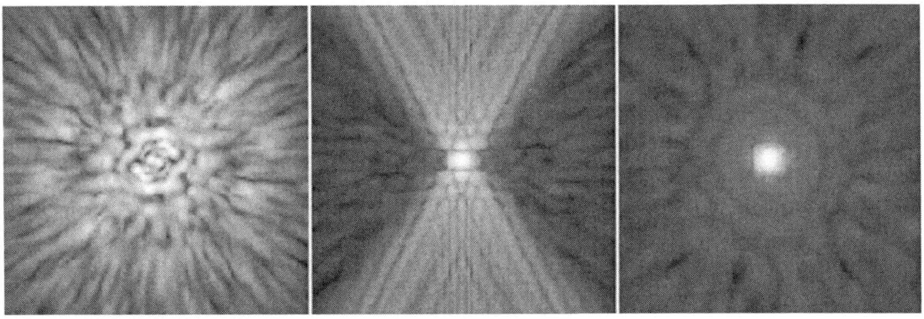

Figure 3. Simulation of 8 m telescope PSFs with different coronagraph apertures. On the left is a Lyot coronagraph, in the middle is a Spergel mask coronagraph, and on the right is a square aperture coronagraph. To show that good diffraction suppression alone is insufficient, we have included a 2 nm wavefront to illustrate the impact of mid-frequency errors in the focal plane. These can be clearly seen as the speckle pattern in the dark region of the Spergel mask coronagraph.

put, increase the PSF width and require more pointings for an observation. In Figure 3, where we show focal plane images each of the three approaches. The simulations include 20 nm of wavefront error to highlight the fact that good diffraction suppression alone is not sufficient. Following the principle of pupil shaping, new approaches to focal plan masks are now appearing and may ultimately simplify manufacturing requirements for occulting masks (Kuchner and Traub 2002).

3. Terrestrial Planet Detection

With the provision of an optimum coronagraph design and mid-frequency wavefront control to $\leq$0.1nm by means of appropriately specified telescope optics and a deformable mirror, a future large aperture UV/visible telescope could achieve the contrast ratios required to detect terrestrial planets. An 8 m aperture size is optimal since it would be able to survey in excess of 100 nearby stars for terrestrial planets (Beichmann et al. 2002), while a smaller aperture of 4 m would be able to survey $\sim$40 nearby stars. Typical exposure times are determined by the final coronagraph design adopted and the distance of the target, but a $S/N = 5$ detection with an 8 m telescope would require $\sim$1 to 2 hours. In order to verify detections it will be necessary to make observations of each target over multiple epochs.

Having achieved the milestone of a terrestrial planet detection, the large aperture UV/visible telescope would be able to conduct a program to characterize the nature of the sources detected. Basic orbital parameters can be determined from the multiple visits to each target obtained for confirmation of the planet detections. The best detections would be the subject of subsequent monitoring on a routine basis. The core strategy for terrestrial planet characterization will be broad-band photometry, which is simple to obtain, and can be used to derive approximate assessments of planet mass and rotational period (Ford, Seager, & Turner 2001; Des Marais et al. 2001). Different surfaces and weather conditions rotating in and out of view can cause flux variations of 100% to 200% providing the opportunity, in combination with low resolution spectra, to probe conditions on the planet.

In order to address the possibility of life, a spectroscopic characterization of detections will be necessary. The coronagraph design will require the capability to obtain low resolution spectra with $R \leq 100$. This could be as simple as a narrow-band filter or a grism capability in combination with the "dark hole" algorithm to reduce the local background. The main spectroscopic features available in the visible spectrum are O_2, O_3, H_2O, CO_2, and CH_4 (Traub & Jucks 2002; Woolf et al. 2002). H_2O is believed to be a biomarker for environments where lifeforms could survive and can be detected in the visible at 940 nm and 720 nm. O_2 in significant amounts is believed to be an indicator for life and can be detected at 760 nm, while O_3 is a non-linear indicator of the presence of O_2.

Earthshine spectra with resolution of $R \sim 50$ (Arnold et al. 2002) clearly show these features and confirm that even a very low resolution spectral analysis will provide clear indication of biomarkers. The earthshine observations of Woolf et al. (2002) and Arnold et al. (2002) also show that Rayleigh scattering can be readily detected just by broad-band photometry offering a simple diagnostic of the planet's atmosphere. A spectral feature known as the red edge (700–800 nm) has been identified as a possible biomarker for vegetation. Both Woolf et al. (2002) and Arnold et al. (2002) tentatively identify this feature in earthshine spectra at the 2% to 10% level, but it can be masked by absorption bands (also biomarkers), so interpretation can be difficult. Exposure times to obtain $R \sim 50$ spectra range from tens to hundreds of hours depending upon the spectral feature of interest and the planetary properties. Another important spectroscopic investigation this instrument could undertake is a detailed investigation of giant planets at higher spectral resolutions of 100 to 1000.

Finally, returning to a technique discussed earlier in the paper, a large aperture UV/visible telescope could also conduct a detailed spectroscopic analysis of terrestrial planets of stars which exhibit transits. Charbonneau et al. (2002) have already detected the presence of sodium in the atmosphere of HD 209458 using HST/STIS to measure very high signal to noise spectra through its transit. With a large aperture telescope, this technique could be extended to obtain spectral diagnostics of terrestrial planets. A 6.5 m telescope could distinguish Venus from Earth in a few tens of hours around stars brighter than $V \sim 7$ (Brown & Gilliland 2002). The number of nearby stars exhibiting transits is likely to remain relatively small, but this technique can be used to conduct detailed spectroscopic studies of these systems.

4. Summary

In summary, a large aperture UV/visible space telescope fitted with an optimized coronagraphic mode would be able to detect terrestrial planets around 50 to 100 stars, for apertures of 4 m to 8 m, respectively. It would also be able to characterize the planets detected and search for simple biomarkers that would provide a very detailed picture of conditions on the planets, completing several key milestones in the search for evidence of life. The incorporation of a coronagraph need not preclude other science programs being undertaken with the telescope. However, the telescope would need to be designed with the coronagraph as a core component of the telescope so that the primary requirements for coronagraphy such as mirror quality, off-axis optics, etc. were met. This approach would also offer benefits to the TPF program. A smaller 4 m-class telescope could be the technology precursor for a larger 8 m visible coronagraph. Alternatively, the discovery space of a 4 m class coronagraph could be complemented by a larger aperture infrared interferometer, an approach which would provide complementary science capabilities, leading to a broader picture of the planetary systems around nearby stars.

References

Arnold, L., Gillet, S., Lardiere, O., Riaud, P., & Schneider, J. 2002, A&A, in press

Beichman, C.A., Coulter, D.R., Lindensmith, C.A., & Lawson, P.R. 2002, Summary Report on Architecture Studies for the Terrestrial Planet Finder, JPL Publication 02-011, Version 1, Rev. A

Brown, T., & Gilliland, R. 2002, private communication

Charbonneau, D., Brown, T.M., Noyes, R.W., & Gilliland, R.L. 2002, ApJ. 568, 377

Clampin, M., et al. 2002, Proc. SPIE, 4860, in press

Clampin, M., et al. 2002, Astrobiology, in press

Des Marais, D.J., et al. 2002, Astrobiology, vol. 2, issue 2, 153

Krist, J.E., Golimowski, D.A., Hartig, G.F., Clampin, M. & Ford, H.C. 2002, Proc. SPIE, 4860, in press

Ford, E. B., Seager, S., & Turner E. 2002, Nature, 412, 885

Golimowski, D.A., Durrance, S.T., & Clampin, M. 1993, ApJ, 411, L41

Grady, C.A., et al. 2001, AJ, 122, 3396

Koch, D., Borucki,W., Webster, L., Dunham, E., Jenkins, J., Marriott, J., & Reitsema, H. 1998, Proc. SPIE, 3356, 599

Kasting, J.F., Whitmire, D.P., & Reynolds, R.T. 1993, Icarus, 101, 108

Krist, J., & Burrows, C.J. 1995, Applied Optics, 34, 4951

Kuchner, M.J., & Traub, W.A. 2002, ApJ, 570, 900

Lyon, R.G., et al. 2002, Proc. SPIE, 4860, in press

Lyot, B. 1932, MNRAS, 99, 580

Malbet, F., Yu, J.W., & Shao, M. 1995, PASP 107, 386

Nakajima, T., Oppenheimer, B.R., Kulkarni, S.R., Golimowski, D.A., Matthews, K., & Durrance, S.T. 1995, Nature, 378, 463

Nisenson, P., & Papaliolios, C. 2001, ApJ, 548, L201

Olivier, S.S., Sommagren, G.E., Clampin, M., Ford, H.C., Illingworth, G.D., Krist, J.E., Krulevitch, P., & Papavasiliou, A. 2002, Proc. SPIE, 4860, in press

Smith, B.A., & Terrile, R. 1984, Science 226, 1421

Spergel, D., & Kasdin, J. 2001, BAAS 199, 86.03

Staplefeldt, K., Colavita, M., Dekany, R. G., Ftaclac, C., Macintosh, B.M., Traub, W.A., Trauger, J., & Woolf, N. 2002, Proc. SPIE, 4835, in press

Traub, W., & Jucks, K.W. 2002, Atmospheres in the Solar System: Comparative Aeronomy, eds. M. Mendilo, A. Nagy & H.J. Waite, AGU Geophysical Monograph 130, 369

Trauger, J.T. 2002, Proc. SPIE, 4854, in press

Weinberger, A.J., Becklin, E.E., Schneider, G., Smith, B.A., Lowrance, P.J., Silverstone, M.D., Zuckerman, B., & Terrile, R.J. 1999, ApJ, 525, L53

Woolf, N., Smith, P.S., Traub, W.A., & Jucks, K.W. 2002, ApJ, in press

Hubble's Science Legacy: Future Optical-Ultraviolet Astronomy from Space
ASP Conference Series, Vol. 291, 2003
K.R. Sembach, J.C. Blades, G.D. Illingworth, R.C. Kennicutt, Jr. (eds.)

Extrasolar Planets and Biomarkers

Wesley A. Traub

Harvard-Smithsonian Center for Astrophysics, 60 Garden St., Cambridge, MA 02138

Abstract. The most important science to be done with a Hubble successor telescope could be the detection of extrasolar terrestrial planets and the search for life on those planets. A telescope which is optimized for terrestrial planet detection will also be a superb facility for ultraviolet-optical astrophysics, with a large collecting area, high angular resolution, and wide-band response. A coronagraphic Terrestrial Planet Finder (TPF) mission should be flown to explore the extrasolar planet frontier around nearby stars, much as Hubble was flown to explore the galactic frontier at the edge of the Universe.

1. Introduction

The search for extrasolar planets and life is a new frontier for astrophysics. Over the past few decades, observations and theory have given us a surprisingly good picture of the origin and evolution of the Universe, galaxies, and stars, notwithstanding numerous remaining mysteries. However, our existence here on Earth tells us that there are additional steps in the hierarchy, viz., planet formation and the origins of life. As yet, no terrestrial extrasolar planets have been found, no signs of life outside the Earth have been seen, and no theory has emerged to explain the known planets or the manifest presence of life on Earth and seeming absence elsewhere.

In this paper I will touch on those topics which may help us make progress on this new frontier. In the following I discuss the probability of finding Earth-mass planets, a model for nearby solar systems, the coronagraphs which could detect Earth-like planets, some optical requirements, a theoretical Earth spectrum, a validation spectrum, a list of planetary properties which we can learn from a spectrum, and finally spectral evidence for signs of life. My conclusion is that TPF can accomplish this task, and therefore TPF would be the ideal successor telescope to Hubble.

2. Mass Distribution Function

What evidence do we have that there are any terrestrial-mass planets outside the Solar System? To date, essentially all of the 80-plus extrasolar planet masses, ignoring orbital inclination factors, are in the range about 0.2 to 20 Jupiters (Schneider 2002). This mass distribution is shown in Figure 1 (left). Superposed curves labeled 2 and 20 m s^{-1} indicate my estimate of the experimental cutoff

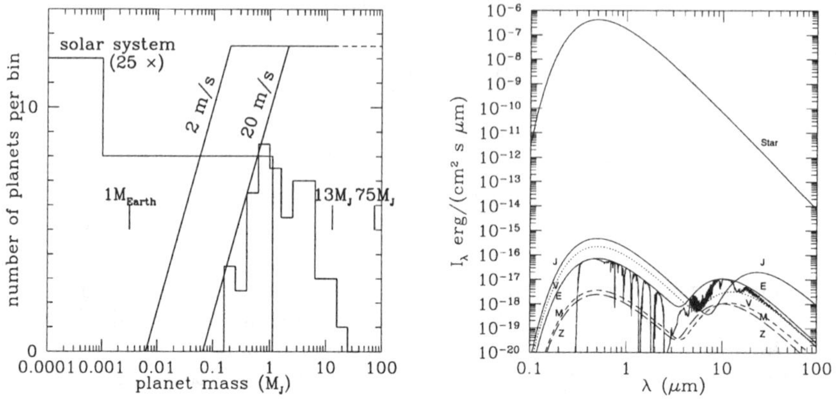

Figure 1. *Left*: Mass distribution function for observed extrasolar planets (lower right histogram), along with a scaled version of the mass distribution in the Solar System, supporting the conjecture that there are many terrestrial mass planets to be found among the nearby stars. *Right*: Spectral flux from the Sun, Jupiter, Earth, Venus, and Mars as seen from 10 pc. The spectra are equivalent blackbody emission and albedo-scaled reflection. A calculated Earth spectrum is superposed, for a clear atmosphere case. The zodiacal curve (Z) is for a 0.1 pc diameter patch at 1 AU from the star.

function imposed by the respective radial velocity uncertainties. The mass of the Earth is well below the range of anticipated radial velocity accuracies, so an Earth-mass planet is unlikely to be measured by this technique.

Our only experimental recourse, at present, is to compare with Solar System masses, shown in Figure 1. From the overlap and trends of the two mass distributions, I conjecture that terrestrial mass planets might be abundant elsewhere. Aside from theory, we have little else to go on, so this admittedly uncertain extrapolation may be our main hope for finding Earth-sized planets.

3. Solar System at 10 Parsecs

How bright would planets be around nearby stars? Using the Solar System at a distance of 10 pc as an example, Figure 1 (right) shows the emitted and reflected fluxes which might be expected, in the approximation that the objects emit as blackbodies and reflect with a flat albedo and a phase function appropriate to maximum elongation (quadrature) from the star. For the Earth, a clear-air spectral curve is drawn for illustration (Traub & Jucks 2002).

In the visible we see that the contrast ratio of an Earth-like terrestrial planet (P) to a Sun-like star (S) is about $P/S \sim 10^{-10}$, and the flux from the planet itself is roughly $P \sim 0.2$ photon m^{-2} $s^{-1}\mu m^{-1}$. As shown below, the visible region has several excellent biomarkers.

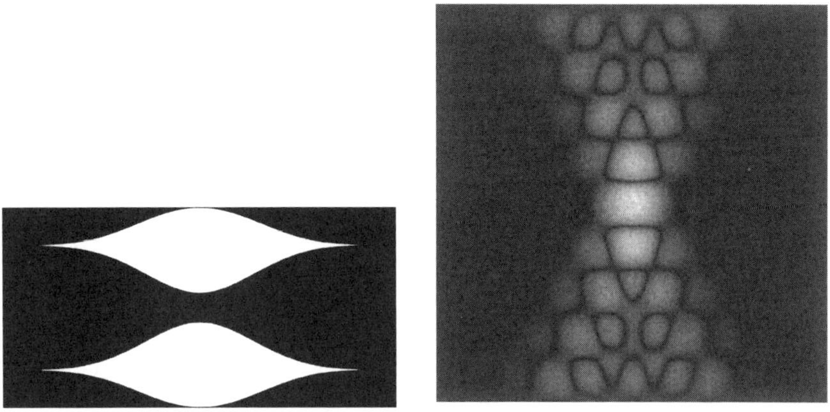

Figure 2. Pupil-plane mask example. *Left*: A pair of prolate-spheroid bounded pupils give minimal diffraction in the direction of the pointed ends. *Right*: Calculated diffraction for this mask, showing very little light is diffracted left and right, thus defining a planet discovery window. From Kasdin, Spergel, & Littman (2002).

4. Coronagraphs

A powerful coronagraph design is needed to isolate an Earth-like planet from its star, a separation of 1 AU / 10 pc = $0.1''$ in our example. There is an entire family of diffraction-minimizing pupil masks that can achieve this separation, an example of which is shown in Figure 2. There is essentially no diffraction from the star toward the left and right, so the planet image will stand out nicely and not be swamped in those positions. Diffracted light does appear above and below, however, so the telescope (or just the pupil mask) must be rotated about 90° in order to search for planets in those regions.

Another type of pupil mask is the apodizing design from Nisenson & Papaliolios (2001), wherein a square mask is further tapered by a graded transmission function. The effect is similar, reducing the diffraction wings along the diagonals. The taper provides another degree of freedom to the family of pupil masks, in addition to shape itself.

Currently both types of designs are being optimized, by varying the number of openings and their shapes in the binary transmission case, or by varying the shape and transmission function in the graded pupil case.

Image-plane masks are an alternative to pupil-plane masks. In fact, the traditional way to design a coronagraph is by placing an opaque mask, or a graded Gaussian mask, in the focal plane, and centering the star on that mask. However, some light is always diffracted around such a mask, and therefore a Lyot stop must be placed in the re-imaged pupil plane, usually with a significant loss of throughput for off-axis objects.

In a new version of this traditional coronagraph, a graded mask is used but its spatial frequency content is limited to a single frequency or a narrow band of frequencies (Kuchner & Traub 2002). As shown in Figure 3, for the case of a sine-squared amplitude transmission mask, the effect is to diffract the on-axis

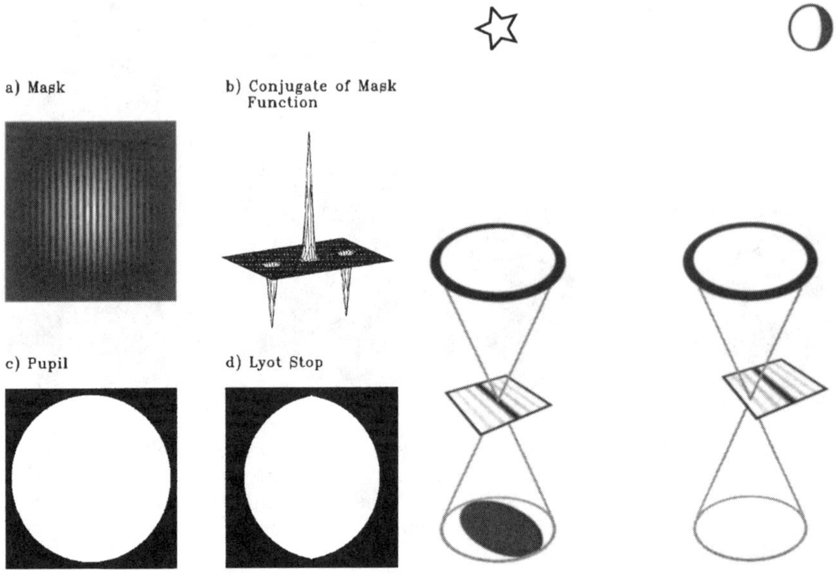

Figure 3. Image-plane mask example. *Left*: A band-limited mask (here $sin^2(\theta)$) is shown, along with the corresponding pupil convolving triplet of near-delta functions, and the Lyot stop. From Kuchner & Traub (2002). *Right*: This cartoon shows the overall mask function, blocking the central star and transmitting the off-axis planet. From Kuchner & Spergel (2002).

star light at such a large angle that it falls in a narrow region at the edge of the re-imaged pupil. The Lyot stop needs to block only a small fraction of the pupil, thus improving throughput. The planet is transmitted at most azimuth angles, and only a small rotation of the telescope is needed to ensure that the planet is detected anywhere in the field.

Hybrid designs mixing pupil and image masks are also possible, but these need further exploration. In any case, it is now clear that diffraction per se is not a fundamental limit to imaging a faint planet near a bright star. Other factors, including manufacturability, scattered light, and stability, are now on a par with diffraction. Planet detection is therefore more an issue of engineering than of fundamental physics.

5. Telescope Requirements

What type of telescope do we need to detect and characterize an Earth at 10 pc? Ultimately we will need to reach to several times this distance in order to obtain a sample goal of, say, 250 suitable target stars, but 10 pc is a convenient touchstone. I assume that pupil diffraction has been reduced to a level below that of the planet, using one of the techniques discussed above.

For general science, the field of view of the telescope itself can be relatively large, subject to classical optical aberrations. However in coronagraph mode, the telescope need only have a small working field of view, since the target planetary systems are small ($\lesssim 1''$). This field is determined by polishing errors in the optics, which in turn generate a wavefront roughness, which can be analyzed in terms of Fourier component "ripples" in the wavefront. Each ripple creates a pair of ghost images of the star in the focal plane. A deformable mirror is employed to reduce the ripples to a tolerable level. If the deformable mirror has N pairs of pistons per diameter, then it can correct N waves of ripple per diameter. The outer working distance is then $OWD = N\lambda/D$, where D is the telescope diameter. The best coronagraph designs will suppress diffraction at about 4 times the diffraction width, so the inner working distance is $IWD = 4\lambda/D$. The IWD determines the mirror diameter, and to reach about 250 stars, a diameter on the order of 8-10 m is required.

If the corrected phase error from the ith ripple is σ_i (rad), and if all ripples are corrected to a common average value, then the total RMS phase error is $\sigma_{rms} = N\sigma_i$. The RMS intensity of each ghost is $(\sigma_i)^2/2$, and each ripple generates two ghosts, at angle $\pm\lambda/l_i$, where l_i is the wavelength of the ripple in the pupil. To make the ghost intensity equal to the planet intensity, we need P/S $= (\sigma_i)^2/2$. Using the values of P/S from above, we find a total RMS wavefront error tolerance of roughly 0.5 Angstrom. This is achievable with a good primary mirror plus a small deformable mirror downstream.

6. Earth's Visible Spectrum

The Earth's reflection spectrum is shown in Figure 4 (left), The dominant bands are water, oxygen, ozone, methane, and carbon dioxide. The latter two are shown for high (early Earth) and low (present) abundances, and are likely to be detectable only in the high-abundance case. The spectra are smoothed for presentation, but are otherwise the result of a full line-by-line calculation using actual vertical profiles of temperature, pressure, and mixing ratios. Figure 4 (right) shows the composite spectrum of atmospheric absorption bands, including the ultraviolet cutoff at about 0.33 μm; a multi-level cloud model is included, diluting the line depths. Additional details of both the visible and infrared models can be found in Traub & Jucks (2002), and Des Marais et al. (2002). The latter includes tables of spectral resolving power and curves of growth.

The narrowest feature in this spectrum is the oxygen band at 0.76 μm, which is matched by a spectrometer resolution of 69. Water bands are formed mostly near the ground in the troposphere. Ozone bands at about 0.6 and 0.3 μm are formed in the ozone layer in the stratosphere, and would be present even if there was a much greater cloud cover. However, since the Earth is not in a runaway condition like Venus, we expect that the cloud cover on an analog planet will always be highly broken, allowing a view to the surface for a large part of the planet's area. The reflection spectra of land plants and Rayleigh scattering are shown in the validation section below.

The integration times to characterize an Earth at 10 pc were estimated by Traub, Jucks, & Noecker (2000). For an 8 m coronagraph and a signal to noise ratio of 5, they find that it would take about 9 days to detect oxygen, 3 days to detect ozone, and 1 day for water.

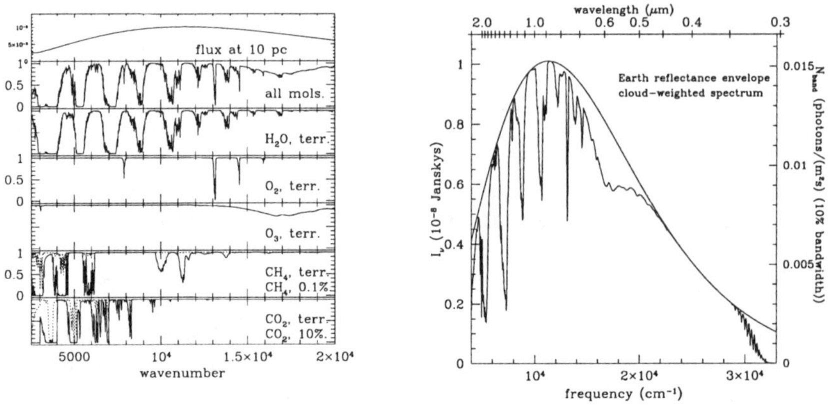

Figure 4. Visible spectra. *Left*: Calculated transmission through Earth's atmosphere for the 5 species shown. *Right*: Composite spectrum for a 4-stream case with some clear-air openings plus 3 layers of reflective clouds.

7. Observational Validation of Spectra

How realistic is our model? We show in Figure 5 (left) a measured reflection spectrum of the Earth, with our model calculation superposed (Woolf et al. 2002). The reflection from the integrated Earth was obtained by observing the "Earthshine" light on the side of the moon away from the sun. Familiar features from the previous section are seen as components here: a high cloud layer over part of the disk, and the composite molecular band spectrum of water, oxygen, and ozone. In addition there is strong Rayleigh (λ^{-4}) scattering at short wavelengths, and the "red edge" reflection from land plant leaves, which reflects strongly at wavelengths longer than about $0.7\,\mu$m. Other components contribute only small amounts, including the blue ocean water, aerosols, and the green pigment in oceanic phytoplankton. The good match validates our model.

8. Planet Characterization

Can we estimate the diameter and other properties of an extrasolar planet? In the infrared we can use the measured flux, observed color temperature, and blackbody equation to estimate a planet radius. However, in the visible we must use another stratagem. In Figure 5 (right) we show albedo measurements of the planets from Irvine et al. (1968) and Earth from Woolf et al. (2002).

Notice that the gas giant curves are distinguishable from those of a runaway cloud case (Venus), a broken cloud case (Earth), and a bare rocky surface with little (Mars) or no (moon, Mercury) atmosphere. By recognizing these different types of reflectors, and making the conjecture that these types (or any other type that can be modeled, e.g., a roaster-type Jupiter) can be represented by their known prototypes, we see how a connection may be made between planet size etc., and its reflection spectrum. My tentative result suggests that mass

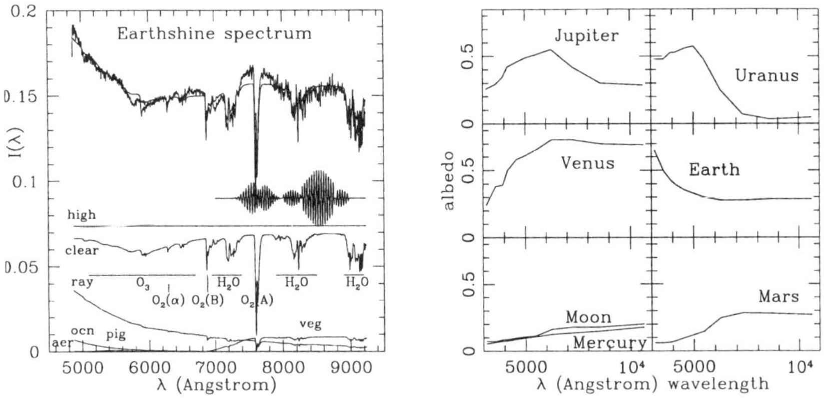

Figure 5. *Left*: Observed Earthshine visible spectrum, with best-fit model. The model components are plotted to scale. The indicated fringing interference was fit and subtracted from the data. *Right*: Albedo measurements of planets, taken with narrow-band filters. Distinctive spectral types can be use to classify planetary types.

could be estimated to about 50% accuracy, and radius to 14%. Knowing the size, albedo, and stellar influx, we can estimate the effective temperature as well.

Other characteristics to be estimated are the total molecular column (from Rayleigh scattering), surface pressure (from column density, radius, and mass), surface temperature (from water bands and a nominal relative humidity), cloud cover, cloud variability, and rotation rate (the latter two from the light curve in various bands, as discussed by Ford, Seager, & Turner 2001). The presence of land plants with leaf structures like those on Earth may be detected through the appearance and light-curve variability of the "red edge".

9. Signs of Life

Yung & DeMore (1999) give an estimate of the effect that life has on the atmosphere of the Earth, drawing on earlier studies by L. Margulis and J.E. Lovelock. On an Earth with and without life, production rates of a number of molecules are quite different. Accounting for the sources and sinks, a steady state abundance can be estimated in each case. For the species discussed here, the result is that the abundances ratios, with/without life, are roughly 100 for oxygen, 1000 for methane, and 1/1000 for carbon dioxide. In Des Marais et al. (2002) a large abundance of oxygen was particularly singled out as being a good marker for life.

Other measurable quantities which would help establish the ability of life to exist are a temperature in the range of liquid water, and the presence of water itself. The presence of the "red edge" of leaves is a clear sign of plant life, although since this reflectivity spectrum is a property of the cell structure of the leaves, and not necessarily a universal spectral signature, caution will be required in interpreting any such reflectivity enhancement.

10. Conclusion

TPF's goal is to search for terrestrial planets around hundreds of nearby stars, and to spectrally characterize those planets. This goal can be met with a visible coronagraph telescope. Key characteristics of a TPF-capable telescope are an 8–10 m diameter, a 0.5 Å RMS wavefront, and wide-band operation at visible (and possibly near-infrared) wavelengths.

These requirements, in terms of telescope diameter, collecting area, and wavefront quality, are necessary to search a large number of stars for terrestrial-type planets. But these same requirements will also permit deeper imaging surveys of the type pioneered by Hubble. Thus, because TPF will figuratively extend the frontier of extrasolar planet observations in much the same way as Hubble extended the search space for distant galaxies, and also because TPF can literally push the Hubble deep field and high-resolution imaging to new levels, it is appropriate to plan that TPF should be the worthy successor to Hubble.

References

Des Marais, D.J., et al. 2002, Astrobiology 2002, 2, 153

Ford, E.B., Seager, S., & Turner, E.L. 2001, Nature, 412, 885

Irvine, W.H., Simon, T., Menzel, D.H., Pikoos, C., & Young, A.T. 1968, AJ, 73, 807

Kasdin, N.J., Spergel, D.N., & Littman, M.G. 2002, Appl.Opt., submitted

Kuchner, M.J., & Spergel, D.N. 2002, ApJ, submitted

Kuchner, M.J., & Traub, W.A. 2002, ApJ, 570, 900

Schneider, J. 2002, http://cfa-www.harvard.edu/planets

Traub, W.A. & Jucks, K.W. 2002, in Atmospheres in the Solar System, Comparative Aeronomy, eds. M. Mendillo, A. Nagy, & J.H. Waite, (Washington D.C.: AGU), 369

Traub, W.A., Jucks, K.W., & Noecker, C. 2000, AAS meeting, 197

Woolf, N.J., Smith, P.S., Traub, W.A., & Jucks, K.W. 2002, ApJ, 574, 430

Yung, Y.L., & DeMore, W.B. 1999, The Photochemistry of Planetary Atmospheres, (New York: Oxford Univ. Press)

Hubble's Science Legacy: Future Optical-Ultraviolet Astronomy from Space
ASP Conference Series, Vol. 291, 2003
K.R. Sembach, J.C. Blades, G.D. Illingworth, R.C. Kennicutt, Jr. (eds.)

Detecting and Characterizing Giant Planets Beyond Hubble

Jonathan I. Lunine

Lunar and Planetary Laboratory, The University of Arizona, Tucson, AZ 85721

Abstract. The current cohort of nearly 100 companions to Sun-like stars, in the Jovian-mass range, is beginning to suggest some planetary systematics whose investigation will be the task of future ground-based and space-based telescopes. We are beginning to understand, from Doppler surveys, the fraction of stars that have giant planets. In the case of one such planet, thanks to the ground-based observation of its transits, the mass and radius — hence the density — are well determined. Transits observed from Hubble Space Telescope provide some constraint on composition and upper atmospheric processes. While the most promising wavelength range for detailed atmospheric studies of extrasolar giant planets is the infrared, there are key features in the optical spectra that are of interest in studies of warmer giant planets.

1. Introduction

It is now eight years since the first announced discovery of an extrasolar giant planet by the Doppler spectroscopic technique (Mayor & Queloz 1995), during which time — after a brief challenge regarding the interpretation of the Doppler shifts in terms of planets — almost a hundred planets have been detected. Little is known about any of these objects, except for the the orbit period and the mass times the sine of the orbital inclination relative to the line of sight to the Earth ($m \sin i$). Assuming a random distribution of such objects in the sky, 90% of these objects should have physical masses within a factor of two of their Doppler masses (Marcy, Cochran, & Butler 2000); as discussed below, direct information on one of these bodies confirms it is like Jupiter in its basic physical properties.

With this foundation, it is possible to construct the statistics of extrasolar giant planets and to consider what future observations will be possible and motivated by the questions we can ask today. In this short review, I address three basic questions with the data available today.

1. What is the fraction of stars that have giant planets?

2. How do we know whether any of these objects have the properties of the giant planets in the Solar System?

3. How can individual planets be characterized?

In terms of future space-borne astronomy, question 3 is of the most interest, but the answers to the first two questions provide the justification for addressing the third.

2. What Fraction of Stars Have Giant Planets?

Doppler spectroscopic surveys have focused on the F, G, and K dwarf stars for the practical reason that the M dwarfs are faint and difficult to observe at the required high signal-to-noise values, and the A stars (and earlier) lack appropriate spectral lines for the Doppler measurements. Within the stellar-type cohort observed, roughly 5% have companions ranging from less than a Jupiter mass up through several Jupiter masses. (As noted above, assuming random inclinations, the physical masses are within a factor of two of the reported Doppler masses for the vast majority of the detections.) The cohort seemed striking, at least initially, because of the bulk of the planets discovered were very close to their parent stars — as close as 0.04 AU. In part, this is an observational artifact because the first planets to be detected after a technique is successfully validated are those whose signal leading to the detection is expressed most rapidly, all else being equal. Since the signal is the periodic oscillation in the position of the stellar line, a consequence of the reflect motion of the star in response to the planet, short-period planets would appear first. Adding to this bias is the issue of the sensitivity of the technique as a function of planetary semi-major axis, which for the Doppler technique is highest for the smallest orbits.

The total cohort of planet detections suggests a discovery process whereby the typical semi-major axes of the newly announced discoveries is increasing with time from 1995 to the present. A histogram of number of planets versus semi-major axis, the latter expressed logarithmically, shows a roughly uniform distribution (D. Trilling, personal communication 2001). Nonetheless, the existence of close-in giant planets as a significant fraction of the total number of detected giant planets is significant, because formation models predict that assembly of stable, largely gaseous planets would be exceedingly difficult within 0.1 AU, and even unexpected within 1 AU. But the discoveries breathed new life into models of inward migration of giant planets interacting with disks (Lin, Bodenheimer, & Richardson 1995) or with each other (Weidenschilling & Marzari 1995). On the outer end of the semi-major axis distribution, the recent discovery of a several-Jupiter-mass planet in a Jovian-type orbit of roughly 5 AU is gratifying to theorists, but is beginning to push the sensitivity of the technique. Given a detection limit (floor) of a few meters per second based on observations of turbulence in the stellar photosphere, a Jupiter-mass planet orbiting 5 AU from a G-dwarf is just barely detectable. More distant Jupiter-like planets or smaller mass planets will require other techniques or some clever means to punch through the stellar photospheric floor.

Given the semi-major axis sensitivity dependence, a good estimate for the total frequency of occurrence of extrasolar giant planets around single F, G, K main sequence stars is about 10%, assuming there is no dramatic rise in the occurrence of such objects beyond 10 AU relative to the 5 to 10 AU region (Trilling, Lunine, & Benz 2003). This is a remarkably high number, considering some dire predictions in the past about the difficulty of making giant planets.

However, the estimate must be a lower limit, because if giant planets migrate inward and dynamically interact with each other, there is a loss of giant planets during and soon after their formation. Application of a simple migration model suggests that two out of three giant planets are lost prior to the star entering the main sequence (Trilling, Lunine, & Benz 2003). If 30% of F, G, and K type stars generate giant planets during formation, then it follows that giant planets may be a key part of the dynamical environment of young terrestrial planets, with important implications for delivery of volatiles to and habitability of other Earth-like planets (Lunine 2001).

3. What Is the Nature of Extrasolar Giant Planets?

Because the Doppler spectroscopic technique provides only a mass and an orbit period, there is little one can do to investigate the nature and origin of planetary companions from those data. Assuming a random orientation of the orbit planes on the sky, roughly 1% of these objects should transit their parent stars as seen from Earth, and indeed one does. The companion to HD209458, a G dwarf, has a 0.05 AU circular orbit with the appropriate inclination. From ground-based observations of the stellar light curves variation, of order 1%, a radius for the planet roughly 30–40% larger than that of Jupiter is derived (Charbonneau et al. 2000). The existence of the transit, of course, immediately yields a physical mass equal to that of the Doppler-derived mass, about 70% that of Jupiter.

The distended size of this sub-Jupiter-mass planet can be understood in terms of the proximity of the planet to the parent star. If the planet absorbs the bulk of the stellar energy it receives (on its day side), models replicate well the planetary radius for the given mass (Burrows et al. 2000). Further, although it is intuitive that downward transport of heat against the outward heat-flow gradient would seem difficult, mechanical mechanisms for moving heat can plausibly explain the transport necessary (Showman & Guillot 2002). In sum, HD209458b seems well explained by models, which were developed to reproduce the properties of Jupiter and Saturn. One can go a step further in considering the origin and evolution of this body. If HD209458b formed at a significantly greater distance from its parent star than its present position, it must have moved inward to its present position within a few tens of millions of years of its formation, otherwise it would have cooled to the point that its radius would be smaller than what is measured today. While downward transport of heat through the planet is possible, re-expansion of the planet in size is extremely difficult, although strong tidal effects might inflate planets to a limited extent (Bodenheimer, Lin & Mardling 2001). Thus, we may have an upper constraint, albeit a weak one, on the time scale over which this planet migrated, or was dynamically scattered, inward. Tens of millions of years is a generous amount of time for either of these processes, based on our current understanding of the early dynamics of such systems.

Further information on the atmospheric composition of HD209458b can be obtained from transits, since the frequency-varying opacity in the atmosphere leads to a wavelength-dependent apparent size of the transiting planet. Dissecting the transit in wavelength requires great photometric stability and high signal-to-noise ratio, and HST observations were the first to accomplish this

(Charbonneau et al. 2002). Examination of theoretical spectra (Seager & Sasselov 2001; Hubbard et al. 2001) indicates that the optical features of sodium should be most apparent, and they were seen in the HST spectra. However, Charbonneau and colleagues inferred less sodium than would be expected in the planet if it were of similar metallicity to that measured for the parent star (which is close to solar). Apart from instrumental effects in what was a difficult measurement, possibilities include ionization of the sodium, formation of clouds or molecules that fall out of the visible upper atmosphere, or problems with the theoretical models. Further work on the existing spectra and the models may elucidate the likely cause, but more significant in the long term is that comparison of theoretical and observational spectra are being made on an extrasolar planetary atmosphere, in much the same spirit as was done for the giant planets of the Solar System three decades ago.

4. How Can Individual Giant Planets Be Characterized?

The transits of HD209458b represent the currently best opportunity to separate the signal of the star from a giant planet, and probably the only approach for close-in giant planets in the foreseeable future. Meanwhile, objects of comparable temperature but larger mass, existing isolated or in very loose orbits around stars, are the subject of detailed spectral analyses. Determination of the appearance and disappearance of broad sodium and potassium features in the optical, and the transition from carbon monoxide to methane in the infrared, have provided an extension of the spectral classification sequence below the main sequence hydrogen fusion limit into a new realm of "L-dwarfs" and "T-dwarfs" (Kirkpatrick et al. 1999). The T-dwarfs, by extension, grade in mass from something in the 10–20 Jupiter mass range down to the giant planets. Whether formation mechanisms differ is an open question; almost certainly sub-stellar mass "brown dwarfs" spectroscopically depicted as L-dwarfs form like stars, by direct collapse. Giant planet formation models allow either direct collapse from a disk around a parent star or a multi-step process in which accretion of rocky and icy planetesimals nucleate the collapse of the surrounding gas. It is unclear whether there are any observational consequences of the difference, for example in the relative metallicities of the planet and the star; direct collapse planets may acquire a large amount of rocky and icy debris after formation if it exists in the local environment, making them look spectroscopically like their nucleated collapse cousins. On the other hand, planets in which the metals are not enriched might be tagged as formed from direct collapse in a disk that, for whatever reason, did not have a large amount of condensed rocky and icy debris. Alternatively, the initial mass function of giant planets might differ for direct collapse versus nucleated collapse mechanisms.

Diagnoses of these possible differences in metallicity or mass function will require good spectra of extrasolar giant planets, akin to those now possible for the isolated T-dwarfs (Burgasser 2002). The reason is clear for the first of these potentially diagnostic differences. For the second, it is assumed that we have no good means for diagnosing the mass of extrasolar planets in large (5 AU or more) orbits, unless we wait for a decade or more for astrometric observations to trace out the slow barycentric wobble of the central star. If direct detection of giant planets is to be followed up by rapid diagnosis, clues to the mass will need to be derived through the presence and appearance of spectral features, which are somewhat sensitive to the surface (photospheric) gravity of the body, and hence provide the mass since the radius might be derived from the brightness (with the ambiguity of the albedo variation due to clouds being a complicating factor). Alternatively, simply deriving a good photospheric temperature from the spectra will, also with a constraint on the albedo, provide the mass through detailed evolutionary models (Burrows et al. 2001). The metallicity would have to be constrained simultaneously, but this is less problematic than constraining the albedo, since cloud modeling remains a difficult theoretical challenge.

With respect to extrasolar giant planets, other than a stand-alone precursor version of Terrestrial Planet Finder, featuring a space-borne interferometer with half-meter mirrors along a 10 m beam (N. Woolf, personal communication 2001), or a specialized coronagraphic small telescope, (Trauger et al. 2000), the future will belong to the largest of adaptive-optics-equipped ground-based telescopes and to Next Generation Space Telescope (NGST). NGST will be equipped with coronagraphic capability, and its sensitivity in the near- and mid-infrared will enable a rich set of water, methane, and ammonia bands to be detected in extrasolar giant planets that are not too close to their parent stars. Assuming the strawman NGST spectrometers and current evolutionary models (Hubbard, Burrows, & Lunine 2002), it is clear that both broad-band and moderate-resolution spectroscopy can be done on a 15 Jupiter-mass body across its near-infrared and infrared spectrum out to ages of billions of years. Isolated Jupiter-mass bodies can be detected and their broad-band colors measured at wavelengths at and beyond 5 μm. Jupiter-like planets younger than a billion years are candidates for moderate resolution spectra across the infrared.

The 5 μm wavelength region is an especially important one for the study of Jovian planets and brown dwarfs. At this wavelength, the atmospheres are significantly cleared of molecular opacity, allowing thermal emission from warmer, deeper levels of the objects to escape directly to space. It has long been known that Jupiter's atmosphere shows broad hot spots in 5 μm images (Ortiz et al. 1998). The spectrum of GL229B also displays a prominent flux enhancement at 5 μm (Oppenheimer et al. 1998). Theoretical spectra of model brown dwarf atmospheres constructed by several groups (Baraffe et al. 2002; Hubbard, Burrows, & Lunine 2002) consistently show the same bright emission near 5 μm. These objects can be many orders of magnitude brighter than an equivalent blackbody radiator at the temperature of the planet. $R = 3$ detection is possible in this window for a nearby Jupiter-mass object at an age exceeding that of the Solar System. The superthermal emission of brown dwarfs and extrasolar giant planets at 5 μm offers a distinct opportunity for planet detection that NGST can exploit.

Transit observations of planets (e.g., HD209458b) at high sensitivity in the near infrared will allow water bands to be detected (Hubbard et al. 2001), and these will aid in interpreting the overall metallicity of the object and the situation with respect to sodium. In the optical part of the spectrum, which is more difficult to access for cool objects except in transit measurements, understanding the variation in the alkali metal bands for warm Jupiter-like planets is important in diagnosing the dynamical and chemical processes occurring in the upper atmosphere. Indeed, transit and reflection spectra can constrain variations in atmospheric properties across the diurnal boundary of the planet, providing the potential for exploring dynamical processes that transport heat. Such observations will be very challenging.

5. Coda

In the short time, less than a decade, since the detection of the first extrasolar giant planet, and (in the same year) that of the first brown dwarf (Oppenheimer et al. 1995), a small but varied menagerie of giant planet and brown dwarf spectra has been obtained. That of Gleise 299B, the first definitive brown dwarf, represents a body whose pedigree-planet or sub-stellar "star-like" body-remains an intriguing problem in its resolution and significance. The spectrum of HD209458b, on the other end of the mass spectrum, is tantalizing but terribly incomplete-more will come with better instrumentation on HST and then bigger space telescopes in the infrared. Spectra of T-dwarfs and their more massive cousins the L-dwarfs provide a chemical sequence that prepares us for the interpretation of the lowest mass T-dwarf members — the giant planets — and perhaps the creation of yet one more, final letter in the stellar-to-planet spectral classification sequence. And then there are the giant planets of the Solar System — remote yet already visited, indeed entered directly in the case of Jupiter and the Galileo probe. The richness of phenomena in, and complexity of interpretation of, their atmospheres beckons us to do likewise with their extrasolar cousins — and we will do so with the next generation of space — and ground-based observatories.

Acknowledgments. The author's work described herein was generously supported by NASA's *Origins* program.

References

Baraffe, I., Chabrier, G., Allard, F., & Hauschildt, P.H. 2002, A&A, 382, 563

Bodenheimer, P., Lin, D.N.C., & Mardling, R.A. 2001, ApJ, 548, 466

Boss, A.P. 1995, Science 267, 360

Burgasser, A. 2002, Dissertation, Caltech

Burrows, A., et al. 2000, ApJ, 534, L97

Burrows, A., Hubbard, W.B., Lunine, J.I., & Liebert, J. 2001, Reviews of Modern Physics, 73, 719

Charbonneau, D., Brown, T.M., Latham, D.W., & Mayor, M. 2000, ApJ, 529, L45

Charbonneau, T. Brown, M., Noyes, R.W., & Gilliland, R.L. 2002, ApJ, 568, 377

Hubbard, W.B., et al. 2001, ApJ, 560, 413

Hubbard, W.B., Burrows, A.S., & Lunine, J.I. 2002, ARA&A, in press

Kirkpatrick, J.D., et al. 1999, ApJ, 519, 802

Lin, D.N.C., Bodenheimer, P., & Richardson, D.C. 1997, Nature 380, 606

Lunine, J.I. 2001, Proc. Nat. Acad. Sci. USA, 98, 809

Marcy, G.W., Cochran, W.D., & Mayor, M. 2000, in Protostars and Planets IV, eds. V. Mannings, S. Russel, & A.P. Boss, (Tucson: University of Arizona Press)

Mayor, M., & Queloz, D. 1995, Nature, 378, 355

Oppenheimer, B., Kulkarni, S.R., Matthews, K., & van Kerwijk, M.H. 1998, ApJ, 502, 932

Ortiz, J.L., Orton, G.S., Firedson, A.J., Steward, S.T., Fisher, B.M., & Spencer, J.R. 1998, JGR, 103, 23051

Seager S., & Sasselov, D.D. 2000, ApJ, 537, 916

Showman, A.P., & Guillot, T. 2002, A&A, 385, 166

Trauger, J.T., et al. 2000, DPS meeting 32, 26.08

Trilling, D.E., Lunine, J.I., & Benz, W. 2003, A&A, in press

Weidenschilling, S.J., & Marzari, F. 1996, Nature, 384, 619

Hubble's Science Legacy: Future Optical-Ultraviolet Astronomy from Space
ASP Conference Series, Vol. 291, 2003
K.R. Sembach, J.C. Blades, G.D. Illingworth, R.C. Kennicutt, Jr. (eds.)

Stellar Populations Studies with an 8 m UV/Optical Telescope in Space

R. Michael Rich

Division of Astronomy, Department of Physics and Astronomy, Math-Sciences 8979, University of California at Los Angeles, Los Angeles, CA 90095-1562

Abstract. The gain in resolution of a diffraction-limited 8 m optical/ultraviolet telescope in space would be roughly equivalent to the gain of Hubble relative to ground-based observations. Among the most exciting programs include a determination of the age and star formation history of the Galactic bulge from the white dwarf cooling sequence, and measurement of the ages for most galaxies (and their dwarf galaxies and globular clusters) in the Local Group, and possibly even at the distance of the Virgo cluster. The 13 milli-arcsecond point spread function may enable astrometry over the Local Group volume and parallaxes to Galactic globular clusters.

1. Introduction

It is indeed a challenge to consider the science program for a space telescope that will begin operations no earlier than 2015. The difficulty of this task increases as we consider the broad subject of stellar populations, a subject area that is admittedly not fashionable at the present time. It is not impossible that by 2015, observations by the Microwave Anisotropy Probe (MAP) and the Planck satellite will have all but settled the cosmological model. The Next Generation Space Telescope (NGST) will have applied its infrared capabilities to the study of galaxy formation at the highest observable redshift and likely will also have had a significant impact in the study of stellar populations. One can legitimately wonder whether so mature an area of astronomy as stellar populations will be worth pursuing. However, if we believe that the direct study of galaxy formation at high redshift is a worthwhile endeavor, then we must acknowledge the value of the detailed information available from the ages, abundances, kinematics, and structures of observable stellar populations at lower redshift. In the case of an 8 m optical/UV space telescope (hereafter, Next-HST or NHST), there is also the real possibility of studying galaxies at moderate redshift ($z > 0.3$) in extraordinary detail.

Figure 1 shows a subraster of Brown's (these proceedings) simulations of the M 31 bulge showing dramatic improvement of an 8 m optical/UV telescope in space relative to HST and predicted NGST performance. In the fullness of hindsight, one easily forgets that the gain of HST over ground-based imaging is of order a factor of 5 (assuming $0.5''$ optical seeing, which is not uncommon). If

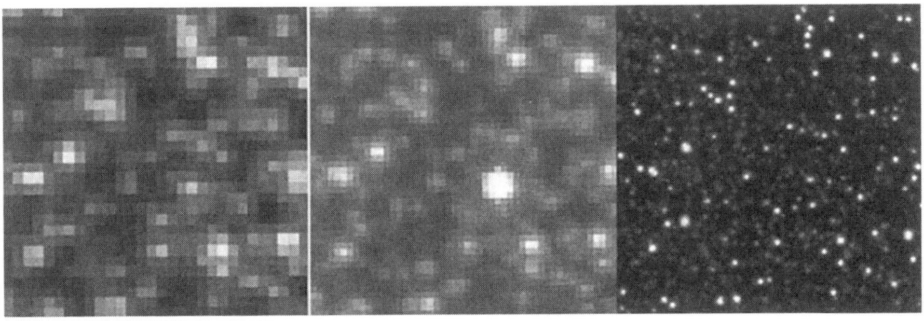

Figure 1. Simulated imagery of the M 31 bulge (courtesy Tom Brown, STScI). This figure shows a $1.5'' \times 1.5''$ subraster of a simulated image of the M 31 bulge with HST/ACS (left panel), NGST (center panel), and an 8 m diffraction-limited optical space telescope: NHST (far right panel). The gain in resolution of the next generation 8 m space telescope is comparable to the gain of Hubble over ground-based imaging. The 8 m also reaches ~ 5 mag deeper in the same length of integration time (see Brown, this volume).

NHST is optimized for high resolution imaging, it would gain a factor of 3 in resolution over HST. Even greater gains are possible if scattered light is suppressed, the point spread function improved, and overall throughput increased relative to HST. Progress leaps when formerly difficult problems become routine, and new capabilities open up new subject areas. One example (from stellar populations) is the impact that HST has had on the measurement of main sequence photometry of star clusters. Prior to the HST era, ground-based imaging had yielded good photometry for a few globular clusters. Well into the HST era, we now have evidence for synchronized formation of the oldest stars over a volume of 100 kpc. The oldest globular clusters have the same age in the old halo (Harris et al. 1997) as in the Magellanic Clouds (Mighell et al. 1996; Johnson et al. 1999).

2. What NGST May Accomplish

The Next Generation space telescope (NGST) will have a significant impact in stellar populations, even though it will be diffraction limited at 1μm, and its actual resolution will likely not significantly better HST in the optical. Nonetheless, the infrared capability will be ideally suited to solving certain classes of problems. For example, the mass-luminosity relationship for stars is such that the faintest stars are best observed in the infrared. At the faint end, a star with $M_K = +12$ ends up at $V = +18$ or fainter, and consequently the 6 m infrared NGST reaches the H-burning limit while the optical NHST cannot attain that goal. The measurement of the slope of the mass function to the hydrogen burning limit is an important problem which must be undertaken in the widest range of stellar systems, sampling the full parameter space in metallicity and system density. Considering the case of the Galactic bulge, one expects to reach the end of the initial mass function at $M_K \approx$ 12-14, or at apparent K mag of 25-28, easily attained by NGST, whereas a ground-based attempt (even with a 10 m telescope)) would be daunting. By contrast, the hydrogen burning limit is at least 6 mag fainter in V and too faint for NHST. The same calculation obviously applies to the Magellanic Clouds and Milky Way dwarf spheroidals.

A summary of the scientific potential of NGST to study stellar populations may be found in my review article in the first NGST science conference (Rich 1998) and in my contributions to the Design Reference Mission (`http://222.stsci.edu/ngst/science/drm/programs.html`).

Yet another area where NGST will excel is in the study of late-type stars in nearby galaxies, perhaps as distant as the Virgo cluster. The most advanced phases of stellar evolution are also infrared bright, and mass is lost during both the first and second ascent of the red giant branch. In an old stellar population, asymptotic giant brnach (AGB) and post-AGB stars are among the most intrinsically luminous stars, but have limited use as age and metallicity diagnostics. At low metallicity, it is well established that detection of stars with $M_{bol} < -4$ is pretty much a guarantee that the population spawning them has some intermediate age fraction, and an age constraint is even possible (Mould & Aaronson 1986). But at the metal rich end even old stellar populations have AGB stars reaching $M_{bol} < -5$ (Guarnieri et al. 1998), which vitiates the age-sensitivity of luminous AGB stars. NGST will also open up a world of mid-IR imaging at high sensitivity and resolution. There will be the prospect of measuring mass loss on the AGB in different populations, and the new additional information offers with it the potential to also constrain population ages, perhaps even at high metallicity.

2.1. Where NGST will Fall Short

Precisely how NGST performs as a high resolution near-IR telescope will depend on presently unknown details of its actual performance. However, the lack of blue sensitivity is a handicap for many problems. In the best case NGST sensitivity begins at 0.6-0.7μm; NGST photometry may be less useful, and require much higher S/N, to discern age and metallicity differences. Because the opacity due to iron lines and molecular bands is greatest in the optical, colors of the turnoff and red giant branch measured in the optical are far more sensitive to age and metallicity. Figure 2 shows that the differences are dramatic. Added to this, red giants are > 1 mag brighter in the IR due to the emergence of flux in the H^- opacity minimum at 1.6μm; flux from the red giant branch (RGB) and AGB stars can overwhelm the interesting horizontal branch (HB) and main-sequence evolutionary phases. These issues are major because much of the attraction of NHST is the possibility of high precision relative photometry of faint stars (relative ages of M 31, Milky Way halos; synchronized star formation of the oldest stars in the Local Group).

For relatively nearby systems (closer than 10 kpc) we may begin to rely heavily on the use of the white dwarf cooling curve as a means of fine age discrimination in the oldest stellar systems. Figure 3 shows what has been accomplished for the nearby globular cluster M4 in a total integration of 123 orbits (combined also with earlier archival data). The right panel of Figure 3 shows Hansen's model fits to the luminosity function of the white dwarf cooling sequence. A recent breakthrough in the understanding of white dwarf cooling (Hansen 1999) demonstrates that collision induced absorption of the H_2 molecule

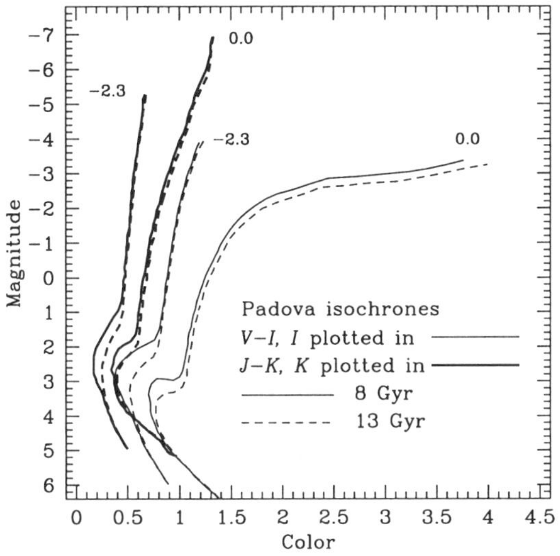

Figure 2. Optical colors give superior results for the measurement of age and abundance. This figure shows Padova isochrones in infrared J, K (1.2 and 2.2 μm) and in the optical V, I. The figures use a common magnitude scale. Notice that the V, I giant branches are far more extended (due to extensive blanketing) and that age differences are also far better defined in optical as opposed to IR colors.

is the most important source of opacity in the coolest DA white dwarfs. The appearance of this molecule has the practical effect of pushing flux from the infrared into the optical; hence the coolest white dwarfs should actually become bluer, not redder as intuition would expect. The predicted absorption in the IR is many magnitudes, and therefore NGST will be disadvantaged in the study of cluster white dwarf luminosity functions. In contrast, NHST would excel in these measurements.

3. The Stellar Populations Science Case for NHST

In the study of resolved stellar populations, spatial resolution is king. One need only examine Figure 1. Image crowding is the most serious problem in the study of any stellar population. The spurious measurement of stars as being too bright leads to underestimates of the age from misplaced main sequence turnoffs and false detection of "bright" AGB stars in bulges (Renzini 1998). Using the fuel consumption theorem (Renzini 1998), we can estimate the number of stars in a given evolutionary phase j as $n_j = B(t)L_T t_j$ where for populations older than 1 Gyr, B (the evolutionary flux) is 2×10^{-11} stars per $L_\odot$ per year. The time in years in stellar evolution phase j is t_j, and L_T is the total luminosity in $L_{bol,\odot}$. In practice, the two controlling factors in whether one can resolve a

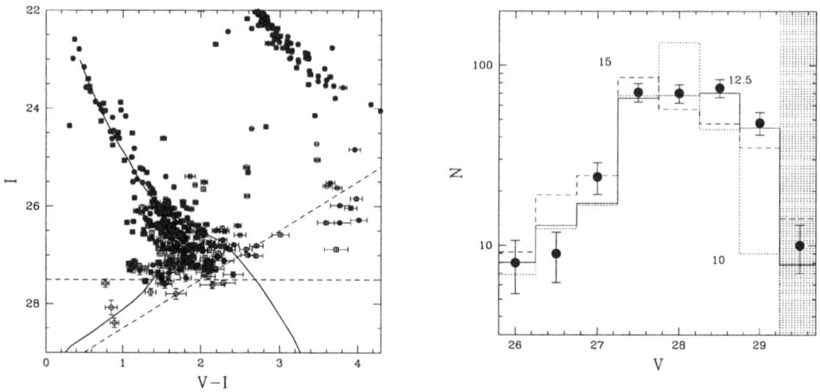

Figure 3. Deep HST/WFPC2 photometry of the white dwarf cooling sequence in the globular cluster M4 (Hansen et al. 2002). *Left Panel:* proper motion selected white dwarf cooling sequence with Hanen's models for DA white dwarfs (solid line extending to the blue) and helium white dwarfs (straight line extending to the red). *Right panel:* Hansen cooling model fits to the data, with ages in Gyr. This 123 orbit observation on HST would take ~ 1 hr on NHST; a 100 hr integration could constrain the age and star formation history of the Galactic bulge.

stellar population are the surface brightness and the distance modulus. Both of these determine the total luminosity (number of stars) per resolution element.

The most secure age indicators are the most difficult to measure. Other then the white dwarf luminosity function (LF) (by far the hardest measurement), the main sequence turnoff point gives the best age constraint. However, main sequence stars live $10^9 - 10^{10}$ yr and therefore outnumber shorter lived, more luminous phases of evolution by factors of 10-100. Their great numbers correspond to serious image crowding. Age estimates for older populations from the main sequence turnoff must also be precise: for a 12 Gyr old population, a change of 0.1 mag in the absolute magnitude of the turnoff point corresponds to an age change of 1 Gyr.

The helium burning phase (horizontal branch) has the potential to give a fine age constraint, but the second parameter problem (which is age plus some other factors) remains unsolved. The presence of a populated blue horizontal branch generally indicates population older than 10 Gyr, while if only red clump stars are present, intermediate age is indicated if [Fe/H] < -1.5. At high metallicity, UV-bright HB stars have interesting potential as a population diagnostic. Helium burning stars are 4 mag brighter than the old main sequence stars, and live only 10^8 yr. Ten to one hundred times fewer in number, and nearly 100 times as bright as old turnoff stars, HB stars are easier to reach; ~ 20 M 31 globular clusters are presently observed to the HB. NHST will reach the HB out to the Virgo cluster. Visible to the greatest distances are the thermally pulsing AGB stars which live only ~ 10^5 yr and reach $M_I < -4$, but provide an strong age constraint only for metal poor, or very young, populations. In somewhat of an analogy to the well-known distance ladder, the "age ladder" has the same problem that the best age indicators are measureable only in the nearest populations.

With a little algebra, the fuel consumption theorem yields the useful form: $surf(V) = (m - M)_0 + const$, where a population at distance modulus $(m - M)_0$ and brighter than $surf(V)$ mag/ sq. arcsec would suffer too much stellar crowding for observations to feasible. For NHST to resolve the main sequence turnoff point of an old population, the limiting surface brightness is $(m - M)_0 - 3.1$, while the constant is -5.1 for red giants and HB stars (the smaller numbers of these stars permit work in more crowded fields (higher surface brightness). At the distances of M 31 and M 33, we are likely to reach the old turnoff point in the disk at 21 mag/sq. arcsec, which is virtually on top of tbe nucleus of M 33, and $20'$ (= 4.5 kpc) into the disk of M 31. Our handy formula suggests that the HB can be studied (distance modulus 24.4) to within $4'$ of the nucleus (the bulge). At the Virgo cluster the old turnoff requires > 100 hours (to reach V=35!) and must involve imaging outside the Holmberg radius at 26.5 mag/sq. arcsec.

3.1. The White Dwarf Cooling Age of the Bulge

The formation of bulges has almost certainly been directly observed at high redshift (Steidel *et al.* 1996). The rapidly star forming Lyman break galaxies (LBGs) are probably the site of bulge formation. Did the bulge of our Milky Way form as rapidly as these observations imply? The bulge giant abundance trends of McWilliam & Rich (1994) and Rich & McWilliam (2000) are most consistent with massive star SNe enrichment over a 1 Gyr timescale. From turnoff photometry, some limits on age spread are possible (notably Ortolani et al. 1995), but it is difficult to argue for bulge formation burst smaller than 2-3 Gyr. The WD LF may give a more powerful constraint, and even the fossil record of the burst history. Our understanding of the old WD luminosity function is presently in its infancy but it will improve dramatically as more clusters are measured and theory is fine tuned to fit observations. Figure 3 shows the potential for the WD luminosity function method to distinguish within an age range of 1-2 Gyr – a precision that main sequence turnoff fitting has not attained. Further, we find that the WD LF gives a distance constraint to 8%, comparable to the best precision available. Even if the globular cluster age scale appears in some part solved by missions such as SIM and GAIA (assuming that these missions which also employ very challenging technology are successful), the WD cooling sequence may be usable to constrain the detailed burst formation history of globular clusters. One of the most interesting applications would be in the Galactic bulge where abundance spread, foreground contamination, and depth effects muddle the main sequence age constraint. A proper motion selected sample of bulge white dwarfs could constrain the age and formation history of the inner halo and bulge, its broad LF insensitive to the detriments troubling the turnoff point. Achieving our M4 result in the bulge requires that we reach 3.2 mag fainter, or to $V \sim 33$, demading a 100 hour integration time, which is lengthy but feasible.

3.2. The Global Star Formation History of Local Group Halos, Disks, and Bulges

Presently, a few M 31 globular clusters have been imaged with HST to the level of the horizontal brach. The preponderance of evidence is that the M 31 cluster system is old and simlar to the Milky Way. Before the end of the HST mission, a handful of M 31 clusters will have photometry measured down to the main sequence turnoff. As Tom Brown's contribution in these proceedings points out, the proposed NHST simultaneously solves the problem of crowding and sensitivity, so that main sequence turnoff photometry will be available in 1-2 hours of integration for each cluster in M 31, and correspondingly longer in more distant galaxies. The < 0.1 mag distance dispersion at M 31 is a huge benefit: photometry will immediately solve a great many problems (cluster age range, age of cluster stars relative to the field, age of the outer M 31 disk relative to the halo). The age dispersion of globular clusters and satellite galaxies may be an interesting constraint in testing galaxy formation models (Johnson et al. 1999).

The relative ages of disks, halos, and bulges remains a debated point among those working on high redshift galaxies, and galaxy formation models. The 8 m space telescope offers the possibility of reaching the old main sequence turnoff point in regions of relatively high surface brightness. The surface photometry of Kent (1987) indicates that for NGC 147, NGC 185, and NGC 205, and M 33, the NHST could reach the old turnoff point at $M_{ABS} = +5.0$ even into the nuclear regions. For M 31 and M32, the old turnoff point could be reached as close as 5-10′ (about 1-2 kpc) from their nuclei. It would therefore be possible to plot the age distribution of the disks of M 31 and M 33 as a function of galactocentric distance (and surface brightness) and to understand whether disk formation has a radial age dependence, in general.

4. Astrometry: the "Killer App?"

If the diffraction limit of 0.013 arcsec (13 mas) at 5000 Å is attained, one may contemplate undertaking problems of fundamental importance using NHST astrometric capabilities. NHST is not alone in having the potential for dramatic astrometric capabilities. SIM and GAIA are well advanced in their planning, and may well be operational before NHST. However, SIM and GAIA both must work brighter than $V = 20$ and have no panoramic capability.

One can imagine NHST measurements of individual stars to 0.01 of the diffraction limit, and a perhaps ten times better estimate for ensembles of 100 or more stars. The latter corresponds to a breathtaking 13 μas, verging on the capability discussed for SIM. Considering that the trigonometric parallax to a typical Galactic globular cluster is 0.2 mas, it should be possible to measure the trig parallax of any globular cluster in the Galaxy, and, with 5 years of observations, the Magellanic Clouds. If this sounds like pie in the sky, consider that I. King and J. Anderson have an active HST program (GO-9058) to obtain a 10% measurement of the 0.25 mas trigonometric parallax of 47 Tucanae. When one considers the capability of NHST to measure the red giant branch tip (and Cepheids) at the distance of the Coma cluster, there is the real possibility that NHST could roll up the problem of the cosmic distance scale, at least out to a volume of 100 Mpc. The obvious implications (precision distances to every nearby galaxy with a type I SN, etc.) are obvious, in terms of settling the question of a Λ-cosmology. Will that problem still be around in 15 years? Considering the complications involved, and the enormous implications, I would very much bet on the question not being conclusively resolved on that timescale.

Other exciting astrometry is possible; at the distance of M 31, 100 km s^{-1} corresponds to 20 mas yr^{-1}. The determination of a secure extragalactic reference frame will be the greatest challenge, but, should it be overcome, one can consider the possibility of measuring transverse velocities for all Local Group galaxies, and actually observing rotation in M 31 and M 33 (the observation that eluded van Maanen). The measurement of space velocities for Local Group galaxies will be a crucial input for timeing models involving the formation of the Local Group (Peebles et al. 2001).

We must design into NHST a high quality point spread function and reliable astrometric capability, with attention given to the careful fabrication of detectors and the focal plane. Will stellar populations still be a living subject in 15 years? One might have thought that, 10 years into HST, the subject would be finished, but such triumphs as the H_0 key project and precision globular cluster age measurements speak otherwise. The observations that I propose in this article will no doubt be superseded by projects we cannot now imagine...at least I hope very much that this will be the case.

Acknowledgments. The author thanks Tom Brown of STScI for providing Figure 1 and beautiful frames used in the talk. The author thanks D. Reitzel for assistance in preparing Figure 2 and Keck observatory at which part of this article was written. The author acknowledges support from grants GO-8679 and GO-9099 from STScI.

References

Guarnieri, M.D., et al. 1998, A&A, 331, 70

Hansen, B.M.S. 1999, ApJ, 520, 680

Hansen, B.M.S., et al. 2002, ApJ, 574, L155

Harris, W.E., et al. 1997, AJ, 114, 1030

Johnson, J., et al. 1999, ApJ, 527 199

Kent, S.M. 1987, AJ, 94, 306

McWilliam, A., & Rich, R.M. 1994, ApJS, 91, 749

Mighell, K.J., Rich, R.M., Shara, M., & Fall, S.M. 1996, AJ, 111, 2314

Mould, J., & Aaronson, M. 1986, ApJ, 303, 10

Ortolani, S., et al. 1995, Nature, 377, 701

Peebles, P.J.E., Phelps, S.D., Shaya, E.J., & Tully, R.B. 2001, ApJ, 554, 104

Renzini, A. 1998, AJ, 115, 2459

Rich, R.M. 1998 in Science with the NGST, ASP conf. ser. 133, p. 129

Rich, R.M., & McWilliam, A. 2000, Proc. SPIE, 4005, 150

Steidel, C.C., et al. 1996, ApJ, 462, L17

Hubble's Science Legacy: Future Optical-Ultraviolet Astronomy from Space
ASP Conference Series, Vol. 291, 2003
K.R. Sembach, J.C. Blades, G.D. Illingworth, R.C. Kennicutt, Jr. (eds.)

Resolved Stellar Populations and the History of Galaxy Evolution in the Nearby Universe

Eva K. Grebel

MPI for Astronomy, Königstuhl 17, D–69117 Heidelberg, Germany

Abstract. What governs galaxy evolution: nature (intrinsic properties) or nurture (external effects)? Resolving stellar populations offers the most detailed information about a galaxy's evolutionary history. A large-aperture, high-resolution, optical/UV telescope would be a perfect tool to search for Population III stars, to date the earliest substantial star formation episodes, to map red giant metallicity distributions, to study the properties of massive stars in clusters and in the field, and to test the universality of the IMF in a variety of galaxies in different environments.

1. Introduction

We still do not know what governs evolutionary processes in galaxies. In particular, it is unclear what impact environment has versus the importance of the intrinsic properties of the galaxies themselves. The Local Supercluster and Local Void provide a rich and diverse environment to investigate these questions. Studies of this kind require a reliable three-dimensional picture of nearby field galaxies, galaxy groups, and galaxy clusters (e.g., Virgo, Fornax, Coma). In order to understand the fundamental underlying processes, we must first turn to galaxies where we can study evolutionary histories in great detail, best accomplished through resolved stellar populations. While this affords the highest degree of accuracy and detail, it is necessarily limited by the achievable resolution and hence by distance, constraining the range of possible targets.

In recent years, an increased observational effort has been put into programs investigating these questions. The most significant breakthroughs in this area have come from Hubble Space Telescope (HST) observations, resulting from its superior angular resolution and low sky background. Recent competition has arisen from ground-based high-resolution near-infrared imaging with adaptive optics. The future holds the option of high-resolution stellar population studies in the infrared through the Next Generation Space Telescope (NGST). In this contribution we identify how stellar population studies in our cosmic neighborhood can benefit from space-based optical and ultraviolet (UV) facilities.

2. The First Episode(s) of Star Formation

A question of primary importance for cosmology and galaxy evolution is when the first stars formed, and how evolution progressed from there. In the nearby universe, we can address these questions by identifying and studying tracers of the most ancient star formation episodes. Questions that we can address through resolved stellar populations include:

- What are the properties of Population III members and how did very early chemical evolution proceed?
- Do galaxies show a common epoch of ancient star formation prior to the epoch of re-ionization, or is there evidence of delayed formation, or even purely "young" galaxies?

2.1. Wide-Field, UV/Blue-Sensitive Surveys for Metal-Poor Stars

While a number of very metal-poor stars have been identified in the Milky Way, no information about potential Population III members is available in other galaxies. A systematic survey for these elusive objects may be feasible through a combination of high resolution, high sensitivity, large fields of view, and broad-band and narrow-band photometry (e.g., Norris, Bessell, & Pickles 1985). By concentrating on prominent blue absorption lines such as Ca II H and K (see Beers, Preston, & Shectman 1985), metal-deficient giants in the outer regions of galaxies within and beyond the Local Group can be identified. This would be a first step towards uncovering the very early chemical enrichment history and to explore the G-dwarf problem in galaxies other than the Milky Way.

2.2. High-Resolution, High-Sensitivity Turn-Off Photometry

In a regime where individual stars are too faint to be studied spectroscopically, photometry offers an accurate tool for the age-dating of sufficiently numerous, old populations: the designation of relative ages via main-sequence turn-offs and subgiant branches. It is likely that Population III stars, if they can be identified, will be too sparse to be dated in such a way. However, the oldest identifiable main-sequence turn-offs provide an excellent means to age-date the earliest substantial episodes of star formation. Interestingly, all Local Group galaxies where these oldest main-sequence turn-offs were resolved share a common epoch of ancient star formation with the Milky Way (see Grebel 2000 for a review). The only exception seems to be the Small Magellanic Cloud, where substantial star formation appears to have started 2 to 3 Gyr later. A number of caveats apply. For instance, the degree of differential age resolution that can be achieved for ancient populations is at best $\sim$ 1 Gyr. Also, comparisons with other galaxies make the implicit assumption that parameters such as the $[\alpha/\text{Fe}]$ ratio are the same as in the Milky Way, an assumption that requires further study (e.g., through spectroscopy with large ground-based telescopes).

In order to extend these kinds of studies to the outer, less crowded regions of more distant galaxies, very high resolution, very high sensitivity, and large-area coverage are needed. The oldest main-sequence turn-offs have an absolute magnitude of $M_V \sim 4$ mag. For an accurate determination of the turn-off location one would ideally aim for a limiting magnitude that is about two magnitudes fainter. At the distance of M31, this would require reaching $M_V \sim 30.5$ mag, which is feasible with HST/ACS and a large number of orbits. For nearby groups, more powerful telescopes are required. For instance, to age-date ancient populations in NGC 5128 (distance $\sim$ 3.8 Mpc), one would need to reach at least $M_V \sim 34.0$ mag. To reach the old main-sequence turn-off outer regions of Virgo cluster galaxies (mean distance $\sim$ 17 Mpc), one would need to resolve stellar populations at $M_V \sim 39.0$ mag.

Blue-optical photometry is preferable since it is optimized for the color range expected in the turn-off and subgiant region and affords the greatest age resolution. A database of the ages of the earliest episodes of significant star formation in nearby galaxies would establish whether there are significant differences from galaxy to galaxy, ultimately imposing constraints on cosmological models and the ubiquity and importance of star formation prior to re-ionization.

2.3. Wide-Field UV Surveys for Horizontal Branch Stars

Horizontal branch stars are unmistakable tracers of old populations since they only occur at ages of $\sim$ 10 Gyr and beyond. In comparison to old main-sequence turn-offs they have the disadvantage of not allowing us to limit their age to a more quantitative statement than "presence of old populations." But they have the advantage of having V-band magnitudes that are 3.5 mag brighter than those of old main-sequence turn-offs. Hence they can be more easily detected in distant but still resolvable galaxies. In the near-ultraviolet wavelength range, the horizontal branch turns into a steep and narrow diagonal line, whose most luminous part is occupied with blue horizontal branch stars. In near-UV color-magnitude diagrams (e.g., using the F225W and F336W filters), horizontal branch stars in old populations are significantly more luminous than red giant branch stars due to their high UV luminosity (see, e.g., Figure 2 in Ferraro et al. 1999).

Ultraviolet observations could be used for systematic searches for old populations in distant galaxies. For galaxies like I Zw 18 it would finally put the question to rest of whether there are genuinely young galaxies that only recently began to form stars (see Östlin 2000 versus Izotov et al. 2001). Differences in the relative numbers of red to blue horizontal branch stars in the field populations of galaxies of different properties and types may provide additional input for solving the second-parameter problem. On larger scales, knowledge of the number and distribution of horizontal branch stars and intermediate-age, UV-luminous post-AGB stars will bear directly on a better understanding of the UV upturn in elliptical galaxies and others. Finally, repeated observations may be useful in monitoring RR Lyrae for distance determinations.

3. Extragalactic Star Clusters and the IMF

As luminous single-burst populations, star clusters are unique and long-lasting chronometers of major star formation episodes. The preferred formation epoch of globular clusters appears to have been at the earliest times of galaxy formation, while open clusters and OB associations are forming (and dispersing) in gas-rich galaxies at all times. Loose clusters and associations are major contributors to the field star populations in galaxies and may indeed represent the dominant origin of field populations. Star formation in massive clusters and associations can have a significant impact on the interstellar medium and may lead to blow-out of much of the star-forming gas in low-mass galaxies. Census and the structural parameters of star clusters as a function of age provide information on the lifetimes of clusters with a range of intrinsic properties and in different galactic environments.

3.1. Blue/UV High-Resolution Studies of Young Clusters

Integrated photometry and spectroscopy of star clusters may be used for age-dating, although these methods are prone to degeneracies for higher ages (beyond ~ 300 Myr, see Figures 10 and 11 in Girardi et al. 1995) unless one uses spectra that extend well into the UV. Higher accuracies can be reached if the clusters can be resolved into individual stars for age-dating via color-magnitude diagrams. Depending on crowding and resolution, this is an easy task for young clusters but an increasing challenge for older clusters.

In young clusters, which are dominated by the energy output from massive blue main-sequence stars and evolved stars, observations at blue and ultraviolet wavelengths including the Balmer discontinuity are particularly useful. At these wavelengths the strongest spectral changes as a function of age are observed in young populations. Moreover, numerous Balmer and metallic absorption lines, nebular and stellar emission lines are included (e.g., Santos et al. 1995). "Young" in this context refers to clusters with ages less than ~ 100 Myr.

Very young (< 10 Myr) clusters hold the clue to massive star formation. In the absence of spatially resolved spectroscopy, photometry in the near UV allows us to place massive stars on evolutionary tracks and to constrain their masses and ages. This enables us to study the properties of the initial mass function (IMF) for massive stars in compact starburst clusters and loose OB associations in a variety of environments, making it possible to address questions such as the universality of the IMF slope, primordial or evolutionary mass segregation, and the nature of possible upper mass limits. These issues provide vital input for theories of massive star formation, stellar evolution, and population synthesis models. Optical photometry alone such as Johnson *UBV* photometry is degenerate for very hot and massive stars (e.g., Massey et al. 1995) as their peak energy output is in the UV. Hill et al. (1995) emphasize the excellent temperature discrimination in the UV. The $(B - V)$ color difference between a B0 and an A0 star is only 0.3 mag, while it is ~ 4 mag when using (160 nm $-V$).

UV observations of young star clusters that are no longer embedded in their natal clouds or otherwise highly obscured provide an important complement to high-resolution near-infrared studies (e.g., with NGST). In turn, NGST will be of paramount importance in the study of very young, highly attenuated clusters and of intermediate-age clusters dominated by asymptotic giant branch (AGB) populations, which are best age-dated through infrared observations.

3.2. Optical/UV High-Resolution Studies of Old Clusters

HST allows us to age-date intermediate-age star clusters with ages of a few Gyr in Local Group galaxies (e.g., Wyder, Hodge, & Zucker 2000) through their resolved stellar populations. With sufficiently high resolution, this should be feasible for old globular clusters in the Local Group as well. Crowding rather than sensitivity would limit these studies. Again, NGST will play an important role here, but for old globulars UV-optical studies have the advantage of providing the largest age resolution along the main-sequence turn-off and subgiant branch, and higher diffraction-limited resolution in the UV than at infrared wavelengths at the same mirror size. Where main-sequence turn-offs cannot be reached, UV studies will be important in order to detect the presence or absence of a horizontal branch as tracer of clusters with ages > 10 Gyr. The potential of UV studies for the detection of blue stragglers, white dwarfs of different ages (e.g., Ferraro et al. 2001), etc. will be covered in Mike Rich's contribution.

4. Spatially Resolved Star Formation Histories of Galaxies

High-resolution, large-scale coverage of nearby galaxies offers the opportunity to study their entire star formation histories as a function of position, revealing

- spatial variations in star formation history,
- population gradients and metallicity distribution/spread, and
- formation and merger histories of galaxy halos.

For old populations, blue-optical studies involving large color baselines such as $U-R$ in the Johnson-Cousins system afford the highest age resolution and the highest metallicity resolution along the main sequence turn-off, subgiant branch, and red giant branch. This resolution exceeds that of the coverage attainable with a color index such as $R-K$ that would be provided by NGST.

For young populations with luminous OB stars, the UV-blue spectral range is optimal as well as discussed earlier.

4.1. Spatial Variations in Star Formation History

Judging from the distribution of integrated colors, most galaxies show variations in their star formation histories and distribution of stellar populations. Old populations tend to be ubiquitous and to extend to large galactocentric radii, whereas younger populations often are more centrally concentrated or confined to disks and spiral arms. But little is known about how star formation proceeded in detail. Wide-field mapping of nearby galaxies with high resolution to obtain deep color-magnitude diagrams would address this question. Spatial variations or even radial population gradients (e.g., Harbeck et al. 2001) could be explored in detail. On time scales where dynamical mixing can be neglected these variations provide information on whether star formation propagated, and where star-forming material was concentrated at different times.

The distribution of special types of stars such as OB and red supergiants (e.g., for the blue-to-red supergiant ratio, Langer & Maeder 1995), Wolf-Rayet stars, or variable stars like Cepheids and RR Lyrae stars could be traced across entire galaxies. Variables would be excellent targets for dedicated distance scale projects and may be useful in resolving the differences between standard candles. The choice of blue colors has the advantage that amplitude variations are greater here than in the infrared. High resolution is crucial to avoid blends, which lead to underestimates of the actual distance (Mochejska et al. 2000).

4.2. Formation of Galaxy Halos

The choice of metallicity-sensitive near UV/blue photometry allows one to use old red giant branch stars for metallicity estimates and derivations of metallicity distribution functions dependent on position (Cole, Smecker-Hane, & Gallagher 2000). This yields important clues of how galaxy halos formed and evolved.

For instance, using optical filters, Harris & Harris (2001 and references therein) studied the halo of NGC 5128 with three WFPC2 pencil beams at 8, 21, and 31 kpc from the center of this giant elliptical galaxy. These pointings cover a tiny fraction of NGC 5128, which extends over more than one degree on the sky. Harris & Harris found a highly homogeneous, metal-rich halo with little evidence of mergers with metal-poor satellite galaxies, which may imply that the halo formed early on through accretion of gas-rich objects and experienced the bulk of its star formation thereafter. This is an interesting result when considering that NGC 5128 experienced a major merger in its past and still exhibits remnant stellar filaments from this event in its halo.

In other galaxies, there is observational evidence for dwarf satellite mergers. For instance, the tidal streams of the Sagittarius dwarf spheroidal galaxy, which is currently merging with the Milky Way (Ibata et al. 1994) wrap around our Galaxy. Ibata et al. (2001) recently found a similarly substantial, metal-rich tidal stream in the halo of M31 through wide-field observations. High-resolution studies of the halos of more distant galaxies are likely to uncover additional streams and constrain the number of such merger events, which in turn constrains hierarchical structure formation models (e.g., Bullock, Kravtsov, & Weinberg 2000). Surveys of this kind also allow us to search for signs of ongoing tidal disruption of satellite galaxies near massive galaxies (e.g., Odenkirchen et al. 2001a).

Not only dwarf companion accretion contributes to the build-up of the halos of more massive galaxies — the disruption of globular clusters does as well, as we recently demonstrated for the Milky Way (Odenkirchen et al. 2001b). Such events should be traceable in nearby galaxies, especially when the photometry reaches the old main-sequence turn-off. Tidal streams mark the orbits of globular clusters (or dwarf satellites) in advanced states of dissolution and thus provide important probes of the underlying galactic potential.

5. Distance and 3-D Structure of the Local Supercluster

To measure recent star formation histories and distances of galaxies within ~ 5 Mpc, we have carried out two HST/WFPC2 snapshot surveys with a total of 200 allocated orbits. These data allow us to quantitatively measure recent star formation events up to several 100 Myr as traced by main sequence, blue loop stars, and red supergiants. We can also evaluate intermediate-age star formation events up to 10 Gyr based on luminous AGB stars. The slope of the red giant branch can be used as a first indicator for the mean metallicity of the old populations. Distances can be determined via the I-band magnitude of the tip of the red giant branch (TRGB), allowing us to measure the three-dimensional location of galaxies within galaxy groups and hence to study the impact of environment on their evolution (see Grebel et al. 2000 for a description of the project goals). The addition of kinematic data enables us to derive dynamical mass estimates for the massive primaries in galaxy groups, the mass of the group as a whole and its zero-velocity surface, and ultimately to measure the local Hubble flow and deviations thereof (Karachentsev et al. 2002a,b,c).

With larger apertures, these studies can be extended to greater distances. To derive recent star formation histories as described above, UV/optical high-resolution imaging will ensure good age segregation. For TRGB distances,

NGST will do a superb job, although a space telescope optimized for optical wavelengths could deliver these distances for free as part of a multi-wavelength program concentrating on the stellar populations in the Local Supercluster. M_I of the TRGB is -4.05 mag. For a robust detection of the TRGB, the limiting magnitude should be at least two magnitudes fainter, putting distances of 40 Mpc ($I \sim 31.0$ mag required for a distance determination) and more within easy reach, provided that the measurements concentrate on uncrowded halos. Measurements of Cepheids may be possible in younger (and more crowded) regions.

6. Final Remarks

For the efficient study of galaxy evolution through resolved stellar populations a telescope/instrument combination should provide the following:

- High sensitivity/low background
- High angular resolution
- Large area coverage

Large-aperture space telescopes with suitable detectors can provide all of the above. Ground-based telescopes with large collecting area and adaptive optics remain limited in sensitivity by the thermal background and provide only small fields of view. Present-day adaptive optics is furthermore limited by low Strehl ratios and the requirement of natural guide stars, but this will likely improve. On the other hand, a 30 m ground-based telescope with spectroscopic capabilities will perfectly complement an 8 m space telescope with excellent imaging capabilities. While NGST will undoubtedly revolutionize infrared astronomy in the coming decade and make major contribution to high-resolution stellar population studies at red/infrared wavelengths, there are a number of important applications that can only be provided by a space telescope optimized in the optical and UV. Moreover, space UV observations have very low background ($\mu_{2000\text{Å}} = 26$ mag arcsec^{-2}; O'Connell 1987), and provide 10 times the diffraction-limited resolution at 2000Å than at 2μm with the same mirror size.

In the preceding sections I listed a number of science cases that I consider of primary interest in resolved stellar population studies in other galaxies, but not all of these require UV/optical wavelengths. In my view, a large-aperture, high-resolution, wide-field UV/optical space telescope would make unique contributions in particular in the following areas:

- Search for extremely metal-poor and Population III stars
- Age-dating of the earliest measurable star formation events and verification of the presence or absence of old populations through surveys for (blue) horizontal branch stars
- Identification and study of massive blue stars and AGB manqué stars in clusters and in the field; constraints on the origin of the UV upturn
- Determination of mass segregation time scales and high-mass IMF slopes
- Recent star formation histories from 1 to ~ 500 Myr out to Coma

Acknowledgments. I am indebted to Jay Gallagher for valuable discussions and suggestions. I gratefully acknowledge travel support from STScI/ NASA and lodging and subsistence support from ESA, without which my participation would not have been possible.

References

Beers, T.C., Preston, G.W., & Shectman, S.A. 1985, AJ, 90, 2089

Bullock, J.S., Kravtsov, A.V., & Weinberg, D.H. 2001, ApJ, 548, 33

Cole, A.A., Smecker-Hane, T.A., & Gallagher, J.S. 2000, AJ, 120, 1808

Ferraro, F.R., Paltrinieri, B., Rood, R.T., & Dorman, B. 1999, ApJ, 522, 983

Ferraro, F.R., D'Amico, N., Possenti, A., Mignani, R.P., & Paltrinieri, B. 2001, ApJ, 561, 337

Girardi, L., Chiosi, C., Bertelli, G., & Bressan, A. 1995, A&A, 298, 87

Grebel, E.K. 2000, 33rd ESLAB Symp. on Star Formation from the Small to Large Scale, eds. F. Favata, A. Kaas, & A. Wilson, (Noordwijk: ESA), 87

Grebel, E.K., et al. 2000, in ASP Conf. Ser. 221, CTIO/ESO/LCO Workshop on Stars, Gas, and Dust in Galaxies: Exploring the Links, eds. D. Alloin, K. Olsen, & G. Galaz, (Provo: ASP), 147

Harbeck, D., et al. 2001, AJ, 122, 3092

Harris, W.E., & Harris, G.L.H. 2001, AJ, 122, 3065

Hill, R.S., Cheng, K.-P., Bohlin, R.C., O'Connell, R.W., Roberts, M.S., Smith, A.M., & Stecher, T.P. 1995, ApJ, 446, 622

Ibata, R.A., Gilmore, G., & Irwin, M.J. 1994, Nature, 370, 194

Ibata, R., Irwin, M., Lewis, G., Ferguson, A.M.N., & Tanvir, N. 2001, Nature, 412, 49

Izotov, Y.I., Chaffee, F.H., Foltz, C.B., Thuan, T.X., Green, R.F., Papaderos, P., Fricke, K.J., & Guseva, N.G. 2001, ApJ, 560, 222

Karachentsev, I.D., et al. 2002a, A&A, 385, 21

Karachentsev, I.D., et al. 2002b, A&A, 383, 125

Karachentsev, I.D., et al. 2002c, A&A, 389, 812

Langer, N., & Maeder, A. 1995, A&A, 295, 685

Massey, P., Lang, C.C., Degioia-Eastwood, K., & Garmany, C.D. 1995, ApJ, 438, 188

Mochejska, B.J., Macri, L.M., Sasselov, D.D., & Stanek, K. 2000, AJ, 120, 810

Norris, J., Bessell, M.S., & Pickles, A.J. 1985, ApJS, 58, 463

O'Connell, R.W. 1987, AJ, 94, 876

Odenkirchen, M., et al. 2001a, AJ, 122, 2538

Odenkirchen, M., et al. 2001b, ApJ, 548, L165

Östlin, G. 2000, ApJ, 535, L99

Santos, J.F.C., Bica, E., Claria, J.J., Piatti, A.E., Girardi, L.A., & Dottori, H. 1995, MNRAS, 276, 1155

Wyder, T.K., Hodge, P.W., & Zucker, D.B. 2000, PASP, 112, 1162

Hubble's Science Legacy: Future Optical-Ultraviolet Astronomy from Space
ASP Conference Series, Vol. 291, 2003
K.R. Sembach, J.C. Blades, G.D. Illingworth, R.C. Kennicutt, Jr. (eds.)

Stellar Population Challenges for the Next Decade

Henry C. Ferguson

Space Telescope Science Institute, 3700 San Martin Dr., Baltimore, MD 21218

Abstract. The high resolution and sensitivity of the Hubble Space Telescope (HST) have revolutionized the study of stellar populations, providing solid constraints on the age and star-formation history of the Milky Way and its neighbors. A successor to Hubble could bring about order of magnitude gains either by improving resolution, sensitivity, or field of view. We outline some scientific goals in stellar population studies that are unachievable by Hubble and discuss the capabilities desirable in its successor.

1. Introduction

The study of stellar populations in the Galaxy and in surrounding galaxies is driven by the desire to understand how galaxies and stars evolve and by the important constraints such studies place on greater issues of cosmology. Early in the twentieth century, observations of the colors and brightnesses of stars in globular clusters led to a general understanding of nuclear fusion and the origin of the elements. At the beginning of the twenty-first century, the same clusters provide the best lower limit to the age of the Universe, and thereby strong independent constraints on the cosmological model.

However, we now think we know (at least generally) how stars evolve, and have a pretty good handle on the geometry of the Universe. Our knowledge of the latter will improve dramatically within a few years with measurements of the redshift-distance relation, cosmic shear, local large-scale structure, the galaxy-cluster mass function, and the fluctuations in temperature of the cosmic microwave background. What, then, becomes of astrophysics that has traditionally provided insight into fundamental physics and more circumspect routes to the cosmological parameters? Do such fields become the province of historians rather than scientists?

There are perhaps two answers to such a question. The first is that better measurement of the cosmological parameters may not necessarily bring deeper understanding. The history of science tells us that breakthroughs in understanding often come from the most unexpected directions. Seemingly innocuous inconsistencies between theory and observation crop up, and are frequently either ignored or disputed until some new insight shows how the whole puzzle fits together. For example, had the inconsistencies between distance-scale and time-scale estimates for the age of the Universe (or between galaxy number counts and simple evolutionary models) been taken seriously 10 years ago, we might have settled on a cosmology with non-zero Λ somewhat earlier. Today we have several examples of issues that may be serious problems, or may be only minor issues for our understanding of cosmology. Inconsistencies between the density profiles of dark-matter halos in cold dark matter (CDM) N-body simulations and the observed density profiles in low-mass galaxies has some in search of exotic solutions (Spergel & Steinhardt 2000) and others re-examining the data (van den Bosch & Swaters 2001). Similarly, the large numbers of low-mass satellite galaxies that appear in N-body simulations appear at odds with estimates of the galaxy mass function (Moore et al. 1999). Understanding galaxy evolution will be the key to deciding whether such problems require revisions in our understanding of dark matter, so this is one area where studies of the properties of nearby galaxies may yield advances in fundamental physics. A second answer to the question of relevance of studies of stars and galaxies in an era of precision cosmology is that we still care about climate, plate-tectonics, and volcanism in an era where we know the diameter, mass, and overall composition of the Earth to better than a few percent. Just as surely as we would like to distill the Universe to its purest physical principles, so also would we like to understand the evolution and behavior of its contents.

The one certainty in today's astrophysical landscape is that the rate of increase in our knowledge (or at least data) is rapid. Today's questions will likely not be tomorrow's questions, and in point of fact, we have almost no idea what the key questions will be ten to fifteen years from now. However, in spite of that we can confidently predict that moving forward in the study of galaxies requires (1) improved angular resolution, (2) continued access to the optical-ultraviolet wavelength range after HST, and (3) a general-purpose facility. Indeed there are many current problems for which a factor of 10-100 in point-source sensitivity would yield breakthroughs.

2. HST Highlights

Before discussing future directions, it is worth pausing to reflect on some of HST's major accomplishments in the field of stellar populations. In the Galaxy one should cite fundamental revisions in our knowledge of the stellar mass function in globular clusters (e.g., Piotto, Cool, & King 1997; de Marchi & Paresce 1997), the bulge (Holtzman et al. 1998; Zoccali et al. 2000), in the disk and halo (e.g., Bahcall et al. 1994; Reid et al. 1996; Gould, Flynn, & Bahcall 1999), and in binary systems (Henry et al. 1999). HST has provided the most precise ages of globular clusters from main-sequence fitting (e.g., Stetson et al. 1999; Piotto et al. 1999) and from the white-dwarf cooling sequence (Hansen et al. 2002). HST also provided some short-lived excitement about the possibility of a dynamically significant population of halo white dwarfs (Ibata et al. 1999; Richer 2001). HST is beginning to provide proper motions for globular clusters (Bedin et al. 2001) and bulge stars (Kuijken & Rich 2002), which will ultimately help to disentangle the structure of the Galaxy. HST has revealed complex star-formation histories for its dwarf-galaxy satellites — which may be related to the missing satellite problem in ways we have yet to understand. HST has resolved individual hot-horizontal branch stars in M32 (Brown et al. 2000), and set constraints on the star-formation history of the disk and halo of M31 (Rich et al.

1996; Ferguson & Johnson 2001; Williams 2002). HST ultraviolet observations of super-star clusters in starburst galaxies have fundamentally revised concepts of how stars form in different environments (Holtzman et al. 1992; Meurer et al. 1995), and studies of star clusters in different types of galaxies (from very young to very old) have begun to yield insight into the formation of early-type galaxies (Whitmore 2001; Kissler-Patig 2001). HST has resolved isolated free-floating stars in the Virgo Cluster (Ferguson, Tanvir, & von Hippel 1998), and has used surface-brightness fluctuations to measure distances to galaxies far enough away that peculiar velocities are not a serious concern (Jensen et al. 2001). Over the next 10 years, with its new instrumentation, we can expect similarly spectacular results from HST.

3. Some Questions

In spite of the many advances in our understanding of stellar populations brought about by HST and other telescopes, a host of relatively simple questions about galaxies and their evolution have as yet no definitive answers. Where, for example, are the remnants of population III, the first population of stars that seeded the Universe with heavy elements? Which came first, galaxy disks or spheroids (halos and bulges)? The Milky Way bulge appears unambiguously older than the disk in the solar neighborhood, but was the bulge formed in a rapid collapse of gas, or in a merger of young disk galaxies at high redshift? If spheroids formed by merging, what are we to make of evidence for high α-element to iron ratios, which suggest rapid enrichment by Type II supernovae rather than slower enrichment by type Ia SNe? Over what span of time did globular clusters form in the Milky Way? Are there still older clusters in other galaxies? Were galaxy halos built by rapid collapse, merging, or accretion? If all three, when and in what proportions? What role do black holes play in galaxy formation? Does feedback from active nuclei have any role in regulating star formation in individual galaxies or in the universe as a whole (e.g., by elevating the metagalactic background)? Was the formation of dwarf galaxies suppressed or delayed as seemingly required by CDM models? How and until when? Studies of resolved stars in nearby galaxies are an essential engine of progress in addressing such questions.

3.1. Discovery Space and Some Examples

In spite of its paltry 2.5 m aperture and relatively inefficient optics and detectors, HST's point-source sensitivity in the optical and near infrared exceeds that of the largest ground-based telescopes in the best of conditions. Many of the advances brought about by HST can be attributed to this great sensitivity. However, even with this great sensitivity it requires major investments of observing time to reach the bottom of the white-dwarf cooling curve in the nearest globular cluster (Hansen et al. 2002), to reach the main-sequence turnoff in M31 (about to be done with the Advanced Camera for Surveys), and to reach far enough down the giant branch in the Virgo Cluster to begin to constrain metallicities of old stellar populations (feasible but not yet in the observing plan). Figure 1 shows sensitivities of an 8 m "Next HST" optical-ultraviolet telescope (NHST), compared to HST, the California Extremely Large Telescope (CELT),

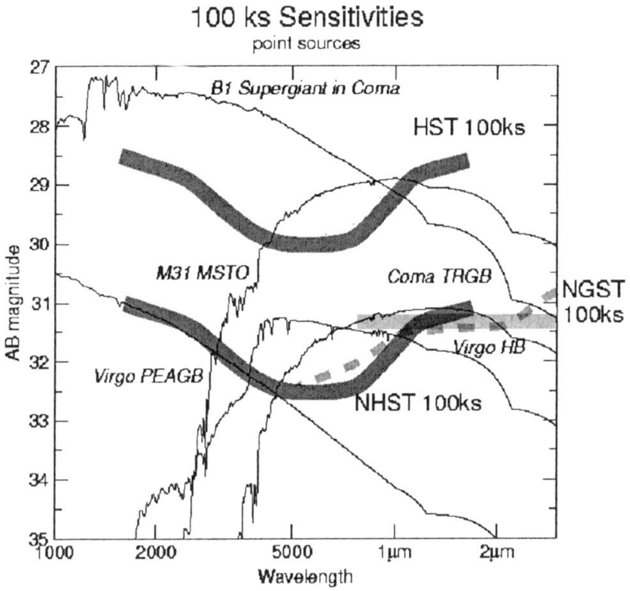

Figure 1. Sensitivity metrics. Thin curves show spectra for a variety of different kinds of stars at different distances. Thick curves show point-source sensitivities for a 100 ks exposure with HST, an 8 m NHST (heavy curves), NGST (light curve), and CELT (dashed). NHST sensitivities are scaled from HST assuming no advances in detector technology or optical throughput. Even with this relatively modest assumption and 8 m NHST makes accessible individual horizontal branch and PAGB stars at the distance of the Virgo Cluster (16 Mpc) and individual red-giant-branch stars at the distance of the Coma Cluster.

and the Next Generation Space Telescope (NGST). The "holy grail" of reaching the main-sequence turnoff in the Virgo cluster is well beyond any of these telescopes. However, the horizontal branch (HB) in the Virgo Cluster is within reach. Directly resolving HB stars in normal ellipticals would provide great insight into the ages, metallicity distribution, and even helium abundance.

Equipped with the same instruments, NHST would gain an order of magnitude in sensitivity relative to HST. In the near infrared it would have competition from both CELT and NGST, but shortward of about 0.6 μm it would be unique.

For many applications, the gain in discovery space brought about by increasing solid angle coverage is as large as the gain brought about by increasing sensitivity. One example of relevance today is the attempt to understand the formation mechanisms of spiral galaxy halos. Recent studies have uncovered disrupted dwarf galaxies (Ibata, Gilmore, & Irwin 1994; Newberg et al. 2002) and tidal streams (Ibata et al. 2001) in the halo of the Milky Way and M31. Such findings have led to a general re-examination of our preconceptions about

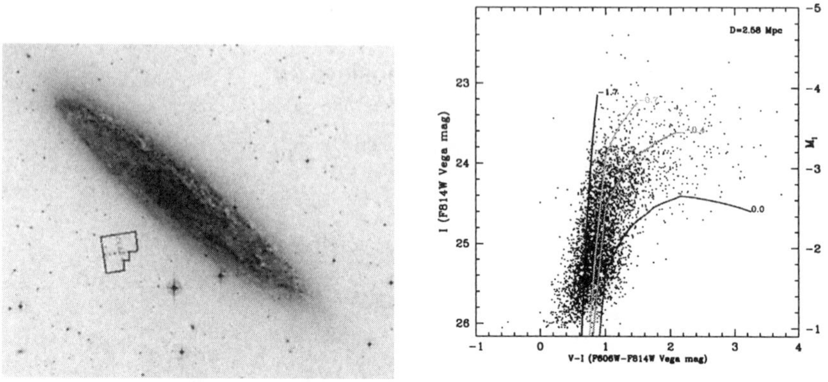

Figure 2. HST study of a spiral galaxy halo. The left panel shows the field subtended by the WFPC2 camera. The right panel shows a preliminary color-magnitude diagram; theoretical isochrones of different metallicities are shown as solid lines. The galaxy evidently has a large spread of metallicities and the position of the tip of the red giant branch (roughly 0.7 mag fainter than the tip of the theoretical isochrones) suggests that the galaxy is further away than typically assumed in the literature (Ferguson 2002).

halo formation, and to speculation that halos might simply be the remnants of accreted dwarf galaxies. Indeed such accretion might help solve the CDM missing satellite problem. However because of the low surface brightnesses of galaxy stellar halos, we don't even know the answers to more basic questions: Do spiral galaxies, in general, have stellar halos? Are they generally metal poor like the Milky Way and M33, or are many of them metal rich like M31's halo? Why are some metal rich and some metal poor? Do their metallicity distributions match a simple model of chemical evolution? Does the mass of the stellar halo scale with the mass of the bulge or the mass of the dark-matter halo? With point-source sensitivity down to $I = 30$ over a wide field of view, we can use the colors and spatial distributions of red-giant-branch (RGB) stars to build a comprehensive picture of halo formation. Models by Johnston et al. show that tidal streamers can exist for more than 1 Gyr, but will typically have surface brightnesses fainter than than $\mu_R = 29$ per square arcsecond. These low surface-brightness features are nearly impossible to detect via surface photometry but are easily detectable with counts of RGB stars. However the predicted density of such streamers for a CDM model is low, about 0.5 per Milky-Way-like galaxy. While HST has the sensitivity to begin to address some of these questions (Figure 2), only a tiny fraction the halo is subtended by the HST field of view for galaxies that are close enough for detailed study. A true attack on the problem will require a large field of view and sensitivity as good or better than HST.

In Figure 3 we show the limiting magnitude versus area surveyed for a 100 ks survey on different facilities (HST, the Supernova Acceleration Probe – SNAP, CELT, and an 8 m NHST). The 100 ks can be divided into a wide-angle shallow survey, or put into a single deep survey covering just the solid angle of the detector. Survey requirements are shown for various illustrative tracers of spiral

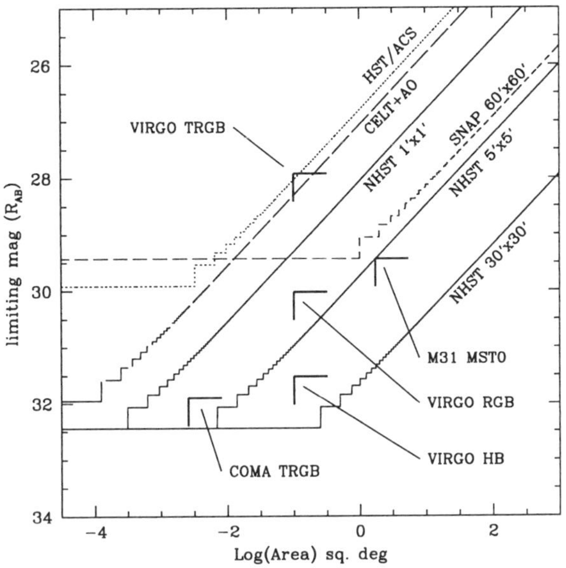

Figure 3. Go wide or go deep? The different lines show the depth and solid angle that could be covered in 100 ks with HST, CELT, SNAP, and NHST with three different FOVs: $1' \times 1', 5' \times 5'$, and $30' \times 30'$. Requirements are shown for five different example surveys. In the Virgo and Coma Clusters we show the area and depth required to survey the entire area within 50 kpc of a single galaxy and measure TRGB, RGB stars, or HB stars ($M_R \approx -3.1, -1.0$, and $+0.5$, respectively). We also show the requirements to survey the entire area within 10 kpc of M31 down to down to $M_R = 5$. While NHST with even a $1' \times 1'$ FOV would provide important gains over HST and CELT, such a small field of view would limit the scope of stellar-population surveys that could be undertaken. NHST becomes unique in this parameter space for a FOV greater than about 25 square arcminutes.

galaxy halos. For example, to measure tidal streamers one would like to survey the entire volume within 50 kpc of a spiral galaxy. Therefore, for Virgo and and Coma Cluster distances we show the area subtended by a single such survey. (To build statistical samples it will be necessary to survey dozens of galaxies.) For deep surveys over very small fields, CELT would be competitive with NHST. For shallower surveys ($R \lesssim 30$), SNAP would be more efficient than NHST unless the NHST field of view (FOV) is greater than about 25 square arcminutes.

Another science example that pushes beyond the limits of HST is the desire to understand the initial mass function. Studies of the IMF in star-forming regions of the Galaxy with HST, SIRTF, and NGST will no doubt bring about great advances in our understanding of star-formation in these environments. However, these environments may or may not be representative of those found in other kinds of galaxies or in the early Universe. There is a variety of indirect arguments for a top-heavy IMF in high-redshift galaxies (Ferguson, Dickinson, & Papovich 2002; Malhotra & Rhoads 2002), and there is evidence for top-heavy initial mass functions in nearby starbursts (Rieke et al. 1993). NHST would be able to resolve stars below 1 $M_{\odot}$ in nearby starburst galaxies such as M32, NGC1705, NGC5253, and would reach $2M_{\odot}$ in I Zw 18. Measurement of both the IMF and the star-formation histories in these starburst systems would provide insight into star formation processes in environments similar to those in high-redshift galaxies — and very different from those in the disk of the Galaxy. This, of course, will tell us much about star formation *today* but what about more direct tests of the IMF at early times? One way to probe this in the fossil record is to search for the remnants of high-mass stars and compare their number densities to the old low-mass stars still living today in the bulges and halos of galaxies. Microlensing may be the only way to do this. For example, a comprehensive microlensing survey of M31 would reveal or set limits on long time scale (100+ day) microlensing events due to neutron stars or black holes. A pixel-microlensing survey near the very center of M31 would be sensitive to putative massive clusters of black holes that ought to form via dynamical processes around the central black hole (Miralda-Escudé & Gould 2000). The observational demands of microlensing searches are not out of the question for a general purpose facility, provided the field of view is large and queue scheduling is efficient. Complementary studies of variable stars could be carried out in tandem and a deep stellar population survey built up over time.

A spectacular program could be constructed taking only a few minutes every few days for several years.

3.2. Telescope Wish List

In summary, there are many problems beyond the reach of HST that cry out for a capable successor at optical-ultraviolet wavelengths. We ought not be shy in our scientific goals or in the implications for the telescope needed to accomplish them. While we cannot be certain what scientific questions will be the most interesting ten or fifteen years from now, we can confidently predict that a capable successor to HST would yield important breakthroughs. A combination of improved sensitivity and a larger field of view would enable systematic studies of resolved stellar populations in ways that are simply not possible with HST. With HST we can study tiny fractions of Local Group galaxies. A capable successor would bring large enough volumes within reach to find representative samples of different types of galaxies (dwarfs, large spirals and ellipticals, starbursts), and allow measurements of spatial distributions, mass-functions, ages, and chemical compositions of a wide range of stellar populations. A worthy successor to HST for stellar populations ought to have (1) at least 10× HST sensitivity at wavelengths 120 – 1000 nm (2) a large field of view (> 100 square arcminutes), and (3) spatial resolution at least twice as good as HST.

References

Bahcall, J.N., et al. 1994, ApJ, 435, L51
Bedin, L.R., et al. 2001, ApJ, 560, L75
Brown, T.M., et al. 2000, ApJ, 532, 308
de Marchi G., & Paresce, F. 1997, ApJ, 476, L19

Ferguson, H.C., Dickinson, M., & Papovich, C., 2002, ApJ, 569, L65
Ferguson, H.C., et al. 2002, in preparation
Ferguson, H.C., Tanvir, N.R., & von Hippel, T. 1998, Nature, 391, 461
Ferguson, A.M.N., & Johnson, R.A. 2001, ApJ, 559, L13
Gould A., Flynn C., & Bahcall, J.N. 1998, ApJ, 503, 798
Hansen, B.M.S. 2002, astro-ph/0205087
Henry, T.J., et al. 1999, ApJ, 512, 864
Holtzman, J.A., et al. 1992, AJ, 103, 691
Holtzman, J.A., et al. 1998, AJ, 115, 1946
Ibata, R., et al. 1999, ApJ, 524, L95
Ibata, R., Gilmore, G., & Irwin, M.J. 1994, Nature, 370, 194
Ibata, R., et al. 2001, Nature, 412, 49
Jensen, J.B. 2001, ApJ, 550, 503
Kissler-Patig, M. 2001, in IAU Symp. 207, Extragalactic Star Clusters, eds. E.K. Grebel, D. Geisler, & D. Minniti, in press
Kuijken, K., & Rich, R.M. 2002, astro-ph/0207116
Malhotra, S., & Rhoads, J.E. 2002, ApJ, 565, L71
Meurer, G.R., et al. 1995, AJ, 110, 2665
Miralda-Escudé, J., & Gould, A. 2000, ApJ, 545, 847
Moore, B., et al. 1999, ApJ, 524, L19
Newberg H.J., et al. 2002, ApJ, 569, 245
Piotto, G., Cool, A., & King I.R. 1997, AJ, 113, 1345
Piotto, G., et al. 1999, AJ, 117, 264
Reid, I.N., et al. 1996, AJ, 112, 1472
Rieke, G.H., et al. 1993, ApJ, 412, 99
Rich, R.M., et al. 1996, AJ, 111, 768
Richer, H. 2001, astro-ph/0107079
Spergel, D.N., & Steinhardt, P. 2000, Phys.Rev.Lett, 84, 3760
Stetson, P.B., et al. 1999, AJ, 117, 247
van den Bosch, F.C., & Swaters, R.A. 2001, MNRAS, 325, 1017
Whitmore, B.C. 2001, in Starburst Galaxies: Near and Far, eds. L. Tacconi & D. Lutz, (Berlin: Springer), 106
Williams, B.E. 2002, MNRAS, 331, 293
Zoccali, M., et al. 2000, ApJ, 530, 418

Hubble's Science Legacy: Future Optical-Ultraviolet Astronomy from Space
ASP Conference Series, Vol. 291, 2003
K.R. Sembach, J.C. Blades, G.D. Illingworth, R.C. Kennicutt, Jr. (eds.)

Ultraviolet Spectroscopy and the Intergalactic Medium

Jill Bechtold

Steward Observatory, University of Arizona, Tucson, AZ 85721

Abstract.
The next generation optical/ultraviolet telescope in space will be indispensable for providing a clear picture of the central Origins theme: how were galaxies and clusters assembled out of the intergalactic gas? I summarize the contributions expected from spectroscopy of quasars, and discuss the technical feasibility of building a very high resolution ($R > 500,000$) spectrograph.

1. Introduction

Ultraviolet spectroscopy of quasar absorption lines has proven to be an indispensable tool for studying the intergalactic medium (IGM) and the interstellar medium (ISM) of galaxies. By combining absorption line spectroscopy with optical imaging and multi-object spectroscopy, it will be possible to have the story of the formation of mature galaxies like the Milky Way, and the "cosmic web" of galaxies and clusters.

The advantage of the absorption line spectroscopy of the ultraviolet (UV) transitions are obvious, but bear articulation. Ultraviolet absorption lines measure the most common elements directly, with observables which are readily converted to physical quantities in a model-free way. The cosmologically important hydrogen, deuterium, and helium have their atomic transitions in the ultraviolet, and the abundances of the main products of stellar fusion – carbon, nitrogen, oxygen, iron, and the other key elements for the origin of life – are readily observed. Abundances of many elements and ions allow us to trace the nucleosynthetic history of galaxies and the depletion history of gas. In a single spectrum, one can measure densities, column densities, and temperatures over a vast range (typically T $\sim 10^2 - 10^5$ K), and transitions of both atomic and molecular species. The most abundant molecule in galaxies, H_2, can be studied directly, in contrast to spectra in the millimeter and sub-millimeter, where trace molecules like CO are the only probes. The ultraviolet transitions are redshifted into the ground-based optical for redshifts $z > 2$, but for most of the age of the Universe, and for the epochs in which most of the star-formation and quasar activity occurred, observations from space are necessary.

Recent reviews of quasar absorption lines are given by Bechtold (2002) and Pettini (2003). Other discussions of the potential science gains from a large, ultraviolet-optimized telescope in space are Bechtold (1999), Tytler (1999), and Martin (1999).

In Figure 1, I show an example of the current state-of-the-art, a quasar absorption line spectrum, obtained with the cross-dispersed echelle of the Space Telescope Imaging Spectrograph (STIS), aboard the Hubble Space Telescope (HST). This spectrum is part of the survey of quasar absorption lines at $z \sim 1$ led by Buell Jannuzi. We observed the extremely bright $V = 14.9$ quasar PG 1634+706, which fortuitously is in the continuous viewing zone of HST.

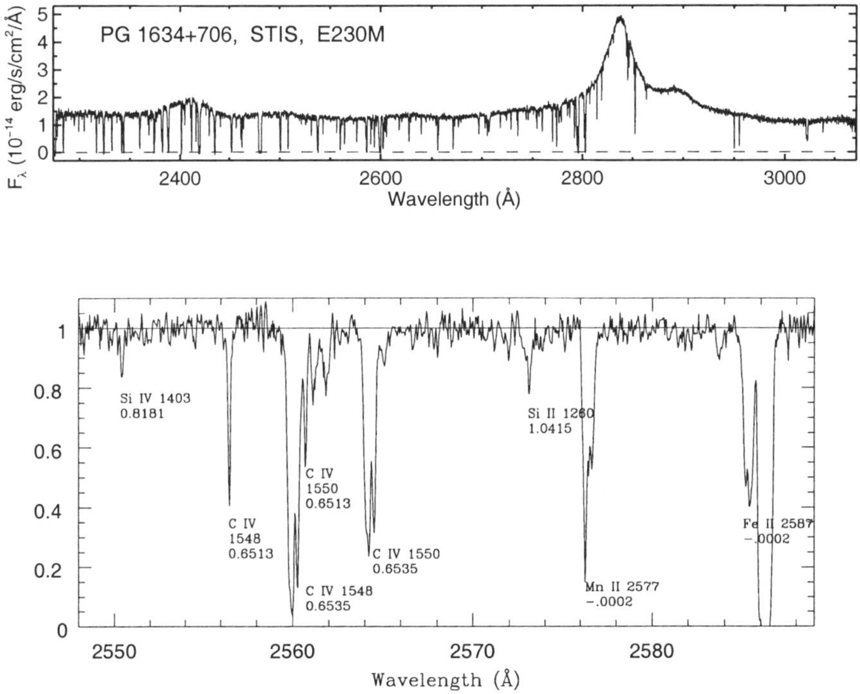

Figure 1. *Top*: HST/STIS spectrum of $z = 1.334$ quasar PG 1634+7061 taken with the E230M echelle grating at 12 km s^{-1} resolution. *Bottom*: Normalized portion of top spectrum, with Voigt profile fits overplotted. Lines are labeled with species, rest wavelength in Å, and redshift. From Jannuzi et al. (2003).

Unfortunately, data of this quality are difficult to obtain with HST. Figure 2 shows a simulated spectrum for a more typical $V = 16.5$ quasar. In 30 orbits, the signal-to-noise, even with the expected improved throughput of the Cosmic Origins Spectrograph (COS), is marginally sufficient.

Since the quasar number density is such a steep function of apparent magnitude, modest gains in sensitivity result in large gains in the number of objects feasible to observe. Now, only a very small number of quasars are bright enough to be targets with HST; tens are needed to assure that cosmic variance is adequately sampled. Moreover, some of the most important observations are of rare objects: fortuitously placed groups of quasars to probe the three-dimensional

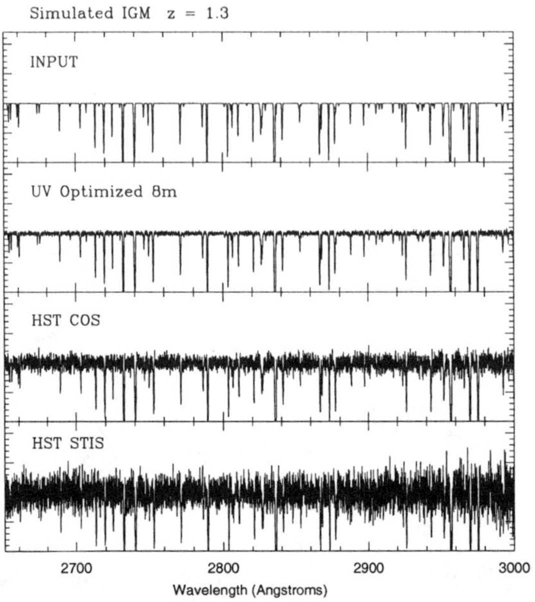

Figure 2. Simulated spectra of the Lyα forest at $z \sim 1.3$, for an $m = 16.5$ quasar, assuming the exposure time obtained with 30 HST orbits. Panels (from top to bottom) show the spectrum with no noise, the spectrum observed with an ultraviolet optimized 8 m telescope, with HST/COS (predicted sensitivity and resolution), and with HST/STIS. The higher throughput of COS compared to STIS is apparent, but only the 8 m spectrum recovers all the input lines.

structure of the IGM; $z > 3$ quasars with no intervening Lyman limit systems to observe He II $\lambda 304$ in the rest frame of the quasar; active nuclei behind nearby galaxies to probe galactic scale winds; quasars reddened by a foreground galaxy to probe fully molecular interstellar gas. The 2 m-class HST can give only tantalizing clues of what could be found by these programs.

2. Science Drivers for Quasar Absorption Line Spectroscopy in Space

Below I describe the main science topics addressed by high resolution UV spectroscopy of quasar absorption lines in the extreme-UV. More detailed discussions of these and other drivers can be found in Shull (1999) and Shull et al. (1999).

(1) What is the cosmological density of baryons, Ω(Baryon)?

Quasar spectra can measure the ratio of deuterium to hydrogen, D/H, in intervening gas in galaxies along the line of sight, and thus test cosmological nucleosynthesis models (see Lemoine et al. 1999 for a recent review). Deuterium is created just after the Big Bang, then destroyed by astration as the Universe evolves. Following D/H as a function of look-back time can test our understanding of the origin of deuterium. Quasar studies of D/H have an advantage over

Galactic studies using stars since photospheric features do not complicate the spectra. The limiting factor for quasar observations is that intervening metal absorbers have a velocity structure comparable to the separation of deuterium Lyα from hydrogen Lyα, 81 km s^{-1}. Definitive measure of D/H therefore requires a statistical analysis of a large sample of relatively quiescent metal absorbers. It is not feasible to observe such a sample with HST.

The second important probe of Ω(Baryon) is the diffuse intergalactic gas, or Lyα forest, out of which galaxies form. Once the column density of H I is measured from spectra of the Lyman series, the key unknown is what ionization correction to apply to derive the total amount of gas. Theoretical expectations are that H I is a trace ion, that the correction factor is large, and that the ionization and metal abundance of the IGM probed by Lyα will vary with galaxy environment. The helium lines in the far ultraviolet are one way to measure ionization, but require the rare line of sight which is clear of intervening Lyman limit systems to rest frame He I λ584 and He II λ304. Data for four objects were obtained with HST, HUT, and FUSE observations that can be described as heroic (Jakobsen et al. 1994; Tytler et al. 1995; Davidson et al. 1996; Reimers et al. 1997; Heap et al. 2000; Kriss et al. 2001; Smette et al. 2002). Observations of highly ionized metals are also of great interest, with most ions having their strongest transitions in the far ultraviolet (see Telfer et al. 2002 for a recent discussion). At $z > 3$, observable from the ground, the 10 m Keck with HIRES was necessary for detections of the very weak C IV and Si IV from Lyα clouds (Tytler et al. 1995; Cowie et al. 1995; Songaila & Cowie 1996), and similar aperture is required for lower redshift systems observable only from space. Extreme-UV observations can be very powerful: for example, oxygen has strong transitions of O VI $\lambda\lambda$1032, 1038, O V λ630, O IV λ788, O IV λ554, O III λ833, O I λ1302, allowing a secure estimate of [O/H] directly from observation. Current estimates, based on very sparse data, are that the IGM probed with the ultraviolet lines constitutes some 90% of the baryons in the Universe (Tripp, Savage, & Jenkins 2000).

(2) How do galaxies and clusters of galaxies form out of the primordial gas?

Observations from space are necessary to measure Lyα for all redshifts less than $z = 1.6$, and the UV transitions of H_2 for all redshifts less than $z = 2.0$. These are sensitive probes of the important constituents of the IGM and galactic interstellar gas at the crucial $z \sim 2$ epoch – the epoch when the quasar population peaked, when the global star-formation rate peaked, and when there was a strong change in the rate of evolution of the Lyα forest lines. The distribution of the gas and its temperature can be measured as a function of redshift. Large scale velocity structure within the gas, and the correlation of the gas with structure in the galaxy distribution, will elucidate whether winds, shocks, or photoionization are important. The three-dimensional structure of the IGM can be determined using adjacent quasar sight lines.

While it is fascinating to study objects at the highest observable redshifts, look-back studies to $z \sim 2$ are sufficient to probe most of the elapsed time in the Universe. Moreover, it has been since $z \sim 2$ that much of the most interesting evolution and development of our current Universe took place. Proto-galactic star-forming fragments merged to form mature galaxies, and somehow the Hubble sequence of galaxies was established. Clusters of galaxies and larger structures formed. The hot intracluster medium was collected in cluster potentials and heated. Gas studies are an essential part of understanding these processes.

(3) How are the heavy elements formed, and how are they dispersed into the intergalactic medium?

The enrichment of the interstellar gas by stellar winds and supernovae is a fairly well understood process, at least in our Galaxy. By contrast the mechanism by which enrichment of the intergalactic medium took place is not well known at present. Did winds from early starbursts carry heavy elements into the intergalactic gas? If so, did these winds also have an important effect on the process of galaxy formation? How did the patterns of metal enrichment seen in present-day galaxies originate? Look-back studies using metal-rich quasar absorbers have provided a rich source of information for $z > 2$ from the ground, but spectroscopy in the ultraviolet is required to probe the redshifts since then. At $z < 2$ it is easier to find the host galaxy and correlate absorber properties with morphology and galaxy environment in the important time period during which the Hubble sequence formed.

3. Instrument Parameters of a High Dispersion UV Spectrograph

There was considerable discussion at the conference on the issue of what spectral resolution would be necessary for realizing the science goals of intergalactic and interstellar work. The real-world trade, of course, is whether one goes for ever higher spectral resolution on bright objects, or more moderate spectral resolution on fainter objects, which are more numerous. Clearly, there will be new ground to explore with either strategy.

Here, I give some rough numbers to show how quantities scale with parameters of the telescope and spectrograph for astrophysical conditions anticipated to be of interest.

The reasons to push to high spectral resolution are the following.

- Resolve line widths, in order to fit profiles and derive temperatures and column densities.
- Resolve velocity structure in the absorption profile arising from what is typically referred to as "velocity components" or turbulent motion: i.e., broadening by the Doppler motions of physically distinct packets of gas, moving in the gravitational potential of the host galaxy or large scale galaxy distribution.
- Detect weak transitions of rare elements, in order to make measurements on the linear part of the curve of growth, and to probe diagnostics of nucleosynthesis and depletion.

Table 1 shows the spectral resolution required to resolve the line widths of several abundant elements as a function of temperature of the gas, for the range of temperatures expected. The line width is usually described in terms of a "b-value" given by

$$b = \left(\frac{2kT}{m}\right)^{1/2} = 0.129 \left(\frac{T}{A}\right)^{1/2} \text{ km s}^{-1} \tag{1}$$

where T = the temperature in degrees Kelvin, and A = the atomic weight of the element. The b-value is related to σ of a Gaussian by

$$b = \sqrt{2}\,\sigma. \tag{2}$$

The spectral resolution, R, meaning the full-width-half-maximum intensity of the instrumental profile, is expressed as

$$R = \frac{\lambda}{\Delta\lambda} \tag{3}$$

Table 1. Thermal Widths of Absorption Lines

Temperature		100 K (diffuse clouds)		6,000 K (WHIM)		100,000 K (Coronal, IGM)	
Element	A	b	R	b	R	b	R
Hydrogen	1	1.29	140,000	9.99	18,000	40.8	4400
Helium	4	0.645	280,000	5.00	36,000	20.4	8900
Carbon	12	0.372	485,000	2.88	62,000	11.8	15,000
Nitrogen	14	0.345	523,000	2.67	67,000	10.9	17,000
Oxygen	16	0.322	560,000	2.50	72,000	10.2	18,000
Iron	56	0.172	1,052,000	1.34	134,000	5.5	33,000

A = atomic weight, Doppler width b in km s^{-1}, R (FWHM) = λ / $\Delta\lambda$

I tabulate the spectral resolution R equal to the thermal width, for three representative temperatures: T = 100 K, typical of diffuse clouds; T = 6,000 K, typical of the warm-hot IGM (WHIM); and T = 100,000 K, typical of coronal gas. Ideally, one would like R to be a factor of two or so better than what is listed in Table 1.

For comparison, the HST/STIS echelle modes have R = 30,000 or 114,000 for $\lambda \sim 2300$ Å, and R = 45,800 or 114,000 for $\lambda \sim 1400$ Å. From Table 1, we see that the HST/STIS choices for spectral resolution are adequate for the WHIM and for coronal lines, but for the cool diffuse clouds, $R = 5 \times 10^5 - 1 \times 10^6$ would be necessary.

The argument is often made that we don't need a bigger space telescope for quasar absorption line observations since we can simply devote large numbers of orbits to observe quasars with HST. However, the exposure times to obtain $R = 10^5$ spectra of quasars with Hubble are prohibitively long.

How long does it take to get a signal-to-noise ratio of 50 for a $V = 17.5$ quasar, with the $R = 10^5$ echelles on HST/STIS (i.e., spectra similar to the kind obtained with the HIRES echelle on the 10 m Keck Telescope)? To estimate exposure times, I used the HST/STIS ETC tool on the STScI web site. I assume 50 minutes per HST orbit exposing on target, which is probably a bit of an overestimate. I calculate that to get the entire spectrum from 1150 to 3000 Å takes 275,000 orbits. Recalling that there are about 5500 orbits per year, observing one $V = 17.5$ quasar would take about 50 years. If you only want one grating setting with the redder echelle, it would take 15 years, with the bluer echelle only about 2 years. I think it is fair to conclude that observing a $V = 17.5$ quasar at $R = 10^5$ is beyond the reach of the 2 m HST, even with the factor of a few increase in sensitivity over HST/STIS that one could gain with newer instruments.

For ground-based 8 m-class telescopes, optical spectroscopy is generally limited to $R < 100,000$, but that is only because ground-based optical spectrographs are seeing-limited. For a 4 m to 8 m in space, with near-diffraction limited images in the UV, very small entrance slits could be used, and therefore $R \sim 5 \times 10^5 - 1 \times 10^6$ spectrographs could be built small enough to launch into space. The limiting technical factors would be (1) the minimum size of pixels in the detector, (2) the ability of the spacecraft to acquire targets and guide, and (3) the minimization of read-noise and sources of background in the detector.

To compute some illustrative numbers, I follow the notation of Schroeder (2000) to write the spectral resolution, R,

$$R\phi = 2\,\frac{d_1 \tan(\delta)}{D}, \tag{4}$$

where ϕ is the width of the slit on the sky, d_1 is the size of the spectrograph beam, δ is the blaze angle of the echelle, and D is the diameter of the telescope primary mirror. If we take ϕ for a diffraction-limited telescope at $\lambda = 2400$ Å, we get

$$\phi = 2.44\,\frac{\lambda}{D} = 0.015\,\left(\frac{\lambda}{2400\,\text{Å}}\right)\,\left(\frac{8\,\text{m}}{D}\right)\ \text{arcsec}, \tag{5}$$

and assuming a 10 inch beam and $\delta = 67°$, we get

$$R = 2 \times 10^6\,\left(\frac{d_1}{254\,\text{mm}}\right)\,\left(\frac{\tan(\delta)}{2.356}\right)\,\left(\frac{2400\,\text{Å}}{\lambda}\right), \tag{6}$$

independent of telescope aperture. The slit projects to the detector with width ω in microns, given by

$$\omega(\mu\text{m}) = 5\ r\ \phi(\text{arcsec})\ D(\text{m})\ f, \tag{7}$$

where r = the anamorphic demagnification (which I take to be unity), and f is the focal ratio of the spectrograph camera. For 2-pixel sampling, and a pixel size of 3 μm, $f \sim 10$, so the camera is about 8 feet long. CCDs and CMOS detectors can have pixels this small, at the cost of losing full-well capability; more commonly, detectors in use at telescopes have have $7 - 27\,\mu$m pixels. Probably a better choice is to have a factor of two larger slit than what is given in Equation (5), and correspondingly smaller f and R.

In summary, a spectrograph capable of spectral resolution $R \sim 10^6$ would be feasible to build for an 8 m-class ultraviolet telescope in space, and would be necessary for probing the $T = 100$ K diffuse ISM. Spectrographs with $R \sim 2 \times 10^4 - 1 \times 10^5$ have adequate resolution for coronal gas and the warm ISM, as well as for studies of the Lyα forest.

References

Bechtold, J. 1999, in ASP Conf. Ser. 164, Ultraviolet-Optical Space Astronomy Beyond HST, eds. J. Morse, J.M. Shull, & A.L. Kinney, (San Francisco: ASP), 41

Bechtold, J. 2002, in Galaxies at High Redshift, eds. I. Pérez-Fournon, M. Balcells, F. Moreno-Insertis, & F. Sánchez, (Cambridge: Cambridge U. Press), 131

Cowie, L., et al. 1995, AJ, 109, 1522

Davidsen, A.F., Kriss, G.A., & Wei, Z. 1996, Nature, 380, 47

Heap, S.R., Williger, G.M., Smette, A., Hubeny, I., Sahu, M.S., Jenkins, E.B., Tripp, T.M., & Winkler, J.N. 2000, ApJ, 534, 69

Jakobsen, P., et al. 1994, Nature, 370, 35

Jannuzi, B., et al. 2003, in preparation

Kriss, G.A., et al. 2001, Science, 559, 1112

Lemoine, M., et al. 1999, New Astronomy, 4, 231

Martin, C.L. 1999, in ASP Conf. Ser. 164, Ultraviolet-Optical Space Astronomy Beyond HST, eds. J. Morse, J.M. Shull, & A.L. Kinney, (San Francisco: ASP), 66

Pettini, M. 2003, Lectures given at the XIII Canary Islands Winter School of Astrophysics, Cosmochemistry: The Melting Pot of Elements, in press, (Electronic copy at http://www.ast.cam.ac.uk/~pettini/canaries13)

Reimers, D., et al. 1997, A&A, 327, 890

Schroeder, D.J. 2000, Astronomical Optics, (San Diego: Academic Press)

Shull, J.M. 1999, in ASP Conf. Ser. 164, Ultraviolet-Optical Space Astronomy Beyond HST, eds. J. Morse, J.M. Shull, & A.L. Kinney, (San Francisco: ASP), 229

Shull, J.M., et al. 1999, The Emergence of the Modern Universe: Tracing the Cosmic Web, White Paper on ST2010, available at http://origins.colorado.edu/uvconf/white-final/UVOWG_final_jun20.html

Smette, A., et al. 2002, ApJ, 564, 542

Songaila, A., & Cowie, L. 1996, AJ, 112, 335

Telfer, R., Kriss, G., Zheng, W., Davidsen, A., & Tytler, D. 2002, ApJ, in press, astro-ph/0207301

Tripp, T.M., Savage, B.D., & Jenkins, E.B. 2000, ApJ, 534, L1

Tytler, D. et al. 1995, in QSO Absorption Lines, Proceedings of the ESO Workshop, ed. G. Meylan, (Berlin: Springer-Verlag), 289

Tytler, D. 1999, in ASP Conf. Ser. 164, Ultraviolet-Optical Space Astronomy Beyond HST, eds. J. Morse, J.M. Shull, & A.L. Kinney, (San Francisco: ASP), 48

Hubble's Science Legacy: Future Optical-Ultraviolet Astronomy from Space
ASP Conference Series, Vol. 291, 2003
K.R. Sembach, J.C. Blades, G.D. Illingworth, R.C. Kennicutt, Jr. (eds.)

The Physics of Extragalactic Gas: One Argument for a Next Generation Space Telescope

Jason X. Prochaska

Obervatories of the Carnegie Institute of Washington, 813 Santa Barbara St., Pasadena, CA 91101

Abstract. Observations of the Galactic interstellar medium have had tremendous impact on our understanding of the physics of galactic gas and the processes of galaxy formation. Similar observations at $z > 2$ reveal the neutral baryonic content of the Universe, trace the evolution of metal enrichment, shed light on process of nucleosynthesis and dust formation, and yield precise measurements of galactic velocity fields. Owing to the limitations of ultraviolet spectroscopy, however, researchers are unable to examine galactic gas at $0 < z < 2$, an epoch spanning $\approx 80\%$ of the age of the Universe. To complement the multitude of ongoing programs to identify and research $z < 2$ galaxies, a next generation space telescope is essential to investigate the gas which feeds and records the history of galaxy formation.

1. What is the Problem?

In the next decade, numerous observational programs will identify and investigate overwhelming numbers of $z < 2$ galaxies. These surveys will characterize the history of star formation, the evolution of galaxy morphology, and the assembly of large-scale structure. Altogether, these efforts are likely to revolutionize our view of stars and galaxies at $z < 2$. In contrast to these achievements, these observations will have minimal impact on our understanding of the gas which feeds star formation. At all redshifts, gas is the major baryonic component of the Universe and, at $z > 1$, probably the dominant baryonic component of most galaxies. The principal challenge associated with examining the physics (e.g., metallicity, density, ionization state, temperature) of this gas is that the majority of diagnostics lie within the ultraviolet (UV) pass-band. To cover this epoch, one requires a next generation space telescope.

With the advent of echelle spectrographs on 10 m-class optical telescopes, researchers have pursued quasar absorption line (QAL) studies at unprecedented levels in the early universe. These observations have revolutionized our understanding of the Lyα forest (e.g., Miralda-Escudé et al. 1996; Rauch 1998), measured the baryonic density of the Universe (Burles & Tytler 1998; Rauch et al. 1997), revealed metals in among the least overdense structures observed (Tytler et al. 1995; Ellison et al. 2000), provided new insight into the chemical enrichment history of the Universe (Pettini et al. 1997; Prochaska & Wolfe 2000), and traced the velocity fields of protogalaxies (Prochaska & Wolfe 1997).

These studies have tremendous impact on our theoretical description of gas in the early universe. For example, aside from the cosmic microwave background, comparisons of cold dark matter predictions with the Lyα forest stand as one of the greatest successes of this cosmological paradigm. Furthermore, the majority of theorists at this workshop stressed one particular point: advances in 'gastrophysics' are essential to addressing the next theoretical frontier. Empirical constraints on 'gastrophysics' can only be made through observations of gas, and QAL observations provide the most efficient avenue of investigation.

In contrast with the $z = 2 - 5$ universe, there are very few diagnostics of the intergalactic medium (IGM) at $z < 2$. This antithesis is easily explained: the majority of physical diagnostics have observed wavelengths below 3000 Å at $z < 1.5$. In particular, the Lyα transition (at 1215 Å) can only be observed at $z > 1.6$ with optical spectrographs. Without coverage of this key transition, one cannot begin to address the preceding list of scientific inquiry.

An extragalactic observer like myself might be prone to nonchalantly dismiss the $z < 2$ universe. After all, this represents only 25% of the redshift path accessible to QAL analysis. This perspective, however, is horribly skewed. The epoch spanning $z = 0 - 1.5$ encompasses roughly 80% of the current age of the Universe. If one considers the temporal evolution of any quantity (e.g., chemical enrichment, galaxy clustering, luminosity functions), then to overlook this epoch is to remain ignorant of the Universe.

In this brief proceeding – written in support of a next generation UV telescope – I will emphasize the preceding introduction through a review of one major area of QAL research: the damped Lyα systems (DLA). These QAL systems are defined to have an H I surface density in excess of $10^{20.3}$ cm^{-2} and they dominate the universal H I content at all epochs following reionization. At $z > 2$, the DLA are believed to be the progenitors of present-day galaxies (Kauffmann 1996; Steinmetz, this volume) and echelle observations of the sightlines penetrating these galaxies provide detailed physical measurements of the protogalactic interstellar medium (ISM). The measurements include chemical enrichment level, star formation rate, dust content, nucleosynthetic enrichment history, as well as clues to the pressure, temperature, and ionization balance of this multi-phase medium. These observations are directly analogous to observations of the Galaxy and Magellanic Clouds carried out by the Hubble Space Telescope (HST) and earlier UV telescopes.

This is the central thesis of this proceeding: current observations reveal the physics of galactic gas locally and at $z > 2$ but very rarely address the ≈ 10 Gyr in between. To probe what amounts to 80% of the Universe, one needs a next generation space telescope. In the following, I will highlight some of the major results of research on damped Lyα systems, primarily those related to my own research with optical echelle spectrographs. To pursue these same research areas at $z < 2$ will require similar instrumentation in the UV pass-band.

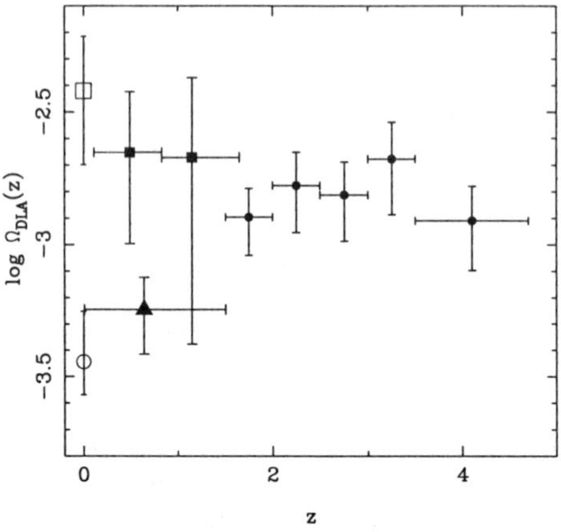

Figure 1. Evolution in the cosmological H I density of the Universe Ω_{DLA} in units of critical density as derived from DLA observations. Note the large uncertainties in Ω_{DLA} at $z < 1.7$ (solid squares and triangle) and the large offset between these two UV surveys (IUE: Lanzetta, Wolfe, & Turnshek 1995; HST: Rao & Turnshek 2000). These uncertainties highlight the inability of current UV spectrographs to address this fundamental measure.

2. What We Would Like to Pursue at z < 2

2.1. H I and Metal-Enrichment

Damped Lyα systems derive their name from the observed quantum mechanical damping of the Lyα transition relating to their very large H I column densities, N(H I). Because the Lyα profile is dominated by this damping, a standard fit to the observed profile has two free parameters: (i) the centroid or z_{abs}; and (ii) N(H I). Accurate measures of N(H I) can be acquired with modest resolution and S/N. Wolfe and his collaborators initiated surveys for these galaxies 20 years ago (e.g., Wolfe et al. 1986; Storrie-Lombardi & Wolfe 2000) and the majority of research has been performed on 4 m-class telescopes. Figure 1 summarizes the principal result from these H I surveys: the cosmic evolution of Ω_{DLA}, the universal mass density of neutral gas in units of the critical density. Focus on the measurements at $z < 2$ which were derived from IUE data (solid triangle; Lanzetta, Wolfe, & Turnshek 1995) and HST follow-up observations of a Mg II-selected sample (solid squares; Rao & Turnshek 2000). The uncertainties in each set of measurements are very large (those are logarithmic error bars!) and one notes a stark disagreement between the central values of the two surveys. These uncertainties emphasize the current challenge of studying H I gas at $z < 2$ with existing UV spectrographs. While HST/COS will enable a modest survey of DLA at $z < 0.6$, the parameter space $z = 0.6 - 1.7$ will require a next generation space telescope.

Aside from the H I content, the most basic measure of the ISM is metallicity. Because of the large H I surface density of DLA, ionization corrections are small and measurements of low-ions like Fe^+, Si^+, and Zn^+ yield accurate measures of the metallicity, i.e., [Zn/H] $\approx$ [Zn^+/H^0]. The only serious systematic error is

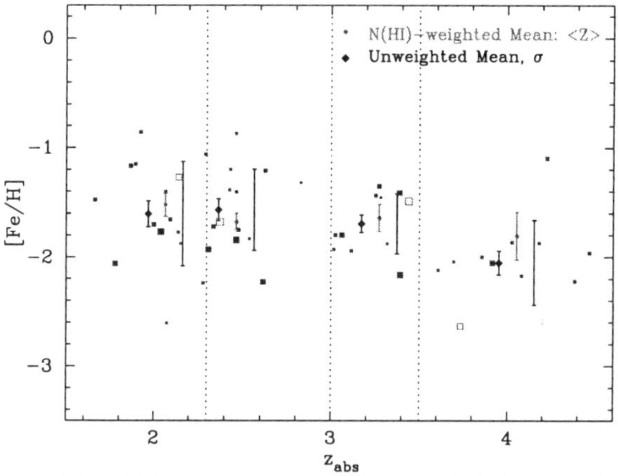

Figure 2. Metallicity measurements for ≈ 50 DLA at redshift $z \approx 2 - 4.5$. Overplotted are the H I-weighted and unweighted means in several redshift bins. Interestingly, one observes minimal evolution in these mean metallicities. Of particular interest, is resolving how the gas chemical enrichment evolves from $z = 2$ to today.

dust depletion; refractory elements like Fe and Si might be depleted from the gas-phase such that Si^+/H^0 and Fe^+/H^0 are lower limits to the true metallicity. In general, the depletion levels of the DLA are small (Pettini et al. 1997; Prochaska & Wolfe 2002) and the basic picture is well revealed by any of these elements at high z. Metallicity observations of a large sample of DLA present two main results: (1) an N(H I)-weighted mean $< Z >$, which is the cosmic mean metallicity of neutral gas; and (2) metallicities for a set of galaxies which presumably span a large range of mass, morphology, and luminosity.

Figure 2 presents over 50 metallicity measurements from $z \approx 2 - 4.5$ (see Prochaska & Wolfe 2002). These are the principal results: (1) the mean metallicity (weighted or unweighted) is significantly sub-solar; (2) there is little evolution in the mean metallicity over this redshift range with the possible exception of a modest decrease at $z > 3.5$; (3) no galaxy exhibits a metallicity lower than 1/1000 solar. These optical observations constrain models of chemical evolution at these epochs (e.g., Pei, Fall, & Hauser 1999) and give the first glimpse into metal production in the early universe. It is crucial, however, to press to lower redshift. The time encompassed by the redshift interval $2 < z < 4.5$ pales in comparison with $z < 2$. Of immediate concern is to determine how the mean metallicity rises to the enrichment level observed today.

2.2. Relative Abundances: Dust and Nucleosynthesis

High resolution ($R > 30,000$), high S/N (> 30 per resoultion element), observations of the damped Lyα systems enablc detailed studies of nucleosynthetic enrichment and dust properties in the early universe. This level of data quality is crucial to achieving the better than 10% precision required by relative abundance studies. Currently, there is an entire "cottage industry" focused on this area (Lu et al. 1996; Prochaska & Wolfe 1999; Molaro et al. 2000; Pettini et al.

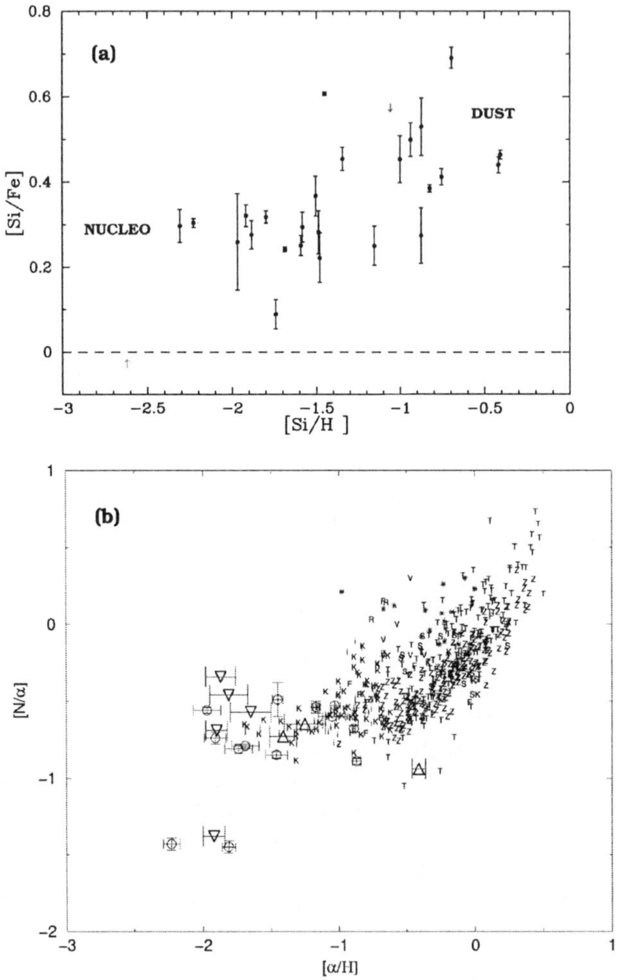

Figure 3. Relative abundances of (a) Si/Fe and (b) N/α versus Si/H and α/H where α refers to either Si or S for the DLA. The upper panel highlights the competing effects of nucleosynthesis and dust depletion in interpreting gas-phase abundances. We currently interpret the plateau of [Si/Fe] at [Si/H] < -1.5 as the result of Type II supernova (SN) nucleosynthesis and are confident the rise in Si/Fe at [Si/H] > -1 is associated with differential depletion (Prochaska & Wolfe 2002). UV observations would allow one to trace these two processes at $z < 2$. The lower panel compares DLA (circles and triangles; the latter are upper/lower limits) against N/α measurements for $z \sim 0$ H II regions and stars (Prochaska et al. 2002; Henry et al. 2000). Although the majority of DLA lie on the N/α plateau observed in metal-poor H II regions, a significant sub-sample is identified at $N/\alpha < -1$. This sub-sample is cautiously interpreted by Prochaska et al. (2002) as evidence for a truncated or top-heavy IMF. N/α observations at $z < 2$ would help reveal the timescales of SF and the nature of the IMF at these epochs.

2000; Ledoux, Bergeron, & Petitjean 2002). Figure 3 presents two of the principal results from our efforts: (a) [Si/Fe] and (b) [N/α] measurements against [Si/H] metallicity.

The super-solar Si/Fe ratios presented in panel (a) highlight the greatest obstacle to interpreting relative abundance measurements from gas-phase abundances: the competing effects of nucleosynthetic enrichment and differential depletion. In terms of dust depletion, one observes Si/Fe enhancements in depleted gas owing to the differential depletion of these two refractory elements. Regarding nucleosynthesis, Si/Fe enhancements suggest Type II SN nucleosynthesis (e.g., Woosley & Weaver 1995), whereas solar ratios would imply Type Ia SN enrichment patterns. Currently, we interpret the plateau of Si/Fe values at low metallicity as the primary result of nucleosynthesis. The mean enhancement matches the Galactic halo-star observations at the same metallicity (e.g., McWilliam 1997) and it would be difficult to understand why differential depletion would imply such a uniform enhancement. In contrast, the rise in Si/Fe at [Si/H] > -1 is highly suggestive of differential depletion. One expects a decrease in Si/Fe from nucleosynthesis at higher metallicity due to the increasing contribution from Type Ia SN and larger depletion levels are sensible in a higher metallicity ISM. Investigating evolution in abundance ratios like these at $z < 2$ would reveal the detailed enrichment history of galaxies and the evolution of dust formation.

Overcoming this dust/nucleosynthesis degeneracy is among the most active areas of DLA research. One avenue is to focus on special pairs of elements which are especially sensitive to only one effect. Panel (b) is an excellent example of this; plotted are N/α pairs from a recent analysis by Prochaska et al. (2002). For N, S, and Si (the latter two are α-elements), depletion effects are small and the results show the nucleosynthetic history of N in the DLA. For comparison, we also plot [N/α], [α/H] pairs for $z \sim 0$ H II regions and stars (see Henry, Edmunds, & Köppen 2000). The majority of DLA observations fall along the locus of local measurements, in particular the plateau of N/α values at [Si/H] < -1. In contrast, a sub-sample of low metallicity DLA exhibit much lower N/α values which Prochaska et al. (2002) interpret these in terms of a truncated or top-heavy initial mass function (IMF). These observations have important implications for the processes of star formation in the early universe and measurements at $z < 2$ would assess the timescale of star formation in these galaxies and further elucidate the nucleosynthesis of nitrogen.

2.3. Star Formation, Velocity Fields, α, etc.

The space allotted for this proceeding precludes a full discussion of the impact of a next generation UV telescope on DLA research. To list a few areas I will neglect: (1) velocity fields: measurements of the gas dynamics in DLA yield mass estimates in a fashion which avoids the luminosity-bias of most traditional approaches (Prochaska & Wolfe 1997). Gas dynamics are sensitive to non-gravitational motions (e.g., SN feedback), however, and interpretation may not be straightforward. Nevertheless, comparisons of QAL velocity widths with stellar measures would lend further insight into measuring galactic masses; (2) H_2: With far-UV spectroscopy, one can examine molecular hydrogen in the DLA (e.g., Bechtold, this volume) and thereby examine the gas which serves as the precursor to star formation. Additionally, one gains insight into dust formation and several physical characteristics of the gas (e.g., temperature, UV radiation field); (3) α: observations of DLA at $z > 2$ have enabled an analysis of the evolution of the fine-structure constant (e.g., Murphy et al. 2001) and have implied possible variations in this fundamental physical constant. UV observations would allow us to perform a similar inquiry at $z < 2$. (4) Star Formation: Observations of the C II* 1335 fine structure transition coupled with detailed ISM modeling yield estimates of the galactic star formation rate of the damped Lyα systems (Wolfe, Prochaska, & Gawiser 2002). The results offer a complimentary assessment of the SFR from emission-line and integrated light techniques and are critical for deciphering the build-up of galactic structure and metal enrichment in the Universe.

3. What We Want or What We Need

The resolution and S/N needed to examine the DLA or perform QAL studies in general is dependent on the specific research area. Nevertheless, we can lean on our extensive experience with high z QAL observations using optical spectrographs. For the majority of scientific applications, $R = 30000$ is a bare minimum. Only at this resolution can one confidently distinguish Lyα clouds from metal-lines in the Lyα forest, obtain abundance measurements to greater than 0.1 dex precision, resolve velocity fields, and investigate a multi-phase medium. Note this value is 50% greater than the $R = 20000$ of HST/COS. I fear this instrument will have limited impact on several important research areas, e.g., relative chemical abundances, kinematic characteristics.

Regarding S/N, a good lower limit is 30 per resolution element (i.e., 15 pix^{-1} for 4 pixel sampling). At this level, one can carefully address systematic errors like continuum placement and also analyse a wealth of very powerful absorption line diagnostics (e.g., Si II 1808, C IV 1550, O VI 1030). As always, one gains by achieving higher S/N and many applications would depend on higher sensitivity.

Now, consider wavelength coverage. For the majority of QAL research, coverage from Lyα (1215 Å) to 2000 Å rest-frame is essential. Many projects (e.g., D/H measurements, photoionization assessment, H_2 observations) need coverage down to 1000 or even 900 Å rest-frame. It is my opinion that an instrument providing coverage from 1100–3200 Å would be ideal with 1200–3000 Å acceptable.

Finally, and perhaps most importantly, observing power. To allow observations of a large enough sample of QSOs at $z < 2$, a UV telescope must achieve the above resolution, S/N, and wavelength coverage for a $V \approx 18$ QSO in a reasonable exposure time (< 10 hr). This would provide enough targets to examine the physical conditions at similar levels as achieved at $z > 2$. Taking the predicted performance of HST/COS as a starting point, the next generation space telescope would need a 5–10× improvement over HST/COS in its highest throughput pass-band (1100–2000 Å) and an increase of 50× over HST/COS at $\lambda > 2000$ Å. Although improvements in coatings and instrument design could make-up 3× of the difference (probably over 10× at $\lambda > 2000$ Å), a larger aperture telescope is unavoidable. After attending this workshop, A 4 m aperture might just satisfy the constraints I have laid out and a $\geq 6\,\mathrm{m}$ aperture would certainly meet the demands. Obviously, this will require terrific funding from NASA and, therefore, considerable support from the astronomical community. I am convinced, however, that the value of examining the gas physics of the Universe at $z < 2$, an effort which complements numerous observational programs within NASA's Origins and SEU Themes, is worth the expense.

Acknowledgments. I would like to thank S. Rao and A. Wolfe who provided the H I and C II* figures. I also wish to apologize to the many folk in the DLA and ISM community whom I failed to reference in this brief proceeding.

References

Burles, S., & Tytler, D. 1998, ApJ, 507, 732

Ellison, S.L., Songaila, A., Schaye, J., & Pettini, M. 2000, AJ, 120, 1175

Henry, R.B.C., Edmunds, M.G., & Köppen, J. 2000, ApJ, 541, 660

Kauffmann, G. 1996, MNRAS, 281, 475

Lanzetta, K. M., Wolfe, A. M., & Turnshek 1995, ApJ, 440, 435

Ledoux, C., Bergeron, J., & Petitjean, P. 2002, A&A, in press (astro-ph/0202134)

McWilliam, A. 1997, ARA&A, 35, 503

Miralda-Escudé, J., Cen, R., Ostriker, J.P., & Rauch, M. 1996, ApJ, 471, 582

Molaro, P., et al. 2000, ApJ, 541, 54

Murphy, M.T., et al. 2001, MNRAS, 327, 1208

Pei, Y.C., Fall, S.M., & Hauser, M.G. 1999, ApJ, 522, 604

Pettini, M., Smith, L.J., King, D.L., & Hunstead, R.W. 1997, ApJ, 486, 665

Pettini, M., Ellison, S.L., Steidel, C.C., Shapely, A.L., & Bowden, D.V. 2000, ApJ, 532, 65

Prochaska, J.X., & Wolfe, A.M. 1997, ApJ, 486, 73

Prochaska, J.X., & Wolfe, A.M. 1999, ApJS, 121, 369

Prochaska, J.X., & Wolfe, A.M. 2000, ApJ, 533, L5

Prochaska, J.X., & Wolfe, A.M. 2002, ApJ, 566, 68

Prochaska, J.X., Henry, R.B.C., O'Meara, J.M., Tytler, D., Wolfe, A.M., Kirkman, D., Lubin, D., & Suzuki, N. 2002, PASP, in press (astro-ph/0206296)

Rao, S.M., & Turnshek, D.A. 2000, ApJS, 130, 1

Rauch, M. 1998, ARA&A, 36, 267

Rauch, M., et al. 1997, ApJ, 489, 7

Storrie-Lombardi, L.J., & Wolfe, A.M. 2000, ApJ, 543, 552

Tytler, D., et al. 1995, in QSO Absorption Lines, ESO Astrophysics Symposia, ed. G. Meylan (Springer, Heidelberg), 289

Wolfe, A.M., Prochaska, J.X., & Gawiser, E. 2002, ApJ, submitted

Wolfe, A.M., Turnshek, D.A., Smith, H.E., & Cohen, R.D. 1986, ApJS, 61, 249

Woosley, S.E., & Weaver, T.A. 1995, ApJS, 101, 181

Hubble's Science Legacy: Future Optical-Ultraviolet Astronomy from Space
ASP Conference Series, Vol. 291, 2003
K.R. Sembach, J.C. Blades, G.D. Illingworth, R.C. Kennicutt, Jr. (eds.)

Studying the Gaseous Phases of Galaxies with Background QSOs

Jane C. Charlton

Department of Astronomy & Astrophysics, The Pennsylvania State University, 525 Davey Laboratory, University Park, PA 16802

Abstract. High resolution rest frame ultraviolet (UV) quasar absorption spectra covering low and high ionization species, as well as the Lyman series lines, provide remarkably detailed information about the gaseous phases of galaxies and their environments. For redshifts less than 1.5, many important chemical transitions remain in the observed ultraviolet wavelength range. I present examples of absorption that arises from lines of sight through a variety of structures, drawn from UV spectra recently obtained with the Space Telescope Imaging Spectrograph (STIS) on the Hubble Space Telescope (HST). Even with the greater sensitivity of the Cosmic Origins Spectrograph (COS) there will be a limit to how many systems can be studied in detail. However, there is a great variety in the morphology of the phases of gas that we observe, even passing through different regions of the same galaxy. In order to compile a fair sample of the gaseous structures present during every epoch of cosmic history, hundreds of systems must be sampled. Multiple lines of sight through the same structures are needed, as well as some probing nearby structures whose luminous hosts have been studied with more standard techniques. Combined with high resolution optical and near-infrared (near-IR) ground-based spectra, it will be possible to uniformly study the gaseous morphologies of galaxies of all types through their entire evolutionary histories.

1. Introduction

The tool of quasar absorption line spectroscopy has several distinct advantages over more traditional imaging studies of distant galaxies. Spectra covering absorption lines from numerous chemical elements in various states of ionization can yield detailed information about the physical conditions in the gaseous components of the universe. This is not limited to only the most luminous components. Dwarf galaxies, low surface brightness galaxies, and even intergalactic structures are probed by quasar lines of sight. Quasar absorption lines can be used to study gas during the birth and death of stars and of entire galaxies. The kinematic information contained in high resolution spectra allows us to study processes. This is not just a still picture snapshot; it is more like a short movie. Finally, the same level of detail is available for our study at all redshifts because the same method can be applied using optical and near-IR spectroscopy.

However, the study of quasar absorption lines also presents some challenges if we are to extract from the quasar spectra all of the detailed information about physical conditions of gas. Imagine if we had to classify a galaxy according to its standard morphological type by zooming in on an image of just a small part of the galaxy. We have to consider carefully what a single line of sight tells us about the global conditions in galaxies. In fact, it should be possible to learn a great deal if we consider the evolution of the ensemble of gaseous structures probed by quasar lines of sight. In order to realize this potential, we must learn to connect absorption signatures to the physical conditions of the phases of gas that produce them, and to the processes that give rise to such signatures in local galaxies.

For studies of the detailed physical conditions along quasar lines of sight, we need high resolution spectroscopy covering all rest-frame UV transitions. For redshifts less than one, a significant fraction of the key transitions still appear in the ultraviolet. In this proceedings, I review examples of various systems that have been studied in detail, focusing on the question of what additional data would yield a significant advance. The goal is to define capabilities for the optimal UV telescope and spectrograph to be used for such a program.

2. A C IV-Deficient Strong Mg II Absorber?

The $z = 0.99$ absorber toward the quasar PG 1634+706 is a strong Mg II absorber, which implies that it is very likely to be within an impact parameter of $35h^{-1}$ kpc of an $\sim L^*$ galaxy (Steidel 1995). It is part of the small subset of Mg II absorbers that have particularly weak C IV absorption features (i.e., it is C IV-deficient relative to other strong Mg II absorbers – Churchill et al. 2000). STIS/E230M spectra of this quasar have been obtained by Jannuzi and by Burles, and Keck/HIRES spectra (covering Mg I, Mg II, and Fe II) have been obtained by Churchill & Vogt (2001). The most important transitions are shown in Figure 1. Detailed results of our modeling of this system have been presented by Ding et al. (2002).

The system can be described by a combination of a minimum of four different phases of gas. A phase is a region or regions with density and temperature within some range, spatially separate from other phases. The Mg II absorption arises in clouds that have densities of $\sim 0.01\,\mathrm{cm}^{-3}$. The bulk of the Mg I absorption cannot arise in the same clouds. Instead, we propose that the Mg I is produced by small (~ 100 AU), cool pockets of the interstellar medium. The metallicity of the Mg II and Mg I clouds is about 0.2×solar. A "broad", low-metallicity ($< 0.01\times$solar) component is required to self-consistently fit Lyα, Lyβ, and the higher-order Lyman-series lines. The low metallicity could relate to why this absorber is C IV deficient. Finally, a strong, smooth Si IV absorption profile suggests an additional collisionally ionized phase with $T \sim 60,000$K, perhaps heated by shocks.

For this system, there are two outstanding issues of particular interest. First, with higher resolution ($R > 100,000$) we could test the narrow Mg I cloud hypothesis. Second, an image of the host galaxy (or those for similar systems) would enable us to see whether an early-type galaxy is responsible for the lack of absorption resembling the Milky Way corona.

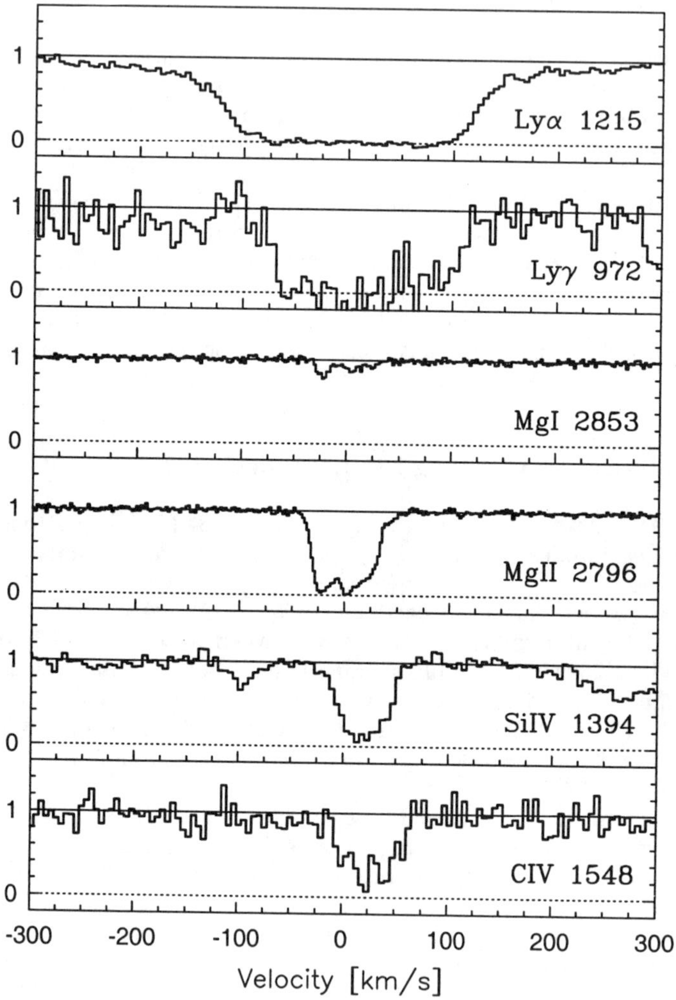

Figure 1. Selected transitions for the $z = 0.9902$ strong Mg II system along the PG 1634 + 706 line of sight, presented in velocity space. The velocity zero-point corresponds to the apparent optical depth centroid for the Mg II profile. The Mg I and Mg II profiles were obtained with Keck/HIRES, with $R = 45,000$ (P.I. Churchill), and the other transitions were observed with HST/STIS, with $R = 30,000$ (P.I. Jannuzi and Burles).

3. A Redshift One Galaxy Group?

The $z \sim 0.93$ systems in the PG $1206 + 459$ line of sight are the subject of a separate contribution in this volume by Jie Ding, and therefore no detail or figures will be presented here. Three distinct systems (two strong Mg II and one weak Mg II) are apparent at $z = 0.925$, $z = 0.928$, and $z = 0.934$, an overall velocity spread of just over 1000 km s^{-1}. This is likely to be a group of galaxies, with the three systems arising in three very different types of galaxy hosts. The study of these systems would be aided by higher S/N since there are suggestions of abundance pattern variations. Understanding of the phases giving rise to high ionization absorption would be much better constrained if a high resolution spectrum covering the O VI transition was obtained.

4. A Pair of Dwarf Galaxies?

The $z = 1.04$ absorber toward PG $1634 + 706$ is a multiple cloud, weak Mg II absorber, with two sets (kinematic subsystems) of low ionization "clouds" separated by $\sim$ 150 km s^{-1}. Each of the subsystems has an O VI absorption profile offset by $\sim$ 50 km s^{-1}. Profiles for key transitions from the HST/STIS and Keck/HIRES spectra are shown in Figure 2. This system was modeled by Zonak et al. (2002), and the results are summarized here.

The Mg II clouds have metallicity $< 0.03\times$solar, constrained by a partial Lyman limit break. Although Si IV has the same kinematic structure as Mg II in the stronger subsystem, the clouds appear to be offset by velocities ranging from 3–5 km s^{-1}. This could be related to an H II region flow, but the interpretation of the multiple cloud structure remains unclear. The Si III profile is smooth, but unsaturated, and is stronger than can be explained by the combination of the Mg II and Si IV cloud phases. This suggests a collisional ionization mechanism for producing Si III. The Lyα profile requires two additional photoionized phases that can also produce the broad O VI profiles.

The low metallicity and the kinematics of the two subsystems suggest that two dwarf galaxies and their halos could be responsible. The lower ionization phases would arise in the inner regions of the dwarfs, and the offset higher ionization phases from their halos.

Very little is known about the expected absorption signature of dwarf galaxies. If quasar absorption line probes through nearby dwarfs were available, this mystery would be quickly solved. However, this would require that some fainter quasars were accessible for high resolution spectroscopy.

5. A Variety of Weak Mg II Absorbers

Figure 3 shows selected transitions for the three single-cloud, weak Mg II absorbers at $z = 0.65$, $z = 0.81$, and $z = 0.90$ along the PG 1634+706 line of sight. Unlike the strong Mg II absorbers, single-cloud, weak Mg II absorbers are generally not found within 50 kpc of L^* galaxies. Rigby, Charlton, & Churchill (2002) found, based on a larger sample for which lower resolution Faint Object Spectrograph (FOS) spectra were available, that they cannot be far below solar metallicity. They also found that at least some of these absorbers are very small

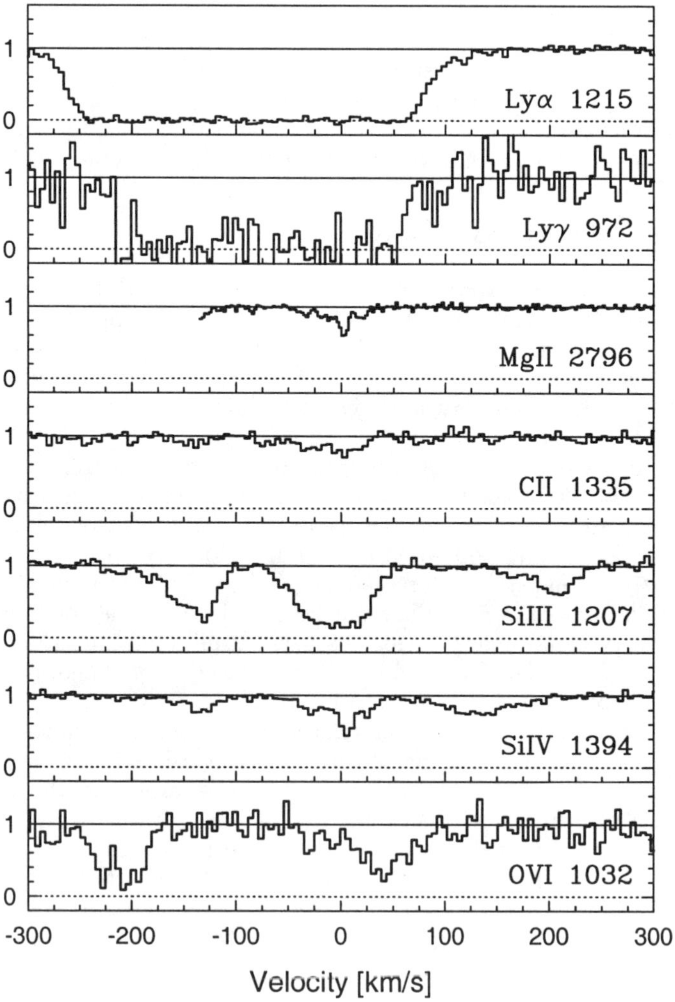

Figure 2. Selected transitions for the $z = 1.0414$ multiple-cloud, weak Mg II system along the PG 1634 + 706 line of sight, presented in velocity space. These are taken from the same spectra noted in the Figure 1 caption.

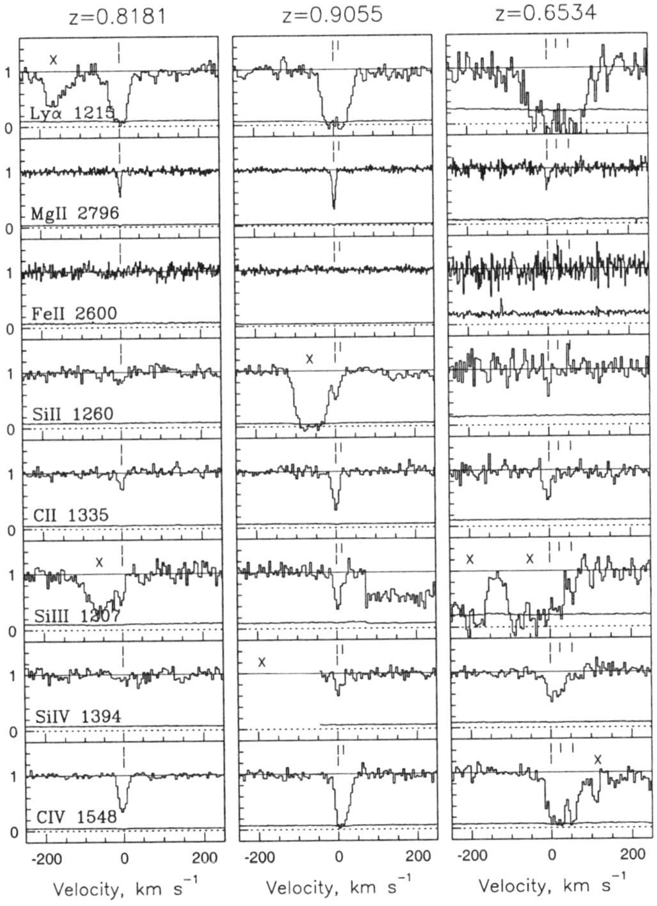

Figure 3. A comparison of selected transitions for the three single-cloud, weak Mg II absorbers along the PG1634+706 line of sight. The spectra from which these were extracted are referenced in the Figure 1 caption.

in dimension. Detailed models of the three PG 1634 + 706 single-cloud, weak Mg II absorbers were presented in Charlton et al. (2002).

The $z = 0.81$ weak Mg II absorber is the kinematically simplest along this line of sight. The C IV absorption is relatively weak. It was not even detected in a low resolution HST/FOS spectrum. The Lyα profile constrains the metallicity to be supersolar, and the ionization conditions indicate that the low ionization phase is sub-parsec scale. Despite the C IV being weak, the broad shape of its profile cannot be fit by the Mg II cloud. The high ionization phase could be collisionally ionized, but photoionization is more likely, with a cloud size of ~ 100 pc. The high ionization phase could surround, or be surrounded by, the low ionization gas.

If the high ionization material is a thicker shell surrounding a small low ionization pocket then we would expect to see many systems with only the high ionization phase. A large statistical sample is needed to determine the geometry. To distinguish between collisional ionization and photoionization, coverage of the O VI transition is key.

In contrast, the $z = 0.90$ weak Mg II absorber has relatively strong C IV. A low ionization phase has supersolar metallicity and a cloud size of ~ 50 pc. The C IV and N V require a separate, broader cloud, which is consistent with photoionization and not with collisional ionization. Also, in this case, an additional offset high ionization cloud is required to fit the blue wing of the C IV and the Lyα, which is not centered on the Mg II. The C IV equivalent width could be large in this case partly because the line of sight passed through a separate cloud at a different velocity, and not because the centered C IV cloud has special physical conditions.

The C IV profile associated with the $z = 0.65$ weak Mg II absorber has distinct subcomponent structure, and therefore a large equivalent width. The Lyα is also quite strong for this system, possibly also because of the large kinematic spread. It is centered on the C IV and not on the single Mg II component. The low ionization phase is poorly constrained, but its metallicity must exceed $0.1\times$solar. The three C IV components arise in high ionization, kiloparsec-scale structures, which are likely to be photoionized.

What are these weak, single-cloud Mg II absorbers? The small size constraints on the low ionization phases implies that there must be huge numbers of the host structures. In fact, to account for the observed number of weak, single-cloud Mg II absorbers, per unit redshift, there must be more than a million hosts for every L^* galaxy in the Universe. Yet, these are not associated with L^* galaxies. They appear to be self-enriched pockets of higher density material embedded in higher ionization, more diffuse gas. Rigby et al. (2002) discuss the origin of the weak Mg II absorbers at length. Their two most likely explanations are small traces of gas in fading remnants of Population III star clusters and fragments of supernova shells in faint blue dwarf galaxies.

A number of new capabilities would help to constrain the properties of this intriguing class of object. Double line of sight observations at a scale of parsecs, though they could not be done for many systems, would provide an essential check on size constraints for these weak Mg II absorbers. Nearby weak Mg II absorbers must be identified so that narrow band imaging can be used to discover if hosts are dwarf galaxies or isolated star clusters. Of course, a large statistical sample would clarify whether there are different classes of weak, single-cloud Mg II absorbers, with different types of hosts.

6. Conclusions

The small sample of systems presented here demonstrates great promise for learning about the detailed physical conditions in a variety of gaseous environments. For these systems, the data have some limitations and, as a result, some questions remain unanswered. These highlight the need for additional UV spectroscopy capabilities. I conclude by presenting a "wish list" if we aim to construct the dynamical history of the ensemble of galaxies and gaseous structures:

- hundreds, or even thousands, of systems
- cover all key chemical species from the rest frame ultraviolet
- study many systems that are at low enough redshift to image their galaxy hosts; these serve as a calibrator for higher redshift systems
- higher spectral resolution ($R = 100,000$ or even higher) for a subset of the systems in order to resolve interstellar medium features
- high signal-to-noise ($S/N > 20$) for a subset to enable abundance pattern studies
- multiple lines of sight through the same objects
- a separate, detailed study of the interstellar medium of the Milky Way and of nearby galaxies using the same techniques

Acknowledgments. Support for this work was provided by the NSF (AST-9617185) and by NASA (NAG 5-6399 and HST-GO-08672.01-A), the latter from the Space Telescope Science Institute, which is operated by AURA, Inc., under NASA contract NAS5-26555.

References

Charlton, J.C., Ding, J., Zonak, S.G., Churchill, C.W., & Bond, N.A. 2002, ApJ, submitted

Churchill, C.W., Mellon, R.R., Charlton, J.C., Jannuzi, B.T., Kirhakos, S., & Steidel, C.C., 2000, ApJ, 543, 577

Churchill, C.W., & Vogt, S.S. 2001, AJ, 122, 679

Ding, J., Charlton, J.C., Bond, N.A., Zonak, S.G., & Churchill, C.W. 2002, ApJ, submitted

Ding, J., Charlton, J.C., Churchill, C.W., & Palma, C. 2002, in preparation

Rigby, J.R., Charlton, J.C., & Churchill, C.W. 2002, ApJ, 565, 743

Steidel, C.C. 1995, in QSO Absorption Lines, ed. G. Meylan (Garching: Springer-Verlag), 139

Zonak, S.G., Charlton, J.C., Ding, J., & Churchill, C.W. 2002, ApJ, submitted

Hubble's Science Legacy: Future Optical-Ultraviolet Astronomy from Space
ASP Conference Series, Vol. 291, 2003
K.R. Sembach, J.C. Blades, G.D. Illingworth, R.C. Kennicutt, Jr. (eds.)

Probing Baryons in Galactic Halos and Gas Near Galaxies

Kenneth R. Sembach

Space Telescope Science Institute, 3700 San Martin Dr., Baltimore, MD 21218

Bart P. Wakker, Blair D. Savage, Philipp Richter, and Marilyn Meade

Astronomy Dept. University of Wisconsin-Madison, 475 N.Charter St., Madison, WI 53706

Abstract. We describe an extensive FUSE survey of highly ionized oxygen in the vicinity of the Milky Way that serves as an example of the type of study that would be desirable for other galactic systems. Understanding the origin of hot gas in the vicinity of galaxies and its relationship to the intergalactic medium presents a major observational challenge. Ultraviolet absorption-line spectroscopy is currently the most direct means for comprehensive investigations of the gas in galactic environments, but even with present (and near-term) facilities the number of background objects available to probe nearby galaxy halos and low-redshift cosmological structures is limited. Studying these structures over a range of impact parameters and angular separations would provide fundamental information about the baryonic content of the hot gas, its physical conditions, and its origins. A large space telescope optimized for high resolution spectroscopy in the 900–3200 Å wavelength region at a sensitivity sufficient to observe faint AGNs/QSOs at angular separations of $< 1^\circ$ would be ideal for such studies.

1. Introduction

Understanding galaxy formation and evolution requires observational information about the hot, highly-ionized gas found in and near galaxies. A complete picture of the relationship between hot gas in galaxies and the intergalactic medium (IGM) does not yet exist because a variety of processes affect the heating and distribution of the interstellar gas in and around galaxies. Galaxy formation, accretion of satellite galaxies, tidal interactions, star-formation, galactic winds, and galaxy-IGM interactions may all contribute to the production of the hot gas observed in the IGM and extended galactic halos.

In this article, we outline a program we have begun with the Far Ultraviolet Spectroscopic Explorer (FUSE) to study the hot gas in the vicinity of the Milky Way. The study is described in detail in a series of three articles devoted to probing the highly ionized oxygen (O VI) absorption along complete paths through the Galactic halo and Local Group. The articles include a catalog of the spectra and basic observational information (Wakker et al. 2003), a study of the hot gas in the Milky Way halo (Savage et al. 2003), and an investigation of the highly ionized high velocity gas in the vicinity of the Galaxy (Sembach et al. 2003). Here, we summarize the high velocity gas results and comment on their implications for understanding other types of highly ionized gas. This comprehensive study serves as a prime example of what can be learned if serious efforts are made to conduct large spectroscopic observing programs with existing space-based instrumentation. Similar types of studies conducted with future generations of instruments on the Hubble Space Telescope (HST) and other large space telescopes will enable us to explore the properties of gas in other groups of galaxies in much greater detail than is presently possible.

In Table 1 we list basic information for some of the best ultraviolet (UV) absorption-line diagnostics of hot gas. The table entries include the atom or ion, rest wavelength of the transition, cosmological redshift required to shift the line(s) to an observed wavelength of 1150 Å, the temperature at which the atom or ion peaks in abundance under conditions of collisional ionization equilibrium, the thermal width of the line at that temperature, and a logarithmic relative line strength that depends on the cosmic abundance of the element (A), the line strength ($f\lambda$), and the peak ionization fraction (ϕ) of the ion in collisionally ionized gas. Larger values of $Af\lambda\phi$ indicate easier detectability.

Table 1. Ultraviolet Diagnostics of Hot Gas at Low Redshift

Species	λ(Å)	z(1150 Å)	T_{CIE} (K)[a]	b_{th}(km s^{-1})[b]	log $[Af\lambda\phi]$[c]
H I Lyα	1215.7	—	$< 10^5$	40.8[d]	−2.06[d]
H I Lyβ	1025.7	0.12	$< 10^5$	40.8[d]	−2.86[d]
H I Lyγ	972.5	0.18	$< 10^5$	40.8[d]	−3.32[d]
C IV	1548.2, 1550.8	—	1.0×10^5	11.8	−1.51, −1.81
N V	1238.8, 1242.8	—	1.8×10^5	14.6	−2.35, −2.65
O IV	787.7	0.46	1.6×10^5	12.9	−1.36
O V	629.7	0.83	2.5×10^5	16.1	−0.97
O VI	1031.9, 1037.6	0.11	2.8×10^5	17.1	−1.65, −1.95
Ne VIII	770.4, 780.3	0.48	5.6×10^5	21.5	−2.36, −2.66
Mg X	609.8, 624.9	0.87	1.1×10^6	27.4	−3.31, −3.61

[a]Temperature of maximum ionization fraction in collisional ionization equilibrium (Sutherland & Dopita 1993).

[b]Thermal line width, $b = (2kT/m)^{1/2}$, at $T = T_{CIE}$ unless indicated otherwise.

[c]Values of $f\lambda$ are from Morton (1991) and Verner et al. (1994). Values of A (abundance relative to hydrogen on a logarithmic scale where H = 12.00, C = 8.55, N = 7.97, O = 8.87, Ne = 8.07, and Mg = 7.59) are from Grevesse & Noels (1993) and Anders & Grevesse (1989). Values of ϕ are from Sutherland & Dopita (1993).

[d]Value at $T = 10^5$ K.

The O VI $\lambda\lambda$1031.926, 1037.617 doublet lines are the best UV resonance lines to use for kinematical investigations of hot ($T \sim 10^5 - 10^6$ K) gas in the low-redshift universe. Oxygen is the most abundant element heavier than helium, and the O VI lines have large oscillator strengths. Lower ionization UV lines observable at high spectral resolution are generally either much weaker than the

O VI lines (e.g., N V $\lambda1238.821, 1242.804$) or are better tracers of collisionally ionized gas at temperatures $T < 10^5$ K (e.g., C IV $\lambda1548.195, 1550.770$, C III $\lambda977.020$, Si IV $\lambda1393.755, 1402.770$, Si III $\lambda1206.500$). This latter set of ions is also considerably more susceptible to photoionization than O VI.

X-ray spectroscopy of the interstellar or intergalactic gas in higher ionization lines (e.g., O VII, O VIII) is possible with XMM-Newton and the Chandra X-ray Observatory for a small number of sight lines toward AGNs and QSOs, but the spectral resolution ($R \equiv \lambda/\Delta\lambda < 1000$) is modest compared to that afforded by FUSE ($R \sim 15,000$). While the X-ray lines provide extremely useful information about the amount of gas at temperatures greater than 10^6 K, the interpretation of where that gas is located, or how it is related to the $10^5 - 10^6$ K gas traced by O VI, is hampered at low redshift by the complexity of the hot ISM and IGM along the sight lines observed.

2. An Example: The Milky Way and its High Velocity Cloud System

We have conducted a study of the highly ionized high velocity gas in the vicinity of the Milky Way using an extensive set of FUSE data. We summarize the results for the sight lines toward 100 AGNs/QSOs and two distant halo stars in this section (see Sembach et al. 2003; Wakker et al. 2003). For the purposes of this study, gas with $|v_{LSR}| \gtrsim 100$ km s^{-1} is typically identified as "high velocity", while lower velocity gas is attributed to the Milky Way disk and halo. A sample spectrum from the survey is shown in Figure 1.

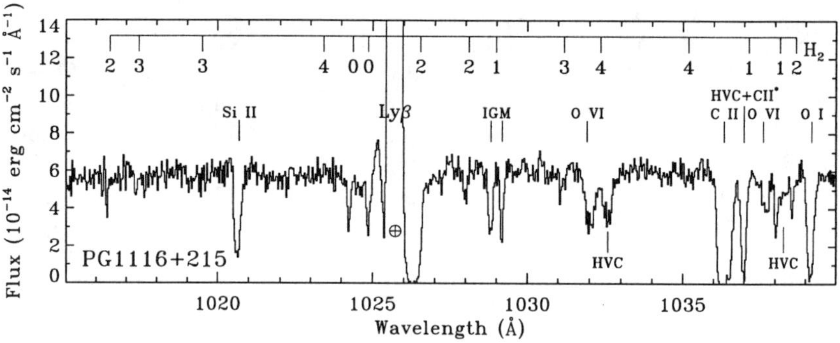

Figure 1. A portion of the FUSE spectrum of PG 1116+215 in the 1015–1040 Å spectral region. Interstellar and intergalactic lines are indicated. The numbers under the top tick marks denote the rotational levels of the H_2 lines ($J = 0 - 2$ lines are prominent). The locations of high velocity O VI features are indicated below the spectrum.

2.1. Observational Results

We have identified approximately 84 individual high velocity O VI features along the 102 sight lines in our sample. A critical part of this identification process involved detailed consideration of the absorption produced by O VI and other species (primarily H_2) in the thick disk and halo of the Galaxy, as well as the absorption produced by low-redshift intergalactic absorption lines of H I and ionized metal species. We searched for absorption in a velocity range of ± 1200 km s^{-1} centered on the O VI $\lambda 1031.926$ line. With few exceptions, the high velocity O VI absorption is confined to $|v_{LSR}| \leq 400$ km s^{-1}, indicating that the O VI features observed are either associated with the Milky Way or nearby clouds within the Local Group.

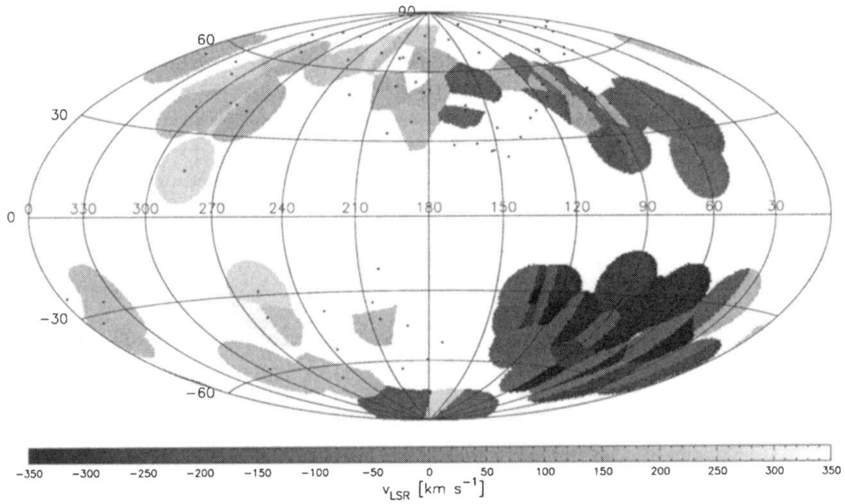

Figure 2. All-sky projection of the high velocity O VI features in the LSR reference frame. The Galactic anti-center is at the center of the plot, and Galactic longitude increases to the left. This map does not include gas at velocities attributed to the thick disk/halo of the Galaxy ($|v_{LSR}| \lesssim 100$ km s^{-1}). The velocities of the O VI features are gray-scale coded and displayed as filled regions of radius 12°. When two features are detected within 12° of each other (either along the same sight line or along adjoining sight lines), the shaded area size is adjusted accordingly. Points with no shading indicate directions where no high velocity O VI is observed.

The high velocity O VI features have velocity centroids ranging from $-372 < v_{LSR} < -90$ km s^{-1} to $+93 < v_{LSR} < +385$ km s^{-1}. There are an additional 6 confirmed or very likely (> 90% confidence) detections and 2 tentative detections of O VI between $v_{LSR} = +500$ and $+1200$ km s^{-1}; these very high velocity features probably trace intergalactic gas beyond the Local Group. We plot the sky distribution of the high velocity O VI in Figure 2, where we code the velocities of the features by shading circular regions around the sight lines for which high velocity O VI is detected.

Most of the high velocity O VI features have velocities incompatible with those of Galactic rotation (by definition). The dispersion about the mean of the high velocity O VI centroids decreases when the velocities are converted from the Local Standard of Rest (LSR) into the Galactic Standard of Rest (GSR) and the Local Group Standard of Rest (LGSR) reference frames. While this reduction is expected if the O VI is associated with gas in a highly extended Galactic corona or in the Local Group, it does not provide sufficient proof by itself of an extragalactic location for the high velocity gas. Additional information, such as the gas metallicity or ionization state, is needed to constrain the cloud locations.

The line widths of the high velocity O VI features range from $\sim$16 km s^{-1} to $\sim$81 km s^{-1}, with an average of $\langle b \rangle = 40 \pm 14$ km s^{-1}. The lowest values of b are close to the thermal width of 17.1 km s^{-1} expected for O VI at $T = 2.8 \times 10^5$ K, while higher values of b require additional non-thermal broadening mechanisms or gas temperatures significantly larger than 2.8×10^5 K.

The high velocity O VI features have logarithmic column densities of 13.06 to 14.59, with an average of $\langle \log N \rangle = 13.95 \pm 0.34$ and a median of 13.97. We show the distribution of the high velocity O VI column densities on the sky in Figure 3. The average high velocity O VI column density is a factor of 2.7 times lower than the typical low velocity O VI column density found for the same sight lines through the thick disk/halo of the Galaxy (see Savage et al. 2003).

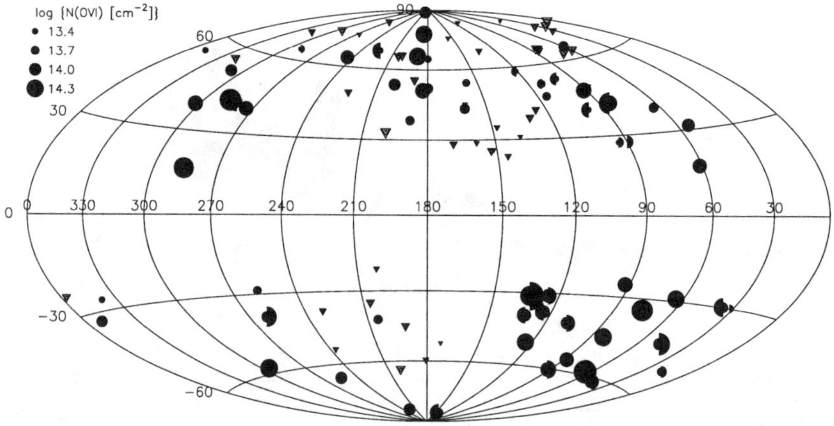

Figure 3. All-sky projection of the high velocity O VI column densities observed in our FUSE survey. The logarithmic column density is coded according to symbol size (see legend), with the symbol split in half if two features are present. Triangles indicate upper limits. Note that large column densities of high velocity O VI ($\log N > 14$) are found in many regions of the sky.

We detect high velocity O VI λ1031.926 absorption with total values of $W_\lambda >$ 30 mÅ at $\geq 3\sigma$ confidence along 59 of the 102 sight lines surveyed. For the highest quality sub-sample of the dataset, the high velocity detection frequency increases to 22 of 26 sight lines. Forty of the 59 sight lines have high velocity O VI λ1031.926 absorption with $W_\lambda >$ 100 mÅ, and 27 have $W_\lambda >$ 150 mÅ. Converting these O VI equivalent width detection frequencies into estimates of $N(\mathrm{H}^+)$ in the hot gas indicates that $\sim$ 60% of the sky (and perhaps as much as $\sim$ 85%) is covered by hot ionized hydrogen at a level of $N(\mathrm{H}^+) \gtrsim 8 \times 10^{17}$ cm^{-2} if the high velocity gas has a metallicity similar to that of the Magellanic Stream ($Z \sim 0.2 - 0.3$). This detection frequency of hot H$^+$ associated with the high velocity O VI appears to be larger than the value of $\sim$ 37% found for high velocity warm gas with $N(\mathrm{H\,I}) \sim 10^{18}$ cm^{-2} traced through 21 cm emission (Lockman et al. 2002).

Some of the high velocity O VI is associated with well-known H I high velocity clouds (HVCs; see Wakker & van Woerden 1997 for a review). These include the Magellanic Stream, possibly Complex A, Complex C, the Outer Arm, and several smaller H I clouds. Some of the high velocity O VI features have no counterpart in H I 21 cm emission. These include discrete high velocity features as well as broad positive-velocity O VI absorption wings that blend with lower velocity O VI absorption in the Galactic thick disk/halo. The discrete features may typify clouds located in the Local Group. The broad, high velocity O VI absorption wings are concentrated mainly in the northern Galactic hemisphere and may trace either tidal debris or thick disk/halo gas that has been accelerated to high velocities by star-formation activity in the Galactic disk.

2.2. Interpretation

High velocity O VI is both widespread and common along complete paths through the Galactic halo. The high velocity O VI traces numerous phenomena, including tidal interactions with the Magellanic Clouds (via the Magellanic Stream), accretion of low-metallicity gas (e.g., Complex C), highly ionized clouds (e.g., the Mrk 509 HVCs), and the outflow of hot gas from the Galactic disk (e.g., the broad positive velocity absorption features). Distinguishing between all of the possible phenomena occurring at large distances from the Galactic plane is absolutely essential for understanding the role of HVCs in Galactic evolution.

One possible explanation for some of the high velocity O VI is that transition temperature gas arises at the boundaries between cool/warm clouds of gas and a very hot ($T > 10^6$ K) Galactic corona or Local Group medium. Sources of the high velocity material might include infalling or tidally disturbed galaxies. Since viscous processes affect the gas but not the stars in interacting systems, tracing the original source of the high velocity gas may prove difficult. A hot, highly extended ($R > 70$ kpc) corona or Local Group medium might be left over from the formation of the Milky Way or Local Group, or may be the result of continuous accretion of smaller galaxies over time. N-body simulations of the tidal evolution and structure of the Magellanic Stream favor a low-density medium ($n < 10^{-4}$ cm^{-3}) for imparting weak drag forces to deflect some of the Stream gas and providing a possible explanation for the absence of stars in the Stream (Gardiner 1999). Moore & Davis (1994) also postulated a hot, low-density corona to provide ram pressure stripping of some of the Magellanic Cloud gas. Hydrodynamical simulations of clouds moving through a hot, low-density medium show that weak bow shocks develop on the leading edges of the clouds as the gas is compressed and heated (Quilis & Moore 2001). Even if the clouds are not moving at supersonic speeds relative to the ambient medium, some viscous or turbulent stripping of the cooler gas likely occurs.

An alternative explanation for the O VI observed at high velocities may be that the clouds and any associated H I fragments are simply condensations within large gas structures falling onto the Galaxy. Cosmological structure formation models predict large numbers of cooling fragments embedded in dark matter, and somc of these structures should be observable in O VI absorption as the gas pass through the $T = 10^5 - 10^6$ K temperature regime. Estimates of the number density of these structures are consistent with the observed IGM O VI detection rate (Tripp, Savage & Jenkins 2000; Fang & Bryan 2001; Savage et al. 2002). This situation is in many ways analogous to the coronal model described above because only about 30% of the hot gas is detectable in O VI absorption, while the remaining $\sim$ 70% is too hot to observe (Davé et al. 2001).

The tenuous hot Galactic corona or Local Group gas may manifest itself through X-ray absorption-line observations of higher ionization species than O VI. For example, the amount of O VII in the hot gas is given by N(O VII) $= (\mathrm{O/H})_\odot\ Z\ \mathrm{f}_{O\,\mathrm{VII}}\ nL$, where Z is the metallicity of the gas, f is the ionization fraction, and L is the path length. At $T \sim 10^6$ K, $f_{O\,\mathrm{VII}} \approx 1$ (Sutherland & Dopita 1993). For $n = 10^{-4}$ cm^{-3}, N(O VII) $\sim 2 \times 10^{16}\ Z\ (L/100$ kpc) (cm^{-2}). Preliminary results (Fang et al. 2003; Nicastro et al. 2002) demonstrate that O VII absorption is detectable near zero velocity at a level consistent with the presence of a large, nearby reservoir of hot gas. A firm association of the X-ray absorption with a hot Galactic corona or Local Group medium will require additional observations since the ionization mechanism and location of the higher ionization gas traced by O VII are still uncertain (see Heckman et al. 2002).

3. Looking Forward: The Need for a Large Space Telescope

Studying the basic physical properties of gas in the vicinity of galaxies is essential for understanding how galaxies evolve over time and quantifying the relationship between galaxies, the IGM, and large-scale cosmic structures. Key quantities to be determined include the elemental abundances, ionization state, and physical properties (n, T, size) of the gas. We have made a start at examining the properties of gas in the Milky Way and Local Group, but extending these studies to other groups of galaxies is required if we are to incorporate the local results into broader descriptions of galaxy evolution and the formation of the "cosmic web" of hot gas expected in the local universe. To make accurate assessments of the kinematics and column densities of a wide range of ionization stages requires high spectral resolution (R $\sim 20,000 - 50,000$), broad UV wavelength coverage (preferably 900–3200 Å so that both zero and low-redshift systems could be observed in C III and O VI), and excellent sensitivity.

Current ultraviolet spectrographs (e.g., FUSE and STIS) are limited to observing QSOs with $m_B < 16$. At this magnitude, the number of QSOs per square degree is $\ll 1$, and it is difficult to observe individual stars in galaxies other than the Milky Way and Magellanic Clouds. The Cosmic Origins Spectrograph (COS), which is scheduled for installation in HST in 2004, will have about an order of magnitude greater sensitivity than STIS at far-UV wavelengths. QSOs with $m_B \approx 17 - 18$ should be observable in reasonable integration times. However, even at this greater sensitivity, the number of QSOs per square degree is still limited. A large space telescope with a diameter of 4 meters (or better still, 8 meters) equipped with a high efficiency spectrograph and detector could easily improve upon the COS sensitivity by a factor of 10 or more and dramatically increase the number of background sources available for spectroscopic studies. At $m_B \approx 20$, the average separation between QSOs is only $\sim 7'$ (see Shull et al. 1999), and there are numerous possibilities for studying multiple sight lines through low-redshift galactic halos, groups, and clusters of galaxies. Furthermore, at these sensitivities lightly reddened B supergiants could be observed out to distances of $\sim$ 2 Mpc, making it possible to study the gaseous content of Local Group galaxies such as M 31 or M 33 in unprecedented detail.

A 4 m to 8 m space telescope optimized for point source spectroscopy would provide an opportunity for an entirely new approach to studying the gaseous halos of galaxies and intragroup gas. Instead of our current methodology of blindly finding low-redshift absorption-line systems spectroscopically and then trying to identify (image) the galaxies responsible for the absorption, one could target specific, well-studied galactic systems for spectroscopic investigations of their gaseous content. A similar pro-active approach may prove fruitful for studying the filamentary structures of the ionized IGM once they are identified through their O VI or X-ray emissions by future observatories such as SPIDR or Constellation-X.

Acknowledgments. We thank Mike Shull and Ed Jenkins for stimulating discussions about the interpretation of the high velocity O VI absorption.

References

Anders, E., & Grevesse, N. 1989, Geochim. Cosmochim. Acta., 53, 197

Davé, R., Cen, R., Ostriker, J.P., et al. 2001, ApJ, 552, 473

Fang, T.T., & Bryan, G.L. 2001, ApJ, 561, L31

Fang, T., Sembach, K.R., & Canizares, C.R. 2003, ApJ, in press

Gardiner, L.T., 1999, in ASP Conf. Ser. 166, The Stromlo Workshop on High Velocity Clouds, eds. B.K. Gibson & M.E. Putman, (San Francisco: ASP), 292

Grevesse, N., & Noels, A. 1993, in Origin of the Elements, ed. N. Prantzos, E. Vangioni-Flam, & M. Cassé, (Cambridge: Univ. Press), 15

Heckman, T.M., Norman, C.A., Strickland, D.K., & Sembach, K.R. 2002, ApJ, 577, 691

Lockman, F.J., Murphy, E.M., Petty-Powell, S., & Urick, V. 2002, ApJS, 140, 331

Moore, B., & Davis, M. 1994, MNRAS, 270, 209

Morton, D.C. 1991, ApJS, 77, 119

Nicastro, F., Zezas, A., Drake, J., et al. 2002, ApJ, 573, 157

Quilis, V., & Moore, B. 2001, ApJ, 555, L95

Savage, B.D., Sembach, K.R., Tripp, T.M., & Richter, P. 2002, ApJ 564, 631

Savage, B.D., Sembach, K.R., Wakker, B.P., et al. 2003, ApJS, in press (May)

Sembach, K.R., Wakker, B.P., Savage, B.D., et al. 2003, ApJS, in press (May)

Shull, J.M., et al. 1999, The Emergence of the Modern Universe: Tracing the Cosmic Web, White Paper on ST2010, available at http://origins.colorado.edu/uvconf/white-final/UVOWG_final_jun20.html

Sutherland, R.S., & Dopita, M.A. 1993, ApJS, 88, 253

Tripp, T.M., Savage, B.D., & Jenkins, E.B. 2000, ApJ 534, L1

Verner, D.A., Barthel, P.D., & Tytler, D. 1994, A&AS, 108, 287

Wakker, B.P., Savage, B.D., Sembach, K.R., et al. 2002, ApJS, in press (May)

Wakker, B.P., & van Woerden, H. 1997, ARA&A, 35, 217

Hubble's Science Legacy: Future Optical-Ultraviolet Astronomy from Space
ASP Conference Series, Vol. 291, 2003
K.R. Sembach, J.C. Blades, G.D. Illingworth, R.C. Kennicutt, Jr. (eds.)

Constructing Physically Realistic Models of Interstellar Gas in Galaxies: Start with Understanding the Physics of the Local ISM

Jeffrey L. Linsky

JILA / University of Colorado and NIST, Boulder, CO 80309-0440

Abstract. High resolution spectra in the ultraviolet and optical regions from the Hubble Space Telescope (HST), the Far Ultraviolet Spectroscopic Explorer (FUSE), and ground-based telescopes are providing important information on the structure and physical properties of the local region of the Galactic disk. This ongoing study is critically important because a detailed understanding of our local region of space provides essential guidance for understanding the interstellar medium elsewhere in the Galaxy, in other galaxies, and even the intergalactic medium. I describe some of the important questions that must be answered and the observational capabilities that are needed in future space telescopes to address these issues.

1. Why Understanding the Local ISM is Important

The low density and dilute radiation fields in the interstellar medium (ISM) require that atoms and ions have significant populations only in their ground states. Atomic physics requires us to study the gas components in the ISM at the wavelengths where their resonance lines are located, which for most species is in the ultraviolet and at shorter wavelengths. To study the ionization, chemical abundance, and dynamics of interstellar gas requires measurement of the line shapes and Doppler shifts of narrow resonance absorption lines along the lines of sight to stars or extragalactic sources. Thus, interstellar gas studies require ultraviolet spectroscopy with the highest spectral resolution of the Space Telescope Imaging Spectrograph (STIS), for which $R = \lambda/\Delta\lambda \approx 100,000$ (3 km s^{-1}).

Before describing in more detail the important scientific questions of interstellar science that should be addressed in future space missions and the observational capabilities needed to address them, I would like to ask two questions:

Question No. 1: Is it possible to construct a physically realistic model of interstellar gas in a starburst galaxy or a low metal abundance galaxy without first understanding the essential physical processes occurring in the galaxy?

Question No. 2: Can one understand the physical processes operating in any galaxy without first understanding the physical processes occurring in the ISM close to home – the local region of the Galactic disk?

I believe that the answer to both questions is a resounding *no*. So, let us take a closer look at the ISM in the solar neighborhood.

2. Important Unanswered Questions Concerning the Local ISM

2.1. What is the Structure of the Local ISM?

There are several identified components of the ISM in the Galactic plane within about 200 pc of the Sun:

Local Bubble: This consists of mostly hot ($\log T = 6.0$–6.1), very low density gas extending 50–200 pc in different directions. A rough outline of its shape is determined from the observed "wall" of Na I absorption (Sfeir et al. 1999). The origin of the hot gas is likely from the winds of massive stars and supernovae in the Sco-Cen Association. The gas flow is from $l = 12.6°$, $b = +11.7°$ at 28.2 km s^{-1} (Frisch 2001).

Local Interstellar Cloud (LIC): This is a warm ($T \approx 7,000$ K) partially ionized cloud with dimensions about 4.7 x 6.8 pc (Redfield & Linsky 2000). The Sun is located just inside the LIC and will leave in < 7000 years. In most directions the measured velocities are consistent with the projected flow of one vector, which is consistent with the flow of neutral interstellar gas inside the heliosphere. However, in the downwind direction the flow is 3 km s^{-1} too slow, and in the upwind direction (usually called the G cloud), the flow is 3 km s^{-1} too fast. Does this mean that the LIC is not behaving as a rigid body but is instead compressed by outside forces?

Other warm clouds: At least 10 other warm clouds located near the LIC have been identified tentatively from HST/GHRS or HST/STIS spectra. They have a range of temperatures and metal depletions. The G cloud (in the upflow direction) and the Hyades Cloud (in the downflow direction) may be in contact with the LIC.

Cold gas: Significant column densities of Na I define the outer edge of the Local Bubble (e.g., Sfeir et al. 1999), indicating the presence of cold gas at this location. Small clouds of cold gas are also present in the local ISM as seen, for example, toward several binary stars, but there has been no systematic survey that could identify these clouds.

Although ultraviolet and optical spectroscopy has revealed some information on the physical properties and structure of the gas in the local ISM, there remain important unanswered questions:

- Is it realistic to talk in terms of discrete clouds with sharp boundaries in which metal depletions, temperature, and density are homogeneous, or should one talk in terms of "clouds" in which all quantities vary spatially?
- Is the LIC surrounded by other warm clouds or by hot Local Bubble gas? Does this make a big difference?

- Why is there a large range of depletions in the local ISM?

2.2. Are We Resolving the Microstructure of Molecular Clouds?

This is an important question because cold clouds are regions of star formation. We do not know how many small cold clouds are present in the local ISM.

- In a high resolution (0.3–1.2 km s^{-1}) survey of interstellar Ca II absorption (formed in warm and cold gas), Welty et al. (1996) concluded that they had identified only about 40% of the velocity components that are likely present. Very narrow velocity components indicate cold gas.

- Thus, most velocity features for heavy elements are unresolved even by HST/STIS, and one cannot accurately measure the gas temperature or non-thermal velocity components without higher resolution spectra. Line saturation is also difficult to determine with the resolution of HST/STIS.

- Are the non-thermal motions subsonic or supersonic? Can they support the gas against collapse?

- The smallest observed structures are sub-parsec in size:

  - Lauroesch et al. (1998) used HST/GHRS to observe both components of the μ Cru binary system. They found large differences between the absorption profiles in the lines of sight to the two stars for neutral species (Na I, Cl I, Mg I) but no significant differences in the profiles of first ions (Cr II and Zn II). They inferred that the extra absorber toward one of the stars is likely a cold interstellar gas cloud with a size of 10–6600 AU (the binary separation) and $n_{HI} < 250$ cm^{-3}.
  - Lauroesch et al. (2000) found changes in the Na I absorption toward HD 32040 over a 4.25 year observing run. Since the star's proper motion was 15–21 AU during this time period, the scale size for the cold interstellar gas must be comparable and $n_{HI} = 20$–200 cm^{-3}.
  - Redfield & Linsky (2001) identified a warm interstellar cloud towards stars in the Hyades Association and extending over a large portion of the sky. This Hyades Cloud could be the "blue cloud" seen toward Sirius (2.6 pc) and thus lie just outside of the LIC. The smallest dimension in the Hyades cloud is ≤ 0.6 pc, but could be much smaller.

Are these small structure sizes controlled by magnetic fields, shocks, thermal instabilities, or something else? Is the LIC filled or does it contain large holes with very low density like Swiss cheese?

2.3. What is the Ionization of the Local ISM, and What is Responsible for the Ionization?

Some fundamental questions:

- Are the ionization states of H, He, and metals in steady-state equilibrium with the incident radiation fields, or are they out of equilibrium following catastrophic events (e.g., supernovae)?

- What are the major sources of ionization: early-type stars, white dwarfs, diffuse ultraviolet (UV) background, shocks, cloud interfaces, or something else?

Some recent progress:

- Lyu & Bruhweiler (1996) computed time-dependent ionization models of the local ISM to explain the observed high ionization of He. They assumed that a supernova (like SN 1987A) in the Sco-Cen Association produced a blast wave that completely ionized H and He, and the gas is now recombining. Their best fit to the observed H and He ionization is for a supernova event occurring 1–3.6 million years ago. Their result does not change significantly when the extreme ultraviolet (EUV) radiation field from ϵ CMa and other hot stars and white dwarfs and the EUV background are included or ignored.

- Jenkins et al. (2000) studied Ar I 1048 Å absorption toward four nearby white dwarfs observed by FUSE. The observed low N(Ar I)/N(H I) ratios favor steady-state photoionization models, because the photoionization cross section of Ar I is larger than H I, but the recombination coefficients are nearly the same. Thus photoionization will make argon more ionized than hydrogen, yielding a lower N(Ar I)/N(H I) ratio than would be predicted for a recombining plasma.

- Slavin & Frisch (2002) computed photoionization models for the local ISM (within $\sim$ 5 pc of the Sun) including radiation fields from hot stars (e.g., ϵ CMa and white dwarfs), the uncertain EUV radiation from the hot gas responsible for the soft X-ray background, and a possible evaporative boundary between hot gas in the Local Bubble and warm clouds. Some of their models can explain the observed ionization fractions, column densities, heating, and cooling in the LIC.

- The recent Wood et al. (2002) empirical study of the first three ionization stages of carbon, nitrogen, and silicon in the LIC toward Capella using HST/STIS and FUSE spectra finds good agreement with the predictions of the Slavin & Frisch (2002) steady-state photoionization model 17.

How to proceed:

- Cloud geometries are needed because clouds absorb the incident EUV radiation field. We need high resolution ultraviolet spectra to separate individual velocity components along a line of sight.

- What is the magnetic field at the cloud/Local Bubble boundary? This is an important question because magnetic fields can suppress thermal conduction at the interface and thus lower the EUV photoionizing radiation.

- Why is helium more highly ionized than hydrogen along some lines of sight? We need to measure accurately the diffuse background at $E > 24.6$ eV which photoionizes neutral helium. The Cosmic Hot Interstellar Plasma Spectrometer (CHIPS) should measure this quantity.

- Significant deviations from steady-state ionization equilibrium are possible because the recombination time for hydrogen is short in the LIC. The case for steady-state ionization equilibrium is not yet proven.

- Why are there large differences in metal depletions over short distances? For example, the depletion of magnesium is 4.4 times smaller in the G cloud toward α Cen than in the LIC.

- We need to test and refine the photoionization models by comparison with accurate measured column densities for a wide range of targets observed with very high resolution ($\sim$ 1 km s^{-1}) in the UV and far ultraviolet (FUV).

2.4. Do we really understand Galactic Chemical Evolution?

One of the pillars of astrophysics is the standard model of galactic chemical evolution. It is generally assumed that the initial chemical abundances in a newly formed galaxy are the primordial abundances from Big Bang nucleosynthesis, and that as stars evolve they destroy deuterium and create metals that are dispersed in the ISM by stellar winds and explosions. New generations of stars form out of this metal-enriched and deuterium-depleted gas and the process continues. It is important to test the predictions of this model to see whether something important is missing. In particular, what are the relevant temporal and spatial scales for mixing of the chemical elements in a galaxy? What roles do turbulence, magnetic fields, starbursts, supernovae, etc. play in establishing these mixing scales? Is the Galaxy well mixed?

Recent results from FUSE concerning D and O: The analysis of FUSE spectra of seven nearby hot stars published in seven papers in the May 2002 issue of the *Astrophysical Journal Supplements Series* and summarized by Moos et al. (2002) provide new information on the abundances of deuterium and oxygen in the local ISM and challenge the standard model for galactic chemical evolution.

- These new results show that D/H = $(1.52 \pm 0.08) \times 10^{-5}$ for five sight lines within the local Bubble (out to $\sim$ 100 pc), consistent with previous HST results towards cool stars (Linsky 1998). Beyond the Local Bubble, the Copernicus and IMAPS data show large variations in the D/H ratio in the range $(0.5\text{–}2.18) \times 10^{-5}$. Why is there such a large range? The most recent value for the lines of sight toward four quasars is D/H = $(3.0 \pm 0.4) \times 10^{-5}$ (O'Meara et al. 2001). This may not be the primordial value of D/H, but it likely refers to gas with low metal abundances and thus only minimal change from the primordial value. Why does the decrease of D/H in the Galactic disk appear to cover the large range between 1.4 and 6?

- Analyses of the FUSE spectra also determine that D/O = $(3.76 \pm 0.20) \times 10^{-2}$ in the Local Bubble but may vary at larger distances.

- Using HST/STIS spectra, Meyer (2001) found that O/H = $(3.43 \pm 0.15) \times 10^{-4}$ and is nearly constant in the Galactic disk out to 5 kpc from the Sun.

- These results indicate large D/H variations but no significant O/H variations beyond the Local Bubble in the disk 100–500 pc from the Sun.

- There is also some evidence that O/H may be correlated (not anti-correlated) with D/H, contrary to the prediction of the standard model for galactic chemical evolution. Additional observations are needed to confirm and quantify or refute this apparent correlation, but the data in hand provide a challenge to our understanding of galactic chemical evolution. My hunch is that a realistic model for the geometrical scale and rate of mixing of gas in the ISM must be added to the theory.

What do we need?

- We clearly need accurate measurements of D/H and D/O over a much larger region of the Galaxy. FUV spectroscopy with a resolution < 3 km s^{-1} is needed instead of the 15–20 km s^{-1} resolution of FUSE.
- UV spectroscopy with resolution ~ 1 km s^{-1} is needed to characterize the physical properties of the gas in which D/H and D/O are measured for the same lines of sight.
- We also need a better theoretical understanding of the sources and sinks of important elements and the mechanisms controlling mixing in the Galaxy.

3. Observational Requirements for a Future Mission

I believe that we will need the following observing capabilities to address these fundamental issues in a systematic way. This table includes science objectives that I have had time to discuss here.

Table 1. Observational Requirements

Science Objective	Resolution (km s^{-1})	S/N	Spectral Range
LISM Structure	1	100	UV, FUV
LISM Microstructure	<1	100	UV, Optical
LISM Components	1	50	UV, FUV, X
Interfaces	1	100	UV, FUV
Ionization	3	100	UV, FUV, EUV
Starburst galaxies	3	50	UV, FUV
Grains	1	100	UV, Optical, IR
Distant ISM	3	50	UV, FUV
Pressure Equilibrium	3	300	UV, FUV
Chemical Evolution	1	100	UV, FUV
Imaging Spectroscopy	10	50	UV, FUV

UV spectral range: 1170–3000 Å.
FUV spectral range: 910–1170 Å.

Other Capabilities Needed in Future Space Missions:

- Suppression of fixed pattern noise is essential for achieving high S/N.

- An onboard wavelength calibration lamp and thermal stability of the spectrograph are essential for an accurate velocity scale.
- Doppler compensation is essential for a satellite in low Earth orbit.
- Simultaneous observations in all spectral bands are not required.
- Synergy is required. Observations in the UV and FUV complement each other beautifully. A FUV + UV dataset is far more valuable than either a FUV or UV dataset alone.
- Sufficient observing time is essential to sample the required number of sight lines is essential.
- Simultaneous observations of several nearby targets (e.g., in a cluster) would be beautiful.

4. Think out of the Box: Compare a Full-Size Multiple Instrument Behemoth to a Series of Half-Size Single Instrument Satellites

Scientific capabilities: To accomplish the important scientific objectives in the 2010–2020 time frame, new UV/optical instruments in space must have many capabilities (e.g., wide-field imaging, low resolution faint object spectroscopy, high resolution spectroscopy, coronagraphy, high UV/optical and FUV throughput, imaging spectroscopy, etc.)

Efficiency: High throughput can be achieved by limiting the types of supported observations and optimizing the telescope/instrument system for just these limited observations. HST/COS will have a factor of 20–30 higher throughput over HST/STIS at 20,000 resolution, effectively increasing HST's aperture from 2.4 m to 10–14 m.

Limiting magnitude: Detector background is often proportional to the pixel size of a resolution element. In this regime, longer integrations can compensate for smaller aperture.

Observing time: Time sharing of the focal plane means each instrument has a low duty cycle. Four focal plane instruments decrease the telescope aperture by a factor of 2 compared to full time use of each instrument.

Cost: Satellite costs are often proportional to weight plus complexity. If all else is equal, doubling the aperture raises the cost by roughly a factor of 4–8. Multiple focal plane instruments increase complexity and cost for construction, testing, and operations. High cost satellites take forever to build and are eventually descoped.

A modest proposal that deserves serious thought and study:

- Design and build a series of satellites each with half the aperture of the behemoth (say 4 m each instead of 8 m). Constellation-X is an example of a multi-satellite program.

- Each satellite will have only one focal plane instrument with limited observing modes that are optimally matched to its telescope optics (f-number, coating, field of view, mirror quality, scattered light suppression, etc.). Make no compromises.

- Each spacecraft will be as identical as possible so that they can be mass produced to reduce cost. Produce a few spares to be used to replace satellites as needed.

- Launch one satellite in the program every year or two and do not design for on-orbit replacement.

- A program of many satellites greatly reduces program risk because failure of a critical component can only end the life of one satellite not the whole program. Less perceived risk means lower costs. The program is designed to live with individual failures.

- If the scaling law is approximately correct, then roughly 4–8 half-size satellites can be built and operated for the cost of one behemoth. This program should produce far more data of similar quality to the behemoth. Even 4 half-size satellites will be more scientifically productive than a behemoth at much lower cost and fewer delays.

References

Frisch, P.C. 2001, in Gaseous Matter in Galaxies and Intergalactic Space, eds. R. Ferlet et al., (Paris: Frontier Group), 19

Jenkins, E.B. et al. 2000, ApJ, 538, L81

Lauroesch, J.T., Meyer, D.M., Watson, J.K., & Blades, J.C. 1998, ApJ, 507, L89

Lauroesch, J.T., Meyer, D.M., & Blades, J.C. 2000, ApJ, 543, L43

Linsky, J.L. 1998, Space Sci. Rev., 84, 285

Lyu, C.-H., & Bruhweiler, F.C. 1996, ApJ, 459, 216

Meyer, D.M. 2001, in Gaseous Matter in Galaxies and Intergalactic Space, ed. R. Ferlet et al., (Paris: Frontier Group), 135

Moos, H.W., et al. 2002, ApJS, 140, 3

O'Meara, J.M., Tytler, D., Kirkman, D., Suzuki, N., Prochaska, J., Lubin, D., & Wolfe, A.M. 2001, ApJ, 552, 718

Redfield, S., & Linsky, J.L. 2000, ApJ, 534, 825

Redfield, S., & Linsky, J.L. 2001, ApJ, 551, 413

Sfeir, D.M., Lallement, R., Crifo, F., & Welsh, B.Y. 1999, A&A, 346, 785

Slavin, J.D., & Frisch, P.C. 2002, ApJ, 565, 364

Welty, D.E., Morton, D.C. & Hobbs, L.M. 1996, ApJS, 106, 533

Wood, B.E., Redfield, S., Linsky, J.L., & Sahu, M.S. 2002, ApJ, submitted

Hubble's Science Legacy: Future Optical-Ultraviolet Astronomy from Space
ASP Conference Series, Vol. 291, 2003
K.R. Sembach, J.C. Blades, G.D. Illingworth, R.C. Kennicutt, Jr. (eds.)

Supermassive Black Hole Research in the Post-HST Era

Laura Ferrarese

Rutgers University, 136 Frelinghuysen Road, Piscataway, NJ 08854

Abstract. Thanks to its unprecedented spatial resolution, the Hubble Space Telescope (HST) has ended a 20 year stalemate by detecting the dynamical signature of nuclear supermassive black holes (SBHs) in a sizeable number of nearby galaxies. These detections have revealed the existence of a symbiotic relationship between SBHs and their hosts, changing the way we view SBH and galaxy formation. In this contribution I review which are the most pressing outstanding issues in SBH research, and what are the technological requirements needed to address them.

1. The Current Status of SBH Searches

The study of supermassive black holes is one of the areas of modern astrophysics which has benefited most from the launch of HST. After two decades of tantalizing but inconclusive ground-based studies, the HST/FOS observations of M87 (Harms et al. 1994) and NGC 4261 (Ferrarese et al. 1996) provided the first firm measurements of SBH masses in galactic nuclei. In the years that followed, HST/FOS and HST/STIS data lead to detections in ten additional galaxies (Bower et al. 1998; van der Marel & van den Bosch 1998; Ferrarese & Ford 1999; Emsellem et al. 1999; Cretton & van den Bosh 1999; Verdoes Kleijn et al. 2000; Gebhardt et al. 2000b; Joseph et al. 2001; Barth et al. 2001; Sarzi et al. 2001). The superiority of HST over ground-based facilities in this field is easily understood. Only dynamical evidence, either from gas or stellar kinematics, can yield compelling proof of the existence of SBHs. With rare exceptions (e.g., M31, M87), ground-based telescopes lack the spatial resolution necessary to resolve the SBH "sphere of influence" (i.e., the region of space within which the SBH gravitational influence dominates that of the surrounding stars):

$$r_h = GM_\bullet/\sigma^2 \sim 11.2(M_\bullet/10^8\ \mathrm{M}_\odot)/(\sigma/200\ \mathrm{km\ s}^{-1})^2\ \mathrm{pc}, \qquad (1)$$

with σ the stellar velocity dispersion and $M_\bullet$ the SBH mass. Resolving r_h is a necessary condition for a SBH detection to be made; not meeting it leads to spurious detections and biased masses (Merritt & Ferrarese 2001a).

With over a dozen secure measurements, it has become possible to search for correlations between $M_\bullet$ and the overall properties of the host galaxies. The first relation to emerge was one between $M_\bullet$ and the blue luminosity L_B of the surrounding bulge (Kormendy & Richstone 1995). A much tighter correlation was subsequently discovered between $M_\bullet$ and the bulge stellar velocity dispersion (Ferrarese & Merritt 2000; Gebhardt et al. 2000c):

$$M_\bullet = \beta\left(\frac{\sigma}{200\ \mathrm{km\ s^{-1}}}\right)^{\alpha}. \tag{2}$$

with $\alpha = 4.58 \pm 0.52$ and $\beta = (1.66 \pm 0.32) \times 10^8\ \mathrm{M_\odot}$ (Ferrarese 2002a). More recently, evidence has emerged that a fundamental relation might exist between $M_\bullet$ and the mass M_{DM} of the dark matter halos in which the SBHs presumably formed (Ferrarese 2002b):

$$\frac{M_\bullet}{10^8\ \mathrm{M_\odot}} \sim 0.10\left(\frac{M_{DM}}{10^{12}\ \mathrm{M_\odot}}\right)^{1.65} \tag{3}$$

The above relations have proven invaluable in the study of SBH demographics (Merritt & Ferrarese 2001b; Ferrarese 2002a; Yu & Tremaine 2002) and have generated intense activity on the theoretical front (Haehnelt, Natarajan, & Rees 1998; Silk & Rees 1998; Cattaneo, Haehnelt, & Rees 1999; Adams, Graff, & Richstone 2000; Monaco et al. 2000; Haehnelt & Kauffmann 2000; Wyithe & Loeb 2002). At the same time, new questions have arisen, and with them the need for further observational constraints. In this contribution, I will address three such questions, and identify the technological requirements necessary to answer them.

- *What are the characteristics of the $M_\bullet - \sigma$ relation? Does the slope and/or normalization of the relation depend on Hubble type, environment, and/or redshift?* As theoretical models are refined, tighter observational constraints will be required. Most of the SBHs detected to date are in the $10^8 \lesssim M_\bullet \lesssim 10^9\ \mathrm{M_\odot}$ range. The $M_\bullet - \sigma$ relation is not sampled below $10^6\ \mathrm{M_\odot}$, and badly sampled for $10^6 \lesssim M_\bullet \lesssim 10^7\ \mathrm{M_\odot}$. There are few spiral galaxies represented, all of which are early type, and only two galaxies well beyond 30 Mpc.

- *Are binary supermassive black holes long lived?* The existence and lifespan of binary SBHs can have dramatic consequences, from shaping the morphology and dynamics of the resulting galaxy (Milosavljevic & Merritt 2001; Milosavljevic et al. 2002; Ravindranath et al. 2002; Yu 2002) to destroying nuclear dark matter halo cusps (Merritt et al. 2002).

- *How small can nuclear BHs be? Are there nuclear BHs in globular clusters?* There is no dynamical evidence for "intermediate mass" black holes (IBHs) in the $M_\bullet \sim 10^2 - 10^6\ \mathrm{M_\odot}$ range, although their existence in the off-nuclear regions of some starburst galaxies is supported by energetic arguments (Fabbiano et al. 2001; Matsumoto et al. 2001). There is also no dynamical evidence that BHs are formed in the nuclei of globular clusters (van der Marel et al. 2000; Gebhardt et al. 2000a). However, whether such black holes exist is critical for our understanding of how SBHs form. In "top-down" self-regulating models that trace the formation of SBHs to the very early stages of galaxy formation, there is a natural lower limit of $\sim 10^6$ $\mathrm{M_\odot}$ to $M_\bullet$ (e.g., Loeb 1993; Silk & Rees 1998; Haehnelt, Natarajan, &

Rees 1998). On the other hand, in "bottom-up" models nuclear SBHs are formed by the merging of IBHs. These are deposited at the galactic center as the globular clusters in which they originally formed spiral in due to dynamical friction (Portegies Zwart & McMillan 2002; Ebisuzaki et al. 2001). In the latter scenario, no physical reason would prevent the formation of SBHs with $M_\bullet \lesssim 10^6$ $M_\odot$.

2. Building the Local Sample

Galaxies come in all flavors, but do all flavors come with a SBH? The sample of galaxies within the reach of HST is remarkably homogeneous. Most are early type galaxies. And although we might expect the number of detections to double or perhaps even triple as new HST data become available (Kormendy & Gebhardt 2001), most will remain in the $10^8 \lesssim M_\bullet \lesssim 10^9$ $M_\odot$ range which is already well sampled by the current data (Merritt & Ferrarese 2002a). Hopes of breaking the 10^6 $M_\odot$ barrier lie in one galaxy only, NGC 205, a compact elliptical already scheduled to be observed with HST (P.I. L. Ferrarese)[1].

But there are other barriers HST cannot break. The low central surface brightness which characterizes giant ellipticals (e.g. Ferrarese et al. 1994; Lauer et al. 1995; Rest et al. 2000) makes stellar dynamical studies with HST prohibitive. For instance, measuring $M_\bullet$ in M87 ($d \sim 15$ Mpc) using stellar dynamics would require over 100 orbits of STIS time[2]. While this problem can be avoided by using gas, rather than stellar, kinematics, only a fraction of ellipticals host the regular NGC4261-like dust disks (Jaffe et al. 1993) which make gas dynamical studies possible. Therefore, ironically, most of the very largest SBHs ($M_\bullet \gtrsim 10^9$ $M_\odot$) are beyond the reach of HST.

Similarly, the vast majority of dwarf elliptical galaxies are beyond HST capabilities. Assuming that the $M_\bullet - \sigma$ relation holds in the $\sigma \sim$ a few $\times 10$ km s^{-1} regime that characterizes these objects, the sphere of influence of the putative SBH at their centers is accessible only in the most nearby systems. In these cases, the stellar population is resolved into individual stars, each too faint to be handled by HST's small mirror. For instance, a constraint on the central mass in NGC 147 ($d \sim 660$ kpc) would require several hundred orbits with STIS. The situation for late type spirals is no better. The upper limit on $M_\bullet$ in M33, the closest Sc galaxy, puts the sphere of influence of the putative BH a factor ~ 20 below the resolution capabilities of HST.

The above statements are quantified in Figure 1. For each galaxy in the CfA redshift sample (Huchra et al. 1990), $M_\bullet$ is calculated from the $M_\bullet - L_B$ relation given by Ferrarese & Merritt (2000), after a correction suitable to each Hubble type is applied to convert total luminosity to bulge luminosity (Fukugita et al. 1998). If only resolution constraints are considered, most late type spirals (Sb - Sc) are expected to host SBHs too small to be resolved by HST. Only a handful of SAs are within reach. It is only with an 8 m class telescope that a complete

---

[1] Based on the $M_\bullet - \sigma$ relation, NGC 205 is expected to host a $\sim 7.5 \times 10^4$ $M_\odot$ black hole; at a distance of 740 kpc, a black hole as small as 6×10^4 $M_\odot$ can be detected by HST/STIS.

[2] The calculation assumes S/N = 50 at ~ 8500 Å.

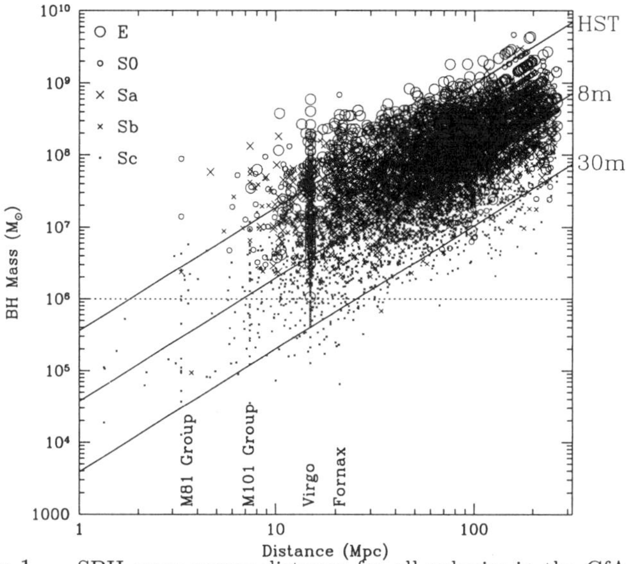

Figure 1. SBH mass versus distance for all galaxies in the CfA Redshift Sample (Huchra et al. 1990). Only for the galaxies which lie above the solid lines the sphere of influence of the putative nuclear SBH can be resolved by HST/STIS, an 8 m and a 30 m diffraction-limited telescope. A few nearby groups and clusters are marked. It should be noted that because of the large scatter in the $M_\bullet - L_B$ relation (a factor of several in $M_\bullet$), this figure has only statistical value.

sample of galaxies spanning the whole Hubble sequence can be collected. Even then, little would be gained below 10^6 $M_\odot$: pushing this limit down by an order of magnitude requires a 30 m diffraction limited telescope.

The above constraints become tighter when exposure time requirements are folded in. Figure 2 shows the detection limits for HST, an 8 m and a 30 m diffraction limited telescopes assuming the equivalent of 3 HST orbits of integration on each galaxy (see footnote 2). The points represent the early type galaxy sample of Faber et al. (1989), with $M_\bullet$ calculated using the $M_\bullet - \sigma$ relation. Again, HST can only see the top of the iceberg, producing enough SBH detections to allow us a glimpse of what lays underneath, but leaving most of the parameter space unexplored. To study the influence of environment on the formation and evolution of SBHs, it would be helpful to probe Coma and the most nearby rich Abell clusters at $d \sim 100$ Mpc. In order to push farther and perhaps study the redshift evolution of the $M_\bullet - \sigma$ relation, reverberation mapping (see Peterson, this volume) will likely remain the only viable method.

3. Binary Supermassive Black Holes

The formation of binary SBHs as a consequence of merging seems unavoidable (Begelman, Blandford, & Rees 1980), however the evolution of the SBH binary after merging is very uncertain (Ebisuzaki, Makino, & Okumura 1991;

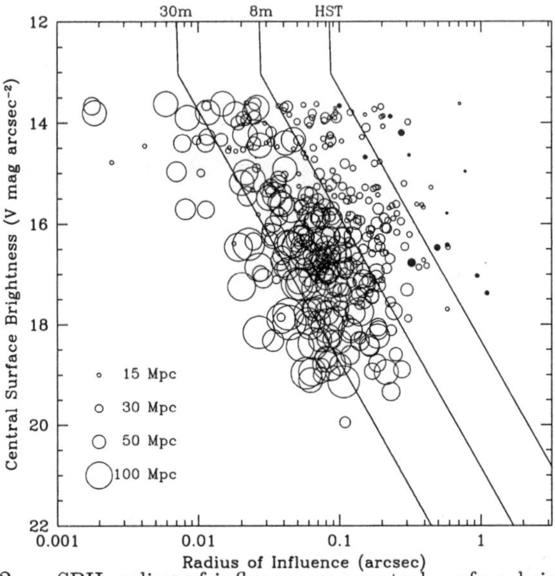

Figure 2. SBH radius of influence vs. central surface brightness for the early-type galaxy sample of Faber et al. (1989). The size of the symbols is proportional to the galaxy distances, as shown in the legend. Spectra with ideal signal-to-noise and spatial resolution for dynamical studies can be collected in the equivalent of 3 HST orbits only for the galaxies to the right of the solid lines (shown for HST/STIS, an 8 m and a 30 m diffraction limited telescopes). The solid circles identify galaxies observed with HST.

Milosavljevic & Merritt 2002; Yu 2002). The recent work of Milosavljevic & Merritt (2002) represents the state of the art in numerical simulations of merging galaxies with SBHs. We will adopt the results of this work to estimate the resolution requirements necessary to detect the dynamical signature of a SBH binary. Milosavljevic & Merritt follow the merging of two low-luminosity ellipticals, characterized by steep power-law stellar density profiles. Well within one Myr after the galaxies merge, the two SBHs form a hard binary with separation between a few hundredths to a few parsecs, depending on the SBHs masses. By exchanging energy with nearby stars, the binary will start to harden; however, Milosavljevic & Merritt (2002) find that the hardening will eventually stall, as a consequence of the depletion of stars with which the binary can interact. At this stage, the separation between the SBHs is $\lesssim 0.2$ pc ($\propto M_\bullet{}^{0.57}$), and the stellar rotational velocity and velocity dispersion differ significantly from the single SBH case (see Figure 15 of Milosavljevic & Merritt 2002). Figure 3 shows the limit at which the "final" distance between the binary SBHs can be resolved with HST, an 8 m and a 30 m diffraction limited telescope, as a function of the binary mass and distance. The solid points show all SBHs detected to date. Rather obviously, a direct dynamical detection of a binary SBH would require a 30 m class telescope, and even then only a few of the most nearby galaxies would be accessible. Unless proof of the existence of binary SBHs can be obtained in

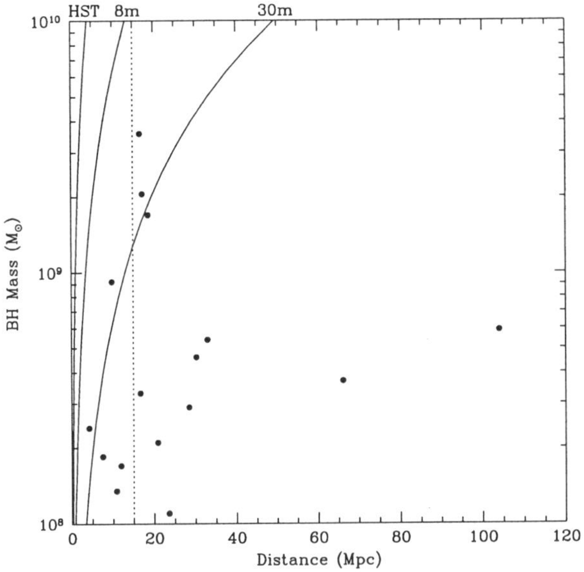

Figure 3. The points represent all SBH masses detected to date (see references in Merritt & Ferrarese 2000). According to the simulations of Milosavljevic & Merritt (2001), a SBH would have to lie to the left of the solid line (drawn for HST, an 8 m and a 30 m diffraction-limited telescope) to be resolved as a binary. The vertical dotted line marks the position of the Virgo cluster.

some other way (with LISA, for instance), it would not be a bad idea to start cheering for OWL (see Gilmozzi, this volume).

4. Black Holes in Globular Clusters

The formation of BHs at the centers of highly concentrated globular clusters (GCs) is suggested by theoretical arguments and numerical simulations (Miller & Hamilton 2001; Mouri & Taniguchi 2002; Portegies Zwart & McMillan 2002). Detecting the signature of a BH in GCs is best done through proper motion, rather than spectroscopic studies. The latter have been pursued in the case of M15, with no conclusive results (van der Marel 2000; Gebhardt et al. 2000). Velocities of several tens to several hundred stars within the sphere of influence of the putative BH must be collected in order to derive an accurate value of the velocity dispersion; unfortunately at the Mg b triplet ($\sim$ 5500 Å), and even more severely at the Ca II triplet ($\sim$ 8500 Å), the fainter turnoff and main sequence stars are drowned in the light of a few nearby giants. In the case of M15, for instance, the sphere of influence of a 10^3 $M_\odot$ BH is $\sim 1''$, and about 90 velocities are needed to produce a value of the velocity dispersion accurate to 10% within this region; using state of the art ground-based instrumentation, Gebhardt et al. (2000) were able to measure velocities for only 5 stars within $1''$.

Table 1. Technical Requirements

Project: Enlarging the Sample and Probing $< 10^6$ and $> 10^9$ $M_\odot$ SBHs

Needed to study systematics, distinguish between "bottom-up" and "top-down" models of SBH formation, constrain the role of feedback in SBH accretion during merging.

Needed: longslit, IFU highly desirable

Resolution	Aperture	FOV	λ range	$\lambda/\Delta\lambda$
<0″.02	> 8m	few×10″	5500−9500 Å	10,000

Project: Resolving SBH Binaries

Needed to contrain dynamical models of galaxy mergers, determine the impact of SBH binaries in the morphological evolution of galaxies, contrain accretion mechanism.

Needed: IFU

Resolution	Aperture	FOV	λ range	$\lambda/\Delta\lambda$
<0″.007	> 30m	few×10″	> 8000 Å	10,000

Project: Detecting BHs in GCs

Needed to constrain dynamical models for SBH evolution, investigate the connection between GCs, nuclear SBHs, and galactic bulges.

Needed: High Resolution Imager, High Dynamic Range

Resolution	Aperture	FOV	λ range	$\lambda/\Delta\lambda$
<0″.05	> 2.5m	few×10″	U or B−band	N/A

The suitability of proper motion studies in constraining the central potential has been demonstrated in spectacular fashion for the Galactic center (Ghez et al. 2000; Genzel et al. 2000), using ground-based facilities. The art of extracting astrometric information from HST/WFPC2 images has been perfected by Anderson & King (2000), who can reach an astonishing 0.005 pixel positional accuracy using properly dithered data. Constraints on the BH mass at the center of Galactic globular clusters can be easily calculated; under appropriate assumptions, with two ten-orbit exposures, taken one year apart, HST could set limits of 2000 $M_\odot$or better on the BH masses at the centers of $\sim$ ten of the nearest Galactic GCs. For a few GCs, the limit is below 1000 $M_\odot$, well in the range of BH masses estimated for the off-nuclear IBHs in the Antennae galaxies and M82 (Fabbiano et al. 2001; Matsumoto et al. 2001). While even tighter constraints could be reached using a larger aperture, the HST/ACS can be very competitive.

Table 1 summarizes the observational requirements needed to answer the questions posed in the introduction of this contribution. While most applications will require higher spatial resolution and a larger collective area, the detection of BHs in GCs could indeed be the next legacy of HST to SBH research.

References

Adams, F.C., Graff, D.S., & Richstone, D. 2000, ApJ, 551, L31
Anderson, J., & King, I.R. 2000, PASP, 112, 1360
Barth, A.J., et al. 2001, ApJ, 555, 685
Begelman, M.C., Blandford, R.D., & Rees, M.J. 1980, Nature, 287, 307
Bower, G. A., et al. 1998, ApJ, 492, L111
Cattaneo, A., Haehnelt, M.G., & Rees, M. J. 1999, MNRAS, 308, 77
Cretton, N., & van den Bosch, F.C. 1999, ApJ, 514, 704
Ebisuzaki, T., Makino, J., & Okumura, S.K. 1991, Nature, 354, 212
Ebisuzaki, T., et al. 2001, ApJ, 562, L19
Emsellem, E., Dejonghe, H. & Bacon, R. 1999, MNRAS, 303, 495
Fabbiano, G., Zezas, A., & Murray, S.S. 2001, ApJ, 554, 1035
Faber, S., et al. 1989, ApJS, 69, 763
Ferrarese, L., et al.. 1994, AJ, 108, 1598
Ferrarese, L., & Ford, H.C. 1999, ApJ, 515, 583
Ferrarese, L., Ford, H.C. & Jaffe, W. 1996, ApJ, 470, 444
Ferrarese, L., & Merritt, D. 2000, ApJ, 539, L9
Ferrarese, L. 2002a, to appear in Current High-Energy Emission around Black Holes, Proc. 2nd KIAS Astrophysics Workshop, ed. C.-H. Lee., (Singapore: World Scientific), (astro-ph/0203047)
Ferrarese, L. 2002b, ApJ in press (astro-ph/0203469)
Fukugita, M., Hogan, C.J., & Peebles, P.J.E. 1998, ApJ, 503, 518
Gebhardt, K., et al., 2000a, AJ, 119, 1268
Gebhardt, K., et al. 2000b, AJ, 119, 1157
Gebhardt K., et al. 2000c, ApJ, 539, L13
Genzel, R., et al. 2000, MNRAS, 317, 348
Ghez, A., et al. 2000, Nature, 407, 349
Haehnelt, M.G., & Kauffmann, G. 2000, MNRAS, 318, L35
Haehnelt, M.G., Natarajan, P. & Rees, M. J. 1998, MNRAS, 300, 817
Harms, R.J., et al. 1994, ApJ, 435, L35
Huchra, J. P., et al. 1990, ApJS, 72, 433
Jaffe, W., et al. 1993, Nature, 364, 213
Joseph, C., et al. 2001, ApJ, 550, 668
Kormendy, J., & Richstone 1995, ARAA, 27, 235
Kormendy, J., & Gebhardt, K., 2001, astro-ph/0105230
Lauer, T., et al. 1995, AJ, 110, 2622
Loeb, A. 1993, ApJ, 403, 542
Matsumoto, H., et al. 2001, ApJ, 547, L25
Merritt, D., & Ferrarese, L. 2001a, in ASP Conf. Ser. 249, The Central Kpc of Starbursts and AGNs, , eds. J.H. Knapen, J.E. Beckman, I. Shlosman, & T.J. Mahoney (San Francisco: ASP), 335

Merritt, D., & Ferrarese, L. 2001b, MNRAS, 320, L30

Merritt, D., et al., 2002, PhysRev, 88, 1301

Miller, M.C., & Hamilton, D.P. 2002, MNRAS, 330, 232

Milosavljevic, M., & Merritt, D. 2001, ApJ, 563, 34

Milosavljevic, M., et al. 2002, MNRAS, 331, L51

Monaco, P., Salucci, P., & Danese, L. 2000, MNRAS, 311, 279

Mouri, H., & Taniguchi, Y. 2002, 566, L17

Portegies Zwart, S., & McMillan S. 2002, ApJ, in press (astro-ph/0201055)

Rest, A., et al. 2001, AJ, 121, 2431

Sarzi, M., et al. 2001, ApJ, 550, 65

Silk, J., & Rees, M.J. 1998, A&A, 331, L4

van der Marel, R.P., & van den Bosch, F.C. 1998, AJ, 116, 2220

van der Marel, R.P. 2000, in Black Holes in Binaries and Galactic Nuclei, Proceedings of an ESO Workshop in September 1999, eds. L. Kaper, E.P.J. van den Heuvel, & P.S. Woudt, (Heidelberg: Springer-Verlag)

Verdoes Kleijn, G.A., et al. 2000, AJ, 120, 1221

Wyithe, S., & Loeb, A. 2002, ApJ, submitted (astro-ph/0206154)

Yu, Q. 2002, MNRAS, 331, 935

Yu, Q., & Tremaine, S. 2002, MNRAS, in press

Hubble's Science Legacy: Future Optical-Ultraviolet Astronomy from Space
ASP Conference Series, Vol. 291, 2003
K.R. Sembach, J.C. Blades, G.D. Illingworth, R.C. Kennicutt, Jr. (eds.)

Integral-Field Spectroscopy of Galactic Nuclei

Tim de Zeeuw

Leiden Observatory, The Netherlands

Abstract. High-resolution imaging and long-slit spectroscopy obtained with the Hubble Space Telescope (HST), combined with ground-based integral-field spectroscopy, provides the kinematics of stars and gas in nearby galactic nuclei with sufficient accuracy to derive the intrinsic dynamical structure, and to measure the mass of the central black hole. This has revealed that many nuclei contain decoupled kinematic components and asymmetric structures, and that nuclear and global properties of galaxies are correlated. Higher spatial resolution and significantly increased sensitivity are required to cover the full range of galaxy properties and types, including the nearest powerful active radio galaxies, and to study the evolution of galactic nuclei as a function of redshift.

1. Integral-Field Spectroscopy

The past decade has seen a revolution in instrumentation for spatially resolved spectroscopy of galaxies and their nuclei. At many observatories, traditional long-slit spectrographs have been replaced by integral-field devices, which produce spectra over an area, fully spatially sampled, in many cases taking advantage of adaptive optics capabilities (see, e.g., Emsellem & Bland–Hawthorn 2002 for a recent summary). These instruments are very efficient in use of telescope time, and allow complete reconstruction of the intensity distribution from the spectra, so that errors in the registration of the spectra relative to the galaxy image, familiar from aperture and long-slit spectroscopy, are avoided.

2. Galactic Nuclei with HST

HST has revealed that most early-type galaxies have centrally cusped luminosity distributions, often containing nuclear stellar and gaseous disks as well as asymmetries. HST/FOS aperture spectroscopy and HST/STIS long-slit spectroscopy of stars and gas have demonstrated that nearly all these nuclei contain a supermassive black hole, with masses of 10^6 to a few times $10^9 M_\odot$ (e.g., Kormendy & Gebhardt 2002). Global and nuclear properties appear to correlate. Examples include a correlation of cusp-slope with total luminosity, and of black hole mass with host galaxy velocity dispersion (e.g., Gebhardt et al. 2000; Ferrarese & Merritt 2000). We discuss four examples of HST studies of nuclei to illustrate the need for two-dimensional spectroscopic coverage, the need for measurement of the absorption-line kinematics, and the need for high spatial resolution.

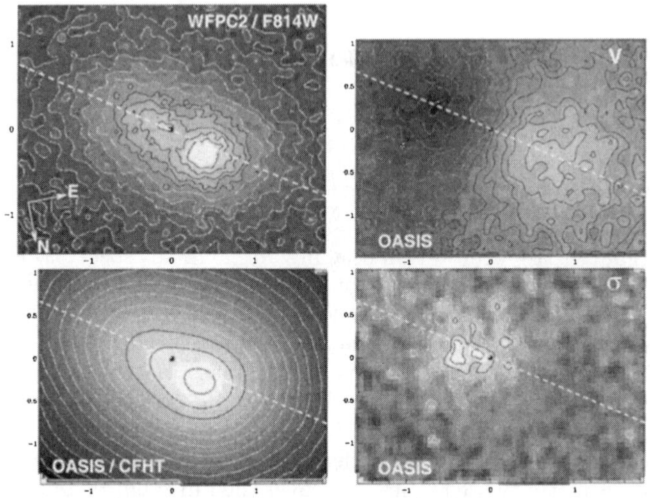

Figure 1. The nucleus of M31. Clockwise from top left: deconvolved HST/WFPC2 I-band image, stellar velocity V, stellar velocity dispersion σ, and intensity I derived from OASIS integral-field spectroscopy near the Ca triplet (Bacon et al. 2001a). The brightest peak in the HST/WFPC2 image is P1; the secondary peak is P2. The dashed line indicates the symmetry axis of the V-field. The field-of-view is 3.″5 by 2.″5.

2.1. The Nucleus of M31

Stratoscope II discovered that the nucleus of M31 is asymmetric (Light, Danielson, & Schwarzschild 1974). HST imaging revealed two peaks in the brightness distribution, labeled P1 and P2 by Lauer et al. (1993), separated by 0.″49. The ground-based integral-field spectroscopy with modest spatial resolution obtained with TIGER showed a nearly symmetric stellar velocity field, with P1 offset from the kinematic center, and a velocity dispersion peak near P2 but offset from it in the direction away from P1 (Bacon et al. 1994). Multiple HST/FOS pointings (Ford, unpubl.), and long-slit spectroscopy with both HST/FOC (Statler 1999) and HST/STIS (GTO 8018, PI Green, see Bacon et al. 2001a) showed an asymmetric velocity curve, and confirmed the velocity dispersion peak near P2, presumably caused by a $7 \times 10^7 M_\odot$ black hole. Adaptive optics (AO) assisted integral-field spectroscopy with OASIS on the Canada-France-Hawaii Telescope (CFHT), with an effective resolution of 0.″45, clarified these puzzling asymmetries (Bacon et al. 2001a). As Figure 1 shows, the velocity field is regular, and the dispersion peaks near P2, but a slit through P1 and P2 (a natural choice based on the HST/WFPC2 image) neither coincides with the kinematic symmetry axis nor intercepts the true σ maximum; without the OASIS data, the HST/STIS profiles are difficult to understand. The nature of the M31 nucleus is not fully understood, but the N-body models of Bacon et al. (2001a), which build on Tremaine's (1995) eccentric disk model, show that a near-Keplerian $m = 1$ density wave provides a good qualitative fit. Future models will have to deal with all the constraints established from two-dimensional spectroscopy.

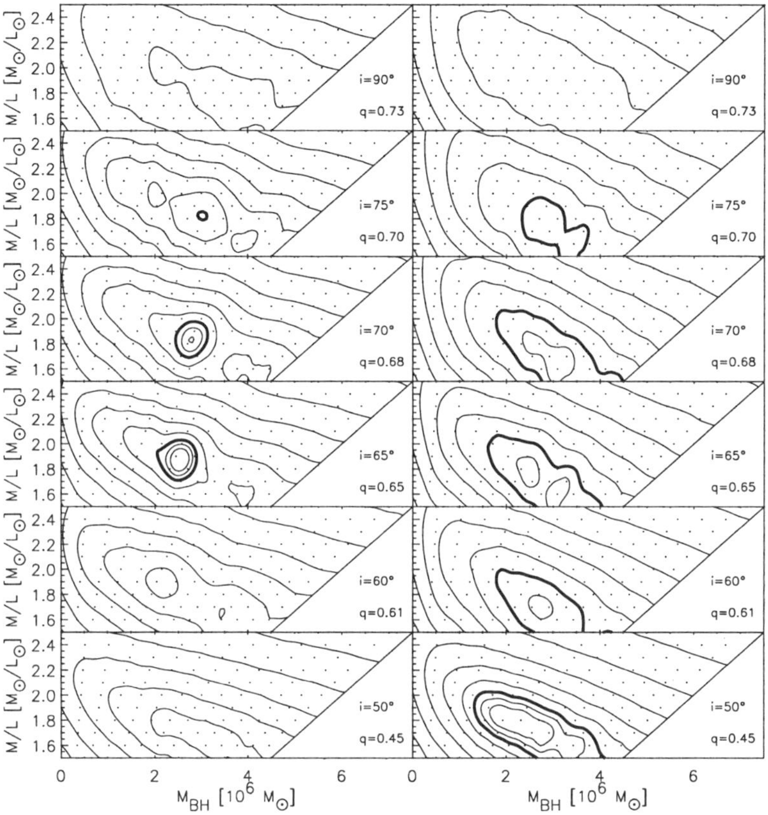

Figure 2. Dynamical models for M32. The panels show contours of $\Delta\chi^2$, which measures the goodness of fit of the dynamical models to the kinematic measurements, as a function of black hole mass $M_{\rm BH}$, stellar mass-to-light ratio M/L and inclination i, taken from Verolme et al. (2002). The thick contour is the 3σ confidence level for three degrees of freedom, at a $\Delta\chi^2$ of 14.2 above the minimum. The intrinsic flattening q of the models is indicated in the lower-right corner of each panel. *Left*: Model fits to the Joseph et al. (2001) HST/STIS major-axis data and the SAURON integral-field measurements over $9'' \times 11''$ sampled with 0.″28 pixels in 0.″95 seeing. This allows accurate measurement of M/L, $M_{\rm BH}$ and i. *Right*: Model fits which include again the HST/STIS major axis-data, but now only the SAURON measurements along 4 slits (major and minor axis, and at $\pm 45°$). This shows that the traditional kinematic coverage provides almost no constraint on i, and that the resulting uncertainties on the inferred values of M/L and $M_{\rm BH}$ are significantly larger.

2.2. The E3 Galaxy M32

M32 is a high-surface brightness inactive E3 companion of M31, and has long been suspected of harboring a central black hole of about $3 \times 10^6 M_\odot$ (e.g., Tonry 1984). This dominates the observed kinematics inside 0.″15, and as a result a reliable mass measurement had to await HST/FOS and the development of general machinery for the construction of axisymmetric dynamical models with the full range of orbital anisotropies (van der Marel et al. 1998). Verolme et al. (2002) compared the HST/STIS major axis kinematics (Joseph et al. 2001), as well as SAURON integral-field spectroscopy of the inner $9'' \times 11''$ of M32 (Bacon et al. 2001b; de Zeeuw et al. 2002) with general axisymmetric dynamical models, varying stellar mass-to-light ratio M/L, black hole mass $M_{\rm BH}$, and the inclination i. All three parameters are well-constrained, with $M_{\rm BH} = (2.5 \pm 0.5) \times 10^6 M_\odot$ (3σ-error), $M/L = 1.85 \pm 0.15\, M_\odot/L_\odot$ in the I-band, and $i = 70^\circ \pm 5^\circ$ (Figure 2, left panel). Experiments showed that the traditional approach of using ground-based kinematics along a few slits significantly degrades the measurement accuracy: it does not allow meaningful determination of i and as a result causes a larger uncertainty on $M_{\rm BH}$ (Figure 2, right panel). Even in an object as dynamically simple as M32, two-dimensional coverage is of key importance.

2.3. Kinematics of Gas or Stars?

Much work has been done on obtaining kinematics of emission-line gas of nearby nuclei (Kormendy & Gebhardt 2002). This is attractive as it allows kinematic measurements with only a few HST orbits at excellent spatial resolution. Often three parallel slits are used to mimic integral-field capability. The results are not always easy to interpret, as the gas is rarely in a simple disk, and even if it is, the kinematics may be asymmetric, or the gas velocity dispersion so large that modeling becomes problematic (e.g., Ho et al. 2002).

Black hole masses and nuclear properties derived independently from the kinematics of gas and stars are available for only a few nearby galaxies. Cappellari et al. (2002) carried out a detailed comparison for the E3 galaxy IC 1459. Dynamical modeling of the stellar kinematics (using HST/STIS and multiple-position angle ground-based data) allows a decomposition/identification of the counter-rotating stellar core in phase space and shows that it is a disk with a mass of about $3 \times 10^9 M_\odot$. The best-fit model has a black hole mass of about $2.6 \times 10^9 M_\odot$, somewhat larger than the value expected from the $(M_{\rm BH}, \sigma)$-relation. Models of the gas velocities suggest $M_{\rm BH} \approx 3.5 \times 10^8 M_\odot$ if the gas is in circular motion in a principal plane, while simple models of the velocity dispersion suggest $M_{\rm BH} \approx 1 \times 10^9 M_\odot$ for a spherical distribution of the gas. This leaves significant uncertainties about the reliability of the gas as a tracer of the galaxy gravitational potential. Studies of NGC4335 (Verdoes Kleijn et al. 2002), and NGC4697 (Gebhardt priv. comm.) reach similar conclusions. While simple modeling of the observed gas kinematics can probably be trusted when the gas has a smooth and regular morphology, symmetric kinematics, and low velocity dispersion (e.g., Barth et al. 2001), in many nuclei this will not be the case.

Much effort has gone into obtaining stellar kinematics with HST/STIS for nearby galactic nuclei (e.g., Pinkney et al. 2002). These studies have typically been restricted to objects with velocity dispersions larger than about 140 km s^{-1}, and cusped luminosity profiles. In some cases the $0\farcs2$ slit is required to reach sufficient signal-to-noise. In combination with multi-position-angle or integral-field ground-based stellar kinematics, these measurements provide reliable black hole masses (e.g., Gebhardt et al. 2002; Verolme et al. 2002). HST/STIS cannot probe the nuclei of the giant ellipticals in Virgo because of their modest surface brightness, and cannot resolve the black hole sphere of influence in low-luminosity ellipticals, because at 15 Mpc its radius is smaller than $0\farcs1$.

2.4. Active Galaxies

Most nearby galaxies contain a quiescent or only weakly active nucleus (Ho et al. 1997). There are few powerful nuclei within 20 Mpc: Cen A at $\approx$3 Mpc (but its nucleus is heavily obscured), as well as classical Seyferts such as NGC1068, and FR I radio galaxies such as M84 and M87 in the Virgo cluster. More powerful active nuclei appear at even larger distances, with Cyg A, the classical FR II at $\sim$200 Mpc. 3C273, the nearest quasar, is another factor 2.5 or so more distant.

HST imaging by Verdoes Kleijn et al. (1999) of a complete sample of 21 northern nearby FR I galaxies (18 of which are beyond 40 Mpc) revealed that all of them have nuclear emission-line gas, as well as dust disks or dust lanes. Comparison with imaging of inactive galaxies shows that the trigger of activity must lie on scales smaller than resolved by HST/WFPC2, even in the nearest active nuclei. HST/STIS emission-line kinematics is now available for these FR I objects but, as we have seen above, the derived $M_{\rm BH}$ will have to be treated with caution, in particular because the gas kinematics may be influenced by non-gravitational motions (in- and outflows). Beyond 40 Mpc the spatial resolution is insufficient to detect all but the largest black holes. Measurement of the stellar kinematics in nearby active nuclei is difficult, as these often reside in low-surface brightness giant ellipticals, and contain a bright nuclear emission-line spike. Increased resolution in both imaging and spectroscopy is required.

3. Requirements for Next Steps

HST studies of nearby galactic nuclei have raised interesting questions, including: Why are nuclear and global properties correlated? What is the role of nuclear disks and asymmetries? Do galaxies evolve dynamically from the inside out on interesting time scales? Why is only a small fraction of the black holes currently active? How does the $M_{\rm BH}$ mass function evolve with time? In order to answer these, we need to extend the studies that are currently possible only for Local Group galaxies to a representative sample. This leads to four requirements:

- Integral-field spectroscopy to obtain the two-dimensional kinematics (and line-strengths), which provides key constraints not only for asymmetric nuclei such as the one in M31, but also in "simple" galaxies such as M32.
- Sufficient sensitivity to be able to measure the stellar absorption-line kinematics over the full range of central surface brightness seen in Virgo galaxies, in order to have a reliable tracer of the gravitational potential.
- Increased spatial resolution, to probe the nuclei of the smaller galaxies in Virgo, and to carry out a representative census to larger distances which also contains the rarer types, in particular the radio-loud galaxies.
- A coronagraphic capability to probe near bright central spikes.

4. Prospects on the Ground

Galactic nuclei are interesting targets for AO-assisted studies from the ground. The OASIS results on M31 in the I-band shown in Figure 1 provide a case in point. This instrument is being upgraded, and will move to the William Herschel Telescope (WHT) on La Palma in 2003. The enhanced throughput, and a laser guide star (planned for 2005) should provide the stellar kinematics in objects like M87 with unprecedented spatial resolution.

Integral-field spectrographs operating at near-infrared wavelengths will soon be available on 8 m class telescopes. Examples include the SINgle Far Object Near-infrared Investigation (SINFONI) on the Very Large Telescope (VLT) and the Near Infrared Integral-field Spectrograph (NIFS) on Gemini, with first light for both expected in 2004. These instruments allow measurement of the stellar kinematics through observations of the CO bandhead at 2.3 μm. With the ten-fold increase in collecting area over HST, and the potential to provide spectra at a spatial resolution of better than $0''\!.1$ at 2μm, this will allow a significant step beyond HST/STIS in the study of normal and dusty objects, including some of the nearest active nuclei.

Adaptive optics in the visible range on 8 m telescopes can in principle improve on the current HST resolution by a factor two or more over a small field of view. This will be hard to achieve in practice, but there is much ongoing effort in this direction, and galactic nuclei are among the natural targets. This will not only involve imaging but also spectroscopy. An example is provided by the proposed second generation VLT Multi Unit Spectrographic Explorer (MUSE, PI Bacon). While its main aim will be multi-conjugate adaptive optics (MCAO) assisted panoramic integral-field spectroscopy of ultra-deep fields in the I and R bands, MUSE will have a mode for spectroscopy at the highest spatial resolution provided by adaptive optics in a small field. Phase A is in progress, and the instrument could be on the VLT by 2007.

Development of interferometry at infrared wavelengths is being pursued actively at Keck and at ESO (VLTI). Interferometry is also an important component of the planned scientific use of the Large Binocular Telescope (LBT). Galactic nuclei (including the Galactic Center) are among the main drivers to go faint and reach K of 20 mag. The anticipated VLTI spatial resolution of order $0''\!.015$ at 10 μm would allow imaging of the dusty broad-line regions of the nearest active nuclei at these wavelengths. Obtaining spectroscopic information on these scales would be very exciting, but is an extremely challenging goal.

Design studies are underway for optical telescopes with filled apertures ranging from 30 m (CELT and GSMT) to 100 m (OWL). If MCAO can be made to work at near-infrared and eventually at optical wavelengths, then these extremely large telescopes will provide the increase in resolution over HST, combined with sufficient sensitivity, to carry out studies of galactic nuclei satisfying the requirements outlined in section 3. A 30 m telescope would allow detailed study of nuclei in most Hubble types. Larger apertures (70–100 m) operating at visible wavelengths would in principle allow detailed studies out to beyond Virgo with the sensitivity and resolution now achievable only for Local Group galaxies. Furthermore, because the angular diameter versus distance relation has a minimum at $z = 1$, these extremely large telescopes could make it possible to estimate black hole masses in large galaxies throughout the Universe from emission-line kinematics.

5. Prospects in Space

The first question to consider is whether significant progress could already be made by equipping HST with an integral-field spectrograph for near-UV and optical wavelengths. Efficient high-throughput designs exist, but this option would require an additional servicing mission. The reward would be the enhancement of the spectroscopic efficiency of HST by about a factor of 50 beyond HST/STIS, including a relaxation of the roll-angle requirements on the observatory. Applications would range from resolved spectroscopy of Io and of proplyds in star-forming regions to resolved kinematics of gravitationally lensed arcs at high redshift. For galactic nuclei, this would provide exquisite capabilities for studying the nuclei of Local Group galaxies, and for measuring the gas kinematics in many nuclei. It would also enlarge the sample for which stellar kinematics can be measured with sufficient resolution.

The current plans have the Next Generation Space Telescope (NGST) as a 6 m class observatory, with spectroscopic capabilities in the near and mid-infrared. An integral-field capability on the Near Infrared Spectrograph (NIRSpec) would allow study of normal and dusty nearby nuclei, provided it has a spectral resolution of a few thousand. The main gain over HST would not be spatial resolution (because of the longer wavelengths) but an order of magnitude in sensitivity, so that one might obtain stellar kinematics of giant ellipticals at $0\farcs1$ resolution. NGST Mid Infrared Instrument (MIRI) integral-field spectroscopy will provide emission-line diagnostics of dusty starbursts and active nuclei, and may probe the nuclear stellar populations of high-z galaxies as the CO bandhead moves to the mid-IR for $z > 1.5$.

A 6-8 m class space telescope with integral-field capabilities in the optical (and perhaps near-UV) would allow stellar absorption-line spectroscopy of many galactic nuclei at $0\farcs1$ resolution or better, and a sufficient gain over HST in sensitivity that the giant ellipticals in Virgo can be studied at this resolution. A coronagraph would allow unique studies of active nuclei (e.g., Seyferts). This would be a truly fantastic resource, which would benefit broad areas of astronomy. However, for galactic nuclei it might not provide a huge gain over the results expected from HST imaging combined with AO-assisted ground-based integral-field spectroscopy on 8 m class telescopes, except in UV diagnostics. A space telescope of significantly larger aperture would provide a major leap in this area, and allow absorption-line studies in a representative sample of galaxy types with sufficient spatial resolution and sensitivity, and emission-line studies to the highest redshifts.

6. Conclusions

The combination of HST imaging and ground-based integral-field spectroscopy is very powerful for the study of the nuclei of normal and active galaxies, and provides a compelling argument for integral-field capabilities in space. Much progress in our understanding of galactic nuclei is expected on the ground in the next decade, and this may continue if near-diffraction limited observations become possible on telescopes with apertures of 30 m or larger. While the study of galactic nuclei is not the main science driver for a 6-8 m class space telescope, even a smaller aperture equipped with an integral-field spectrograph would yield a rich scientific harvest.

Acknowledgments. It is a pleasure to thank Eric Emsellem for providing Figure 1, and to thank him, Roland Bacon, Wilfried Boland, Michele Cappellari, Gijs Verdoes Kleijn, and Ellen Verolme for comments on the manuscript. ESA support for participation in this conference is gratefully acknowledged.

References

Bacon, R., Emsellem, E., Monnet, G., & Nieto, J.-L. 1994, A&A, 281, 691

Bacon, R., Emsellem, E., Combes, F., Copin, Y., Monnet, G., & Martin, P. 2001a, A&A, 371, 409

Bacon, R., et al. 2001b, MNRAS, 326, 23

Barth, A., et al. 2001, ApJ, 555, 685

Cappellari, M., et al. 2002, ApJ, 578, ApJ, 578, 787

de Zeeuw, P.T., et al. 2002, MNRAS, 329, 513

Emsellem E., & Bland–Hawthorn, J. 2002, in ASP Conf. Ser., Galaxies, the Third Dimension, eds M. Rosado, L. Binette, & L. Arias, in press (astro-ph/0206244)

Ferrarese, L., & Merritt, D.R. 2000, ApJ, 539, L9

Gebhardt, K., et al. 2000, ApJ, 539, L13

Gebhardt, K., et al. 2002, ApJ, in press

Ho, L.C., Filippenko, A.V., & Sargent, W.L.W. 1997, ApJ, 487, 568

Ho, L.C., et al. 2002, PASP, 114, 137

Joseph, C.L., et al 2001, ApJ, 550, 668

Kormendy, J., & Gebhardt, K. in 20th Texas Symposium on Relativistic Astrophysics, eds H. Martel & J.C. Wheeler, AIP in press (astro-ph/0105230)

Lauer, T.R., et al. 1993, AJ, 106, 1436

Light, E.S, Danielson, R.E., & Schwarzschild, M. 1974, ApJ, 194, 257

Pinkney, J., et al. 2002, ApJ, submitted

Statler, T.S. 1999, ApJ, 524, 87

Tonry, J.L. 1984, ApJ, 283, L27

Tremaine, S.D. 1995, AJ, 110, 628

van der Marel, R.P., Cretton, N., de Zeeuw, P.T., & Rix, H.W. 1998, ApJ, 493, 613

Verdoes Kleijn, G.A., Baum, S.A., de Zeeuw, P.T., & O'Dea, C.P. 1999, AJ, 118, 2592

Verdoes Kleijn, G.A., et al. 2002, AJ, in press (astro-ph/0208132)

Verolme, E.K., et al. 2002, MNRAS, 335, 517

Hubble's Science Legacy: Future Optical-Ultraviolet Astronomy from Space
ASP Conference Series, Vol. 291, 2003
K.R. Sembach, J.C. Blades, G.D. Illingworth, R.C. Kennicutt, Jr. (eds.)

The Structure and Energetics of Active Galactic Nuclei

Bradley M. Peterson

Department of Astronomy, The Ohio State University, 140 West 18th Ave., Columbus, OH 43210

Abstract. The black-hole/accretion-disk paradigm for active galactic nuclei (AGNs) is now reasonably secure, but there are still important unresolved issues, some of which will require the capabilities of an 8 to 10 m class ultraviolet/optical space-based telescope. Imaging spectroscopy with a diffraction-limited large telescope will be required to measure AGN black-hole masses from stellar dynamics for direct comparison with reverberation mapping-based masses. High spectral resolution in the ultraviolet (UV) is required to determine the mass and kinetic energy of the outflows observed in the absorption spectra of AGNs and to understand the energetics of the accretion process. As with ground-based astronomy, however, effective use of a large UV/optical space telescope requires complementary smaller facility instruments; a meter-class UV spectroscopic telescope, for example, can fit into a Medium Explorer budget.

1. Key Questions in AGN Astrophysics

Not long after the discovery of quasars, it was realized that the fundamental source of energy for these objects must in fact be gravitation, and fairly straightforward arguments led to the long-standing paradigm of a supermassive black hole (SMBH) surrounded by an accretion disk. Observational evidence has been rather ambiguous, although there were early strong clues, such as the near-UV/optical "big blue bump" (Shields 1978; Malkan & Sargent 1982) and the rapid X-ray variability that supported the model if only because they defied other explanation. But it is only within the last several years that the circumstantial evidence has accumulated to the point that few doubters remain. While there is now general, though not unanimous, agreement about the fundamental nature of AGNs, we cannot claim any real understanding of the quasar phenomenon until we successfully address a number of key questions, including the following:

1. *What are the masses of AGN black holes?* As described below, we are making progress, but there are still areas where our understanding is dangerously incomplete, especially with regard to the magnitude of possible systematic errors.

2. *What are the energetics of the accretion process?* In particular, for the various types of AGNs what are the accretion rates and radiative efficien-

cies and how do these scale with luminosity? How much of the output is in the form of kinetic energy (e.g., jets and absorbing gas) as opposed to radiation?

3. *How does the AGN mass function evolve over time?* Does the accretion process contribute significantly to black-hole growth, and how do black-hole demographics evolve with time?

4. *What is the nature of the line-emitting and absorbing gas in AGNs?* There is good reason to believe that these are somehow related to the accretion process, but there is no standard paradigm for the origin and role of these components in AGNs. This is one of the remaining outstanding mysteries in AGN structure.

2. Masses of Black Holes in AGNs

Ironically enough, the first reasonably secure SMBH masses were not measured in AGNs, but in quiescent galaxies, in which stellar or nuclear gas dynamical methodologies were employed (see Kormendy & Richstone 1995). The first and still highest precision AGN SMBH mass was determined from megamaser motions in NGC 4258 (Miyoshi et al. 1995). It was the fortuitious combination of source geometry and inclination that made this possible and, unfortunately, the method is not generally applicable to a broader range of AGNs. Most SMBH mass determinations in AGNs are based on emission-line reverberation mapping (Blandford & McKee 1982; Peterson 1993, 2001), which at this stage remains a rather crude tool; these masses are systematically uncertain to at least a factor of a few, and more accurate determination of these masses is an important current problem in AGN astrophysics.

2.1. Reverberation-Based Black-Hole Masses

Reverberation mapping makes use of the natural variability of AGNs to probe the central structure. The broad emission lines that dominate the UV/optical spectra of AGNs vary in response to changes in the continuum flux with a time delay, or lag, τ that reflects the light-travel time across the broad-line region (BLR). If gravity dominates the dynamics of the BLR, then by combining the size of the BLR $c\tau$ (as measured in a particular emission line) with the Doppler width V of the line, we can make a virial estimate of the central mass

$$M_{\rm BH} = \frac{kc\tau V^2}{G}, \tag{1}$$

where the constant k depends on the structure, kinematics, and projection (inclination) of the line-emitting region. It is the factor k, and thus the level of possible systematic uncertainties in the reverberation method, that remains unknown.

While the fundamental assumption that the gravitational field of the SMBH determines the BLR gas dynamics is unproven, there are nevertheless good reasons to believe that the reverberation-based black-hole masses are meaningful and that the systematic errors are not large enough to render the method useless. First, different emission lines within an individual source give consistent virial mass estimates, i.e., the emission-line time lags and line widths show a virial relationship, $\tau \propto V^{-2}$. This has now been demonstrated for four AGNs (Peterson & Wandel 1999, 2000; Onken & Peterson 2002), and constitutes the best evidence to date that the BLR motions are determined primarily by gravity. The relationship is surprisingly tight given that the line-emitting regions for the high-ionization and low-ionization lines are not necessarily expected to have similar geometries. Second, AGNs follow the same relationship between black-hole mass and stellar bulge velocity dispersion that is seen in quiescent galaxies (Gebhardt et al. 2000b; Ferrarese et al. 2001), constituting a strong demonstration that reverberation-based masses are comparable to stellar-dynamical masses in accuracy.

There are two important things that need to be done to secure our understanding of black-hole masses in AGNs and their relationship to the black holes in quiescent galaxies:

1. More intensive reverberation-mapping experiments that will allow determination of detailed two-dimensional transfer functions for multiple emission lines. This will lead to an understanding of the geometry and kinematics of the BLR and thus allow us to assess the magnitude of possible errors in reverberation-based masses from less well-sampled experiments. The data requirements for such experiments are well-understood (Horne et al. 2002) and there is still much that can be done with smaller facilities.

2. Direct comparison of stellar-dynamical and reverberation-based masses by applying both methods to common AGNs. This is in fact a very challenging exercise that can be attacked with an 8 m class diffraction-limited optical telescope, as described below.

2.2. AGN Black Hole Masses from Stellar Dynamics

A useful criterion for obtaining an accurate stellar-dynamical black-hole mass is spatially resolving the black-hole radius of influence

$$r_* = \frac{GM_{\rm BH}}{\sigma^2}, \tag{2}$$

where σ is the host galaxy bulge velocity dispersion. Especially for Type 1 AGNs, the best stellar absorption features for velocity dispersion measurement are the Ca II triplet lines around 8600 Å, in no small part because of the favorably low contrast between the AGN light and starlight in this spectral region. To resolve r_* at wavelength λ, we require

$$\mathrm{FWHM} = \frac{1.22\lambda}{D} < \frac{r_*}{d}, \tag{3}$$

where D is the telescope aperture and d is the distance to the AGN. We can eliminate σ from Equation (2) by using the now well-known relationship between black-hole mass and bulge velocity dispersion (Ferrarese & Merritt 2000;

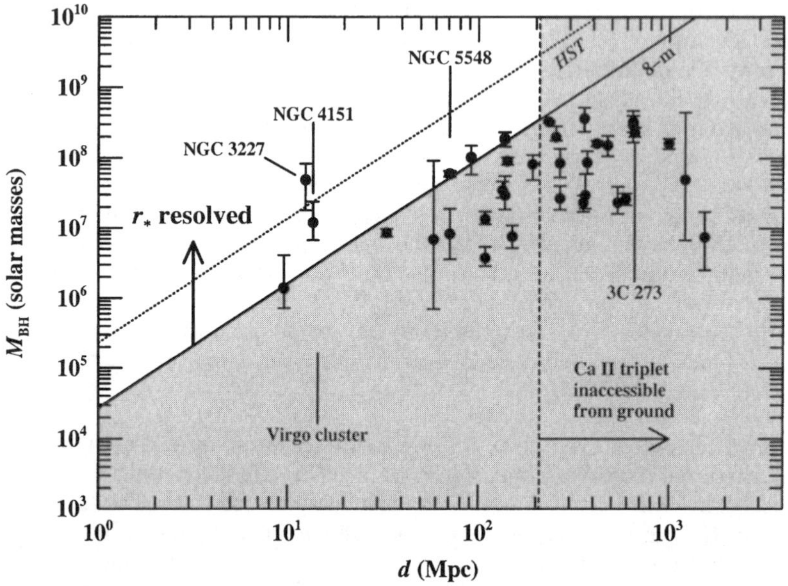

Figure 1. The diagonal lines show the minimum detectable black-hole mass as a function of distance for HST (dotted line) and an 8 m diffraction-limited telescope (solid line), based on stellar dynamical measurements of the Ca II triplet lines. At redshifts $z > 0.06$, the Ca II triplet lines are inaccessible from the ground due to water vapor in the Earth's atmosphere. The individual points show AGNs for which reverberation-based mass estimates are available.

Gebhardt et al. 2000a). The parameterization of Ferrarese (2002) gives

$$\frac{M_{\rm BH}}{M_{\odot}} = 1.66 \times 10^8 \left(\frac{\sigma}{200\,{\rm km\,s^{-1}}} \right)^{4.58}, \tag{4}$$

and by combining Equations (2) and (4), we obtain the criterion

$$\frac{M_{\rm BH}}{M_{\odot}} > 1.1 \times 10^6 \left[\frac{d({\rm Mpc})}{D({\rm m})} \right]^{1.78}. \tag{5}$$

This is the smallest black-hole mass that can be measured by stellar-dynamical methods as a function of distance d and telescope diameter D. The strong dependence on distance favors measurement of nearby lower-mass black-holes over higher-mass distant black holes.

We plot Equation (5) in Figure 1 for two cases, (a) for Hubble Space Telescope (HST) with $D = 2.4$ m, and (b) for a diffraction-limited 8 m telescope. SMBH masses above these respective lines are large enough to be measurable by stellar-dynamical methods. Also plotted on this diagram are masses and distances for reverberation-mapped AGNs. This diagram shows that no more than two reverberation-mapped AGNs, NGC 3227 and NGC 4151, are even possible candidates for measurement of their central masses by stellar-dynamical methods with HST. On the other hand, with a diffraction-limited 8 m telescope, a half-dozen or so AGNs are amenable to stellar-dynamical measurements of their SMBH masses. This is a marginally large enough sample to compare the stellar-dynamical and reverberation-based masses in a definitive way and to allow a reasonable assessment of the currently unknown systematic errors in the reverberation-based masses.

We note that attempts have been made to observe both NGC 3227 and NGC 4151, the two reverberation-mapped AGNs in which the black-hole radius of influence might be resolvable with HST (Gary Bower, PI). Neither attempt has been successful, in both cases because at the time the AGN was observed it was in an anomalously bright state, swamping the Ca II triplet absorption lines in the host-galaxy spectrum. Spatially resolved spectroscopy of both of these sources should be re-attempted with HST, but perhaps on a target-of-opportunity basis when the active nucleus is in a relatively faint state.

An obvious question to ask is why must these measurements be made from space, since the Ca II triplet is accessible from the ground? The main reason is that the strong contamination by the point-like nuclear source requires a truly diffraction-limited optical system. Ground-based adaptive optics systems are inadequate because of their low Strehl ratios (ratio of core intensity relative to a diffraction-limited point-spread function). Even in the low-luminosity AGNs, the weak stellar absorption features are completely washed out by the bright nuclear light. A second reason to observe from space is that at a fairly low redshift ($z \gtrsim 0.06$), the Ca II triplet lines are redshifted into the strong telluric water vapor lines, thus making an already very difficult observation virtually impossible.

It is also worth noting that it is only Type 1 AGNs (i.e., those with prominent broad lines in the UV/optical) that are amenable to reverberation mass measurements; on the other hand, of course, this is an important method that needs further development since it can be applied to Type 1 AGNs at arbitrary distance. Black-hole masses can be measured by megamaser motions in only very special cases. Thus direct measurement of the central masses of most local galaxies, including many AGNs and related objects (e.g., Type 2 AGNs and LINERs), will require stellar-dynamical studies. These, we see from Fig. 1, are capable of measuring SMBH masses for $M_{\rm BH} > 10^8\,M_\odot$ out to a distance of $\sim 100\,$Mpc.

Comparison of stellar-dynamical and reverberation-based masses is of critical importance. Once this has been effected, we will be able to employ (a) reverberation-mapping methods to distant AGNs and (b) secondary methods that are tied to reverberation (e.g., Wandel, Peterson, & Malkan 1999; Vestergaard 2002) to estimate with confidence masses of the black holes in distant quasars and thus address how the quasar mass function evolves with time.

3. X-Ray/UV Absorption

During the last decade, largely because of moderate-resolution, high-sensitivity UV spectroscopy with HST and soft X-ray spectroscopy with the Advanced Satellite for Cosmology and Astrophysics (ASCA), it has been recognized that UV and X-ray absorption features are ubiquitous properties of low-luminosity AGNs (e.g., Crenshaw et al. 1999). The origin of the resonance-line and ionization-edge absorption features is poorly understood. In many instances, a strong case can be made that the UV and X-ray features arise co-spatially. The absorbing gas is always blueshifted relative to the emission lines, and multiple velocity components are often identifiable in the UV lines. The large column densities indicate that the flows are massive[1], and in some cases can involve kinetic energy fluxes similar to the radiative output of AGNs. These properties suggest that they are the analogs of the polar outflows seen in young stars, i.e., they are a by-product of the accretion process. They may be somehow related to the much more massive outflows seen in about 10% of high-luminosity AGNs, those known as broad absorption line (BAL) QSOs. Perhaps most importantly, outflows from AGNs may have a profound feedback effect on star-formation processes in the host galaxies (e.g., Silk & Rees 1998).

The basic questions that need to be addressed are:

1. How much mass and kinetic energy is involved in these outflows, and how does this compare to the radiative output of AGNs?

2. How do the properties of the absorbers vary with other AGN properties, especially mass, luminosity, and radio loudness?

Only a very large space-based telescope can address these questions, as the sources are faint and high spectral resolution observations are required. Spectral resolution must be high enough to resolve the individual velocity components at their thermal width ($\sim 10\,\text{km s}^{-1}$, or $R = 30,000$), and the most important features are the UV resonance lines of abundant elements, notably C IV, Si IV, N V, and O VI. It is only in the lower-luminosity, lower-redshift objects that the individual velocity components are distinguishable from one another; we cannot address the same question simply by observing similar systems at higher redshift, as the absorption systems seen in the higher-redshift, higher-luminosity BALs are virtually continuous in velocity, making physical analysis extremely difficult and model dependent.

Variability of absorption lines in lower-luminosity AGNs, which occurs on time scales as short as a day, and detection and measurement of weak fine structure lines afford useful probes of physical conditions (mainly particle density) in the absorbing gas. Large collecting apertures are key to acquiring data of sufficient quality to utilize these tools.

---

[1]This conclusion is dependent on the assumed "covering factor", but this must be of order unity since absorption is present in virtually all nearby well-studied AGNs.

4. Space Astronomy Infrastructure

A generally recognized principle in ground-based astronomy is that to make efficient use of new-technology very large telescopes, we must off-load essential work that can be done with smaller telescopes. During the early years of HST, the International Ultraviolet Explorer (IUE) served in this capacity. At the present time, only HST and the Far Ultraviolet Spectroscopic Explorer (FUSE), which do not have overlapping capabilities, are operational. By no later than 2010, there is likely to be *no* UV spectroscopic instrument generally available to the community. Even optimistic launch dates for an 8 m class UV telescope might mean a long hiatus during which no UV observations are possible and hard-won expertise in UV spectroscopy begins to evaporate. Moreover, once an 8 m UV/optical telescope becomes operational, either it will spend some fraction of its time doing truly critical UV observations that might equally well be done on a smaller, cheaper telescope, or this important work simply will not get done.

NASA's contribution to astrophysics has been fantastically successful; in many areas of astronomy, and active galactic nuclei constitute only one example, UV data are simply too critical to do without. Large areas of astrophysics opened by UV astronomy cannot be allowed to die or languish when HST reaches the end of its lifetime. Certainly the case put forward at this meeting tells us that there are many frontiers in which we can progress significantly with an order-of-magnitude increase in collecting area and a factor of a few improvement in angular resolution. However, in our planning, we need to recognize the need for smaller workhorse facilities as part of the space astronomy infrastructure. These do not be need to be glamorous telescopes ladden with superlatives or science programs designed to address in a definitive way one of the handful of current big mysteries of the Universe. Neither would they be expensive; indeed, a 1 m class UV spectroscopic telescope fits into a Medium Explorer funding envelope ($\sim$\$200M) and expendable launch vehicle (e.g., Delta II with a 3 m fairing). In other words, we are fully capable of solving this problem essentially with existing resources because new funding lines would not necessarily be required. If we fail to solve this problem, we have no one but ourselves to blame.

5. Summary

The most dramatic impact that an 8 m space-based UV/optical telescope would have on AGNs would be to enable stellar-dynamical mass measurements out to a distance of $\sim$ 100 Mpc, a volume large enough to include several AGNs with reverberation-based mass measurements. The increase in collecting area relative to HST will enable far more detailed studies of the poorly understood, but energetically important, massive outflows seen in AGN spectra. The order-of-magnitude improvement in spatial resolution afforded by HST relative to ground-based observations has provided us with a wealth of information on the inner structure of AGNs (e.g., Pogge & Martini 2002) and on the evolution of AGN host galaxies. Certainly, another factor of a few improvement in resolution and the much larger collecting area will allow us push these frontiers forward.

Acknowledgments. I wish to thank my colleagues D.M. Crenshaw, L. Ferrarese, F. Hamann, L.C. Ho, A. Laor, P.T. O'Brien, P.S. Osmer, R.W. Pogge, J.C. Shields, and M. Vestergaard for helpful discussions and suggestions.

References

Blandford, R., & McKee, C.F. 1982, ApJ, 255, 419

Crenshaw, D.M., Kraemer, S.B., Boggess, A., Maran, S.P., Mushotzky, R.F., & Wu, C.-C. 1999, ApJ, 516, 750

Ferrarese, L. 2002, in Current High-Energy Emission Around Black Holes, ed. C.-H. Lee (Singapore: World Scientific), in press (astro-ph/0203047)

Ferrarese, L., & Merritt, D. 2000, ApJ, 539, L9

Ferrarese, L., Pogge, R.W., Peterson, B.M., Merritt, D., Wandel, A., & Joseph, C.L. 2001, ApJ, 555, L79

Gebhardt, K., et al. 2000a, ApJ, 539, 13

Gebhardt, K., et al. 2000b, ApJ, 543, L5

Horne, K., Peterson, B.M., Collier, S.J., & Netzer, H. 2002, submitted to PASP (astro-ph/0201182)

Kormendy, J., & Richstone, D. 1995, ARAA, 33, 581

Malkan, M.A., & Sargent, W.L.W. 1982, ApJ, 254, 22

Miyoshi, M., Moran, J., Herrnstein, J., Greenhill, L, Nakai, N., Diamond, P., & Inoue, M. 1995, Nature, 373, 127

Onken, C.A., & Peterson, B.M. 2002, ApJ, 572, 746

Peterson, B.M. 1993, PASP, 105, 247

Peterson, B.M. 2001, in Advanced Lectures on the Starburst–AGN Connection, ed. I. Aretxaga, D. Kunth, & R. Mújica, (Singapore: World Scientific), 3

Peterson, B.M., & Wandel, A. 1999, ApJ, 521, L95

Peterson, B.M., & Wandel, A. 2000, ApJ, 540, L13

Pogge, R.W., & Martini, P. 2002, ApJ, 569, 624

Silk, J., & Rees, M.J. 1998, A&A, 331, L1

Shields, G.A., 1978, Nature, 272, 706

Vestergaard, M. 2002, ApJ, 571, 733

Wandel, A., Peterson, B.M., & Malkan, M.A. 1999, ApJ, 526, 579

Hubble's Science Legacy: Future Optical-Ultraviolet Astronomy from Space
ASP Conference Series, Vol. 291, 2003
K.R. Sembach, J.C. Blades, G.D. Illingworth, R.C. Kennicutt, Jr. (eds.)

Measuring the Influence of Supernovae at High Redshift

Kurt Adelberger

Harvard-Smithsonian Center for Astrophysics, 60 Garden St., Cambridge, MA 02138

Abstract. Supernovae play a large but poorly understood role in our attempts to explain the evolution of the baryons in the Universe. Numerous observations throughout astronomy cannot be explained if we neglect their influence, yet our quantitative understanding of the ways in which supernovae affect the Universe remains remarkably poor. This is one of the most embarrassing gaps in our knowledge of the cosmos, and planned telescopes and surveys will probably not do much to fill it. The problem is that these surveys will be optimized to observe galaxies and intergalactic material independently of each other, while (in the author's view) by far the best information will come from simultaneous surveys of galaxies and the intergalactic material (IGM) in their vicinity. Only this will show directly how galaxies affect their surroundings and provide a rough energy scale for supernova-driven winds. Redshifts $1 \lesssim z \lesssim 3$ are ideal for the joint galaxy/IGM surveys we advocate, because the comoving density of star formation is near its peak, because the Lyα forest is thin enough for QSO spectra to reveal the locations of the dominant metallic species, and because bright background QSOs are common. But a new ultraviolet (UV) spectrograph in space will be required.

1. Introduction

Fifteen years from now we will be awash in galaxies. The 2dF, Sloan Digital Sky Survey, and maybe a Sloan successor will have given us redshifts for more than a million galaxies in the nearby Universe. The Deep Extragalactic Evolutionary Probe (DEEP) and the Visible and Infrared Multi-Object Spectrograph VLT Deep Survey (VVDS) will have added $\sim 10^5$ galaxies out to redshift $z \sim 5$. The Next Generation Space Telescope (NGST) and 30 m optical/infrared telescopes on the ground may have detected thousands of galaxies to $z \gtrsim 10$ — some of the first sources of light in the Universe. These surveys and others will teach us a tremendous amount about galaxy and structure formation. Our communal efforts so far will seem little more than a prologue to the vast literature on galaxies that will exist in 15 years.

Nevertheless, I suspect that one of the most fundamental questions in galaxy formation (and in all astronomy) will remain largely unanswered. This is the role that supernovae played in shaping the baryonic Universe. The influence of supernovae is thought to account for a wide range of observations throughout our field. The disruption of star-formation by supernova explosions is the favored explanation for why so few baryons are found in stars today (e.g., White & Rees 1978; Springel & Hernquist 2002). Numerical simulations cannot reproduce the large disk galaxies that we observe around us unless they include substantial heat input from supernovae (e.g., Weil, Eke, & Efstathiou 1998). The material between galaxies at high redshift is hotter than would be expected if gravity and the background radiation field were the only sources of heating; another source, presumably supernovae, appears to be required (Cen & Bryan 2001). It is difficult to explain why the soft X-ray background is so faint and so dominated by active galactic nuclei without asserting that supernovae blew apart dense clumps of baryons that would otherwise have produced copious free-free emission (e.g., Pen 1999). The shape of the galaxy cluster X-ray temperature/luminosity relationship differs from naive expectations in a way that suggests that supernovae may have imparted ~ 1 keV of energy to each of the nucleons in the young Universe (e.g., Kaiser 1991; Ponman, Cannon, & Navarro 1999).

These examples are only a few of many. We are unable to account for much of what we observe around us without invoking the indistinct notion of strong supernova "feedback", and our understanding of how the Universe evolves will remain seriously incomplete until we comprehend quantitatively how this feedback works.

The standard picture is that the numerous supernova explosions in a young galaxy create an enormous blast-wave (or wind) that rips through the galaxy and lays waste to its surroundings. But working through the details of this picture remains challenging even after 30 years of theoretical studies. There are many complications, but the central problem is that we have little idea of the characteristic energy scale to associate with the blast-waves that supernovae drive. The energy released by a single supernova, $\sim 10^{51}$ erg, is known, but it is unclear how large a fraction of the energy released by supernova explosions is imparted to nascent winds. Much of it may be harmlessly radiated away by the dense gas it heats. Physical arguments and numerical simulations are at present incapable of estimating a priori the energy of galactic winds to within even an order of magnitude; the energy of the winds is largely what determines how large an impact they have on the evolution of baryons. For this reason it is still unclear what types of galaxies were responsible for seeding the intergalactic medium with metals, or what effect blast-waves have on galaxy formation and evolution, or even whether realistic blast-waves would be physically capable of filling the large role that they are assigned in the standard lore.

Whether numerical calculations in 2017 will be able to reliably estimate the energy of galactic winds from first principles is anyone's guess, but in any case we will certainly want empirical support for the numerical results. Properties of galaxies (e.g., disk sizes) and of the IGM (e.g., metal content) are affected by the strength of supernova winds, and I can imagine convoluted and uncertain chains of reasoning that would provide an estimate of the strength of winds from observations of one or the other. Surely the most straightforward way to estimate the strength of galactic winds is to observe galaxies and the intergalactic medium simultaneously and see how much winds have disturbed the surroundings of the galaxies. Winds lose energy as they climb out of galactic potential wells and crash into nearby intergalactic matter, and a robust (if crude) measurement of the typical energy of galactic winds can be obtained by seeing how far they are

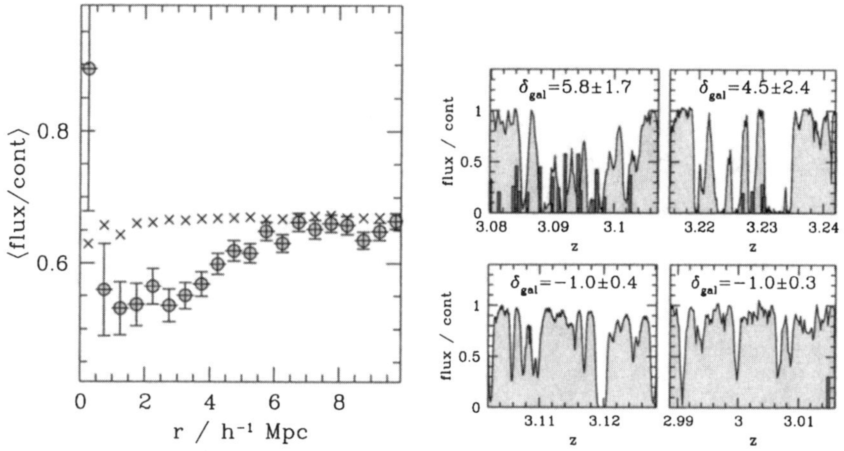

Figure 1. *Left*: The mean Lyα transmissivity of the IGM as a function of comoving distance from Lyman-break galaxies. Points with error bars show our measurements; crosses show the result if we randomize the galaxy redshifts. Since Lyα transmissivity decreases as the H I content of the IGM is increased, the plot shows that Lyman-break galaxies tend to be surrounded by large amounts of H I at comoving radii $1 \lesssim r \lesssim 5h^{-1}$ Mpc but little H I at smaller radii. *Right*: The H I content of the IGM along skewers through our two most significant galaxy overdensities (top) and underdensities (bottom). The shaded region shows the Lyα absorption in the QSO spectrum; vertical bars mark the measured redshifts of galaxies, which are typically a few comoving Mpc away from the QSO line-of-sight. The overdensities, presumably young Abell clusters, contain large amounts of H I — a result consistent with the idea that gas between the galaxies has been roiled by supernova-driven winds (see Adelberger et al. 2002; Adelberger 2003).

able to propagate. Simultaneous observations of galaxies and the intergalactic material that surrounds them are easy to obtain, at least in principle; one need only conduct a galaxy redshift survey in a field that contains background QSOs whose absorption spectra reveal the locations of H I and metals in the IGM. These sorts of observations, present and future, are the subject of my talk.

2. Current Data

Figures 1 and 2 show examples of what one can learn about galactic winds at high redshift from data that are currently available. These data, described more thoroughly by Adelberger et al. (2002), consist of measured redshifts for 431 Lyman-break galaxies in 6 fields at $z \sim 3$ together with high-resolution spectra of a bright background QSO in the middle of each field. The leftmost panel of Figure 1 shows the mean Lyα transmissivity of the IGM as a function of distance from Lyman-break galaxies at redshift $z \sim 3$. Increases in the H I content of the IGM lead to more Lyα absorption and hence to a lower Lyα transmissivity. (Here, Lyα transmissivity in the IGM at redshift z is quantified as the ratio of observed flux in the QSO spectrum at wavelength $\lambda = (1+z)\lambda_{Ly\alpha}$ to the expected

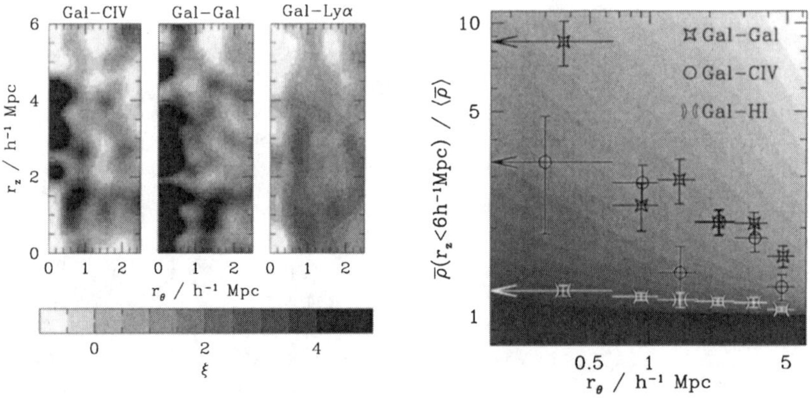

Figure 2. *Left*: The two-dimensional correlation functions of galaxies with galaxies, with C IV absorption systems, and with Lyα transmissivity decrements. Distances along the line-of-sight (r_z) and along the plane of the sky (r_θ) are comoving. The Ly-α panel was multiplied by -10 to make its structure visible. *Right*: A marginalization over redshift of the data in the left panel. Points with error bars are labeled. Striations in the background show how this plot would appear if the unmarginalized density had the form $\rho(r) \propto 1 + (r/r_0)^{-1.6}$.

QSO flux if there were no absorption from the IGM.) The figure shows that as one approaches a Lyman-break galaxy from afar, the density of intergalactic H I at first begins to rise. This is in accord with the simple view that galaxies ought to be found where the density of matter (including H I) is highest. But at small separations ($r \lesssim 0.5h^{-1}$ comoving Mpc) something else happens; intergalactic H I largely disappears. One interpretation is that the galactic winds have largely driven away all material within this radius; the competing hypothesis that the galactic light has ionized the H I is ruled out by the large size of regions with lowered H I content (Adelberger et al. 2002).

The right panel of Figure 1 shows the difference between the neutral hydrogen content of the intergalactic medium in young galaxy clusters and in young voids. Far more neutral hydrogen is found in the young clusters. The trend appears stronger than would be expected solely from the fact that galaxy clusters ought to contain more matter of all sorts (Adelberger et al. 2002). A possible explanation is that the roiling of the young intracluster medium by galactic blast-waves creates density inhomogeneities and increases the neutral hydrogen fraction (Adelberger 2002, 2003).

If Mpc-scale winds are responsible for the H I results just discussed, one might expect to find an increase in the density of metals near galaxies. This appears to be the case, as can be shown in a variety of ways. Consider first the triptych in the left half of Figure 2. Shown are the two dimensional correlation functions of galaxies with galaxies, of galaxies with C IV absorption systems, and of galaxies with Lyα transmissivity decrements. The panels show how the densities of H I, C IV, and other galaxies vary with spatial separation from a typical Lyman-break galaxy at redshift $z \sim 3$; one should imagine that the typical galaxy is at the origin and that the shadings indicate the amount of material at various angular and redshift separations. $\xi(r)$ is defined so that the mean density at any separation r is $\rho(r) = \langle\rho\rangle(1+\xi(r))$ with $\langle\rho\rangle$ the global mean. Positive values of ξ correspond to overdensities, and negative values correspond to underdensities. One may immediately see that the intergalactic medium near galaxies at $z \sim 3$ contains a large overdensity of C IV systems.

Now consider the right panel of Figure 2, which shows the same data marginalized in the redshift direction, $\bar{\rho}(r_\theta, < r_z) \propto \int_0^{r_z} dr'_z \rho(r_\theta, r'_z)$, where the marginalization helps remove dependence on redshift measurement errors or peculiar velocities. Contemplation of this figure reveals (a) that the ratio of C IV to H I density increases near galaxies, suggesting perhaps that the intergalactic metallicity is highest close to galaxies, and (b) that on large scales the galaxy-C IV cross-correlation function is similar to the galaxy-galaxy auto-correlation function, suggesting that C IV absorption systems and galaxies may be similar objects. Readers will find a slightly less glib discussion in Adelberger et al. (2002).

3. Ideal Data

I will not pretend that the data just presented are entirely convincing. In fact the entire point of my talk is to advertise their inadequacy! Some might worry that the statistical significance of any one of these results is low, but that does not concern me much. The flaw is easily fixed; existing telescopes will shrink our error bars by a factor of a few in a relatively short time. I am more concerned with the fundamental limitations of the data. There are several. I will mention two.

Measuring the content of the IGM along a single skewer through the galaxy distribution can tell us something about the characteristic stalling radii of the winds, and hence about their typical energies, but it tells us almost nothing about their geometry. Are they isotropic or bipolar? This has significant implications but cannot be determined if the IGM is observed only along a single skewer. Ideally, one would have several background QSOs whose light pierced each galaxy's wind at a range of angles and impact parameters.

More serious is the fact that working at $z \gtrsim 3$ forces us to rely almost exclusively on the C IV content of the IGM as a proxy for its metal content. The density of C IV is not related in a simple way to the metal content of the IGM (see, e.g., Adelberger et al. 2002). Inferences about intergalactic metals based solely on C IV will always be somewhat suspect, to say the least. But metals are the most direct signature of supernovae, and measuring the dispersal of metals out of galaxies and into the IGM will be central to our attempts at understanding supernova-driven winds. Ideally one would be able to measure the density of several metallic species, not merely C IV. This would provide constraints on the ionization parameter and let one say something definitive about metallicity. The relative abundances of different elements would also (in principle) provide information about their nucleosynthetic origin. But the vast majority of the strongest absorption lines at intergalactic temperatures and densities have wavelengths similar to or shorter than Lyα. This means that at redshifts $z \gtrsim 3$ they are buried in the thick Lyα forest and are extraordinarily difficult to detect. Only at redshifts $z \lesssim 3$, where the decreasing density of the Universe slows recombination and thins the forest, can these metallic absorption lines be straightforwardly found.

So this, then, is my picture of the ideal survey for quantifying empirically how supernova winds have affected the Universe. Find a field with a very large number of QSOs at redshifts $1 \lesssim z \lesssim 3$ and obtain high-S/N, high-resolution spectra of each. (The upper redshift bound comes from our desire that several intergalactic metal lines be easily detectable; the lower bound comes from the fact that the comoving density of star-formation, and presumably the prevalence of powerful winds, declines rapidly at $z < 1$.) Derive from the QSO absorption lines a detailed picture of the distribution of H I and metals in the intergalactic medium. Conduct a redshift survey of galaxies at $1 \lesssim z \lesssim 3$ in the same field and discover where star-forming galaxies lie amid the jumbled distribution of intergalactic matter. Then compare the relative spatial distributions of galaxies and intergalactic material. To the extent that galaxies send winds into their surroundings, one should find that the intergalactic medium near them is disturbed. Perhaps most of the intergalactic gas will have been driven away. Perhaps large amounts of metals will be found at the stalling radii of the winds. In any case, one should be able to answer the following questions (and others): What fraction of the energy released by supernovae appears to have been imparted to galactic winds? What sorts of galaxies drive winds? Are winds from dwarf galaxies really the most important, as many have argued? Are the winds more isotropic or bipolar? What is their dominant effect on the the large fraction of the baryons that lie outside of galaxies?

Now let us think how such a survey might be constructed. Obtaining galaxy redshifts in this redshift range is trivial with 8 m-class telescopes (e.g., Steidel et al. 1996; Adelberger 2002). That part of the problem is solved. The difficulty lies in obtaining quality spectra of the background QSOs. Over much of the redshift range $1 \lesssim z \lesssim 3$, Lyα and (especially) the strongest metal lines lie in the observed near- or far-UV and can only be observed from space. Unfortunately only a small number of QSOs are bright enough to be observed with current and planned space-based UV spectrographs. As an illustration, the left panel of Figure 3 compares the rough high-resolution, high-S/N magnitude limits of HST/STIS and HST/COS to the apparent magnitude distribution of all known QSOs in the (arbitrary) redshift range $1.0 < z < 1.6$. Only a small fraction of these QSOs are bright enough to observe with either spectrograph. The few hundred sources accessible to COS may seem like more than enough, but this is not true, at least not if the goal (as it should be) is to obtain information about the intergalactic distribution of hydrogen and metals along numerous skewers through the foreground galaxy distribution. This is illustrated by the right panel of Figure 3, which shows, as a function of assumed magnitude limit, the number of unique pointings in the entire sky that contain 5 or more NED QSOs at $1.0 < z < 1.6$ within a circle of radius $r \sim 20'$. The number of QSOs and the circle radius were selected arbitrarily; the plot is designed only to illustrate how much easier it is to find fields with large numbers of background sources as the limiting magnitude of the QSO becomes fainter. Because the NASA/IPAC Extragalactic Database (NED) is increasingly incomplete towards fainter magnitudes, the true gains from a fainter magnitude limit are far larger even than those indicated by this plot.

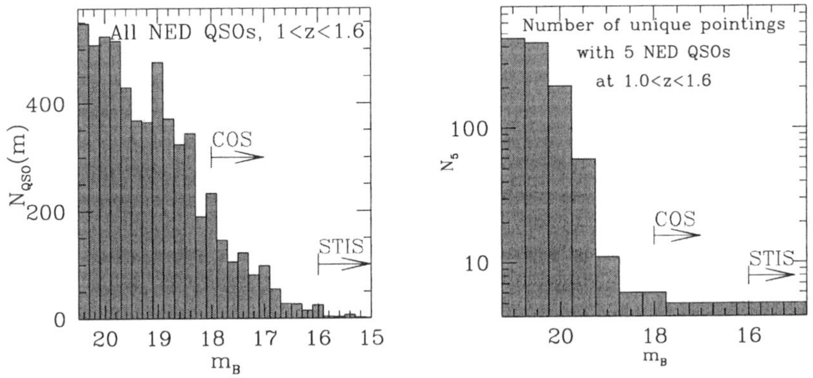

Figure 3. *Left*: The number of QSOs with $1.0 < z < 1.6$ in NED (including the Sloan early data release) as a function of apparent magnitude. Also shown are the rough magnitude limits for high-S/N high-resolution spectroscopy with HST/STIS and HST/COS. NED is seriously incomplete at the faintest magnitudes, and the true number density of QSOs rises much more rapidly at faint luminosities than this plot suggests. *Right*: The number of unique $r = 20'$ pointings that contain 5 or more of the NED QSOs in the left panel as a function of the imposed QSO magnitude limit. STIS and COS will let us sample this density of H I and metals in the IGM for only a handful of pointings across the entire sky — and the ideal number of QSOs in a field this size is much higher than 5. With a somewhat deeper magnitude limit (e.g., $m \sim 20$), dense sampling of the IGM would be vastly easier. The gain is much larger even than this figure suggests because of NED's incompleteness at fainter magnitudes.

There are evidently a few known pointings with 5 or more QSOs suitable for HST/COS within a $r = 20'$ circle. Might these pointings be sufficient for the project we are advocating? The answer, sadly, is no. Five QSOs per $20'$ circle works out to roughly one QSO per ~ 250 square arcminutes. For an $\Omega_M = 0.3$, $\Omega_\Lambda = 0.7$ cosmology at $z = 1.5$, the distance along the plane of the sky between the QSOs in these pointings will typically be more than $10h^{-1}$ comoving Mpc. For comparison, the winds (apparently) detected around galaxies at $z \sim 3$ have a radius smaller than $1h^{-1}$ Mpc. If our goal is to have a few QSOs behind an average galaxy's expanding blast-wave (in order to say something about the blast-wave's geometry, or learn about the connection between galaxy properties and blast-wave properties on a case-by-case basis), we will need a density of background QSOs that is at least 100 times higher.

A crude estimate of the required spectrographic sensitivity comes from the following calculation, which could and should be improved. Assume the QSO luminosity distribution at the bright end has the shape $n(l) \propto l^{-\alpha}$ with $\alpha \simeq 3.7$ (e.g., Pei 1995). Then the number density of QSOs brighter than magnitude limit m is proportional to $10^{0.4(\alpha-1)m}$, and an improvement in sensitivity of 2 magnitudes over HST/COS would lead to the desired factor of 100 increase in the density of background QSOs.

The good news is that a gain in sensitivity of this size is hardly inconceivable. The required increase in sensitivity over HST/COS is similar to the increase that will come when STIS is replaced by COS, for example. Obtaining the data set that I have described, and settling at last the vexing question of how supernovae have changed the evolution of the baryons in the Universe, is not much beyond our reach. How long will we have to wait? This is the bad news. The necessary QSO absorption lines are almost all at wavelengths that can only be observed from space; the atmosphere is an obstacle that will not change. Under current NASA plans these data will not be available until most of us are retired or dead.

Acknowledgments. I would like to acknowledge the major contributions of my collaborators Chuck Steidel, Alice Shapley, and Max Pettini; the support and patience of this meeting's organizers; and the hospitality of Bukowski's, in Boston, where much of this polemic was composed.

References

Adelberger, K.L. 2002, PhD Thesis

Adelberger, K.L., Steidel, C.C., Shapley, A.E., & Pettini, M. 2002, ApJ, submitted

Adelberger, K.L. 2003, ApJ, submitted

Cen, R. & Bryan, G.L. 2001, ApJ, 546, 81

Kaiser, N. 1991, ApJ, 383, 104

Pei, Y.C. 1995, ApJ, 438, 623

Pen, U.-L. 1999, ApJ, 510, 1

Ponman, T.J., Cannon, D.B., & Navarro, J.F. 1999, Nature, 397, 135

Springel, V., & Hernquist, L. 2002, MNRAS, in press

Steidel, C.C., Giavalisco, M., Pettini, M., Dickinson, M., & Adelberger, K.L. 1996, ApJ, 462, 17

Weil, M.L., Eke, V.R., & Efstathiou, G. 1998, MNRAS, 300, 773

White, S.D.M., & Rees, M.J. 1978, MNRAS, 183, 341

Hubble's Science Legacy: Future Optical-Ultraviolet Astronomy from Space
ASP Conference Series, Vol. 291, 2003
K.R. Sembach, J.C. Blades, G.D. Illingworth, R.C. Kennicutt, Jr. (eds.)

Studying Galaxy Formation with Hubble's Successor

Rachel S. Somerville
Department of Astronomy, University of Michigan, Ann Arbor, MI 48109

Abstract. In this paper, I discuss the capabilities and limitations of an 8-10 m ultraviolet/optical telescope in space, the proposed successor to the Hubble Space Telescope, in the context of galaxy studies. The exquisite spatial resolution and excellent sensitivity of such a facility would open up new possibilities for the study of nearby dwarf galaxies ($z \lesssim 0.5$), and for studying the internal structure and kinematics of more luminous galaxies at high redshift ($z \gtrsim 2$). These applications are of particular importance because they would address areas in which the popular Cold Dark Matter theory is in potential conflict with observations.

1. What Do We Need to Know About Galaxy Formation?

The "Cold Dark Matter" (CDM) or hierarchical paradigm of structure formation provides a useful framework for attempting to understand galaxies and cosmology. As the values of the cosmological parameters within this framework have become more and more tightly constrained by non-galaxy-based observations (like the cosmic microwave background and supernovae), most of the major uncertainties have to do with the messy "gastrophysics" that connects dark matter with the gas and stars that we can observe directly. Some of these issues include:

- **Cooling** — observations indicate that less of the gas in the Universe has cooled than is predicted by simulations. In particular, cooling in clusters seems to have been "shut off" by some unknown process.
- **Star formation** — what determines how efficiently a galaxy can turn cold gas into stars? How does this efficiency scale with redshift and galaxy properties? What is the physical basis of empirical scaling laws such as the "Kennicutt Law" (Kennicutt 1989; 1998), and are they universal? What factors determine the duration and efficiency of the bursts of star formation seen in interacting galaxies? Is the stellar initial mass function universal in space and time, and why does it have its observed shape?
- **Stellar feedback** — how does the energy from massive stars and supernovae affect the interstellar medium (ISM), intergalactic medium (IGM), and intracluster medium (ICM), and how does this impact future generations of stars? Is the main effect predominantly thermal or kinetic? How important are turbulence and large-scale galactic winds and outflows?

- **Heavy elements and dust** — how efficiently are metals expelled from the potential wells of galaxies? How did the ICM and IGM get so uniformly polluted with heavy elements? Is there a universal dust-to-metal ratio in all galaxies? What determines the degree of optical/ultraviolet extinction a galaxy experiences? Do the properties of dust in galaxies (composition, temperature, etc.) differ dramatically from one galaxy to another, or as a function of redshift?

While the observations obtained with the new generation of space-based facilities and the large number of 6-10 m-class ground-based telescopes now coming on line will doubtless bring us a much better understanding of many of these issues, I expect that even 10–15 years from now, some of them will not be completely put to rest. In the remainder of this paper, I focus on a few specific examples of observations that would utilize the unique capabilities of the proposed Hubble Space Telescope successor (the "Next HST" – NHST) to address important theoretical questions about galaxy formation.

2. What Could We Do with a 10 m Optical Telescope in Space?

At the risk of stating the obvious, I shall go through some of the capabilities and limitations of the proposed NHST facility for studying galaxies. The ultraviolet/optical wavelength sensitivity is optimal for studying objects with young stellar populations. There are many indicators of star-formation activity in this wavelength range, although they are suspect due to the uncertain effects of dust. Because of the large aperture and the location in space, NHST would have exquisite spatial resolution. Space-based observations also have lower background and so lower flux limits. This is useful for looking at high redshift or intrinsically faint and/or low surface brightness objects (a familiar lesson from HST). However, by $z \simeq 6$, the Lyman break has redshifted into the I-band, so a purely optical telescope will not be useful for observing galaxies at $z \gtrsim 6$ or so (but that is what NGST is for).

Quantitatively, what would the proposed parameters of NHST allow us to do? The following calculations were done assuming a flat geometry, $\Omega_m = 0.3$, $\Omega_\Lambda = 0.7$, and $H_0 = 70$ km s^{-1} Mpc^{-1}. Figure 1 (left) shows the apparent magnitude (AB system) or flux of several representative objects placed at different redshifts (no k-correction or evolution). The top line represents a large spiral galaxy similar to the Milky Way ($M_B = -20$), the middle line represents a typical dIrr galaxy like the SMC ($M_B = -16$), and the bottom line represents a very faint dwarf galaxy such as Sculptor ($M_B = -10$). Assuming a flux limit on the order of a 0.5–1 nJy, which should be easily obtainable with NHST, we see that luminous $L \simeq L_*$ galaxies will be easily visible out to $z \sim 6$. Even much fainter objects like the SMC (about three magnitudes below L_*) will be visible to $z \sim 2$. Objects like the very faintest dwarf galaxies known in the Local Group (9–10 magnitudes below L_*) will be very difficult to see at truly impressive redshifts — but they should be visible out to redshifts of $z \sim 0.5$–1, while currently these extremely tiny objects can only be studied very locally (mostly within the Local Group). Of course, surface brightness dimming will make these objects even more difficult to detect at large distances.

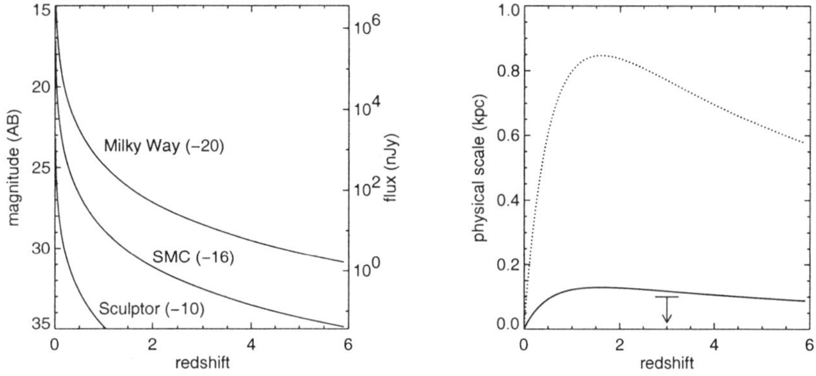

Figure 1. *Left*: The apparent magnitude or flux of several representative objects as a function of redshift. *Right*: Physical scale corresponding to the approximate angular resolution of the Wide Field Camera on HST (top line) and a diffraction limited 10m NHST (bottom line), both at V_{606}. The horizontal line shows the approximate upper limit on the required resolution necessary to study the internal kinematics of galaxies at $z \sim 3$ (see text).

The right panel of Figure 1 shows the physical scale corresponding to the angular resolution of a diffraction limited 10 m telescope in the V-band (bottom curve). For comparison, the top curve shows the same quantity for the approximate angular resolution of the Wide Field Camera on HST (about 0.1 arcsec). This function is nearly flat at $z \gtrsim 1$, so that physical scales on the order of 0.1 kpc would be resolved at all redshifts of interest ($z \lesssim 6$). In the following sections, I discuss some science goals that require this kind of resolution.

3. Dwarf Galaxies in the Nearby Universe

Dwarf galaxies have long presented a particular challenge to CDM. Most people are now familiar with the problem that CDM models tend to produce steep luminosity functions and too many low-luminosity galaxies (e.g., White & Frenk 1991; Kauffmann, White, & Guiderdoni 1993). It is usually assumed that this can be cured by invoking supernova (SN) feedback, but most models for SN feedback are currently quite ad hoc, and it is important to obtain better direct observational constraints on this process. This is one of the reasons that the study of nearby dwarf galaxies is so important.

Another form of this problem that has received a lot of attention recently is known as the "substructure problem"; virialized halos in high-resolution CDM simulations have large amounts of surviving substructure, from previous generations of dark matter halos that have merged to form a larger halo (Klypin et al. 1999; Moore et al. 1999). There are several hundreds of bound substructures with $V_c \gtrsim 20$ km s^{-1} within Local Group sized halos in the simulations, at least an order of magnitude larger than the number of dwarf galaxies observed in the Local Group (at most around 40).

To solve these problems within the context of standard CDM, clearly we must invoke some mechanism to either destroy these small halos or to prevent them from collecting gas and from forming stars. Supernova feedback may help, but would have to be strongly differential, much more effective in small mass objects than in large mass ones, in order to solve this problem without making star formation too inefficient in larger mass galaxies. Another possibility is that the presence of a photoionizing background could suppress gas infall and cooling in very small halos ($V_c \lesssim$ 30–50 km s^{-1}). A number of recent studies using semi-analytic methods (Somerville 2002; Benson et al. 2002; Bullock, Kravtsov & Weinberg 2001) concur in the conclusion that if this effect (sometimes known as "squelching") is as efficient as suggested by recent hydrodynamic simulations, then it can probably cure the substructure problem in the Local Group.

If the "squelching" picture is correct, there are a number of observational consequences that should be tested. There have been several recent proposals for ways to probe the substructure mass function in other galaxies using gravitational lensing (e.g., Metcalf & Madau 2001; Moustakas & Metcalf 2002). We might expect that an additional consequence of this picture would be that some small galaxies, which formed before the Universe became reionized, would experience a burst of early star formation but then be "squelched" — these objects should then have uniformly old stellar populations. This seems to be in conflict with the star-formation histories derived for dwarf galaxies in the Local Group — however, theory predicts that the halos that can collapse before reionization should be predominantly found in cluster environments (Tully et al. 2002). Another implication of the squelching picture is that as one goes down the mass function of dark matter halos, the mass-to-light ratio of halos should greatly increase. Since squelching is expected to be a somewhat gradual and stochastic process, one might be able to find very faint, surviving dwarf galaxies which could be used as dynamical tracers of the density field. Indeed, several candidates for these "dwarf groups", with mass-to-light ratios about five times larger than those of normal groups, have been identified (Tully et al. 2002). A direct probe of squelching could also be obtained by measuring a "baryonic Tully-Fisher relation" (or the analog for spheroidals) for extreme dwarf galaxies ($M_B \leq -10$).

4. Structure and Kinematics of Galactic Disks

Another potentially serious problem for CDM-based models is the set of issues surrounding the internal structure of disk galaxies. It is now well-known that the density profiles of dark matter halos formed in dissipationless CDM simulations are "cuspy" – i.e., they rise fairly steeply in the center, with a slope of about r^{-1} (Navarro, Frenk, & White 1997 (NFW); Bullock et al. 2001; Power et al. 2002). It is difficult to reconcile the observed rotation curves of some dwarf and low surface brightness galaxies with these NFW halos, which seem to predict too much dark matter in the central part, leading to rotation curves that rise too abruptly. There is considerable disagreement, however, about whether there really is a problem or whether the discrepancies can be explained by selection effects or observational errors. All of the problematic objects are from a subclass of extremely late type, dwarf and/or low surface brightness (LSB) galaxies.

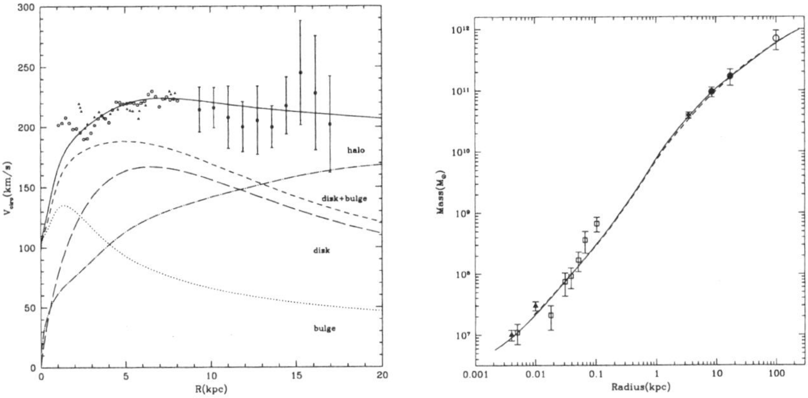

Figure 2. *Left*: Rotation curve of the Milky Way. Symbols represent observational data. Curves represent various components from the favored model of KZS02, as indicated on the figure. In this model, transfer of angular momentum from the baryons to the dark matter has flattened the dark matter "cusp" near the center of the halo. *Right*: Mass profile of the Milky Way, from pc to $\sim$ 100 kpc scales. Symbols represent observational data from stellar kinematics (inner points), H I kinematics (intermediate points) and satellite velocities (outer point). Observational references are given in KZS02, from which both figures are reproduced.

Perhaps this population can be accounted for if they are formed within halos drawn from an extreme tail of low concentrations or flat inner profiles. In order to lay this debate to rest, we need high resolution kinematics for a statistically complete sample of galaxies, extending to very low luminosities and surface brightnesses. The existing samples are sparse because these tiny, faint objects are so difficult to detect, and even more difficult to obtain good spectra for. The resolution, sensitivity, and high S/N that could be obtained with NHST could provide the necessary breakthrough to determine whether this is indeed a "crisis" for CDM.

If there really is a problem, perhaps it can be solved by appealing to baryonic processes. It is generally argued that this is unlikely for the objects in question, which are clearly dark matter dominated in the center. To go this route, one must appeal to an event that occurred at an early stage, leaving no directly observable trace. For example, cusps could be destroyed by an early, intense burst of star formation which simultaneously drove all the gas out of the halo, perhaps combined with subsequent tidal heating (Gnedin & Zhao 2001; Dekel & Devor 2002). Weinberg & Katz (2002) proposed that a large and massive rotating bar in the high-redshift progenitors of present day galaxies could destroy the inner dark matter cusps. These solutions all involve events which should produce rather dramatic observational signatures but which must occur at high enough redshift that they are safe (for the moment) from direct confrontation with observations. Perhaps the best way to test these ideas is to look directly at the high redshifts where these processes are assumed to occur. NHST would be very useful for studying the internal structure and dynamics of the progenitors of present day galaxies, which may provide constraints on these kinds of scenarios.

It has also been suggested that there may be too much dark matter in the centers of luminous galaxies, like our own Milky Way (e.g., Binney & Evans 2001). Klypin, Zhao, & Somerville (2002; KZS02) found, however, that if there is moderate transfer of angular momentum from the baryons to the dark matter (about a factor of two), the detailed inner structure and kinematics of the Milky Way and M31 can be understood within the context of NFW halos (see Figure 2). The analytic solution of KZS02 is reminiscent of the process seen by Weinberg & Katz (2002) in their simulations.

One of the main uncertainties in interpreting galaxy rotation curve data is due to the difficulty in connecting the virial mass of the dark matter halo (typically, a few hundred kpc for bright galaxies) with the observed rotation curves, which probe only about the inner tenth of the halo. Especially for luminous galaxies, the gravitational force of the baryons is expected to significantly modify the profile in the central parts, where the baryons comprise about half of the mass. For the unique case of the Milky Way, we can reconstruct the mass profile over a huge range of scales (see Figure 2) by using different dynamical tracers: stellar kinematics on small scales (pc), the H I rotation curve on intermediate scales (kpc), and satellite velocities on large scales (100 kpc). As shown in KZS02, these combined data produce very strong constraints on the distribution of baryons and dark matter in the Milky Way. NHST could be used to obtain very high resolution kinematic data for the central parts of galaxies, and kinematic tracers of the outermost reaches of the halos of galaxies such as satellite galaxies, planetary nebulae, or globular clusters. These data could be combined with CO rotation curves from the Atacama Large Millimeter Array (ALMA) to obtain very tight constraints on the distribution of mass out to the virial radii of galaxies. This will be invaluable in understanding the complex process of disk formation and what determines the surface density profile and rotation curve of spiral galaxies. Similar exercises may be done for early type galaxies.

5. High Redshift Galaxies: What Are They?

We now know of the existence of well over a thousand galaxies at $z \gtrsim 2$. While this is an enormous achievement, in many respects we still do not know what these objects are. Are they disks or spheroids? Are their star-formation rates so much higher than present-day galaxies because they have been brightened by collisions, or because their disks have higher gas surface densities? Do they follow the same fundamental scaling relations as nearby galaxies? What objects in the lower redshift Universe are their descendants? Arguably, we will not really know what these objects are until we image them at considerably higher resolution, and ideally, obtain spatially resolved kinematic data for them. Combined with information on the gas distribution from ALMA, this could allow us to test whether empirical star-formation laws like the Kennicutt law (1989; 1998) are universal.

What kind of resolution would this require? Figure 3 shows the predictions of a CDM-based semi-analytic model of galaxy formation for the structural

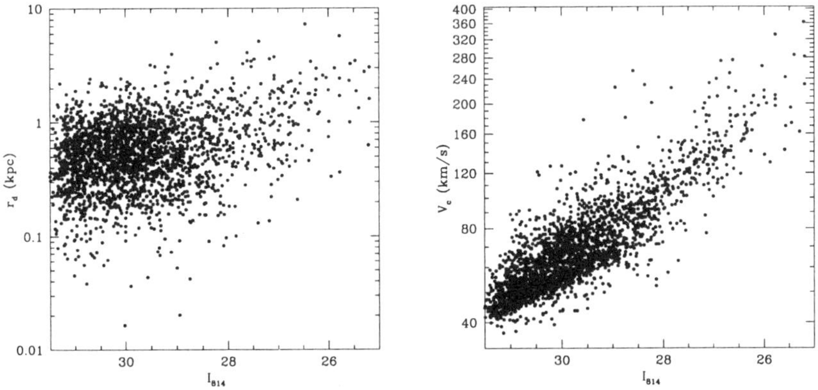

Figure 3. *Left*: exponential disk scale length versus apparent magnitude (observed frame) for model disk galaxies at $z \sim 3$. *Right*: Tully-Fisher relation for model disk galaxies at $z \sim 3$.

properties of galaxies at $z \sim 3$. This model is similar to those described in Somerville & Primack (1999) and Somerville, Primack, & Faber (2001; SPF), but includes a more detailed model of disk formation within cosmological "NFW" halos, including the effects of adiabatic contraction of the baryons using an approach similar to the one described in Mo, Mao, & White (1998). The left panel of Figure 3 shows the disk scale length (in kpc) versus the I_{814} magnitude (AB). A couple of things are worth noting. While the predicted sizes of the brighter objects are consistent with those measured for Lyman break galaxies in the Hubble Deep Field (see SPF), the models predict that these objects are only the tip of the iceberg; there is a large population of objects too small and too faint for HST to resolve. The right panel shows the Tully-Fisher relation for the same model galaxies at $z = 3$ (note that the quantity plotted is the observed frame I-band, corresponding to rest far-ultraviolet, so it is not surprising that there is a large scatter in this relation). Note the break in this relation at around $I_{814} > 28$, which is caused by the inclusion of "squelching" in the models.

6. Summary

I have suggested several applications of the proposed NHST 8-10 m ultraviolet/optical space telescope to the study of galaxies. A concise summary of these suggestions follows:

- Study stellar populations of dwarf galaxies (constrain their star-formation histories, and test "squelching" scenario).
- Look for "dwarf groups" with high mass-to-light ratios (the tail of the "squelched" population).
- Collect high resolution kinematic data for a complete sample of nearby galaxies, including dwarf and LSB galaxies (assess "cusp" crisis).

- Constrain the dark matter around galaxies on scales of $\sim$ 100 kpc using satellite galaxies, planetary nebulae, and/or globular clusters as kinematic tracers (understand disk and spheroid formation).
- Obtain high resolution imaging and spatially resolved kinematics of $z \gtrsim 2$ galaxies.
- Obtain star-formation indicators and gas surface densities for a broad range of environments and redshifts, to study the universality of empirical star-formation scaling laws.

References

Benson, A.J., Frenk, C.S., Lacey, C.G., Baugh, C.M. & Cole, S. 2002, MNRAS, 333, 177

Binney, J., & Evans, W. 2001, MNRAS, 327, 27

Bullock, J.S., Kravtsov, A.V., & Weinberg, D.H. 2000, ApJ, 539, 517

Bullock, J.S., Kolatt, T.S., Sigad, Y., Somerville, R.S., Kravtsov, A.V., Klypin, A., Primack, J.P., & Dekel, A. 2001, MNRAS, 321, 559

Dekel, A., & Devor, J., 2002, preprint, astro-ph/0204452

Gnedin, O.Y., & Zhao, H.S. 2002, MNRAS, 333, 299

Kauffmann, G., White, S.D.M., & Guiderdoni, B., 1993, MNRAS, 264, 201

Kennicutt, R.C. 1989, ApJ, 344, 685

Kennicutt, R.C. 1998, ApJ, 498, 541

Klypin, A.A., Zhao, H.S., & Somerville, R.S. 2002, ApJ, 573, 597

Klypin, A.A., Kravtsov, A.V., Valenzuela, O., & Prada, F. 1999, ApJ, 522, 82

Metcalf, R.B., & Madau, P. 2001, ApJ, 563, 9

Mo, H.J., Mao, S., & White, S.D.M. 1998, MNRAS, 295, 319

Moore, B., Ghigna, F., Governato, F., Lake, G., Stadel, J., & Tozzi, P. 1999, ApJ, 524, L19

Moustakas, L.A., & Metcalf, R.B. 2002, preprint, astro-ph/0206176

Navarro, J.F., Frenk, C.S., & White, S.D.M. 1997, ApJ 490, 493

Power, C., Navarro, J.F., Jenkins, A., Frenk, C.S., White, S.D.M., Springel, V., Stadel, J., Quinn, T. 2002, preprint, astro-ph/0201544

Somerville, R.S. 2002, ApJ, 572L, 23

Somerville, R.S., & Primack, J.R. 1999, MNRAS, 310, 1087

Somerville, R.S., Primack, J.R. & Faber, S.M. 2001, MNRAS, 320, 504

Tully, R.B., Somerville, R.S., Trentham, N., & Verheijen, M.A.W. 2002, ApJ, 569, 573

Weinberg, M.D., & Katz, N. 2002, preprint, astro-ph/0110632

White, S.D.M., & Frenk, C.S. 1991, ApJ, 379, 52

Hubble's Science Legacy: Future Optical-Ultraviolet Astronomy from Space
ASP Conference Series, Vol. 291, 2003
K.R. Sembach, J.C. Blades, G.D. Illingworth, R.C. Kennicutt, Jr. (eds.)

Galaxy Formation Now and Then

Matthias Steinmetz

Astrophysikalisches Institut Potsdam, An der Sternwarte 16, 14482 Potsdam, Germany

Steward Observatory, University of Arizona, 933 N. Cherry Ave., Tucson, AZ 85721

Abstract.
I review the current state of our understanding of the galaxy formation and evolution process from the modelers perspective. With the advent of the cold dark matter model and the support of fast computers and advanced simulation techniques, there has been considerable progress in explaining the growth of structure on the largest scales and in reproducing some of the basic properties of galaxies and their evolution with redshift. However, many properties of galaxies are still only poorly understood or appear to be in conflict with the prediction of the cold dark matter model. I discuss in what direction the next generation of galaxy formation models may go and why a large space-based optical-ultraviolet telescope could be critical for the calibration and testing of these advanced models.

1. Introduction

The past couple of years have witnessed a dramatic increase in the quantity and quality of observations on the formation and evolution of galaxies. Galaxies are routinely identified at redshifts exceeding three and high resolution imaging allows us to study their internal structure. These data are complemented by high resolution spectroscopy of QSO absorption systems that provide further clues on the evolution of baryons in the Universe. In fact, this increase has been so rapid that observations have outgrown their theoretical framework. Traditional approaches, which rely heavily on the morphological classification of galaxies and which intend to disentangle the star formation history of galaxies, seem outdated if compared with the much richer structure seen in galaxies at different redshifts.

Motivated by the increasing body of evidence that most of the mass of the Universe consists of invisible "dark" matter, and by the particle physicist's inference that this dark matter consists of exotic non-baryonic particles, a new and much more fruitful approach has been developed; rather than to model the formation and evolution of galaxies from properties of present-day galaxies, it is attempted to prescribe a set of reasonable initial conditions. The evolution of galaxies is then modeled based on physical processes that are considered to be relevant, such as gravity, hydrodynamics, radiative cooling, and star formation. The outcome at different epochs is then confronted against observational

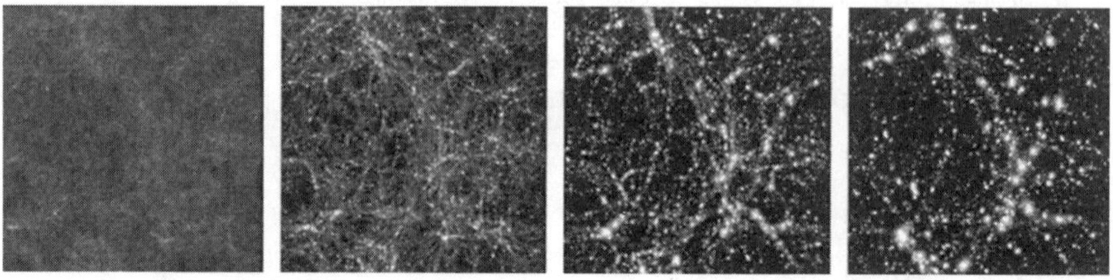

Figure 1. Time sequence of structure formation in a hierarchical clustering universe, here for the so-called ΛCDM model. The four snapshots correspond (from left to right) to redshifts of 9, 3.5, 1, and 0, respectively. The simulation box is 50 Mpc (comoving) on the side.

data. One scenario that has been extensively tested in this way is the model of hierarchical clustering, currently the most successful paradigm of structure formation.

In this contribution I review the main successes and some of the generic problems of models in which structure forms by hierarchical clustering. I briefly compare the current state of the field with that 10-15 years ago followed by some speculations in which direction the field may develop in the next decade and how a large optical-ultraviolet telescope in space may support such developments.

2. The State of the Field

Hierarchical clustering is at present the most successful model for structure formation in the Universe. In this scenario, structure grows as objects of progressively larger mass merge and collapse to form newly virialized systems (Figure 1). Probably the best known representative of this class of models is the cold dark matter (CDM) scenario. The initial conditions consist of the cosmological parameters (Ω, $\Omega_{\rm baryon}$, Λ, H_0) and of an initial density fluctuation spectrum, such as the CDM spectrum. The remaining free parameter, the amplitude of these initial fluctuations, is calibrated by observational data (e.g., the measured anisotropies of the microwave background). Since the CDM model was introduced in the early 1980s, the values of these parameters have been revised and tuned to match an ever growing list of observational constraints, from the $\Omega_0 = 1$, $H_0 = 50\,{\rm km\ s^{-1}\ Mpc^{-1}}$, and $\sigma_8 = 0.6$ of the former "standard" cold dark matter model to the currently popular "concordance" ΛCDM model. This ΛCDM model envisions an eternally expanding universe with the following properties (Bahcall et al. 1999): (i) matter makes up at present less than about a third of the critical density for closure ($\Omega_0 \approx 0.3$); (ii) a non-zero cosmological constant restores the flat geometry predicted by most inflationary models of the early Universe ($\Lambda_0 = 1 - \Omega_0 \approx 0.7$); (iii) the present rate of universal expansion is $H_0 \approx 70$ km s^{-1} Mpc^{-1}; (iv) baryons make up a very small fraction of the mass of the Universe ($\Omega_b \approx 0.04 \ll \Omega_0$); and (v) the present-day root-mean-squared mass fluctuations on spheres of radius 8 h^{-1} Mpc is of order unity ($\sigma_8 \approx 0.9$). The hierarchical structure formation process in this ΛCDM scenario is illustrated in Figure 1, which depicts the growth of structure within a 50 Mpc box between redshifts nine and zero. The ΛCDM model is consistent with an impressive array of well-established fundamental observations such as the age of the Universe as measured from the oldest stars, the extragalactic distance scale as measured by distant Cepheids, the primordial abundance of the light elements, the baryonic mass fraction of galaxy clusters, the amplitude of the cosmic microwave background fluctuations measured by COBE, BOOMERANG, MAXIMA, and DASI, the present-day number density of massive galaxy clusters, the shape and amplitude of galaxy clustering patterns, the magnitude of large-scale coherent motions of galaxy systems, and the world geometry inferred from observations of distant type Ia supernovae, among others.

The hierarchical build-up is also thought to determine the morphology of a galaxy, most noticeably the difference between disk–like systems such as spiral galaxies (some of them barred) and spheroidal systems such as elliptical galaxies and bulges. This picture envisions that whenever gas is accreted in a smooth fashion, it settles in rotationally supported disk-like structures in which gas is slowly transformed into stars. Mergers, however, convert disks into spheroids. The Hubble type of a galaxy is thus determined by a continuing sequence of destruction of disks by mergers, accompanied by the formation of spheroidal systems, followed by the re-assembly of disks due to smooth accretion (Figure 2). This picture of a hierarchical origin of galaxy morphology has been schematically incorporated in so-called semi-analytical galaxy formation models used to study the evolution of the galaxy population, but its validity in a cosmological setting has only just recently been directly demonstrated (Steinmetz & Navarro 2002).

Numerical simulations have been an integral part in the detailed analysis of the virtues of the CDM scenario. Only numerical techniques can account for the highly irregular structure formation process and for at least some of the complicated interaction between gravity and other relevant physical processes such as gas dynamical shocks, star formation, and feedback processes. Simulations also provide the required interface to compare theoretical models with observational data and are able to link together different epochs. While simulations of structure formation on the larger scales have mainly used large massively parallel supercomputers, studies of how individual structures such as galaxies or clusters of galaxies form in the ΛCDM scenario have heavily used special purpose hardware like the GRAPE (=GRAvity PipE) family of hardware N-body integrators (Sugimoto et al. 1990).

Although gas dynamical simulations were considerably successful in explaining some details of the galaxy formation process, the largest impact so far has been in the field of QSO absorption systems. Numerical simulations can reproduce the basic properties of QSO absorbers covering many orders of magnitude in column density (Cen et al. 1994; Zhang, Anninos, & Norman 1995; Hernquist et al. 1996; Haehnelt, Steinmetz, & Rauch 1996). Indeed, gas dynamical simulations were even responsible for a paradigm shift, as QSO absorbers are no longer considered to be caused by individual gas clouds. Absorbers of different column density (Lyα forest, metal line systems, Lyman-limit systems, and damped Lyα absorption systems) are rather reflecting different aspects of the large-scale structure of the Universe. While the lowest column density systems ($\log N \approx 12 - 14$) arises from gas in voids and sheets of the "cosmic web", sys-

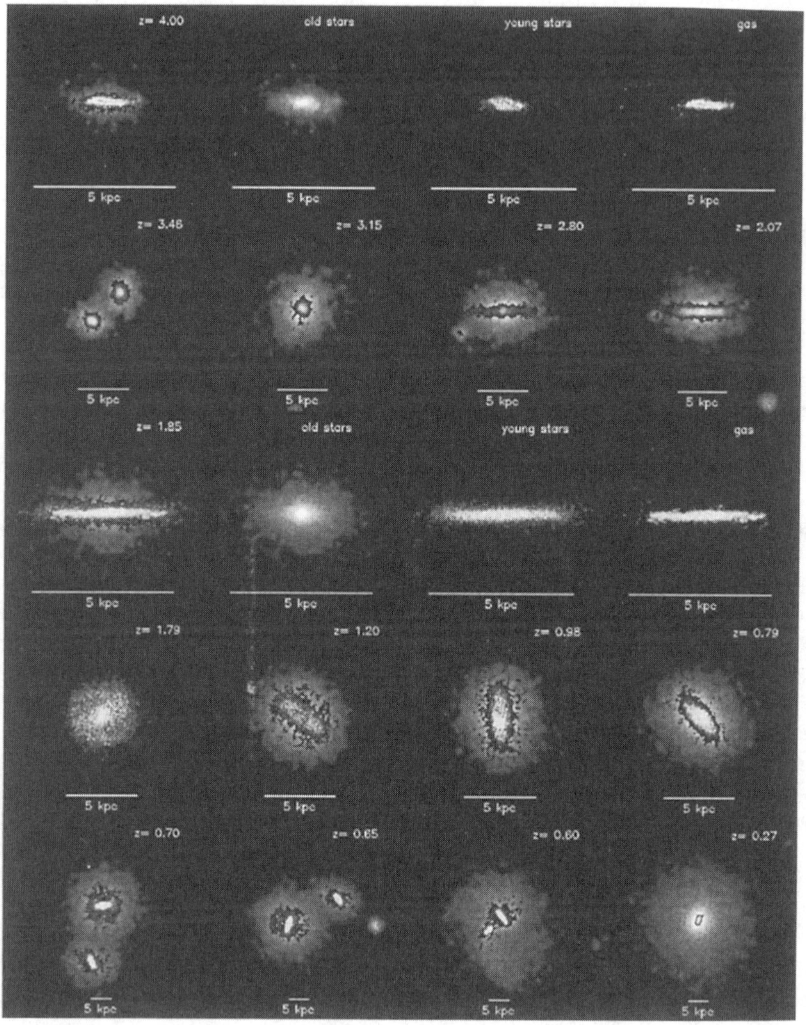

Figure 2. Surface mass density of the gaseous and stellar components of ΛCDM halos at various epochs. Horizontal bars in each panel are 5 (physical) kpc long and indicate the scale of each figure. Rows 2, 4, and 5 show time sequences near some key evolutionary stage. Rows 1 and 3 decompose a galaxy at a particular redshift (left) into its constituents: old stars, young stars, and gas (from left to right). Top row: The most massive progenitor at $z = 4$, seen edge-on. Second row: The formation of a bulge and the rebirth of a disk. Third row: The appearance of the galaxy at $z = 1.8$, seen edge-on. Fourth row: The tidal triggering of bar instability by a satellite resulting in the emergence of a rapidly rotating bar. Bottom row: A major merger and the formation of an elliptical galaxy.

tems of higher column density are produced by filaments ($\log N \approx 14 - 17$) or even by gas that has cooled and collapsed in virialized halos ($\log N > 17$).

Even though the above list of achievements appears quite impressive, it mainly addresses structures on scales exceeding a few hundred kpc. On smaller scales, theoretical models have at best provided some qualitative insights into the physics of the galaxy formation process, but we are still far from being able to make quantitative predictions of the properties of the galaxies now or at higher redshifts on those small scales. For example, even though we have some qualitative insight into how the Hubble sequence has formed, we are still unable to account in detail for the mix of morphologies at different redshifts. To a lesser extent we still do not have the computing power to study the small scale properties for a cosmologically representative sample of galaxies, but the more important factor is likely our rather poor understanding of the astrophysical processes acting on such scales, such as star formation, and energetic feedback from supernovae and stellar winds.

Furthermore, findings have surfaced in the past few years that seem to be at odds with the model predictions: (i) Because of their negligible primordial velocity dispersion, cold dark matter particles can achieve enormous phase space densities. As a result, numerical simulations have consistently shown that near the centers of halos the density profiles of virialized CDM halos diverge as r^{-1} (Navarro, Frenk, & White 1996) or perhaps even as steeply as $r^{-3/2}$ (Moore et al. 1999a). These divergent profiles are at odds with the usual interpretation given to the "solid-body" H I rotation curves reported for some low surface brightness (LSB) dwarf galaxies (Flores & Primack 1994; Moore 1994). (ii) Another generic prediction of CDM models is that virialized galactic halos will typically be triaxial and have dense cores. Yet in a number of galaxies where detailed stellar and gas dynamical observations are possible, the dark matter contribution within the optical radius appears to be rather small: many disks, perhaps including the Milky Way, are "maximal" (see, e.g., Debattista & Selwood 1999). (iii) High resolution N-body simulations indicate that, if the dark halo of our own Milky Way is made up of CDM, it should contain several hundred dark matter sub-condensations, a number that climbs to roughly one thousand if all halos within the Local Group are considered (Klypin et al. 1999; Moore et al. 1999b). On the other hand, observations of the Milky Way surroundings and of the Local Group reveal an order of magnitude fewer galaxies than expected in this picture (Mateo 1998). (iv) A difficulty indirectly related to the substructure problem concerns the angular momentum of gaseous disks assembled in hierarchical clustering scenarios. In the absence of heating, most of the mass of a galactic disk forming within a CDM halo is accreted through mergers of proto-galaxies whose own gas components have previously collapsed to form centrifugally supported disks. Numerical simulations (Navarro & Benz 1991; Navarro, Frenk, & White 1995; Navarro & Steinmetz 1997) show that most of the angular momentum of the gas is transferred to the surrounding halos during mergers. As a result, the spin of gaseous disks formed by hierarchical mergers is much lower than those of observed spirals.

It should be clear from this list that a considerable concerted effort is required by the community in order to promote our cosmological concordance model to a concordance model of galaxy formation. Considering the complexity of the processes involved, progress is likely to continue to be drive by observations.

3. Where Did We Come From ?

Before speculating what the state of the field may be like in 10–15 years, it may be illustrative to reconsider how the current state described above compares to the state of the field 10–15 years ago, i.e. before the advent of the Hubble Space Telescope and ground-based 8 m-class telescopes. Cosmological parameters were only very poorly constrained; the Hubble constant was known to only within a factor of 2, Ω at best to within a factor of ten. Hierarchical clustering in the form of a $\Omega = 1$ CDM model was already favored by most theorists, but it was certainly not accepted within the larger community as it is today. Similar statements can be made concerning the more basic concept that the morphological type of a galaxy may primarily reflect its merging history. From the observational side, barring some episodic evidence of very high redshift objects (radio galaxies, QSOs) and some indirect evidence for evolution (e.g., the Butcher-Oemler effect), whose relationship to regular galaxies was quite unclear, galaxy properties were mainly determined only at very low redshift. In fact, most galaxy models did not include, or included only very schematically, a cosmological context. Similarly, galaxy populations were mainly discussed in terms of non-evolution versus passive evolution models. It should be clear from this listing how rapidly the field of galaxy formation has changed in the past decade.

4. Prospects

Considering this rapid progress it seems (and probably is) impossible to make sound predictions of how the field may change over the next ten or even fifteen years. Nevertheless, the exercise may be entertaining, especially if one comes back for an evaluation after ten years. Assuming that the field continues to progress at the same pace and that basic concepts (like ΛCDM as a cosmological concordance model) remain unchanged, the following appear to be "fair bets":

- Microwave background experiments, in particular satellite missions like MAP and PLANCK, will accurately determine the cosmological parameters. Extragalactic astronomy will therefore be dominated by the quest to understand the formation and evolution of galaxies and the astrophysics behind it, and will no longer be just a tool to measure cosmological parameters.

- The Next Generation Space Telescope (NGST) will provide us with a look at the so-called dark ages of the Universe (i.e., the Universe at redshifts larger than six). Furthermore, NGST will allow us to study the evolution of the stellar component between redshifts 3 and 1, the epoch in which galaxies appear to have developed from some irregular bright clumps into the Hubble sequence.

- Astrometric missions like GAIA will provide a detailed dynamical and chemical record of how the Milky Way has been built up and how it transformed its gas into stars.

- In ground-based astronomy, the second generation of instrumentation on 8 m-class telescopes will be routinely used, and (from an optimist's point of view) the first data may already come in from 30 m optical telescopes.

- Galaxy formation models will largely benefit from the increase in computing power. According to Moore's law, the speed of computers doubles every 18 month. If this trend continues to hold, then standard PCs in 15 years will have a speed comparable to the largest supercomputers today. By then numerical simulations should be capable of simulating cosmologically representative volumes (a few hundred Mpc) of the Universe at a resolution better than a kpc. The understanding of astrophysics rather than the availability of sufficient CPU power will limit our insight into the details of the galaxy formation process.

What will be the role of an 8 m optical/ultraviolet space telescope in this framework? Two potential applications that depend critically on the availability of large ultraviolet-sensitive instruments are:

- *Observe the development of the Hubble sequence.* Deep high-resolution imaging of ultraviolet dropout galaxies between redshifts 3 and 1 will unravel the building blocks of typical L* galaxies at the present day epoch. Furthermore, the transition from clumps at $z \approx 3$ to Hubble-type galaxies at $z \approx 1$ can be observed. Using low-resolution spectroscopy, the kinematics and thus the development of scaling relations like the fundamental plane or the Tully-Fisher relation can be investigated.

- *Create a 3D-map of the baryons in the Universe between $z = 3$ and $z = 0$.* Using fainter QSOs (as obtained, e.g., by the Sloan Survey) as background sources, several patches of the night sky can be probed along many lines of sight. High-resolution spectroscopy of these background sources will map out the distribution of baryons in the intergalactic medium and eventually provide detailed 3D maps of the gas density, temperature, metallicity, and ionization state. The current paradigm that galaxies preferentially form at the intersections of filaments in the cosmic web can be probed directly. Studies of the transverse proximity effect measure in detail the evolution of the cosmic ultraviolet background and thus on the reionziation history of the Universe. Furthermore, they can constrain the amount of beaming in the ultraviolet flux of QSOs.

Should these developments be accompanied by a corresponding increase in our understanding of how stars form, how they release energy to the interstellar medium, and how star formation interacts with the evolution of galaxies, we may indeed make a major step toward a standard model of galaxy formation by 2020.

References

Bahcall, N., Ostriker, J.P., Perlmutter, S., Steinhardt, P. 1999, Science, 284, 1481

Cen, R., Miralda-Escudé, J., Ostriker, J.P., & Rauch, M. 1994, ApJ, 437, 9

Debattista, V.P., & Sellwood, J. 1999, ApJ, 493, L5

Flores, R.A., & Primack, J.R. 1994, ApJ, 427, L1

Haehnelt, M., Steinmetz, M., & Rauch, M., 1996, ApJ, 465, L95

Hernquist, L., Katz, N., Weinberg, D.H., & Miralda–Escude J. 1996, ApJ, 457, L51

Klypin, A., Kravtsov, A.V., Valenzuela, O., & Prada, F. 1999, ApJ, 552, 82

Mateo, M. 1998, ARA&A, 36, 435

Moore, B. 1994, Nature, 370, 629

Moore, B., Quinn, T., Governato, F., Stadel, J., & Lake, G. 1999a, MNRAS, 310, 1147

Moore, B., Ghigna, S., Governato, F., Lake, G., Quinn, T., & Stadel, J. 1999b, ApJ, 524, L19

Navarro, J.F., & Benz, W. 1991, ApJ, 380, 320

Navarro, J.F., Frenk, C.S., & White, S.D.M. 1995, MNRAS, 275, 56

Navarro, J.F., Frenk, C.S., & White, S.D.M. 1996, ApJ, 462, 563

Navarro, J.F., & Steinmetz, M. 1997, ApJ, 478, 13

Steinmetz, M., & Navarro, J.F. 2002, New Astronomy, 7/4, 155

Sugimoto, D., Chikada, Y., Makino, J., Ito, T., Ebisuzaki, T., & Umemura, M. 1990, Nature, 345, 33

Zhang, Y., Anninos, P., & Norman, M.L. 1995, ApJ, 453, L57

Hubble's Science Legacy: Future Optical-Ultraviolet Astronomy from Space
ASP Conference Series, Vol. 291, 2003
K.R. Sembach, J.C. Blades, G.D. Illingworth, R.C. Kennicutt, Jr. (eds.)

The Whys and Hows of Finding 10,000 Lenses

C.S. Kochanek

Smithsonian Astrophysical Observatory, Center for Astrophysics MS-51, 60 Garden St., Cambridge, MA 02138

Abstract. Very large samples of lenses are ideal for studying the relations between dark matter halos and astronomical objects and the evolution of these relations with cosmic epoch. Samples of 10^3 lenses are likely to be available in 10 years, but obtaining a sample of 10^4 lenses probably requires a wide-field, high resolution space telescope.

1. Introduction

Studies of gravitational lenses are undergoing a major change driven by the rapidly rising number of known systems. We now know of approximately 80 systems, and the discovery rate is over 10 systems per year and accelerating. Somewhat more than 1 in 1000 objects with $z > 1$ are multiply imaged by a foreground galaxy, so the only limitation on finding more gravitational lenses is the number of sources you can survey at a high enough resolution to identify the lens. For compact sources with surface brightnesses well above that of galaxies (e.g., QSOs at any wavelength), only moderate resolution ($0\farcs5$) is necessary to find most lenses. For extended sources or sources with surface brightnesses similar to galaxies, higher resolution ($0\farcs1$) is needed. Since technology is simultaneously advancing the sensitivity, resolution, and size of astronomical surveys, the discovery of lenses will continue to accelerate for the foreseeable future.

In considering the use of a next generation optical space telescope to find lenses, we will first discuss how lenses are found today (§2) and their application to current astrophysical problems (§3). However, most of these present day problems will have been solved by the time the telescope is launched (we hope!), so in §4 we discuss reasons for finding very large lens samples which should still be valid 10–20 years from now. Finally, in §5 we discuss the technological developments that will lead to these discoveries.

2. How Do We Find Lenses Today?

Unfortunately, a complete one page treatment of finding lenses is impossible. Indeed, even a 100 page treatment would be difficult. Systematic imaging surveys have found approximately 2/3 of the 80 lenses, with the remaining 1/3 due to serendipity and experimental surveys. Of the lenses found in imaging surveys about 1/2 are due to radio surveys and 1/2 to optical quasar surveys.

The first systematic radio lens survey, the MIT - Green Bank survey (MG, see Lehar et al. 1997), found lenses by obtaining Very Large Array (VLA) snapshot images of flux-limited samples of 5 GHz radio sources. While the MG survey found several very interesting systems, such as the first Einstein ring lens MG1131+0456, the complexity of the intrinsic radio morphologies made it difficult to identify the lenses. The Jodrell Bank VLA Astrometric Survey (JVAS; Patnaik et al. 1992), Cosmic Lens All Sky Survey (CLASS; Myers et al. 2001), and Parkes-MIT-NRAO Survey (PMN; Winn et al. 2001) solved this problem by examining only flat spectrum radio sources, which generally consist of a single compact core. Further radio imaging, particularly very long baseline interferometry (VLBI) imaging, can generally confirm candidates as lenses. Since the candidates are selected using only radio data, the resulting lens samples are unbiased by either emission or absorption from the lens galaxy. While efficient and productive, the surveys have nearly run out of bright sources visible from the VLA. Approximately 1 in 500 radio sources is a lens.

Quasar imaging surveys (see the summary by Kochanek 1996 or Gregg et al. 2001 for a current survey) have focused on samples of bright (15-18 mag) quasars where magnification bias can raise the lens fraction to 1 in 100 sources. For fainter quasars, the magnification bias weakens and the lensing rate should be comparable to that for radio sources. Until recently, the inability of ground-based telescopes to reliably provide high resolution ($0\farcs5$ FWHM) images made finding lensed quasars a relatively painful process. Several surveys solved the problem using Hubble Space Telescope (HST) snapshot images (e.g., Maoz et al. 1993; Gregg et al. 2001), but ground-based surveys are becoming significantly easier due to the steady improvement in image quality.

The remaining systems are found by a huge variety of methods, including good luck. Two interesting possibilities for the future are finding lenses through the detection of emission lines from the source in the spectrum of a lower redshift galaxy (e.g., Mortlock & Webster 2001), and looking for extended radio lobes with optical counterparts (Lehar et al. 2001). Both of these are experimental survey methods at present, but may become enormously productive as a result of large surveys like the Sloan Digital Sky Survey (SDSS).

Increasingly, the bottleneck in using the lenses as tools is the follow-up observations rather than the discovery. The first bottleneck is the problem of missing redshifts. Roughly 1/3 of lenses lack both lens and source redshifts, and another 1/3 lack one of the two redshifts. The rapid increase in the numbers of large telescopes, improvements in adaptive optics (AO) systems and infrared spectrographs, and (as a last resort) the Next Generation Space Telescope (NGST) should solve this problem. The second bottleneck is that only HST can provide the images needed to use lenses as astrophysical tools. Most modern applications of the lenses need precise component positions, fluxes, colors, and the physical parameters of the lens or host galaxy. While we have tried some experiments with ground based AO systems, the energy in the point spread function (PSF) wings makes it very difficult to quantitatively study systems that are a mixture of very high (AGN cores) and very low (lens, AGN host galaxy) surface brightness components separated by $1''$.

Finally, we note that two qualitatively different types of data will become common. First, for the brighter lenses it is possible to make direct stellar dynamical measurements for comparison to the lens data (5 at present, see Koopmans

& Treu 2002). Second, monitoring programs will determine many more time delays between lensed images (9 at present, see Kochanek 2002a).

3. Why Study Lenses Today?

We cannot review all the present applications of lenses here, so we will simply illustrate their uses with a few recent examples. Each of these applications is critically dependent on obtaining HST imaging data to provide the necessary astrometry and photometry used in the calculations.

The Hubble constant remains an interesting problem for two reasons. First, the best current local estimate of 72 ± 8 km s^{-1} Mpc^{-1} (Freedman et al. 2001) is subject to enough systematic uncertainties to make an independent check of the results desirable. Second, most degeneracies in cosmological parameter estimation for cosmic microwave background (CMB) experiments can be eliminated given an accurate independent estimate of H_0. Lens time-delay measurements constrain the variable combination $H_0(1 - \langle\kappa\rangle)$ where $\langle\kappa\rangle = \langle\Sigma\rangle/\Sigma_c$ is the average surface density in the annulus between the images used to measure the delay (Kochanek 2002a). In cold dark matter (CDM) halo models $\langle\kappa\rangle$ is determined by the cold baryon fraction, f_b, the fraction of the halo mass which has cooled to form the visible galaxy (Kochanek 2002b). The best fit mass model for lenses, a singular isothermal profile, corresponds to the halo model we would expect for galaxies with the locally observed baryon fraction of $f_b \sim 0.01$. As shown in Figure 1, current results for the 4 simplest time delay lenses show a contradiction between the delay measurements, the halo model, and local estimates of H_0.

Gravitational lenses can also be used to measure the evolution of galaxies with cosmic epoch. They have the important advantage over other samples, which we return to in §4, of being selected based on mass rather than on luminosity or surface brightness. Current results are obtained by using the fundamental plane (FP) for early-type galaxies to determine the difference between the observed surface brightness and the surface brightness a galaxy with the same effective radius and velocity dispersion has at redshift zero. The average change in the rest frame B-band mass-to-light ratio for the sample of 29 lenses shown in Figure 2 is $d\log(M/L)_B/dz = -0.55 \pm 0.04$ (Rusin et al. 2002). This rate matches that found for clusters (e.g., Kelson et al. 2000) and the van Dokkum et al. (2001) sample of field galaxies. It is significantly slower than the Treu et al. (2002) estimate for field galaxies. Semi-analytic models (e.g., Kauffman & Charlot 1998) predict significant age differences between cluster and field early-type galaxies that we do not seem to observe. The present results are limited by the need for HST observations of more lenses.

In the last few years there has been considerable debate about the compatibility of dark matter halo models with the observed structure of galaxies. One problem is that the Galaxy seems to have too few satellites compared to CDM models (e.g., Moore et al. 1999). Proposed solutions are to make them invisible by suppressing star formation, destroy them with interacting dark matter, or avoid forming them by changing the power spectrum. Substructure in lenses has the common observable consequence of producing anomalous flux ratios, images which should have the same brightness by symmetry but do not (Mao & Schneider 1998; Metcalf & Madau 2001). Dalal & Kochanek (2002) developed

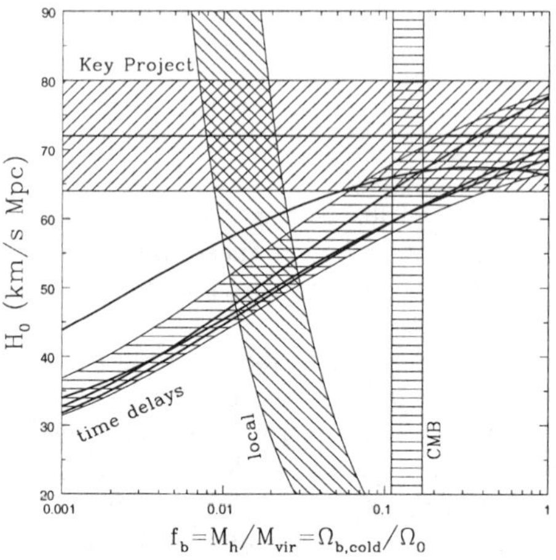

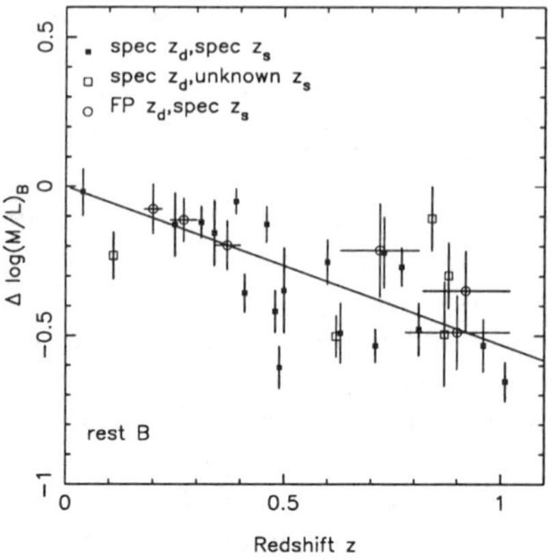

Figure 1. [*Above, left*] The Hubble constant for CDM halo models. The solid curves show H_0 inferred for the four simple time delay lenses as a function of the halo mass fraction, f_b, represented by the visible galaxy. The hatched region tracking the curves is their joint estimate. The baryon fraction is bounded by estimates of Ω_b/Ω_m from the CMB and local inventories of known baryonic populations. The Key Project estimate of H_0 is consistent with the time delays only as we approach constant M/L models where $f_b = 1$ and the lenses have no dark matter.

Figure 2. [*Above, right*] Evolution of M/L of early-type lens galaxies. Each point is a lens, coded by the available redshift information. The error bars and scatter are comparable to similar studies using dynamical observations of intermediate redshift galaxies. The line shows the average rate of change of the B-band mass-to-light ratio.

a method to estimate the substructure mass fraction. Figure 3 shows that the results almost exactly agree with typical CDM predictions. Larger samples and close studies of lenses with complicated VLBI structures should improve the constraints and allow an accurate estimate of the substructure mass scale.

Our final application is the use of lenses to study the interstellar medium in the lens galaxy. Lenses show a wide range of differential extinctions between the lensed images with rest-frame $\Delta E(B-V)$ values ranging from 0 to 2.5 with a median of 0.04 mag (Falco et al. 1999). Differential extinction studies in lenses have a precision comparable to what can be achieved locally by studying individual stars because they use differential photometry of the same quasar. The total extinction is far more poorly constrained because it depends on a model for the intrinsic spectrum of the quasar. If the lens has just the right amount of differential extinction [$\Delta E(B-V) \simeq 0.1$ mag], then it becomes possible to trace the absorption as a function of wavelength to estimate the extinction curve associated with the dust. By combining a high redshift lens (to shift the 2175 Å absorption feature into the optical) with a low redshift source (to keep the Lyman break at shorter wavelengths than the absorption feature), we can trace the ultraviolet extinction curve of the lens galaxy using optical

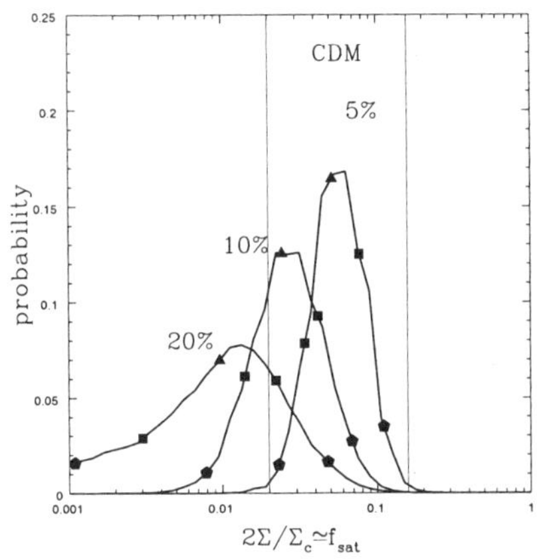

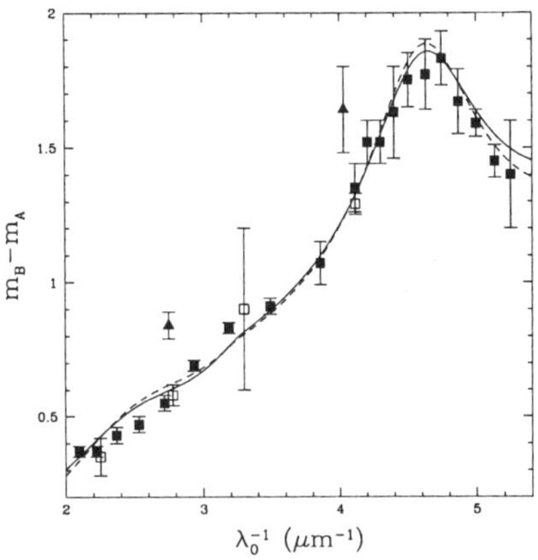

Figure 3. [*Above, left*] Estimates of the substructure mass fraction. The curves show the estimated mass fraction in substructure for image flux errors of 5% (too low), 10% (about right) and 20% (too high). Standard CDM models lie in the band between 2% and 20%, while the observed Milky Way satellites correspond to $f_{sat} \sim 10^{-3}$-10^{-4}.

Figure 4. [*Above, right*] The extinction curve of SBS0909+532 at $z_l = 0.83$. Solid squares are from spectra, open squares from images, (offset) triangles are for emission lines. Dashed line is a standard Galactic $R_V = 3.1$ extinction curve, solid is the best fit $R_V = 2.1 \pm 0.9$ model.

spectra. Figure 4 shows the result for the $z_l = 0.83$ lens SBS0909+532 (Motta et al. 2002). The lens galaxy shows a clean 2175 Å feature, and the shape of the extinction curve is almost exactly that of a standard $R_V = 3.1$ Galactic extinction curve.

4. Why Study Lenses Tomorrow?

For the purposes of this section, we assume that over the next 10 years the basic parameters of the cosmological model and the power spectrum are known and that the focus of cosmology is to measure quantitatively the relationships between halo mass functions and astronomical objects as a function of time.

The most important difference between lenses and all other astronomical objects is that they are detected by mass rather than light. This distinction will become steadily more important as the lens sample size increases. Other methods of finding halos suffer from three important problems.

1. They cannot find dark halos. While there is no absolutely convincing example of a massive, dark halo (although see Miralles et al. 2002), the related debate about the prevalence of low surface brightness systems persists. Moreover, at the low mass end, the differing slopes of standard mass functions ($\sim 1/M^2$) and luminosity functions ($\sim 1/L$) means that a steadily increasing fraction of $M \lesssim 10^{10} M_\odot$ halos are not found as visible galaxies. The substructure problem of §3 is one facet of this problem.

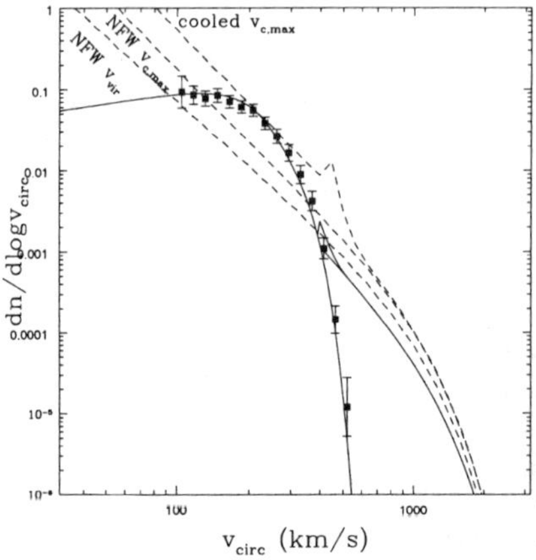

Figure 5. The solid line shows the estimated velocity function of galaxies and groups/clusters. The dashed curves show the velocity function of CDM halos using either the virial velocity (NFW v_{vir}), the peak circular velocity (NFW $v_{c,max}$), or the peak circular velocity including a crude model for baryonic cooling (cooled, $v_{c,max}$). Note the problems combining galaxies and clusters, and the difference between galaxies and halos at low velocity. The cold baryon fraction used in these models, $f_B \simeq 0.02$, reproduces the observed separation distribution of gravitational lenses and matches the models in Figure 1 giving low H_0 but roughly the observed isothermal mass distributions of gravitational lenses.

2. They have trouble studying halos at the boundary between groups and galaxies. Optical observations are well-suited to studying individual galaxies, and X-ray observations are well-suited to rich clusters, but neither works well in the transition region.

3. Selection methods based on emission (or absorption) will be difficult to match across different cosmic epochs because of evolution. This can be an obvious effect, like changes in surface brightness, or a more subtle effect, like comparing the velocity functions of galaxies between epochs when the relative numbers of early-type and late-type galaxies have changed.

Figure 5 illustrates these problems based on an admittedly crude attempt to piece together a global velocity function (Kochanek & White 2001).

Unlike any other method, lenses provide a uniform method for detecting halos on all mass scales. It is not independent of baryonic physics because the efficiency of a halo as a lens depends on the density profile of the halo as well as the mass, but the same search method works on all mass scales, the density profiles can be measured using the lens data, and density profiles evolve slowly with cosmic epoch.

For known populations, the rough distribution of the lenses is understood. Massive galaxies dominate the lens population because the high central densities produced by cooling the baryons increase enormously their lensing efficiency over more massive halos (Kochanek & White 2001). As a result, the overwhelming majority ($\simeq$ 80%) of lenses are massive early-type galaxies. Spiral lenses are rarer (10–20%) because they are less massive even if they are more numerous, and group/cluster lenses are rare (5–10%) because the failure of the baryons to cool gives them far lower central surface densities than galaxies. Lenses with velocity dispersions below 100 km s^{-1} are both intrinsically rare and filtered out by survey resolution limits. With the present sample of 80 lenses, only some of which have complete follow up observations, the only lens population we can study in detail is the dominant, early-type galaxy lens population. At present there are too few spiral lenses to examine their evolution (as in Figure 2), and we are just beginning to find lenses genuinely produced by groups.

Thus, extending the use of lenses to studying the relationship between astronomical objects and the halo mass function over a broad range of masses requires a much larger lens sample. To be able to study spiral lenses or group lenses in as much detail as we now study early-type lenses requires $\sim 10^3$ lenses. A uniform survey with good statistics on the mass scale of rich clusters would require $\sim 10^4$ lenses; while rich cluster lenses are easy to find by looking at rich clusters, they represent a negligible fraction of the lenses. On cluster scales, finding dark lenses is no harder than finding luminous lenses because neither (presumably) has benefited from the enormous increase in lens cross section produced by cooling the baryons. On the scales of galaxies, dark halos in which the baryons failed to cool or were reheated will have far lower lensing cross sections than normal galaxies. Finding or limiting a population of dark halos on the mass scales of galaxies requires surveys which produce large samples (10^3-10^4) of normal lenses and very good angular resolution ($< 0\farcs1$).

5. How Will We Find Lenses Tomorrow?

The first new generation survey is the SDSS. Roughly speaking, if the survey finds 10^5 quasars, we would expect to find ~ 100 lensed quasars. Recognizing them will be hard work because of the poor resolution of the survey ($1\farcs0$).

On a 10 year time scale the most important new development is the Expanded VLA (EVLA), which will increase the sensitivity and resolution of the VLA. Radio selection remains the cleanest approach to finding uniform samples of lenses which are unbiased by the properties of the lens galaxy. The FIRST survey of 10^4 deg^2 at 1.4 GHz and 1 mJy sensitivity contains $\sim 10^6$ sources (Becker et al. 1995), so a sample of $\sim 10^3$ lenses can be found by re-imaging most of the FIRST survey at high resolution. Performing snapshots of individual sources is probably the most efficient approach, but performing the search as part of a new, high resolution survey might serve a broader range of scientific purposes. In either case, it would require roughly a year of observing time to obtain such a sample.

The problem with radio surveys, or any other survey based on finding lensed AGNs, is that their surface density is relatively low. Surveys to find really large samples of lenses, $\sim 10^4$ lenses, probably have to use high-redshift galaxies as their source population. Optical surveys are intrinsically more sensitive to emission and absorption by the lens galaxies, making them "dirtier" surveys than radio surveys. This becomes particularly acute when the background sources are fainter galaxies whose central surface brightnesses are lower than the foreground lenses. The only way to find such systems is with high resolution imaging from space so that the sources can be separated from the lens. The Medium Deep Survey (MDS) survey, for example, found 10 plausible lens candidates with an effective density of 15 deg^{-2} (Ratnatunga et al. 1999). Thus, the Supernova Acceleration Probe (SNAP), which will survey $\sim 10^2$ deg^2 with resolution and depth comparable to the MDS, can find $\sim 10^3$ lenses. Neither the HST Wide Field Camera (used for the MDS) nor the SNAP design is ideal for strong lenses because of poor PSF sampling.

It is easy to write the words for the telescope which could find 10^4 lenses. Keep the field of view of SNAP, but put it on a larger aperture telescope (to go deeper faster and to improve the resolution to $0''.05$), and use a well-sampled near-infrared sensitive camera (to improve the contrast between the lens and the images). Essentially, it is the SDSS or Large Synoptic Survey Telescope (LSST) project done with better resolution than HST. Practicality and affordability are harder issues to address, but the science would be great!

Acknowledgments. CSK is supported by NASA ATP grant NAG5-9265.

References

Becker, R.H., White, R.L., & Helfand, D.J. 1995, ApJ, 450, 559

Dalal, N., & Kochanek, C.S. 2002, ApJ, 572, 25

Falco, E.E., Impey, C.D., Kochanek, C.S., et al. 1999, ApJ, 523, 617

Freedman, W.L., Madore, B.F., Gibson, B.K., et al. 2001, ApJ, 553, 47

Gregg, M.D., et al. 2001, in ASP Conf. Ser. 237, Gravitational Lensing: Recent Progress, Future Goals, eds. T. Brainerd & C.S. Kochanek, (Provo: ASP), 41

Kauffmann, G., & Charlot, S. 1998, MNRAS, 294, 705

Kelson, D.D., et al. 2000, ApJ, 531, 184

Kochanek, C.S. 1996, ApJ, 466, 638

Kochanek, C.S., & White, M. 2001, ApJ, 559, 531

Kochanek, C.S. 2002a, ApJ, in press [astro-ph/0205319]

Kochanek, C.S. 2002b, ApJ, in press [astro-ph/0206006]

Koopmans, L.V.E., & Treu, T. 2002, ApJ, 568, L5

Lehar, J., Burke, B.F., Conner, S.R., et al. 1997, AJ, 114, 48

Lehar, J., Buchalter, A., McMahon, R.G., et al. 2001, ApJ, 547, 60

Mao, S., & Schneider, P. 1998, MNRAS, 295, 587

Maoz, D., Bahcall, J.N., Schneider, D.P., et al., 1993, ApJ, 409, 28

Metcalf, R.B., & Madau, P. 2001, ApJ, 563, 9

Miralles, J.-M., Erben, T., Hammerle, H., et al. 2002, A&A, 388, 68

Moore, B., et al. 1999, ApJ, 524, L19

Mortlock, D.J., & Webster, R.L. 2001, MNRAS, 321, 629

Motta, V., et al. 2002, ApJ, 575, 719

Myers, S.T., et al. 2001, in ASP Conf. Ser. 237, Gravitational Lensing: Recent Progress, Future Goals, eds. T. Brainerd & C.S. Kochanek, (Provo: ASP), 51

Patnaik, A.R., et al. 1992, MNRAS, 254, 655

Ratnatunga, K.U., Griffiths, R.E., & Ostrander, E.J. 1999, ApJ 117, 2010

Rusin, D., Kochanek, C.S., Falco, E.E., et al. 2001, ApJ, submitted

Treu, T., Stiavelli, M, Casertano et al., 2002, ApJ, 564, L13

van Dokkum, P.G., et al. 2001, ApJ, 553, L39

Winn, J.N., et al. 2001, in ASP Conf. Ser. 237, Gravitational Lensing: Recent Progress, Future Goals, eds. T. Brainerd & C.S. Kochanek, (Provo: ASP), 61

Hubble's Science Legacy: Future Optical-Ultraviolet Astronomy from Space
ASP Conference Series, Vol. 291, 2003
K.R. Sembach, J.C. Blades, G.D. Illingworth, R.C. Kennicutt, Jr. (eds.)

Dark Matter and Dark Energy: The Critical Questions

Michael S. Turner

Dept. of Astronomy & Astrophysics, Dept. of Physics, Enrico Fermi Institute, and Center for Cosmological Physics, The University of Chicago, Chicago, IL 60637-1433

NASA/Fermilab Astrophysics Center, P.O. Box 500, Fermi National Accelerator Laboratory, Batavia, IL 60510-0500

Abstract. Stars account for only about 0.5% of the content of the Universe; the bulk of the Universe is optically dark. The dark side of the Universe is comprised of: at least 0.1% light neutrinos; 3.5%±1% baryons; 29%±4% cold dark matter; and 66%±6% dark energy. Now that we have characterized the dark side of the Universe, the challenge is to understand it. The critical questions are: (1) What form do the dark baryons take? (2) What is (are) the constituent(s) of the cold dark matter? (3) What is the nature of the mysterious dark energy that is causing the Universe to speed up.

1. Introduction

The past five years have witnessed great progress in identifying the basic features of our Universe. It is spatially flat and thus has the critical density ($\rho_{\rm crit} = 3H_0^2/8\pi G \approx 10^{-29}\,{\rm g\,cm^{-3}}$). The expansion is speeding up, not slowing down (i.e., $q_0 < 0$). The mass/energy density is distributed as follows:

- Bright stars: 0.5%
- Baryons (total): $4\% \pm 1\%$
- Nonbaryonic dark matter: $29\% \pm 4\%$
- Neutrinos: at least 0.1% and possibly as large as 5%
- Dark Energy: $66\% \pm 6\%$

The evidence for flatness comes from measurements of the anisotropy of the cosmic microwave background (CMB) on angular scales of about 1 degree. The position of the first acoustic peak at multipole number 200, as determined by the BOOMERANG, MAXIMA, DASI and CBI experiments, implies that $\Omega_0 = 1 \pm 0.04$ (Sievers et al. 2002). This means that the curvature radius of the Universe is greater than about 5 times the Hubble radius since $R_{\rm curv} = H_0^{-1}/|\Omega_0 - 1|^{1/2}$. In the next sections I will discuss the evidence for the accounting of the various dark components.

I did not mention the cosmic microwave background, which today accounts for $\Omega_{\rm CMB} = 2.47h^{-2} \times 10^{-5}$ (about 0.005%), or the relativistic neutrino backgrounds which account for $\Omega_\nu = 0.56h^{-2} \times 10^{-5}$ per relativistic neutrino species. Though unimportant to the energy budget today, at very early times relativistic particles were dominant.

The existence of (at least) three components that evolve differently with redshift divides the evolution of the Universe into (at least) three epochs: 1. Early ($z > 10^4$ and $t < 10^4$ yrs) radiation-dominated era (photons, neutrinos, and a spectrum of relativistic particles that grows with temperature); 2. Matter-dominated era ($10^4 > z > 0.2$, $10\,{\rm Gyr} > t > 10^4$ yrs) during which cosmic structure grew; and 3. Dark-energy dominated era ($z < 0.2$, $t > 10$ Gyr) characterized by accelerated expansion and cessation of structure formation.

I note that the precision of the present accounting still allows for an unidentified component that contributes perhaps as much as 10% of the critical density.

Now that we can enumerate the components of the Universe, the task is to understand them. Three critical questions arise:

1. What form do the dark baryons take?
2. What is (are) the constituent(s) of the nonbaryonic dark matter?
3. What is the nature of the dark energy?

2. Dark Baryons

For many years the theory of big-bang nucleosynthesis (BBN) and a lower limit to the primordial deuterium abundance were used to argue for a low baryon density (see e.g., Schramm & Turner 1998), $\Omega_B \lesssim 0.1$. The measurement of the primordial deuterium abundance in high-redshift hydrogen clouds has turned the upper limit into the most precise determination of the baryon density, $\Omega_B h^2 = 0.020 \pm 0.001$, or $\rho_B = 3.8 \pm 0.7 \times 10^{-31}\,{\rm g\,cm^{-3}}$ (Burles et al. 2001; Tytler et al. 2000; O'Meara et al. 2001). When combined with knowledge of the Hubble constant, $h = 0.72 \pm 0.07$ (Freedman et al. 2001), this implies

$$\Omega_B = 0.04 \pm 0.008 \tag{1}$$

Within the past year the credibility of the BBN baryon density got a tremendous boost from the DASI, BOOMERANG, MAXIMA, and CBI CMB experiments (Sievers et al. 2002). Based upon the ratios of the heights of the odd to even acoustic peaks, these experiments imply: $\Omega_B h^2 = 0.022 \pm 0.003$. The agreement with BBN is striking, and the underlying physics is very different: gravity-driven acoustic oscillations at 400,000 years versus nuclear physics at a few seconds. Not only does this give one confidence in a low baryon density, but it also provides a powerful consistency test of the whole framework.

The absorption of light from $z \sim 3-4$ quasars by Lyman-alpha clouds provides a lower limit to the fraction of critical density in intergalactic gas which is consistent with BBN and the CMB: $\Omega_B h^2 > 0.018[\Omega_M h^2/0.17]^{1/4}$ (McDonald et al. 2001). This determination of the baryon density depends upon a number of assumptions, including the flux of ionizing radiation, the physics of the Lyman-alpha forest being well described by CDM, and the bulk of the baryons at $z \sim 3-4$ not being in stars or stellar remnants.

Today, baryons in stars account for only about 10% of all baryons; most of the baryons are still unaccounted for. In rich clusters, the census appears to be complete: The ratio of baryons in the hot, intracluster gas to that in stars is close to 10. While clusters provide an environment where it is easier to carry out a baryon census because of their deep gravitational potential, clusters only account for a few percent of the total matter density and a minority of the baryons.

Where then are the bulk of the baryons today? The conventional wisdom is that they reside in hot intergalactic gas or warm gas more closely associated with galaxies (see e.g., Cen & Ostriker 1999). While unlikely, it is possible that the bulk of the baryons, or even a significant fraction, exists as the remnants of an early generation of stars.

Finally, it is worth noting that at redshifts $z \sim 3, 1000$, and 10^{10} the baryon census is complete and consistent. At BBN the baryons were in neutrons and protons that were in the processing of being fused into the light elements; at last scattering they were in ionized H and He gas; and at $z \sim 3$, they were in intergalactic gas. To complete the picture, we need a baryon census today (see Fukugita et al. 1998 for a preliminary estimate).

3. Nonbaryonic Dark Matter

The total matter density can now be determined by measurements that do *not* depend upon the connection between mass and light (see e.g., Turner 2002). They include CMB anisotropy (the heights of the acoustic peaks determine $\Omega_M h^2$ and $\Omega_B h^2$), the power spectrum of inhomogeneity (its shape determines $\Omega_M h$ and Ω_M/Ω_B), the cluster baryon fraction (Ω_M/Ω_B) and the primordial deuterium abundance ($\Omega_B h^2$). From these a robust and "unbiased" value of the matter density follows (Turner 2002):

$$\Omega_M = 0.33 \pm 0.035 \qquad \Omega_B = 0.004 \pm 0.008 \;\Rightarrow\; \Omega_{NB} = 0.29 \pm 0.04 \tag{2}$$

With a high degree of confidence, we can say that almost 90% of the matter is in a form other than baryons and yet to be identified! This conclusion receives support from another important, though indirect, argument: Absent nonbaryonic dark matter, there is no model for structure formation that can begin with the level of matter inhomogeneity indicated by the anisotropy of the CMB ($\delta\rho/\rho \sim 10^{-5}$) and explain the structure seen in the Universe today.

For almost 20 years there has been a working hypothesis: the bulk of the dark matter exists in the form of stable (or longlived) relic particles left over from the earliest moments of creation. For almost as long, the leading particle candidates have been: one or more light neutrinos, the axion and the neutralino. Significant progress has been and is being made toward testing all three candidates (see e.g., Turner 2000a; Griest & Kamionkowski 2000).

3.1. Neutrinos

Results from the SuperKamiokande (SuperK, Fukuda et al. 1999; Fukuda et al. 2000) and Sudbury Neutrino Observatory (SNO, Ahmad et al. 2002a, 2000b) experiments have provided strong evidence for neutrino oscillations, with two mass differences identified: $\Delta m_{12}^2 \simeq 10^{-4}\,\mathrm{eV}^2$ (SNO/solar neutrinos) and $\Delta m_{23}^2 \simeq 3\times 10^{-3}\,\mathrm{eV}^2$ (SuperK/atmospheric). Oscillation experiments only probe relative masses, and not the absolute mass scale. With the simple assumption of a mass hierarchy, $m_3 \gg m_2 \gg m_1$, this implies $m_3 \sim 0.05\,\mathrm{eV}$ and $m_2 \sim 0.01\,\mathrm{eV}$, corresponding to $\Omega_\nu \sim 0.1\%$ (note, $\Omega_\nu = m_\nu/93.2h^2\,\mathrm{eV}$). Hierarchy or not, this provides a lower limit to what neutrinos contribute to the mass budget – in the same ball park as stars – and validates the concept of particle dark matter. Cosmology has long since made its judgment on neutrinos – hot dark matter cannot account for the structure we see today and even a modest of amount of hot dark matter ($\Omega_\nu \gtrsim 0.05$) leads to an unacceptable deficiency of small-scale structure (Croft et al. 1998; Elgaroy et al. 2002).

In setting the absolute scale of neutrino mass, cosmological observations may play an important role. CMB anisotropy and large-scale structure measurements have sensitivity at the tenths of an eV level (see e.g., Hu et al. 1998).

3.2. Cold Dark Matter

Cosmology has also made its judgment on cold dark matter – the cold dark matter scenario has much, if not all, the truth about structure formation. The two leading CDM particles are the axion, a very light particle ($m \sim 10^{-6}\,\mathrm{eV} - 10^{-5}\,\mathrm{eV}$), which is cold by virtue of being born in a Bose condensate, and the neutralino, a heavy particle ($m \sim 50\,\mathrm{GeV} - 500\,\mathrm{GeV}$) and the lightest of the supersymmetric partners of the known particles, which is cold by virtue of its large mass.

Experiments to detect the axions or neutralinos that may comprise our own halo have reached sufficient sensitivity to begin seriously testing both possibilities. In addition, the neutralino can be produced at an accelerator or detected by its annihilation products (see e.g., Griest & Kamionkowski 2000). (Neutralinos that accumulate in the Sun annihilate and produce high-energy neutrinos; neutralinos in the halo can annihilate and produce photons, antiprotons, or gamma rays.)

Like the neutrino, the interactions of the axion and neutralino with ordinary matter are very weak (mean free path $\gg H_0^{-1}$). So far as cosmology goes, axions and neutralinos are particles that only interact via gravity. This fact has made simulation of cosmic structure formation relatively simple (compared to the typical astrophysical simulation). Unfortunately, it also means that axionic cold dark matter and neutralino cold dark matter cannot be distinguished by purely cosmological observations.

There are some indications that CDM does not have all the truth on small scales, namely, the halo cusp problem and the overabundance of substructure (see e.g., Sellwood & Kosowsky 2000). While plausible astrophysical explanations exist for both problems (see, e.g., Benson et al. 2001; Bullock et al. 2000; Klypin et al. 2001; Merritt & Cruz 2001; Milosavljevic & Merritt 2001; Milosavljevic et al. 2001; Somerville 2001; Weinberg & Katz 2001), the solution could involve an unexpected property of the dark-matter particle, e.g., large self-interaction cross section (Spergel & Steinhardt 2000), large annihilation cross section (Kaplinghat et al. 2000), or mass of around 1 keV (i.e., warm dark matter).

At present, there is the tantalizing possibility that further study of cosmic structure on small scales could reveal a new property of the dark-matter particle; in any case, this is the arena where CDM needs to be tested.

4. Dark Energy

Dark energy is my term for the causative agent of the current epoch of accelerated expansion. According to the second Friedmann equation,

$$\ddot{R}/R = -4\pi G\left(\rho + 3p\right)/3; \tag{3}$$

thus, stress-energy with large negative pressure can produce accelerated expansion (R is the cosmic scale factor).

Dark energy has the following defining properties: (1) it emits no light; (2) it has large, negative pressure, $p_X \sim -\rho_X$; and (3) it is approximately homogeneous (more precisely, it does not cluster significantly with matter on scales at least as large as clusters of galaxies). Because its pressure is comparable in magnitude to its energy density, it is more "energy-like" than "matter-like" (matter being characterized by $p \ll \rho$). Dark energy is qualitatively very different from dark matter.

4.1. Two Lines of Evidence for an Accelerating Universe

Two independent lines of evidence point to an accelerating Universe. The first is the measurement of type Ia supernova light curves by two groups, the Supernova Cosmology Project (Perlmutter et al. 1999) and the High-z Supernova Team (Riess et al. 1998). The teams used different analysis techniques and different samples of high-z supernovae and came to the same conclusion: the Universe is speeding up, not slowing down.

The recent discovery of a supernova at $z \simeq 1.76$ bolsters the case significantly (Riess et al. 2001) and provides the first evidence for an early epoch of decelerated expansion (Turner & Riess 2002). SN 1997ff falls right on the accelerating Universe curve on the magnitude – redshift diagram, and is a magnitude brighter than expected in a dusty open Universe.

The second, independent line of evidence for the accelerating Universe comes from measurements of the composition of the Universe, which point to a missing energy component with negative pressure (Turner 2000b). The argument goes like this. CMB anisotropy measurements indicate that the Universe is flat, $\Omega_0 = 1.0 \pm 0.04$ (Sievers et al. 2002). In a flat Universe, the matter density and energy density must sum to the critical density. However, matter only contributes about 1/3rd of the critical density, $\Omega_M = 0.33 \pm 0.04$ (Turner 2002), leaving two thirds of the critical density missing.

In order to have escaped detection this missing energy must be smoothly distributed. In order not to interfere with the formation of cosmic structure, the energy density in this component must change more slowly than matter, so that it was subdominant in the past. For example, if the missing two-thirds of the critical density were smoothly distributed matter ($p = 0$), then linear density perturbations would grow as $R^{1/2}$ rather than as R. The shortfall in growth since last scattering ($z \simeq 1100$) would be a factor of 30, leading to far too little growth to produce the structure seen today.

The pressure associated with the missing energy component determines how it evolves:

$$\rho_X \propto R^{-3(1+w)} \quad \Rightarrow \quad \rho_X/\rho_M \propto (1+z)^{3w} \tag{4}$$

where w is the ratio of the pressure of the missing energy component to its energy density (here assumed to be constant). Note, the more negative w, the faster the ratio of missing energy to matter goes to zero in the past. In order to grow the structure observed today from the density perturbations indicated by CMB anisotropy measurements, w must be more negative than about $-\frac{1}{2}$ (Turner & White 1997), and since $q_0 = \frac{1}{2} + \frac{3}{2}w\Omega_X \sim \frac{1}{2} + w$, this implies $q_0 < 0$ and accelerated expansion.

4.2. Gravity Can Be Repulsive in Einstein's Theory

In Newton's theory mass is the source of the gravitational field and gravity is always attractive. In general relativity, both energy and pressure source the gravitational field, cf. Eq. 3. Sufficiently large negative pressure leads to repulsive gravity, and so accelerated expansion can be accommodated. Of course, that does not preclude that the ultimate explanation for accelerated expansion lies in a fundamental modification of Einstein's theory.

Repulsive gravity is a stunning feature of general relativity. It leads to a prediction every bit as revolutionary as black holes – the accelerating Universe. If the explanation for the accelerating Universe fits within general relativity, it will be a major new triumph for Einstein's theory.

4.3. The Biggest Embarrassment in Theoretical Physics

Einstein introduced the cosmological constant to balance the attractive gravity of matter in order to create a static cosmological model. He quickly discarded the cosmological constant after the discovery of the expansion of the Universe.

Quantum field theory makes the consideration of the cosmological constant obligatory not optional. The only possible covariant form for the energy of the (quantum) vacuum,

$$T^{\mu\nu}_{\rm VAC} = \rho_{\rm VAC} g^{\mu\nu},$$

is mathematically equivalent to the cosmological constant. It takes the form for a perfect fluid with energy density $\rho_{\rm VAC}$ and isotropic pressure $p_{\rm VAC} = -\rho_{\rm VAC}$ (i.e., $w = -1$) and is precisely spatially uniform. Vacuum energy is the *almost* perfect candidate for dark energy.

Here is the rub: the contributions of well-understood physics (say up to the 100 GeV scale) to the quantum-vacuum energy add up to 10^{55} times the present critical density. This is the well known cosmological-constant problem (see e.g., Weinberg 1989; Carroll 2001).

While string theory currently offers the best hope for a theory of everything, it has shed precious little light on the problem; in fact, it has raised some new issues. The deSitter space associated with the accelerating Universe poses serious problems for the formulation of string theory (Witten 2001).

The cosmological constant problem leads to a fork in the dark-energy road: one path is to wait for theorists to get the "right answer" (i.e., succeed in showing that indeed quantum vacuum energy contributes $\Omega_\Lambda \simeq 2/3$); the other path is to assume that even quantum nothingness weighs nothing and something else with negative pressure must be causing the Universe to speed up. Of course, theorists follow the advice of Yogi Berra: where you see a fork in the road, take it.

4.4. Parameterizing Dark Energy: For Now, It's w

Theorists have been very busy suggesting all kinds of interesting possibilities for the dark energy: networks of topological defects, rolling or spinning scalar fields (quintessence and spintessence), influence of "the bulk", and the breakdown of the Friedmann equations (Carroll 2001; Turner 2000a).

The uniformity of the CMB testifies to the near isotropy and homogeneity of the Universe. This implies that the stress-energy tensor for the Universe must take the perfect fluid form. Since dark energy dominates the energy budget, its stress-energy tensor must be to a good approximation

$$T_X{}^\mu_\nu \approx \mathrm{diag}[\rho_X, -p_X, -p_X, -p_X] \tag{5}$$

where p_X is the isotropic pressure and the desired dark energy density is

$$\rho_X = 2.8 \times 10^{-47}\,\mathrm{GeV}^4$$

(for $h = 0.72$ and $\Omega_X = 0.66$). This corresponds to a tiny energy scale, $\rho_X^{1/4} = 2.3 \times 10^{-3}\,\mathrm{eV}$.

The pressure can be characterized by its ratio to the energy density (or equation-of-state):

$$w \equiv p_X/\rho_X,$$

which need not be constant; e.g., it could be a function of ρ_X or an explicit function of time or redshift.

For vacuum energy $w = -1$; for a network of topological defects $w = -N/3$ where N is the dimensionality of the defects (1 for strings, 2 for walls, etc.). For a minimally coupled, rolling scalar field, w is time dependent and can vary between -1 and $+1$.

I believe that for the foreseeable future getting at the dark energy will mean trying to measure its equation-of-state, $w(t)$.

4.5. The Universe: The Lab for Studying Dark Energy

Dark energy by its very nature is diffuse and a low-energy phenomenon. It probably cannot be produced at accelerators; it isn't found in galaxies or even in clusters of galaxies. The Universe itself is the natural lab – perhaps the only lab – in which to study it.

The primary effect of dark energy on the Universe is on the expansion rate. The first Friedmann equation can be written as

$$H^2(z)/H_0^2 = \Omega_M(1+z)^3 + (1-\Omega_M)\exp\left[3\int_0^z [1+w(x)]d\ln(1+x)\right], \tag{6}$$

where Ω_M is the fraction of critical density contributed by matter today, a flat Universe is assumed, and the dark-energy term follows from energy conservation, $d(\rho_X R^3) = -p_X dR^3$. For constant w the dark-energy term becomes $(1-\Omega_M)(1+z)^{3(1+w)}$. In a flat universe $H(z)/H_0$ depends upon only two parameters: Ω_M and $w(z)$.

While $H(z)$ is probably not directly measurable (however see, Loeb 1998), it does affect significantly two observable quantities: the (comoving) distance to an object at redshift z,

$$r(z) = \int_0^z \frac{dz}{H(z)}, \tag{7}$$

and the growth of (linear) density perturbations, governed by

$$\ddot{\delta}_k + 2H\dot{\delta}_k - 4\pi G\rho_M = 0, \tag{8}$$

where δ_k is the Fourier component of comoving wavenumber k and overdot indicates d/dt.

The comoving distance $r(z)$ can be mapped out with the aid of standard candles (e.g., type Ia supernovae) by measuring luminosity distances $d_L(z) = (1+z)r(z)$. It can also be probed by counting objects of a known intrinsic comoving number density, through the comoving volume element, $dV/dzd\Omega = r^2(z)/H(z)$.

Both galaxies and clusters of galaxies have been suggested as objects to count (Newman & Davis, 2000; Holder et al. 2001). For each, their comoving number density evolves (in the case of clusters very significantly). However, it is believed that much, if not all, of the evolution can be modelled through numerical simulations and semi-analytical calculations. In the case of clusters, evolution is so significant that the number count test probe is affected by dark energy through both $r(z)$ and the growth of density perturbations, with the latter being the dominant effect.

The various cosmological approaches to ferreting out the nature of the dark energy have been studied extensively. Based largely upon my work with Dragan Huterer (Huterer & Turner 2001), I summarize what we know about the efficacy of the cosmological probes of dark energy:

- Present cosmological observations prefer $w = -1$, with a 95% confidence limit $w < -0.6$ (e.g., Perlmutter, Turner & White 1999).
- Because dark energy was less important in the past, $\rho_X/\rho_M \propto (1+z)^{3w} \to 0$ as $z \to \infty$, and the Hubble flow at low redshift is insensitive to the composition of the Universe, the most sensitive redshift interval for probing dark energy is $z = 0.2 - 2$.
- The CMB has limited power to probe w (e.g., the projected precision for Planck is $\sigma_w = 0.25$) and no power to probe its time variation.
- A high-quality sample of 2000 SNe distributed from $z = 0.2$ to $z = 1.7$ could measure w to a precision $\sigma_w = 0.05$. If Ω_M is known independently to better than $\sigma_{\Omega_M} = 0.03$, σ_w improves by a factor of two and the rate of change of $w' = dw/dz$ can be measured to precision $\sigma_{w'} = 0.16$.

- Counts of galaxies and of clusters of galaxies may have the same potential to probe w as SNe Ia. The critical issue is systematics (including the evolution of the intrinsic comoving number density, and the ability to identify galaxies or clusters of a fixed mass).

- Measuring weak gravitational lensing by large-scale structure over a field of 1000 square degrees (or more) could have comparable sensitivity to w as type Ia supernovae. However, weak gravitational lensing does not appear to be a good method to probe the time variation of w (Huterer 2001). The systematics associated with weak gravitational lensing have not yet been studied carefully and could limit its potential.

- Some methods do not look promising in their ability to probe w because of irreducible systematics (e.g., Alcock – Paczynski test and strong gravitational lensing of quasars). However, both could provide important independent confirmation of accelerated expansion.

References

Ahmad, Q.R., et al. (SNO Collaboration) 2002, nucl-ex/0204008
Ahmad, Q.R., et al. (SNO Collaboration) 2002, nucl-ex/0204009
Benson, A.J., et al. 2001, astro-ph/0108218
Bullock, J., Kravtsov, A., & Weinberg, D. 2000, ApJ, 539, 517
Burles, S., et al. 2001, ApJ, 552, L1
Carroll, S. 2001, http://www.livingreviews.org/Articles/Volume4/2001-1carroll
Cen, R., & Ostriker, J.P. 1999, ApJ, 514, 1
Croft, R.A.C., Hu, W., & Dave, R. 1999, Phys.Rev.Lett, 83, 1092
Elgaroy, O., et al. 2002, Phys.Rev.Lett, 89, 061301
Fukuda, Y., et al. (SuperK Collaboration) 1999, Phys.Rev.Lett, 82, 2644
Fukuda, Y., et al. (SuperK Collaboration) 2000, Phys.Rev.Lett, 85, 3999
Fukugita, M., Hogan, C.J., & Peebles, P.J.E. 1998, ApJ, 503, 518
Griest, K., & Kamionkowski, M. 2000, Phys. Rep., 333-4, 167
Holder, G.P., et al. 2001, ApJ, 553, 545
Hu, W., Eisenstein, D.J., & Tegmark, M. 1998, Phys.Rev.Lett, 80, 5255
Huterer, D. 2001, astro-ph/0106399
Huterer, D., & Turner, M.S. 2001, Phys.Rev.D, 64, 123527
Kaplinghat, M., Knox, L., & Turner, M.S. 2000, Phys.Rev.Lett, 85, 3335
Klypin, A., Zhao, H.-S., & Somerville, R.S. 2001, astro-ph/0110390
Krauss, L., & Turner, M.S. 1999, Gen. Rel. Grav., 31, 1453
Loeb, A. 1998, ApJ, 499, L111
McDonald, P., et al. 2001, ApJ, 570, 457
Merritt, D., & Cruz, F. 2001, ApJ, 551, L41
Milosavljevic, M., & Merritt, D. 2001, ApJ, 563, 34
Milosavljevic, M., et al. 2001, astro-ph/0110185

Newman, J., & Davis, M. 2000, ApJ, 534, L11

O'Meara, J.M., et al. 2001, ApJ, 552, 718

Perlmutter, S., Turner, M.S., & White, M. 1999, Phys.Rev.Lett, 83, 670

Perlmutter, S., et al. 1999, ApJ, 517, 565

Riess, A., et al. 1998, AJ, 116, 1009

Riess, A., et al. 2001, ApJ, 560, 49

Schramm, D.N., & Turner, M.S. 1998, Rev. Mod. Phys., 70, 303

Sellwood, J., & Kosowsky, A. 2000, astro-ph/0009074

Sievers, J.L., et al. 2002, astro-ph/0205387

Somerville, R.S. 2001, astro-ph/0107507

Spergel, D.N., & Steinhardt, P.J. 2000, Phys.Rev.Lett, 84, 3760

Turner, M.S. 2000a, Physica Scripta, T85, 210

Turner, M.S. 2000b, in Proceedings of Type Ia Supernovae: Theory and Cosmology, eds. J.C. Niemeyer & J.W. Truran (Cambridge: Cambridge Univ. Press), 101

Turner, M.S. 2002, ApJ 576, L101

Turner, M.S., & Riess, A. 2002, ApJ, 569, 18

Turner, M.S., & White, M. 1997, Phys.Rev.D, 56, R4439

Tytler, D., et al. 2000, Physica Scripta, T85, 12

Weinberg, M.D., & Katz, N. 2001, astro-ph/0110632

Weinberg, S. 1989, Rev. Mod. Phys., 61, 1

Witten, E. 2001, hep-th/0106109

Hubble's Science Legacy: Future Optical-Ultraviolet Astronomy from Space
ASP Conference Series, Vol. 291, 2003
K.R. Sembach, J.C. Blades, G.D. Illingworth, R.C. Kennicutt, Jr. (eds.)

A Large, Serviceable HST Successor

Roger Angel and Johanan L. Codona

Steward Observatory, University of Arizona, Tucson, AZ 85721

Abstract. A large optical space telescope with a wide field-of-view, high resolution, and high sensitivity is needed to complement the future capabilities of the Next Generation Space Telescope (NGST) and adaptively-corrected ground-based telescopes. We describe a long-term, evolutionary observatory concept, to be built and operated in an accessible low-Earth orbit using low-risk, space-proven methods. The plan provides for incremental upgrades to the optics as well as the instruments. Initially, the telescope's 10 m aperture would be populated with six 2.5 m diameter mirror segments, giving images 4 times sharper than the Hubble Space Telescope (HST) over a 4 arcminute field of view, and a 1.7 mag improvement in sensitivity. Later, two elliptical mirrors, sized to fit the largest available vehicle fairings, would be added, increasing the gain to 2.7 mag and allowing for the development of powerful exoplanet imagers.

1. Introduction

A strong scientific case for a successor to the HST has been made at this meeting. As we consider different possible implementations, it is instructive to examine the circumstances that led to the enormous success of HST itself. Its huge scientific productivity has depended on its long life, its accessibility to astronomers worldwide, and its unique observational capabilities — most especially its tenfold increase in resolution compared to seeing-limited imaging. But, despite their lower resolution, ground-based telescopes have played a crucial complementary role — new objects found by HST and imaged at high resolution could only be studied spectroscopically with the largest-aperture telescopes on the ground. On the technical side, the success of HST was crucially dependent on servicing and repairs by astronauts. Its optics relied on relatively mature optical fabrication capabilities developed for lightweight mirrors in military systems.

The scientific capability that has remained untouched by the HST is for infrared wavelengths $>2\,\mu$m, where the scientific potential is great and the natural sky background is low. This gap will be filled by NGST, a new large telescope with cryogenic optics much larger than the Space Infrared Telescope Facility (SIRTF).

At shorter wavelengths, there is the potential for a larger optical space telescope to make another major improvement in the depth and resolution of imaging — going beyond HST by a factor comparable to its advance over previous imaging capability. However, this is a domain where there will be competition from ground-based telescopes with adaptive optics (AO) systems that correct atmospheric blurring. Unlike the situation in the infrared, where atmospheric absorption and emission by both the telescope and atmosphere provide fundamental limitations to ground performance, at optical wavelengths there is little reduction in sky brightness in space. Thus, for those types of observations that allow use of adaptive optics, space telescopes offer little advantage. In planning for a successor a decade in the future, we need to consider and understand what are likely to be the areas of complementary strengths.

2. The Potential for Adaptive Optics

The many 8 m-class telescopes now in operation will be joined in 2005 by the Large Binocular Telescope (LBT), a coherent 8 m pair with an effective aperture of 8×23 m (Hill & Salinari, 2000), and a filled-aperture 20–30 m (or even larger) telescope may well be built within the next ten years. Whenever operation at the diffraction limit is possible, such telescopes will have far higher resolution and light-gathering power than HST. The primary limitation for ground-based telescopes in achieving this limit is atmospheric distortion.

Atmospheric distortion is well characterized, and we understand, in principal, the way to correct fields of view of an arcminute or so. The method requires multiple deformable mirrors conjugated to turbulent layers at different heights in the atmosphere, the different wavefront aberrations corresponding to different paths through the atmosphere being simultaneously corrected. Deformable mirrors suitable for this purpose have been developed, but their use has been prevented by the difficulty of determining what corrections to apply corresponding to the different atmospheric layers.

This information will be derived by tomographic methods from wavefront measurements of a constellation of laser beacons covering the area to be corrected and separated on the scale of the isoplanatic angle. The characteristic corrected field size for multi-conjugate adaptive optics (MCAO) laser correction is D/h, where D is the telescope diameter and h the height of the beacon. This is ~ 1 arcminute for either a 30 m telescope used with sodium beacons ($h \approx 90$ km) or an 8 m telescope used with Rayleigh beacons ($h \approx 30$ km). Calculations by Lloyd-Hart et al. (2001) and De La Rue & Ellerbroek (2000) show that for both cases, $\sim$5 beacons are sufficient to yield correction in the near infrared over ~ 1 arcminute with 50% Strehl, and natural guide stars needed to complement the laser constellation by correcting image motion are available over most of the sky. The minimum wavelength that can be corrected depends strongly on the number of discrete turbulent layers that contribute to the distortion, and the number and power of the laser beacons used. Wavelengths $\geq 1.2\,\mu$m appear tractable with currently available YAG lasers for Rayleigh scattering ($\sim$10 W), while wavelengths down to $0.6\,\mu$m should be possible with more lasers of higher power. Wavelengths shorter than $0.6\,\mu$m are unlikely to be practical.

For the special case of imaging extrasolar planets, a different type of adaptive optics is needed. In this case a bright star is always available, and reduced photon noise allows for a much more accurate measurement of wavefront errors. In theory, 6.5–8 m telescopes should be able to detect Jupiter in a Solar System twin at 10 pc in a single night of integration (Angel 1994; Stahl & Sandler

1995), and a 30 m telescope should be able to detect an Earth-like planet in the same system, also in a single night (Angel 2002). Spectroscopy should also be possible from a high dry site, though because of the extremely faint signal from an exo-Earth, an integration of $\sim$ 100 hours would be needed to detect even the strong spectral features of water at 1-2 μm. A space telescope would avoid the complication of absorption of the same features in the Earth's atmosphere.

3. The Role for an Optical HST Successor

Based on the above considerations, we see that a large, room temperature telescope in space holds a unique optical potential for:

- Any observation requiring a large collecting area for $\lambda < 0.3\,\mu$m.
- Diffraction-limited imaging with higher resolution than HST over any field size for $\lambda < 0.6\mu$m.
- Diffraction-limited imaging with higher resolution than HST over fields $>$ 1 arcmin anywhere in the optical.
- Diffraction-limited imaging over any field in the range 1–2 μm, where ground-based telescopes are hampered by a sky background from atmospheric OH emission lines $\sim$ 50 times brighter than in space, as well as limited fields of view.
- Very high sensitivity imaging, as is needed for spectroscopy of Earth-like planets. The main requirements are: large continuous apertures approaching 10 m in one dimension, freedom from mechanical vibration, and the capability to learn by experiment how to reach the exquisite accuracy required for a 10^{-10} contrast ratio at 0.1 arcsec.

4. A Permanent Observatory to Complement NGST

NGST must be run passively at cryogenic temperature, almost certainly far from the Earth with no access for repair or servicing. These requirements present very challenging technical problems. Even the current descoped primary mirror of $\sim$6.5 m is too large to be launched in one piece. It is planned that lightweight mirror segments will be deployed, relying on automated systems to establish and maintain optical alignment. Further descoping and simplification may yet be required to achieve acceptable risk and cost.

However, even with a 6.5 m aperture, NGST's image sharpness in the unique thermal spectral region of $\lambda > 2\mu$m will be no greater than that of optical HST images. Its optics and detector pixels are scaled for these longer wavelengths, and will not reach the diffraction limit in the optical. Thus, a space telescope with a larger aperture and diffraction-limited imaging in the optical and ultraviolet over a wide field is needed to complement NGST and ground-based telescopes.

The large aperture and high stability needed for such a telescope would be difficult and expensive to achieve in another remotely-deployed system, even when NGST's cryogenic requirement is dropped. Moreover, for a remote system, the instrument upgrades that will almost certainly be needed to realize sensitivity to Earth-like planets would not be possible. But an extremely powerful permanent observatory to satisfy all these needs could be built in low Earth orbit, taking advantage of the wealth of experience in space assembly and refurbishment developed through the construction of the Space Station and the HST program. Astronaut capabilities would be valuable, not only for instrument upgrades, but in the initial construction and possible evolution of the telescope optical assembly itself. Telescope construction in space will minimize risk because it allows the initial primary mirror to be assembled from space-proven, rigid, round substrates figured and tested by established methods.

As an illustration, we consider here a possible 10 m successor to the HST. Covering the same wavelength range as HST, the proposed larger diameter would yield a large increase in resolution and light-gathering power. Its relationship to current and other proposed facilities is shown in Figure 1.

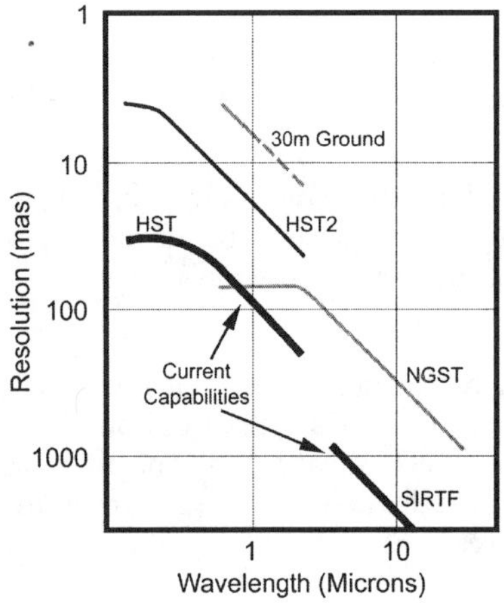

Figure 1. Resolution and spectral range for current or impending telescopes (HST and SIRTF) and planned future larger telescopes. Our proposed 10 m successor to HST (marked HST2) has 10 times the resolution of the longer wavelength NGST, 10 times the field of view of AO-corrected ground telescopes at red and near-infrared wavelengths, and unique access and resolution in the visible and ultraviolet.

The basic structure uses low-risk technology and methods, and would be assembled in space, either at the International Space Station or in a more convenient orbit using techniques developed in the assembly of the Station. The telescope consists of a rigid 10 m mirror support structure built as a rigid truss from carbon composite, incorporated into a basic spacecraft including sunshade, power, and data handling. The geometry would be configured for an f/1 primary, with mirror elements installed on orbit. The whole 10 m aperture does not have to be completed in one go, but could be upgraded over time. We show as an example in a concept for an initial configuration very similar to the Golay-6 configuration (Golay 1970) slightly modified to allow a larger filling factor (Figure 2). The design includes six 2.5 m segments, to be upgraded later to a near fully-filled aperture.

4.1. Initial Golay Configuration

The six-mirror Golay configuration would be built very conservatively from rigid honeycombed glass mirror elements with a 2.5 m diameter. These are similar in size to the HST primary, but would now be constructed with significantly lighter

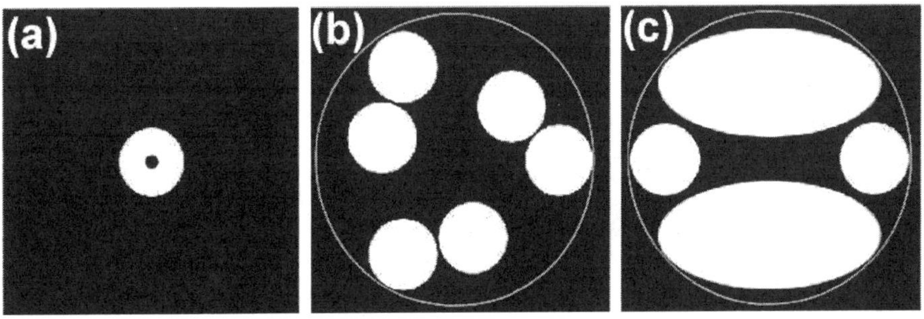

Figure 2. HST (a) and proposed pupils "Golay" (b) and "Upgraded" (c). Pupils are shown to the same scale.

weight. The configuration of off-axis elements has a collecting area of 29.5 m^2, nearly 7 times the 4.3 m^2 of HST. With the segments phased and filling 37.5% of the 10 m envelope, the images show a strong central core with the full resolution of a 10 m aperture. The central peak itself contains 2.9 times more energy than HST's peak, and is 4.2 times sharper. This is a performance advance comparable to that made by HST over ground-based telescopes. Because the aperture is not fully filled, not all the energy collected is centrally concentrated. However, the relatively smooth halo is no larger than an HST image. The 6.8-fold increase in total flux would thus be available for deep spectroscopy over wide fields and into the far ultraviolet.

4.2. Upgrade with Two Large Segments

At a later stage, four of the Golay elements would be replaced by the two elliptical 3.8 × 8 m segments to largely fill the 10 m aperture. This would bring the filling factor and central core energy up to 73%. The additional segments are chosen to be the largest that could be reasonably launched in current vehicle fairings (Angel et al. 1996), with the rationale of providing the capability for exoplanet imaging. To be effective, a space telescope for this purpose must very efficiently apodize the largest possible continuous aperture. The two ellipsoids could accommodate pupil geometries favorable for exoplanet detection (Nisenson & Papaliolios 2001; Spergel 2002).

Figured mirrors of 8 m size are already made for ground-based telescopes. They can be lightweighted for space by a process developed at the University of Arizona in which a finished figured mirror is ground away from the back to reduce the thickness to as little as 1 mm. The thin shell is then mounted to a rigid carbon fiber composite support via adjustment actuators to restore the original figure (Burge et al. 2001). This process has yielded a 2 m mirror with 2 mm thick meniscus and could be further developed to make 8 m ellipsoids. With a glass thickness of 1 cm, for example, each glass meniscus would weigh 700 kg, and two complete mounted mirrors would weigh no more than 5 tons. Thus, both mirrors could be carried in a single launch to refurbish the telescope. Such mirrors are much more ambitious than the 2.5 m elements, but could be developed in time for the telescope upgrade. Current mirror figure accuracy of 15 nm RMS would be adequate even for imaging Earth-like planets, since the meter-scale Fourier components that control scattered light at 0.1 arcsec would in any case have to be under continuous active control to exquisite accuracy.

4.3. Optical Design and Performance

The optical design must provide a large diffraction-limited field, since this is uniquely available from space. Our basic parent design is an axisymmetric Ritchey-Chretien (RC) with a 10 m $f/1$ primary and a 1 m secondary. The large fields at different scales will be obtained with auxiliary optics near the RC focus. Relays of this type, with a powered tertiary mirror and off-axis fields of 4×4 arcminutes, have been worked out and could be adapted (Burge et al. 2001). Large detector mosaics that could fully sample such large fields, at least for the longer optical wavelengths, are likely to be commonplace in a decade. Large mosaics are already proven in ground-based imaging systems, and a 30,000×30,000 element mosaic is planned for use in the ground-based Large Synoptic Survey Telescope. We will assume that several cameras with mosaic arrays of 10^9 pixels will be used, with sensitivity and sampling optimized for different wavelength bands. Such detectors may be too ambitious for the initial launch, but the optics and focal plane arrangements should make provision for their addition at some later date. The resolutions and fields of view will then range from 4 mill-arcseconds over 1 arcminute at 200 nm to 32 mill-arcseconds over 8 arcminutes at 1.6 μm. Such imaging capability represents a large gain over HST, with 50-1000 times the number of pixels sampling the sharper images. Images from the new telescope will each contain the same wealth of detail as photographic Schmidt plates.

Because the Golay configuration is not fully filled, the images will show a sharp core with the full resolution of 10 m aperture set in a halo the size of an HST image core. Azimuthally-averaged point spread functions (PSFs) for each of the configurations are plotted on linear and log scales in Figure 3. Table 1 gives the FWHM and fraction f of the total star flux contained within the central peak for the HST and new telescope configurations. A useful performance measure is the signal/noise (S/N) ratio for an unresolved point source seen against a uniform brightness background, such as the zodiacal background, which is proportional to $f\sqrt{A}/FWHM$. Table 1 also gives this measure.

The Golay configuration yields a 4.66 times improvement in S/N ratio, and thus will reach 1.7 magnitudes deeper than HST for the same length exposure, as well as giving 4 times sharper images. The upgraded aperture increases the S/N ratio to 12.5 times, reaching 2.7 magnitudes deeper than HST with the same sharpness advantage.

Table 1. Telescope Configuration Comparison

Telescope	Primary Mirror Area (m^2)	Flux Fraction in Central Peak	FWHM (HST units)	S/N (HST units)
HST	4.3	0.77	1	1
Golay	29.4	0.33	0.24	4.6
Upgrade	57.2	0.61	0.23	12.5

Imaging and sensitivity parameters for the HST, Golay, and upgraded telescope configurations. The FWHM and S/N ratios are normalized to the values for HST.

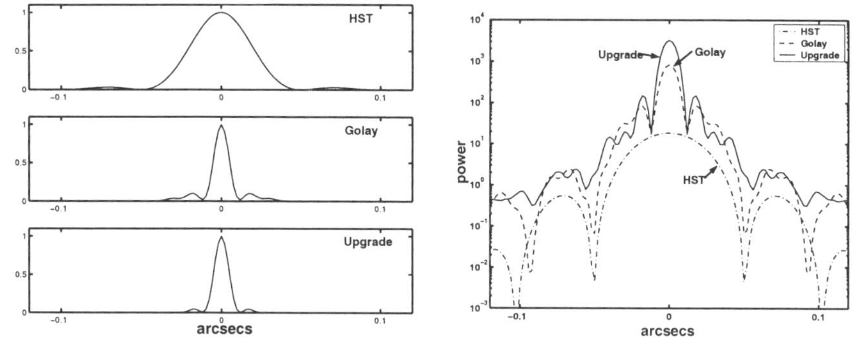

Figure 3. *Left*: Normalized azimuthally-averaged PSFs for the HST, Golay, and upgraded configurations, compared on a linear power scale. *Right*: The unnormalized PSFs with the HST (dash-dot), Golay (dash), and upgraded (solid) configurations on the same absolute log scale. The resolution is calculated for $\lambda = 0.5\,\mu$m.

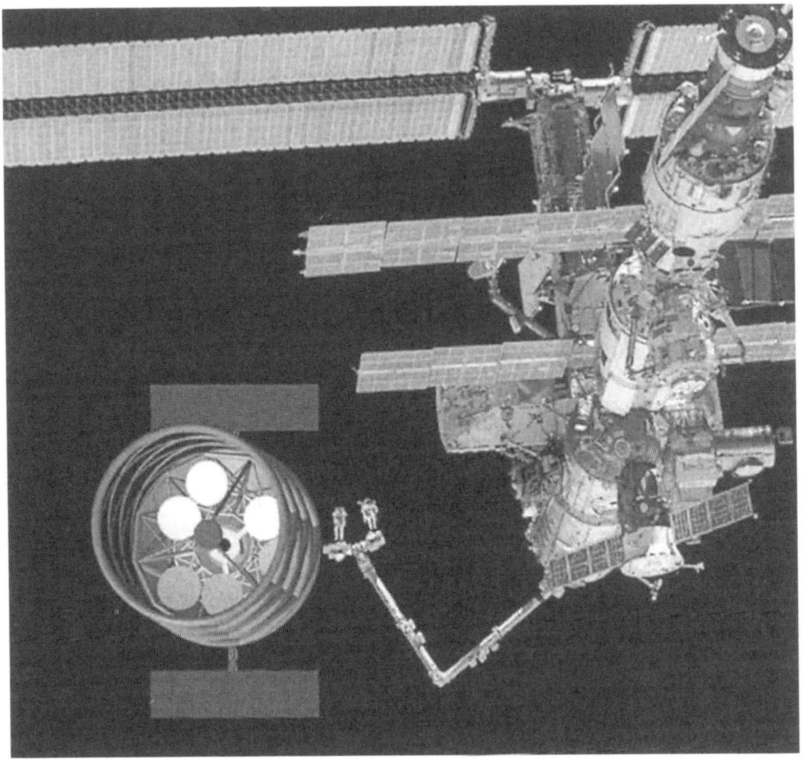

Figure 4. A computer model of the 10 m telescope to scale against a photograph of the International Space Station.

4.4. Construction, Servicing, and Operation in Low-Earth Orbit

The four main elements of the telescope are the telescope structure, the primary mirror segments, the focal plane/spacecraft unit, and an inflatable/rigidized sunshield. We envisage that the structure, including hexapod mirror mounts, will be launched as sub-units to be bolted to each other and to the spacecraft unit on-orbit. The mirror segments will then be attached and the sunshield deployed. Alignment of the mirror segments by the hexapods will be guided by measurements of a bright star. After check-out, the telescope would be moved to its operating orbit, where it will be accessible for continued development (Figure 4).

A low-Earth orbit has the well-known disadvantages of Earth occultation and thermal cycling. We feel that these will be outweighed by the advantages of accessibility for construction, maintenance, and upgrades that have already proved invaluable for HST. Astronauts were slated for a rather limited role on HST, swapping out new instruments for old. In fact, they have been able to make repairs of systems in ways never envisioned (e.g., installing a refrigerator to revitalize HST/NICMOS). Taking advantage of this capability, we can plan for an observatory of great robustness and longevity, a worthy successor to HST.

Acknowledgments. We are grateful to Tom Connors for providing the telescope concept drawing and the composite image with the Space Station. Johanan Codona acknowledges support from the Air Force Office of Scientific Research under the grant AFOSR #F49620-01-1-0383.

References

Angel, J.R.P. 1994, Nature, 368, 203

Angel, R., et al. 1996, Proc. SPIE, 2807, 354

Angel, J.R.P. 2000 in Science with the LBT, eds. Tom Herbst, Bavaria, 21

Angel, R. Lesser, M. Sarlot, R., & Dunham, E. in Imaging the Universe in ASP Conf. Ser. 195, Three Dimensions: Astrophysics with Advanced Multi-Wavelength Imaging Devices, eds. W. Bruegel & J. Bland-Hawthorn, (San Francisco: ASP), 81

Angel, J.R.P. 2002, Nature, submitted

Burge, J., & Angel, R. 1999 in ASP Conf. Ser. 164, Ultraviolet-Optical Space Astronomy Beyond HST, eds. J.A. Morse, J.M. Shull, & A.L. Kinney, (San Francisco: ASP), 234

Burge, J.,et al. 2001, Proc. SPIE, 4451, 27

De La Rue, I.A., & Ellerbroek, B.L. 2002, Proc. SPIE, 4494, 290

Golay, M.J.E. 1970, J. Opt. Soc. Am., 61, 272

Hill, J.M., & Salinari, P. 2000, Proc. SPIE, 4004

Lloyd-Hart, M, Goerges, J. Angel, R. Brusa, G. & Young, P. 2001, Proc. SPIE, 4494, 259

Nisenson, P., & Papaliolios, C. 2001, ApJ, 548, L201

Spergel, D.N. 2002, Applied Optics, in press

Stahl, S.M., & Sandler D. 1995 ApJ, 454, L153

Hubble's Science Legacy: Future Optical-Ultraviolet Astronomy from Space
ASP Conference Series, Vol. 291, 2003
K.R. Sembach, J.C. Blades, G.D. Illingworth, R.C. Kennicutt, Jr. (eds.)

Ultraviolet and Visible Detectors

Bruce E. Woodgate

Laboratory for Astronomy & Solar Physics, Code 681, NASA/GSFC, Greenbelt, MD 20771

J. Chris Blades

Space Telescope Science Institute, 3700 San Martin Dr., Baltimore, MD 21218

Abstract. In order to build new ultraviolet observatories in space with substantially better performance than those currently flying or in near term development, more sensitive ultraviolet detectors with large formats are required. New opportunities for technological developments exist to substantially improve the performance of ultraviolet detectors, particularly with new photocathodes and silicon microchannel plates.

1. Introduction

To make substantial scientific advances beyond the current ultraviolet (UV) space missions, future missions need to observe fainter objects than current ones such as HST (STIS and COS), FUSE, and GALEX. This can be achieved by increasing the collecting area and the sensitivity (Shull et al. 1999). The cost to build space telescopes of 6–8 m diameter is expected to greatly exceed that of improving the quantum efficiency (QE) of detectors, assuming that substantial detector improvements are technically feasible. We show below that new photocathode materials being investigated will likely lead to much higher QE and low-noise imaging detectors in the far-ultraviolet (FUV) and near-ultraviolet (NUV), if they are developed.

A NASA Office of Space Science study of the requirements, status, and prospects of UV and optical detectors was recently undertaken by an Ad-Hoc UV-Visible Detectors Working Group, chaired by one of us (JCB), and its final report was available at the Chicago meeting, entitled "Ultraviolet and Visible Detectors for Future Space Astrophysics Missions" (Blades et al. 2002). One of its key findings was: "The sensitivity of ultraviolet solar-blind detectors needs improvement. We find opportunities through development of a silicon-based MCP technology, new photocathode materials, and use of AlGaN based detectors."

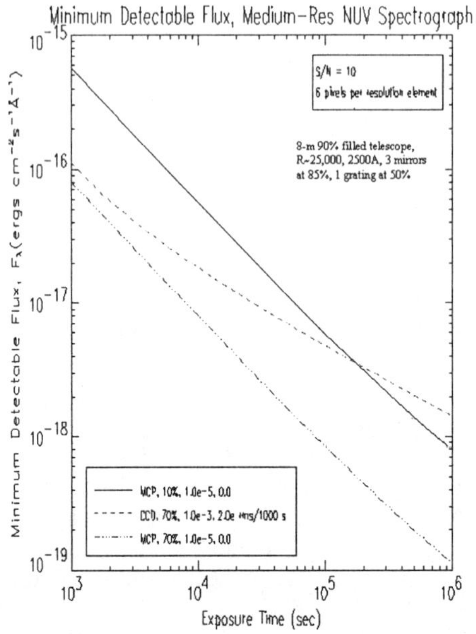

Figure 1. The effect of cumulative read noise on sensitivity. Comparative simulations of the minimum detectable flux as a function of exposure time are shown for a NUV spectrograph with R=25000 at 250 nm behind an 8-m 90%-filled telescope with three mirrors at 85% efficiency and one grating at 50% efficiencies, for S/N=10 at 6 pixels per resolution element. For exposures greater than 200,000 sec, a 10% zero-read-noise photon counter is more sensitive than a 70% efficient CCD with 2 e^- pixel^{-1} RMS read noise read out every 1000 sec.

2. UV Detectors

Photon-counting detectors remain the standard for numerous applications in UV/EUV/X-ray astronomy, atmospheric science, solar physics, and space physics and will for the foreseeable future. No other technology offers the same powerful combination of large format (in area and pixel count), low background, photon-counting sensitivity (zero read noise), solar blindness, radiation tolerance, and room temperature operation. Compared with integrating detectors such as CCDs, for example, photon counters often prove to be the detector of choice based on either:

- their superior "faint-end" performance (a low background photon-counter eventually achieves a fainter limiting sensitivity than a system with finite read noise, even if starting with a QE disadvantage),
- the need for time-resolved observations,

- the strong long-wavelength rejection of photocathode-based systems, crucial to numerous short-wavelength applications (as visible/UV flux ratios can be as large as 10^9), or
- the poor radiation tolerance of CCDs.

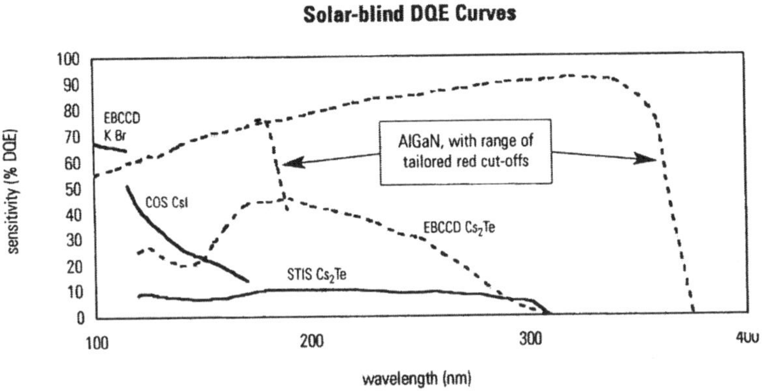

Figure 2. Comparison of sensitivities for solar-blind detectors. Solid curves are measured. The EBCCD KBr and COS CsI curves are for opaque photocathodes, and the STIS Cs_2Te curve is for a semi-transparent photocathode. Dashed curves are predicted for a future sealed-tube opaque Cs_2Te photocathode EBCCD, and for theoretical photoconductive AlGaN.

2.1. Low Background

The "limiting sensitivity" advantage of photon counters is particularly crucial, because so much frontier science is carried out near the faint-end limit of an instrument's capabilities. For detector-background-limited observations (a situation often encountered in UV spectroscopic applications, where sky backgrounds are negligible), it is easily shown that the signal-to-noise ratio obtainable in a given observing time scales as $QE/\sqrt{B}$ (and the minimum detectable flux at a given S/N scales inversely with that quantity), where QE is the quantum efficiency of the detector and B is the background rate per pixel. For integrating detectors such as CCDs or other proposed solid state sensors (such as AlGaN arrays), this background term includes both dark current and the statistical effects of non-zero read noise. The strong performance advantage for low-background photon counters is illustrated in Figure 1.

2.2. High QE

The QE of current UV detectors leaves significant room for improvement, as shown in Figure 2. The solid curves have been achieved, and the dashed curves are predicted. In this plot, the predicted AlGaN curve is for a photoconductive device, which requires a readout system which adds noise, and currently the dark noise of photoconductive AlGaN is many orders of magnitude too high for astronomical detectors. If the intrinsically high internal QE of AlGaN can be converted to an external QE by ejecting the photoelectrons, forming a photocathode, a high-QE low-noise detector could be made. Current progress in QEs achieved with cesiated AlGaN photocathodes is shown in Figure 3, starting with the simpler GaN limit (Ulmer et al. 2002; Korotokov et al. 2001). One key factor in obtaining high QE is to increase the amount of p-doping with Mg to bend the band structure near the surface to increase the probability of electron escape. The 50% QE measurement at 310 nm by Machuca et al. (2000) was for a Mg doping level of 5×10^{17} cm^{-3}. The "theoretical limit" shown in Figure 3 does not include internal electric field effects, which could allow higher QEs, similar to those already achieved in the FUV. Improvements are also possible in the FUV, as achieved in photomultipliers by Niigaki et al. 1999 using alkali oxide negative electron affinity treatments (Uebbing & James 1970) in diamond, with QEs up to 60% at 170 nm cutoff, a factor >4 higher than the currently used CsI (see Figure 4).

These photocathodes could be used in an imaging detector, such as the Electron Beam CCD (EBCCD) used as the IMAPS detector, provided that a sealed tube with clean materials could be made to preserve the sensitive negative electron affinity coatings. They can also be applied to microchannel plates (MCPs), which can be combined with a variety of imaging readout systems, such as crossed delay-line detectors and their derivatives (Siegmund et al. 2001) and intensified advanced pixel sensor (APS) centroiding detectors (Kimble et al. 2002).

2.3. Large Format

Both spectroscopic and imaging applications require large pixel formats to be competitive. For imaging, wide area coverage is needed to obtain data on a large number of sources, either for good statistics, as in galaxy surveys or weak lensing, or to pick out rare objects such as in supernova or eclipsing planets searches. For spectroscopy, a very long format is needed to cover a wide spectral range at medium resolution in a low order spectrum, and a wide area is needed to cover multiple echelle orders or for a multi-object spectrograph. The number of effective pixels is a critical factor in deriving a figure of merit in comparing detector technologies.

2.4. Microchannel Plates

Drawn glass MCPs have been used as electron intensifiers for about 40 years, primarily for night vision, but they are also the workhorse technology for most recent and current NASA UV missions: HST/STIS (FUV and NUV), HST/ACS (FUV) and HST/COS (NUV) with MAMA readouts; EUVE, SOHO, FUSE, IMAGE, GALEX, and HST/COS (FUV) with delay line readouts. While these glass MCPs are the preferred technology currently available, time consuming and expensive developments are usually encountered in obtaining flight quality scientific grade MCPs. While glass MCP fabrication should be a mature and long developed technology, the production processes continue to have inadequate control for custom scientific grade MCPs, and introduce considerable geometric and performance variability, non-uniformity and defects, resulting in low yields

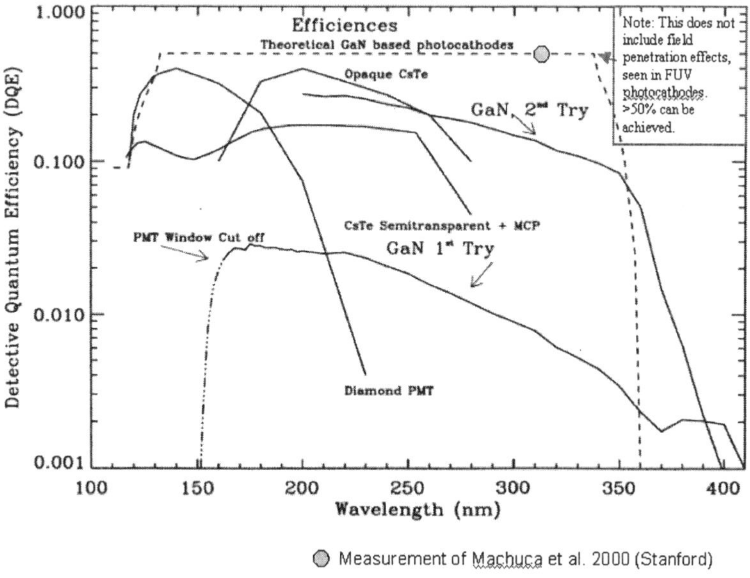

Figure 3. Measured GaN and diamond quantum efficiencies. Initial measurements by Ulmer et al. (2002) show an order of magnitude improvement between the first and second tries, partly due to increasing the amount of Mg doping, to $\sim 2 \times 10^{17}$ cm^{-3}, and another factor 4 is obtained by Machuca et al. (2000) with doping of 5×10^{17} cm^{-3}.

and programmatic impacts. In addition, their radioactivity due to potassium in the glass can cause significant dark rates.

Even more seriously, they have not yet enabled the use of high QE photocathodes at wavelengths above Lyα. A few standard photocathodes have been used with the glass MCPs, Cs_2Te for the NUV, CsI for the FUV and other alkali halides for shorter wavelengths. In the NUV, the QEs are very poor. A net peak QE of only 9.6% was obtained for the HST/STIS NUV MAMA in spite of many Cs_2Te deposition attempts (Figure 3). In that case a semi-transparent photocathode was used on the detector window. Attempts to obtain a higher QE by depositing an opaque Cs_2Te photocathode on the MCP (as is done for most FUV CsI photocathodes, such as on the HST/STIS and COS FUV detectors) have never achieved higher QE than the semi-transparent photocathodes This has been attributed to chemical poisoning of the photocathodes by materials ejected from the chemically complex glass MCPs during the electron multiplication process.

However, a promising new opportunity lies in coupling new photocathodes with the new MCP technology that has been emerging, based on the Si lithography methods and materials developed for computer chips. Nanosciences Inc. of Oxford CT, has succeeded in deep electrochemical etching down Si crystal

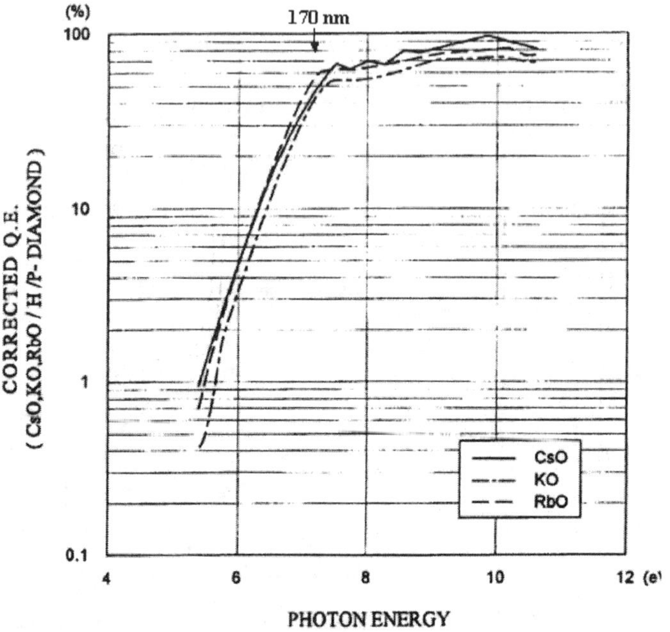

Figure 4. Measured QE of negative electron affinity diamond. Alkali oxide coatings with CsO, KO, or RbO on hydrogenated diamond show QEs above 60% at 170 nm, a factor 4 higher than CsI, and above 80% near Lyman alpha.

planes to create MCPs with hole separations of 5–8 microns and length-to-diameter ratios of up to 300:1 (Figures 5 and 6). The Si MCPS have been made with square and hex patterns. They can be made with smaller pores for higher spatial resolution. Their semiconducting and secondary emitting surface processes down the pores have been rapidly improving, resulting in MCP gains of 5000 at 1000 volts, similar to that of glass MCPs. We expect them to be used in chevron pairs to obtain suitable gain for current readout systems. These Si MCPs are chemically much simpler and cleaner than the glass MCPs, and can be processed at higher temperatures for further cleaning and for application of more refractory photocathode materials. However, much work is required to develop the processes for matching photocathodes such as AlGaN to the front of the MCPs.

Nanosciences Inc. is also developing a process of boron-doped diamond, initially as a secondary electron emitter with very high gain near the front of the pores for the first and second strike electrons. They have succeeded in making continuous layers on the MCP's web, and have deposited it down the pores to two pore diameters. The secondary emission can be further enhanced by cesiation. These same properties that make B-doped diamond an excellent secondary emitter also make it an excellent photocathode in the FUV (Niigaki et al. 1999, Figure 4; Tremsin & Siegmund 2000).

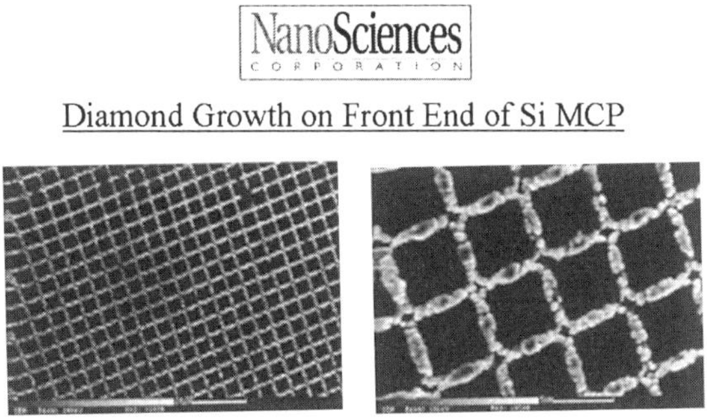

Figure 5. Silicon MCP front surface showing an even surface without the bundle interface defects of glass MCPs, and diamond particles on the MCP web. More recently, diamond has been deposited continuously and into the pores.

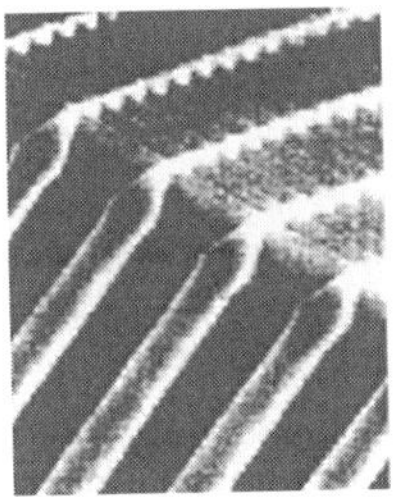

Figure 6. Isometric view of a silicon MCP with 6 micron spacing, and a tapered end, increasing the open area ratio.

2.5. Energy Resolving Detectors

New energy-resolving detector such as superconducting tunneling junctions (STJ) and transition edge sensors (TES) and their variants offer exciting longterm potential, and several groups are pursuing the development of such devices. Their potential and progress are discussed in the Ad-Hoc UV-Visible Detectors Working Group report (Blades et al. 2002), and so will not be repeated here. However, for the UV, the 3-D detectors suffer from: 1) pixel formats that will remain severely limited in comparison with MCP detectors for a very long time to come, 2) the necessity for complex, costly cryogenic operations, 3) the difficult problem of long-wavelength rejection in some applications, and 4) the contamination risk of cryogenic optical surfaces in a UV system. We therefore conclude that energy-resolving cryogenic detectors will not supplant large-format, room-temperature, MCP-based detectors over the broad range of photon-counting applications. In the UV/optical in particular, the energy resolving power of such devices is relatively low, while the efficiency of conventional dispersive spectroscopy is high. The tradeoff between energy-resolving detectors and non-resolving detectors with vastly greater pixel formats is thus completely different at UV/optical wavelengths than in the X-ray, where the spectral resolution of the 3-D detectors is high and the efficiency of dispersive spectroscopy is low.

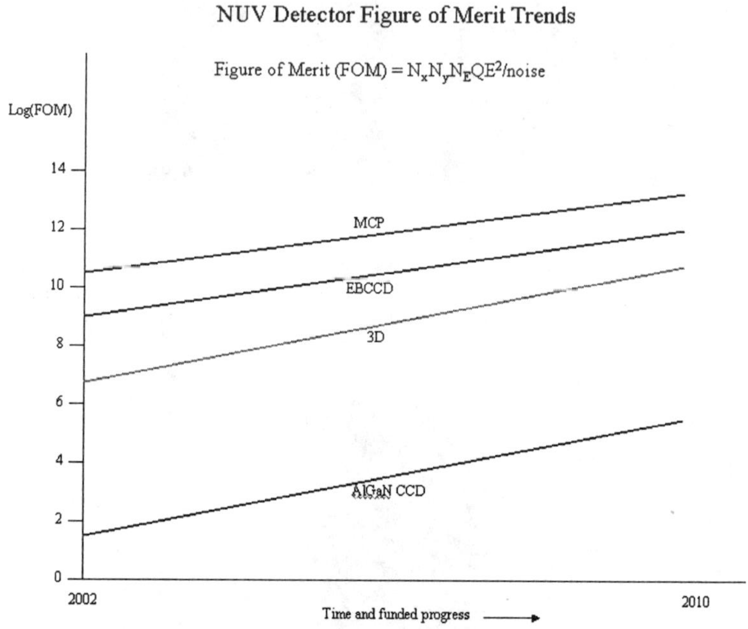

Figure 7. Near-UV figure of merit (FOM) trend predictions for MCP and EBCCD photon counters, energy resolving detectors, and high band-gap photoconductive detectors, if each technology were to be adequately funded. The FOM combines potential improvements due to QE increases and number of pixel increases. After the assumed improvements, all the detector types should have QEs above 50%. The energy resolving (3D) detector lag because of their smaller formats, while the photoconductive devices (AlGaN CCDs) lag because of noise.

2.6. Figure of Merit Comparison of Detectors

We can compare the performance of various detector types by comparing their relative sensitivities and coverage. A critical measure of the utility of an instrument is the observing time required to perform a specific observation, since for a given mission lifetime, the overall development and operations cost can be divided by the lifetime to derive the cost per unit observing time. For the long exposures expected for limiting faint object observations which set the cost drivers for the observatory, the setup overheads should be minimal. For a detector background limited observation, the observing time for a given source, S/N, per pixel is inversely proportional to QE^2/noise, so we adopt a figure of merit per pixel of QE^2/noise. The number of objects or spectral lines observable is proportional to the number of pixels, or $N_x N_y N_E$, where N_E is the number of resolution elements for an energy resolving detector. We combine these to form an overall figure of merit $N_x N_y N_E QE^2$/noise. Rough estimates of this measure are shown in Figure 7 for several detector types now and 8 years in the future, on the assumption that each technology were funded adequately to achieve its technological potential. The 2D photon counters could maintain their lead with plausible improvements in QE and format, with MCP based detectors able to have large 2D formats while the EBCCDs will be limited in their second spatial dimension because of their magnet structure. The 3D detectors should improve more because they are at an early stage, but difficulties in providing very large formats will prevent them from catching up. The AlGaN CDDs as photoconductive devices suffer from enormous dark currents, and will remain limited to brighter object applications even with rapid materials property improvements.

3. Visible Detectors

Currently CCD detectors dominate in the visible because of their high QE and relatively low readout and dark noise. Only for the highest spectral resolution and long exposure times requiring the addition of a large number of images would zero read noise be superior if they had similarly high QE, and currently they have much lower QE.

Efforts are being made to reduce the read noise of CCDs, most notably the avalanche readout CCDs by Marconi in the UK. A serious concern is the supply of scientific grade CCDs. After 3 decades of leadership in the development and production of high quality CCDs in the USA, the supply from US industry is almost non-existent, partly due to transfer of research and production to CMOS devices, and party due to the limited and sporadic astronomical and space market.

Also a technical problem exists for CCDs in space, due to radiation damage from the trapped radiation belts in low Earth orbit, and from solar flares outside of the Earth's magnetic field. After a few years, the performance of the CCDs on HST has deteriorated significantly with loss of charge transfer efficiency and increase of hot pixels and dark noise (e.g., Kimble et al. 2000). This problem becomes more serious as CCD formats get larger, and charge has to be transferred across more radiation-induced traps in the silicon. Two potential solutions exist, deserving of development support. P-channel CCDs, such as those being developed by Lawrence Berkeley Labs for the SuperNova Acceleration Probe (SNAP) mission, have less susceptibility to radiation damage (Holland 2002; Bebek et al. 2002). The CTE loss for the p-channel CCDs is at least a factor 10 less than that for the regular n-channel CCDs. An alternative solution is to use CMOS devices, alternatively called advanced pixel sensors, due to their having amplifiers in each pixel, so that the original charge does not need to be transferred across the detector area to the amplifier. These devices currently have much higher read noise and lower QE than CCDs, but research is underway to include backside thinning to raise the QE, and to improve amplifiers to lower the read noise (e.g., Stirbl et al. 2002).

References

Bebek, C.J., et al., 2002, http://snap.lbl.gov/pubdocs/pubdocs.html

Beetz, C.P., Boerstler, R., Steinbeck, J., Lemieux, B., & Winn, D. R., 2000, Nuc. Instr. Methods, Phys. Res. A 442, 443

Blades, J.C., et al. 2002, "Ultraviolet and Visible Detectors for Future Space Astrophysics Missions: A Report from the Ad-hoc, UV-Visible Detectors Working Group of NASA's Office of Space Science," (Baltimore: Space Telescope Science Institute)

Holland, S. 2002, Presentation at R and D Review of SNAP, LBL, Berkeley.

Kimble, R.A., Goudfrooij, P, & Gilliland, R.L. 2000, Proc. SPIE, 4013, 532

Kimble, R., Norton, T., Pain, B., Ortiz, M., & Haas, P. 2002, Proc. SPIE, in preparation

Korotokov, R.Y., Gregie, J.M., Han, B., & Wessels, B.M. 2001, "Optical study of GaN co-doped with Mn and Mg grown by metalorganic vapor phase epitaxy," International Conference on Defects in Semiconductors

Machuca, F., Sun, Y., Liu, Z., Ioakeimidi, K., Pianetta, P, & Pease, R.F.W. 2000, J. Vac. Sci. Technol. B 18(6), 3042

Niigaki, M., Hirohata, T., Kan, H. and Yamada, M. 1999, "Electron tube with polycrystalline diamond photocathode," U.S. patent 5,982,094

Shull, J.M., et al. 1999, "The Emergence of the Modern Universe: Tracing the Cosmic Web", White Paper of the UV-Optical Working Group, (Boulder: University of Colorado)

Siegmund, O.H., Tremsin, A.S., Abiad, R., & Vallerga, J.V. 2001, Proc. SPIE, 4498, 131

Stirbl, R.C., Pain, B., Cunningham, T.J., Hancock, B.R., Yang, G., Heynssens, J.B. & Wrigley, C.J. 2002, Proc. SPIE, 4547, 1

Tremsin, A.S., & Siegmund, O.H.W., 2000, Proc. SPIE, 4139, 16

Uebbing, J.J., & James, L.W. 1970, J. of Appl. Phys. 41, 11, 4505

Ulmer, M.P., Wessels, B.W., Shahedipour, F., Korotokov, R.Y., Joseph, C.L., & Nihashi, T. 2001, Proc. SPIE, 4288, 246

Hubble's Science Legacy: Future Optical-Ultraviolet Astronomy from Space
ASP Conference Series, Vol. 291, 2003
K.R. Sembach, J.C. Blades, G.D. Illingworth, R.C. Kennicutt, Jr. (eds.)

Detecting Extrasolar Planets with HST-II

David N. Spergel, Jeremy Kasdin, and Edwin Turner

Princeton University, Princeton, NJ 08540

Robert Brown

Space Telescope Science Institute, 3700 San Martin Dr., Baltimore, MD 21218

Steve Kilston

Ball Aerospace Corporation, Boulder, CO 80306

Wes Traub

Harvard-Smithsonian Center for Astrophysics, 60 Garden St., Cambridge, MA 02138

Abstract. A properly designed HST-II should be capable of detecting extrasolar planets. This will require building a very stable telescope with excellent optics and a properly designed coronagraph. Properly equipped, a 3×6 meter telescope should be capable of detecting Earth-Sun twins out to 20 parsecs.

1. Extrasolar Planet Detection: Science Goals

1.1. Planet Detection

The direct imaging of an Earth-like planet will be an exciting event with significant cultural and scientific impact. Direct imaging is a superior approach to any indirect technique for gathering both qualitative and quantitative information. It is also most cost-effective in terms of science because of its access to photometric properties, its broad applicability to physical studies, and its mission compatibility with other instrumentation, such as spectrographs, for general astronomy and astrophysics. Direct detection complements ground-based radial velocity surveys by detecting planets in wide orbits. Indirect detection is biased towards massive planets and/or close-in orbits.

Direct detection discovers immediately all the planets not hiding right next to the star. It finds them in the first picture (and finds the hidden ones after they move). By contrast, indirect observations must span most of an orbital period to enable a discovery, which raises mission risk relative to direct techniques. Confusion from multiple-planet systems does not further slow the process, as it does for indirect techniques. Most important for the community, direct detection enables prompt observational and theoretical follow-on studies to clarify the nature of candidate planets and systems.

As reviewed by Brown et al. (2002a), direct detection adds great value to existing information from indirect searches. One measurement of the apparent separation of the star and planet, learned from a picture, when combined with the true separation from a radial velocity discovery, determines the inclination angle of the orbit and therefore the true mass of the planet. In the case of longer orbits, where incomplete data from astrometry or radial velocity show only an acceleration, direct detection can not only find but also characterize the suspected planet photometrically; in addition, the measured separation can be used to compute a valuable probability distribution for the planet mass.

Direct detection provides a wealth of orbital and physical information about planets, whereas indirect techniques yield only the orbit and the mass. Direct detection yields a rough mass estimate that is useful for most scientific questions. Spectrophotometry can determine the presence and basic nature of the atmosphere, which can be used both to classify the planet according to Solar System templates and to estimate the geometric albedo, from which the planet radius can be estimated.

Only direct detection can address the question of life on planets orbiting other stars. By analyzing the light from planets, we can learn whether conditions are compatible with life or even learn of the presence of life.

1.2. Planet Characterization

An optical Terrestrial Planet Finder (TPF) detects visible starlight reflected from the planet rather than thermal infrared emission from the planet. As members of the Ball TPF Architecture Study team, we have shown that visible light is the right choice for myriad reasons, including superior signal-to-noise per unit time, immunity to exo-zodiacal light, and superior science. We have shown that a single-aperture visible-light telescope has many technical advantages over a multi-aperture infrared interferometer, including the multiplex advantage and fewer technology and systems-engineering concerns. Direct imaging is sovereign: it is faster and better.

While mass is an important characteristic of planets, direct imaging will enable much more detailed studies of planetary properties. The mass, together with orbital distance, is an important constraint in Solar System formation studies (although a rough mass estimate is really all that can be compared with the current state of theory). The surface gravity, surface temperature, and orbital and rotational parameters are arguably more important for habitability and life. Given optical images, we will determine the orbit, and can probably fix the rotational period, of an Earth-like planet.

The ability to reliably detect 10% changes in photometry will open a whole new window into Earth-like planet characterization (Ford, Seager, & Turner 2001). Earth is very different from the other terrestrial planets (and, in fact, from all Solar System planets except Pluto-Charon, which has 30% variations due to Charon) in that its brightness varies dramatically. Our theoretical light curves of the unresolved Earth have 10-20% variation on a daily (1 rotation period) basis. This intensity variation is due to the contrast between clouds and land or ocean (< 0.1 for ocean, 0.3 to 0.4 for land, > 0.6 for clouds) rotating in and out of view. Thus, a time series monitoring of terrestrial planets will tell us if our new planet has diverse components (e.g., ocean, land, ice) and meteorological activity.

With a rotational period, observations can be folded into the period and a higher accuracy can be reached. With this higher accuracy, a more detailed study of the planetary surfaces and atmospheric features is possible. For example, seasonal changes in the cloud pattern and planet albedo can be detected for a planet like Earth. Even for a planet like our Earth with varying cloud cover, the rotational period can still be extracted because cloud patterns on Earth are coherent over several days. Brightness variations could be much higher than the 10-20% of our own Earth. If an Earth-like terrestrial planet had much less cloud cover but still had land, ocean, and ice, its brightness could vary by up to 200%.

In addition to studying "geological" surface features we might learn about biological features encoded on the planet's surface. On Earth, vegetation has a dramatic sudden rise in albedo by almost an order of magnitude around 750 nm, known as the red edge. Vegetation has evolved this strong reflection to prevent overheating, which would cause chlorophyll to degrade. While the albedo of sand also increases towards the infrared, the red edge vegetation signature is more rapid and may be detectable from the unresolved images. While we cannot necessarily expect to find Earth-like vegetation on extrasolar terrestrial planets, photometric measurements in different colors may be able to detect a unique signature different from any known surface features or atmospheric constituents on Earth or other Solar System planets.

2. Telescope Requirements

The wavefront requirements, technical performance, and feasibility of TPF for a range of aperture sizes has been reviewed by Brown et al. (2002b).

2.1. Think Elliptical

Our ability to detect planets around extrasolar systems will be limited by the inner working distance of the telescope. This inner working distance will be roughly $4\lambda/D$, where D is the long axis of the telescope's primary mirror.

There are significant advantages to an elliptical telescope. With a deployable secondary, we will be able to fit a larger telescope into a cylindrical launch shroud. Since planet finding is limited by the long-axis, a 3×6 meter telescope operating at 0.4 microns will be capable of detecting an Earth twin around stars out to 20 parsecs, while a 4 meter telescope can only detect Earth twins out to 12 parsecs. For ordinary imaging, a 3×6 meter telescope observing at several orientations will deliver a higher resolution image than a 4 meter telescope. For spectroscopy, the elliptical mirror does not offer a significant advantage; however, it will not degrade system performance significantly.

2.2. Controlling Diffracted Light

The role of the coronagraph is to control diffracted light. This section reviews the basics of coronagraphy and then outlines several different approaches to coronagraph design. (See Kuchner & Traub 2002 for a recent discussion of coronagraphy.)

Light enters the coronagraph through an entrance pupil P_1 and is then focused to an image plane. A mask, M_1, at the first image plane blocks some (or perhaps nearly all) of the starlight. The light distribution after the mask is the product of mask with the Fourier transform of the transmission through the first pupil, P_1:

$$I_1(\vec{\xi}) = M_1(\vec{\xi}) \int d\vec{x} \exp(ik\vec{\xi} \cdot \vec{x}) P_1(\vec{x}) \tag{1}$$

The signal that leaves through the exit pupil is the product of the Lyot stop, P_2, with the Fourier transform of the electric field energy distribution at the first image plane. This light is then refocused at a second image plane where it is detected. For an ideal coronagraph, the light distribution from the star at the detector is

$$I_2(\vec{\xi'}) = \int d\vec{x'} \exp(ik\vec{\xi'} \cdot \vec{x'}) P_2(\vec{x}) \int d\vec{\xi} I_1(\vec{\xi}) \exp(-ik\vec{\xi} \cdot \vec{x}) \tag{2}$$

Since we are free to specify the transmission properties of both the pupils, P_1 and P_2, and the masks, M_1, there are a number of different coronagraphic solutions. We will divide them into pupil plane coronagraphs and classical coronagraphs.

In a pupil plane coronagraph, most of the work is done by the entrance pupil. It focuses the starlight so that most of the light lands in a confined region on the first image plane. The Nisenson & Papaliolios (2001) apodized square aperture design controls the transmission by apodizing the transmission (i.e., specifying particular functional forms for $P_1(x)$). At Princeton, we have been exploring binary pupil masks (Spergel 2001; Kasdin et al. 2001). These masks are either transparent or completely opaque and have the advantage of ease of manufacture and higher throughput. If this pupil mask was used as part of a coronagraph, then an X-shaped image plane mask would block the residual starlight and the remaining region could be searched for a planet. By rotating the telescope, we can search the full stellar system.

In a classical coronagraph, most of the work is done by the image mask, M_1. In a Lyot coronagraph, a hard spot at the position of the star blocks most of the star light and then diffracts much of the residual starlight so that it is blocked at P_2. This classical design, however, does not suppress enough starlight to enable planet detection at $4\lambda/D$. A gaussian spot coronagraph replaces the hard spot with a mask whose transmission smoothly varies across the image plane. This design, which is incorporated into several ground-based coronagraphs, can in principle achieve the needed contrast ratios.

Kuchner & Traub (2001) have recently introduced a promising variant on the classical coronagraph: a band-limited mask coronagraph. This approach has a number of advantages: for an on-axis star, the coronagraph completely blocks the starlight. The design has a higher throughput than the classical Gaussian coronagraph and allows a smaller inner working distance.

Can we build these various coronagraph designs? The apodized systems are likely to be the most challenging. Kuchner and Traub note that the transmission needs to be controlled to better than 1 part in 10^8 in any apodized mask.

We can avoid these manufacturing difficulties by building a binary mask whose transmission is either 0 or 1. Kuchner & Spergel (2002) have introduced a series of these "notch filter" maps that combine the advantages of band-limited masks with the ease of manufacturability of a binary mask.

2.3. Controlling Scattered Light

In the previous section, we assumed that the wavefront entering the coronagraph was "perfect". Unfortunately, this is not true. The greatest challenge for a coronagraphic telescope is the need to sense, correct, and maintain a high quality wavefront (Brown & Burrows 1990).

Let's imagine that there is a small error in the surface of the primary mirror of amplitude $\epsilon\lambda$ and wavelength D/n. This distortion produces a speckle in the image plane at position $n\lambda/D$ whose intensity is $\sim \epsilon^2$. These speckles are the bane of planet detection.

The first step towards correcting these wavefront errors is sensing their amplitude and position. The coronagraphic camera is well-suited for the task; in the absence of diffracted light, the position and intensity of the speckle immediately yields the amplitude and wavelength of the mirror distortion. If we change the set-up of the coronagraph by either altering the masks or by moving the actuators in the deformable mirror [see e.g., the Coronagraph Optimized for Detection and Exploration (CODEX) proposal, R. Brown, PI], then by measuring the interference between the diffracted and scattered light, we can infer the phase of the mirror error.

John Trauger's group at JPL has been played a leading role in developing arrays of PbMgNb actuators for wavefront correction. Their Eclipse test bed (Trauger, Hull, & Redding 2001) has all of the elements needed for a system-level test of the full coronagraphic system.

Another possible source of speckle are variations in reflectivity in the optical system. These variations also produce speckles that can confound planet detection. The CODEX study has proposed canceling phase and amplitude errors in one half of the image plane over a modest bandpass (10%). As long as the variations in reflectivity are small ($< 10^{-3}$), this approach is viable. The need for nearly uniform reflectivity will be an important driver for the design of HST-II and may conflict with the desire to use more exotic coatings on the primary mirror to improve ultraviolet sensitivity.

2.4. Detectors

What are the detector needs of a coronagraphic telescope? The minimal requirements are low read-noise, high quantum efficiency, and linearity. The current generation of CCDs satisfy these requirements.

A detector (or camera) with energy resolution ($\delta E/E \sim 0.1$) would be highly beneficial. This energy resolution would enable us to separate planet light from stellar speckles. The position of the planet on the image plane is wavelength independent, while the speckle position moves outward from the star. The energy resolution would also assist our efforts at wavefront control by separating the effects of amplitude error (which is likely gray) from phase error. The amplitude of speckles produced by phase errors in the mirror scales as λ^{-2}

There are several possible ways of achieving this energy resolution. There are currently several classes of 3-dimensional detectors under development. Since we would only need to cover a relatively small field of view, a $128 \times 128 \times 10$ detector would be well suited for TPF. Alternatively, we could build a dichroic camera that simultaneously observes the star and its planetary system at several discrete wavebands. This approach has two advantages: (1) it enlarges our search area, as we can search the inner system at blue wavelengths and the outer system at redder wavelengths, and (2) its energy resolution will help discriminate against speckles. While the dichroic cameras may not achieve the energy resolution of the 3-dimensional detectors, they are lower risk.

2.5. Stability, Stability, Stability

Planet detection requires very accurate control of the wavefront entering the coronagraph. Since we will only be able to detect wavefront errors after long integrations (tens of minutes), the optical system must be stable on long timescales. Since we are assuming currently wavefront errors at the $\lambda/10,000$ level, the primary mirror (and subsequently optics) cannot vary at this level.

This stability requirement implies that the satellite will need to be in a very stable thermal environment and be dynamically stable. In our Ball TPF study, we envisioned that TPF would be in either an L2 or Earth-trailing orbit and would employ a sunshield to further isolate the mirror from variable heating. Dynamical stability may likely require active control via actuators on the primary mirror.

While these requirements will not alter the ability of the telescope to carry out other scientific programs, they will likely significantly increase mission cost and complexity.

3. Conclusions and Other Science Capabilities

An optical TPF will be capable of carrying out a wide ranging scientific program. With proper instrumentation, it will be able to carry out the science program outlined in the SUVO study (Shull et al. 1999). The excellent mirror needed for planet finding will enable diffraction limited seeing into the ultraviolet.

Any telescope capable of planet detection will have superb wavefront control. This will enable very accurate narrow angle astrometry. It will likely be able to position stars to better than 10^{-3} of its resolution. For the 3×6 meter elliptical telescope, this implies that a one hour integration will enable 15 micro-arcsecond astrometry on stars down to 26th magnitude. Once the Space Interferometry Mission (SIM) and GAIA have established an accurate global grid, this narrow angle astrometry can be tied to a global reference frame. This precision astrometry opens up a large region of new parameter space and will enable a number of interesting science programs ranging from the astrometric detection of distant planets to measure the masses of stars and stellar remnants through gravitational lensing (Paczynski 1998).

If properly designed, the follow-on to HST should be capable of detecting Earth-like planets. This "killer application" should be designed in from the start.

References

Brown, R.A., et al. 2002a, PASP, submitted

Brown, R.A., Burrows, C.J., Kasdin, N.J., Kilston, S., Kuchner, M.J., Spergel, D.N., Traub, W.A., Trauger, J.T., & Turner, E.L., 2002b, Nature, submitted

Brown, R.A., & Burrows, C.J. 1990, Icarus 87, 484

Brown, R.A., Ford, E.B., Seager, S., & Turner, E.L. 2001, Nature, 412, 885

Ford, E.B., Seager, S., & Turner, E.L. 2001, Nature, 412, 885

Kasdin, J., Spergel, D.N. & Littman, M., 2001, BAAS, 199, 8603

Kuchner, M., & Spergel, D.N. 2002, ApJ, submitted

Kuchner, M., & Traub, W. 2002, ApJ, 570, 900

Nisensen, P., & Papaliolios, C. 2001, ApJ, 548, L201

Paczynski, B. 1998, ApJ, 494, L23

Shull, J.M., et al. 1999, UVOWG White Paper, (astro-ph/9907171)

Spergel, D.N. 2001, astro-ph/0101142

Trauger, J.T., Hull, A.B.,& Redding, D.A. 2001, BAAS, 199, 8604.

Hubble's Science Legacy: Future Optical-Ultraviolet Astronomy from Space
ASP Conference Series, Vol. 291, 2003
K.R. Sembach, J.C. Blades, G.D. Illingworth, R.C. Kennicutt, Jr. (eds.)

The Darwin-GENIE Experiment: An ESA-ESO Partnership

Ph. Gondoin, O. Absil, C. Erd, M. Fridlund, R. den Hartog, and N. Rando

Research and Scientific Support Department, European Space Agency - Postbus 299, 2200 AG Noordwijk, The Netherlands

Abstract. Darwin is one of the most challenging space projects ever considered by the European Space Agency (ESA). Its principal objectives are to detect Earth-like planets around nearby stars and to characterize their atmospheres. Within the framework of the Darwin program, ESA and the European Southern Observatory (ESO) intend to build a ground-based technology demonstrator called Darwin-GENIE. Such a demonstrator built around the Very Large Telescope Interferometer (VLTI) in Paranal (Chile) will test some of the key technologies required for the Darwin Infrared Space Interferometer. The paper describes the objectives of the Darwin-GENIE project.

1. Introduction

Understanding the principles and processes that generated the Earth, and allowed the evolution of life forms to take place is a major scientific objective which deserves the utmost attention. In particular, the successful detection of Earth-like planets possessing environments benign to life would answer central questions such as "How unique is the Earth as a planet?" and "How unique is life in the Universe?". To achieve these objectives, the IRSI-Darwin mission of the European Space Agency (Fridlund 2000, 2002) will survey a large sample of nearby stars and search for Earth-size planets within their "habitable zone". IRSI-Darwin will measure their spectra in order to infer the presence of an atmosphere and search for the presence of biomarkers.

Detection of Earth-size bodies circling nearby stars is extremely difficult because of the weakness of the planetary signal emitted within a fraction of an arcsecond from an overwhelmingly bright star. A solar type star outshines an Earth size planet by a factor of more than 10^9 in the visible wavelength range. In the infrared spectral range, where the planet's thermal emission increases and the star's emission decreases, the contrast is still higher than 10^6. Only the planetary signal, a millionth of the stellar light, should remain in the input feed of a spectrograph in order to register a planet spectrum in a reasonable time. To accomplish such an extinguishing of light at the relevant spatial scales, the technique of "nulling interferometry" has been selected for IRSI-Darwin. By applying suitable phase shifts between different telescopes in an interferometric array, destructive interference can be achieved on the optical axis of the

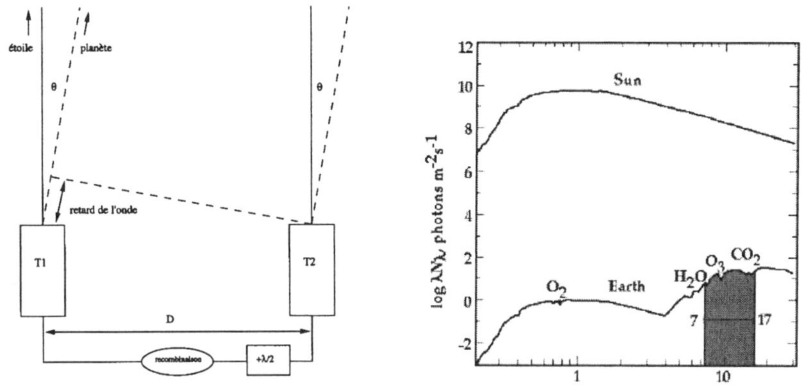

Figure 1. *Left*: Principle of a two-telescope Bracewell interferometer. *Right*: Contrast between the solar flux and the Earth as a function of wavelength in microns. Adapted from Angel et al. (1986).

system in the combined beam while interference is constructive for small off-axis angles. The principle of nulling interferometry for a simple two-telescope Bracewell interferometer (Bracewell 1978) is described in Figure 1. The equivalent transmission map of the nulling interferometer is a set of interference peaks with a sharp null in the center. By placing the central star under this null, and adjusting the interferometer baseline to the required angular resolution, planets can be detected in the "habitable zone". The actual shape and transmission properties of the pattern around the central null depend on the configuration and the distance between the telescopes.

2. The IRSI-Darwin Development Program

To date, the IRSI-Darwin mission concept (see Figure 2) consists of six 1.5 m telescopes, each of which is a free flying spacecraft transmitting its light to a central beam-combining unit (see ESA-SCI/2000/12, 2000). Using the nulling interferometry technique, the beam combiner could detect and analyze the light from Earth-like planets at distances of up to 25 parsecs. The mission concept is based on the ability to co-phase telescopes on independent spacecraft to an accuracy better than 20 nm and to perform nulling interferometry with a rejection factor of $\approx 10^6$ in a wide spectral band extending from 5 to 18 μm. In-orbit co-phasing of the free-flying telescopes could be performed in successive steps combining first a local radio-frequency positioning system with milli-Newton propulsion devices, then inter-satellite laser metrology with micro-Newton propulsion devices, and finally fringe sensors with optical delay lines. These items will be developed within the frame of the ESA Technology Research Program (TRP). They will then be tested in-orbit within the frame of the ESA Small Missions for Advanced Research in Technology (SMART) program.

In addition to the co-phasing of free-flying telescopes, a second key issue for IRSI-Darwin is the nulling interferometry technique. Hence, the TRP activi-

Figure 2. *Left*: Artist view of the IRSI-Darwin Space Interferometer orbiting at Sun-Earth L2 point (courtesy Alcatel Space Division). *Right*: The ESO Very Large Telescope Interferometer.

ties also include the development of Darwin specific optical components, namely achromatic phase shifters, wavefront filtering devices and infrared (IR) single mode fibers, integrated optics, IR detectors, electronics and coolers, optical delay lines and fringe sensors, as well as other components for interferometry. Laboratory nulling breadboards with star-planet simulators are currently designed that will test these components before the end of 2002. These innovative designs use the narrow telecommunication band around 1.65 μm where off-the-shelves components are available. Based on their performance, a follow-up activity has been identified that shall adapt their design to the mid-IR band and test the nulling interferometry technique on astronomical targets in IRSI-Darwin representative operating conditions.

The Very Large Telescope Interferometer (VLTI) at the ESO is the most appropriate infrastructure on the ground to perform such a test. Hence, ESA and ESO decided to initiate a definition study for a Groundbased European Nulling Interferometry Experiment which shall operate in the central laboratory of the VLTI at Mount Paranal (Chile). This experiment is called Darwin-GENIE.

3. Darwin-GENIE: A Technology Demonstrator

The primary objective of the Darwin-GENIE nulling experiment is to gain experience on the design, manufacture, and operation of a nulling interferometer using the IRSI-Darwin representative concept and technology. Nulling tests with the highest rejection factor on single stars or close binaries in broad mid-IR spectral bands will achieve this objective with the limitations imposed by the turbulence and infrared background of the Earth's atmosphere. The Darwin-GENIE experiment will combine all optical functions foreseen into the future Darwin Infrared Space Interferometer. It will benefit from the existing VLTI infrastructure, including the telescopes, the chopping schemes, the adaptive optics, the delay lines, the fringe sensors, and the beam combiner laboratory. The overall performance of the instrument will depend heavily on the performance of all VLTI subsystems and in particular on the adaptive optics and co-phasing subsystems. The GENIE optical bench within the VLTI laboratory will provide the functions specific to the nulling interferometry technique, namely photometry and amplitude control, polarization matching, phase shifting, beam combination and internal modulation, spatial filtering, spectrometry, detection, electronics, and cryogens. The detailed architecture of Darwin-GENIE will be defined during the definition study and will take into account the ESO VLTI interface characteristics and the output of the ESA TRP activities.

4. Darwin-GENIE: Preparation for the Darwin Science Program

A second objective of GENIE is to prepare the IRSI-Darwin science program through a systematic survey of IRSI-Darwin candidate targets. The solar zodiacal cloud, a sparse disk of 10–100 μm diameter silicate grains, is the most luminous component of the solar system after the Sun. Its optical depth is only $\approx 10^{-7}$, but a patch of the solar zodiacal cloud 0.3 AU across has roughly the same emitting area as an Earth-sized planet. Similar and even brighter clouds may be common in other planetary systems and may present a severe obstacle for the direct detection of extra-solar terrestrial planets. A systematic survey of IRSI-Darwin candidate targets will screen out those stars for which circumstellar dust prevents the detection of Earth-like planets. Bright exo-zodiacal clouds are easier to detect than extra-solar terrestrial planets, but finding an exo-zodiacal cloud is still difficult. The total emission from our zodiacal cloud is no more than 10^{-4} of the Sun's at any wavelength. Photometric surveys like the Infrared Astronomical Satellite (IRAS) survey can only detect exo-zodiacal clouds that are $>$ 500 times as optically thick as the solar clouds (Backman & Parece 1993). Attempts to spatially resolve faint exo-zodiacal clouds with single-dish telescopes in the mid-infrared and near infrared (Kuchner & Brown 2000) have not yielded better detection limits.

A Bracewell interferometer using two ESO VLTI 8 m Unit Telescopes can provide better performance. However, the spatial resolution of the stellar disk by a two telescope interferometer does not allow a complete null of the starlight. Erratic fluctuations of this stellar leakage due to residuals in the control of the optical path difference (OPD) between the two arms of the interferometer are the major limitation to the achievable signal-to-noise ratio for the detection and spectroscopy of exo-zodiacal dust clouds with two VLT telescopes. The effect can be minimized by operating at long wavelengths in the N band and by using the shortest VLTI baselines (e.g., 46 or 56 meters). Assuming a conservative 20 nm OPD control accuracy, it is anticipated that an exo-zodiacal dust disk only fifteen times brighter than the solar zodiacal cloud could be detected around a G2 V star at 10 pc within 750 s. An assessment study of this scientific case for the Darwin-GENIE experiment will be further conducted, defining the analysis methods of the interferometric data that separate the circumstellar signal from the residual starlight, and inferring the morphology of circumstellar clouds.

5. Darwin-GENIE: Direct Detections of Exo-Planets?

Since 1995, more than 80 extra-solar planets have been discovered as companions to solar-type stars (Marcy, Cochran, & Mayor 2000). All of these planets were discovered indirectly in high-precision Doppler surveys. These surveys measure the radial velocity Doppler shift of the parent star due its periodic motion around the center of mass of the star plus planet system. This technique is most sensitive to massive planets with short orbital periods. The extra-solar planets that have been discovered today have mass ranges from $M \times sini \approx 0.15$ to 10 Jupiter masses (where i is the orbit inclination onto the observer line of sight) with orbital periods ranging from 3 days to a few years. Because of their proximity to the parent star (<<1 AU), the atmospheres of some of these planets are much hotter than that of Jupiter (and of the Earth), hence their name of hot Jupiters. With temperatures sometimes higher than 1000 K, their thermal emission is maximum in the infrared K and L bands.

With baselines ranging from 46 to 130 meters, the VLTI is optimized for the detection of planets in the L band at about 5-15 milli-arcsec from their parent star, which corresponds to 0.05-0.15 AU for a nearby star at 10 parsecs. This is especially fortunate since these planets are precisely the hot Jupiters of which a significant sample is known. The contrast between a hot Jupiter and its parent star in the L band is reduced to less than 10^4 compared to 10^6 in the N band for Earth-like planets. The starlight leakage through a two-telescope Bracewell interferometer increases as the square of the baseline and decreases as the square of the wavelength. The requirements on the OPD control accuracy between the two arms of the interferometer are slightly relaxed in the L band compared with the K band. They are within the limitation imposed by the brightness of the reference central star for nearby objects. Three-telescope configurations can in principle achieve a broader null of the star signal in the K and L bands but are not robust to residual atmospheric OPD fluctuations. Preliminary calculations indicate that a 4-VLT double-Bracewell configuration using internal modulation between two Bracewell nulling interferometers could subtract the IR background and the starlight leakage from the overall nulled signal with an accuracy sufficient to retrieve the signal from the brightest exo-Jupiters. Thus, direct detection of the hottest exoplanets might be possible on the ground using the Darwin nulling interferometry technique. The measurement of the spectrum of an extra-solar planet would be an outstanding achievement providing unique data to planetary physicist and a major step for exo-planetology.

6. Darwin-GENIE: A General User Instrument

The nulling interferometry technique in the infrared is well suited to spectroscopic studies of any faint cool object located in the immediate environment of a bright astrophysical source. A wide range of studies could potentially be conducted using the Darwin-GENIE experiment, including infrared spectroscopy of cool dwarfs in spectroscopic binaries, protoplanetary disks around T Tauri stars and Herbig Haro objects, and dust tori in the nuclei of active galaxies. Such programs will be important scientific by-products of the Darwin-GENIE experiment.

7. Conclusion

The prime objective of Darwin-GENIE is to gain experience on the design, manufacture, and operation of a nulling interferometer. The ESO Very Large Telescope Interferometer offers a unique opportunity to fulfill this objective. Secondary objectives of Darwin-GENIE are to prepare the IRSI-Darwin science program through a systematic survey of IRSI-Darwin candidate targets, to perform Darwin related science achievable from ground including spectroscopy of protoplanetary disks and low-mass companions around nearby stars, and to provide the science community with a scientifically useful instrument. ESA and ESO will initiate an initial definition study to be conducted by European industries and scientific institutes. The study will establish a preliminary design of the instrument including the specifications of its subsystems and a description of their implementation using technologies developed within the framework of the ESA Technology Research Program. Upon completion of the definition study, ESA and ESO will decide whether to proceed in the manufacturing and testing of Darwin-GENIE.

References

Angel, J.R., Cheng A.Y., & Woolf, N.J. 1986, Nature, 322, 341

Backman, D.E., & Parece, F. 1993, in Protostar and Planet III, eds. E.H. Levy & J.I. Lunine, (Tucson, Univ. of Arizona Press), 1253

Bracewell, R.N. 1978, Nature, 274, 780

ESA-SCI/2000/12, 2000, Darwin: the Infrared Space Interferometer: Concept and Feasibility Study Report

Fridlund, M. 2000, in Proceedings of the Conference "Darwin and Astronomy – the Infrared Space Interferometer", Stockholm, Sweden, 17–19 November 1999, ESA SP-451

Fridlund, M. 2002, in Proceedings of the Conference "Stellar Structure and Habitable Planet Finding", Cordoba, 11–15 June 2001, ESA SP-485

Kuchner, M.J., & Brown, M.E. 2000, PASP, 112, 827

Marcy, G.W., Cochran, W.D., Mayor, M. 2000, in Protostars and Planets IV, eds. V. Mannings, A.P. Boss, & S.S. Russel, (Tucson: Univ. of Arizona Press), 1285

Hubble's Science Legacy: Future Optical-Ultraviolet Astronomy from Space
ASP Conference Series, Vol. 291, 2003
K.R. Sembach, J.C. Blades, G.D. Illingworth, R.C. Kennicutt, Jr. (eds.)

Large-Optics Technology Lessons from Ball's NGST and TPF Programs

Steven Kilston

Ball Aerospace & Technologies Corp., M/S T-3, P.O. Box 1062, Boulder, CO 80306-1062

Abstract. Ball Aerospace & Technologies Corporation and its partners have been considering many large space-astronomy design concepts, especially through our work on the Next Generation Space Telescope (NGST) and Terrestrial Planet Finder (TPF) programs. During those efforts our teams strove to be open-minded and creative about design approaches and suitable technology elements. Our evaluations of the candidate concepts were based on performance, costs, and technology risks and required development. We assessed many performance factors relevant to an optical-UV successor to the Hubble Space Telescope (HST), such as system sensitivity, angular resolution, and usable fields of view. Our teams investigated key technology issues such as attitude control, thermal and vibrational disturbances, ease of launch and deployment, and operational reliability beyond Earth orbit (if that is required). In the case of TPF we considered several large-optics designs operating at optical wavelength bands, with wavefront quality more than adequate for UV operation too. We will discuss here some of our work's potential implications for choosing an HST optical-UV successor's technologies.

1. Introduction

For the past several years Ball Aerospace & Technologies Corporation has been part of teams studying and developing concepts, technologies, and plans for the Next-Generation Space Telescope (NGST) and the Terrestrial Planet Finder (TPF) programs. These programs have very challenging mission performance goals well beyond those of current space telescopes. To achieve those goals we must develop and incorporate into the system designs the best technologies suited to the program timelines at affordable cost and risk levels. This discussion will highlight lessons from our TPF study, as the present volume contains an extensive discussion of NGST lessons in the paper by John Mather. For both NGST and TPF the key system performance capabilities include wavelength coverage, angular resolution, system sensitivity and dynamic range, and usable field-of-view, as shown in Figure 1. For the TPF visible-light coronagraph concept, the TPF option relevant to this discussion, there is also the special requirement to achieve high sensitivity to faint planets separated by very small angles from their extremely bright host stars. To design systems with the required capabilities, including a potential large-optics follow-on ("Hubble-2") to

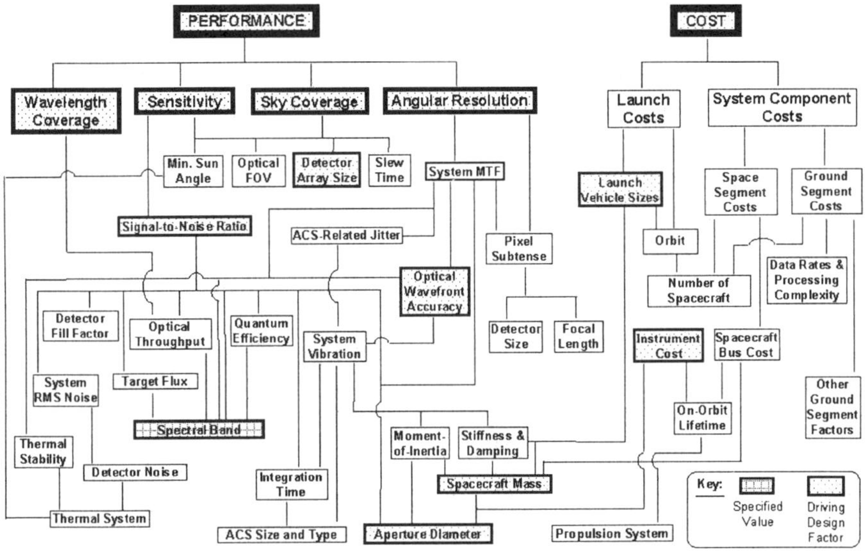

Figure 1. System performance dependencies.

the Hubble Space Telescope (HST), as compared in Table 1, we have addressed instrument and spacecraft technologies, launch and deployment options, and operational reliability far from Earth. We evaluated system options in terms of their robustness, risks, and costs, and constructed appropriate technology development roadmaps for the preferred options.

The major performance features of Ball's visible-light coronagraph concept for TPF include its capability to directly detect Earth-like planets in the habitable zones surrounding about 150 solar-type stars within 20 pc of our Solar System, to observe the signatures of chemical biomarkers in the spectral band extending from 0.4 to 1.0μm, and to perform general astrophysics research utilizing our largest telescope mirror in space. The design concept includes an optical train with a 10×4 meter near-monolithic off-axis primary mirror only 4 cm thick, supported by shape-controlling actuators attached to a backplane, with a mass totaling 940 kg. This incorporates the part of the active optics for preliminary wavefront control, with a deformable mirror (DM) inside the coronagraph instrument supplying extremely high-level final control of the wavefront in order to suppress light scattered by optical surface imperfections. In addition, the instrument permits shaped-pupil options to produce very low levels of diffracted starlight at locations in the focal plane where the planet signal is to be detected. The TPF 3-axis-stabilized spacecraft is to be launched in 2015 into either an Earth-trailing drift-away or an L2 orbit by a Delta IV H launch vehicle; this would be a direct insertion launch. The spacecraft, with a

Table 1. NGST, TPF, and Hubble-2 (SUVO – Shull et al. 1999) Comparison

	NGST	Visible TPF	Hubble-2
Aperture Size	~ 6 m	4×10 m	8 m
Wavelength Coverage	0.6–28 μm	0.3–1.0 μm	0.11–1.0 μm
Angular Resolution	30 mas	10×25 mas	15 mas
Wavefront Quality	$\lambda/14$@0.75 μm	$\lambda/140$@0.5 μm	$\lambda/14$@0.5 μm
System Sensitivity	$M_{AB} = 31$	$V = 34$	$M_{AB} = 32$
Usable FOV (arcmin)	5×6	$> 2 \times 2$	12×12

launch mass near 6,000 kg supplying significant ($> 25\%$) mass margin, would have a 5 year design life (but would carry 10 years worth of expendables) and, as shown in Figure 2, will include a large thermal shield providing an extremely stable thermal environment for the optics and instrument.

Our team's colleagues at Eastman Kodak have been working on technology development to enable segmented optics construction of a near-monolithic primary mirror big enough for TPF and other large space telescopes. The Kodak process fuses together separate core and faceplate segments to form the monolithic primary mirror blank, using existing Kodak facilities and capabilities to produce the segments. The primary mirror would be built from segments derived from Advanced Mirror System Demonstrator (AMSD) program technology developed for NGST, which has the goal of achieving a total mirror/actuator system areal density of $15\,\mathrm{kg\,m^{-2}}$. The TPF design does not require that low a density; the availability of a heavy-lift launch vehicle permits us to allocate $35\,\mathrm{kg\,m^{-2}}$ to the primary mirror element, helping to yield a mirror with adequate stiffness and stability. Currently a proof-of-concept 30 cm mirror blank, shown in Figure 3, made of ultra-low-expansion (ULE) glass has been constructed, with a segmented 19 mm deep core with support struts 0.02 inch (0.5 mm) thick, a segmented back plate, and a monolithic structure formed by low-temperature fusion of the segments. The TPF monolithic primary mirror similarly will have a segmented front and back plane. The TPF primary mirror is currently sized at a 4×10 m baseline, but the launcher may offer sufficient capacity to carry a 4×13 m primary. Either mirror size would require new facilities to assemble and test the full primary mirror. A technology plan to develop these capabilities and facilities is shown in Figure 4.

2. Active Optics

One stage of wavefront correction is used for NGST, but two are needed for a TPF coronagraph, since strong suppression of the starlight requires wavefronts correct to about 0.0001 of a wavelength. Both telescopes will employ actuators between their primary mirror's back surface and its backplane, typically about 10 actuators per m^2. These actuators will have a stroke capability of tens of mm, with a setting resolution near 7 nm. In addition, for the TPF coronagraph,

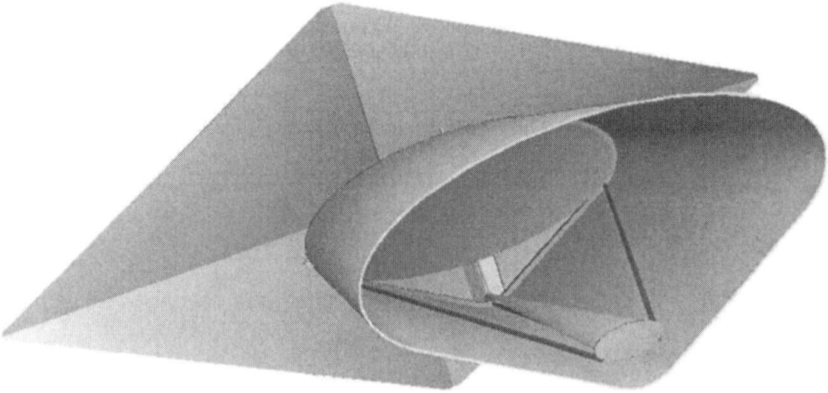

Figure 2. Ball TPF visible light coronagraph.

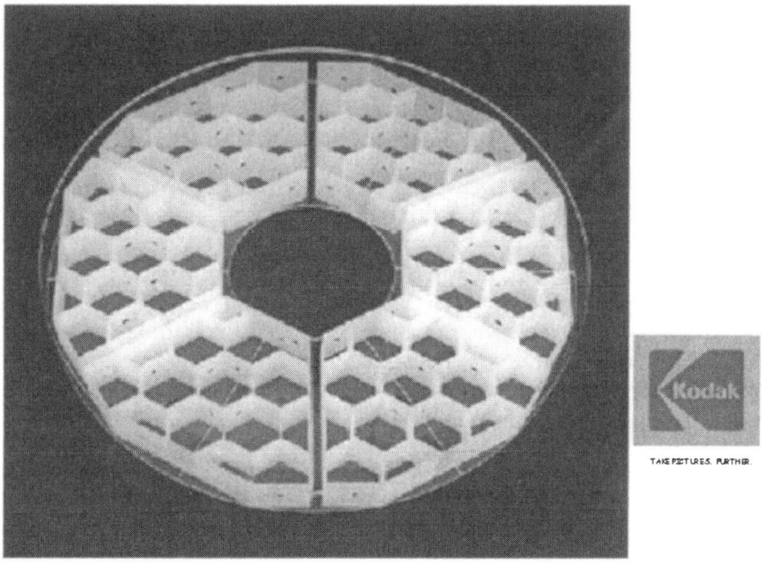

Figure 3. Kodak demonstration mirror.

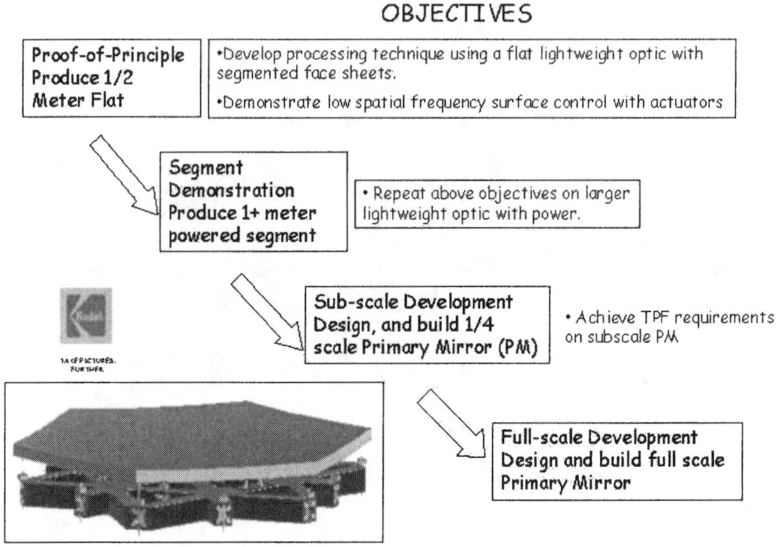

Figure 4. TPF primary mirror technology development plan.

a DM with about 250×100 elements will be able to correct the wavefront at mid spatial frequencies. Its stroke need be only about 100 nm, to compensate for primary mirror actuator inaccuracy and to provide a much finer setting resolution of 25 pm (which Xinetics DMs have already demonstrated). Iterative setting avoids the need for the DM to provide absolute setting accuracy. Temporal and thermal stability requirements are very stringent for the DM, and experiments have shown that Xinetics DMs, as shown in Figure 5, can be held stable to levels of 100 pm over weeks; in order to do this a 10 mK temperature stability is needed. Also, for use in TPF, such DMs will need accelerated life tests to prove that they will be durable in space. Also, since other types (for example, some based on MicroElectroMechanical Systems, or MEMS, technology) and suppliers are entering the DM arena, comparative projected performance capabilities will need evaluation, including the potential of any DM to effectively reduce image speckle due to amplitude non-uniformities.

3. Mirror Coatings and Spectral Passbands

Particularly for use as a versatile UV/optical telescope, there will be a system tradeoff of throughput versus spectral range. Silver coatings give high reflectivity, but no ultraviolet (UV) performance, while aluminum allows operation in the UV and provides a reflectivity in the range of 92-96% per mirror across the visible band. However, in the complex coronagraphic optical train, throughput

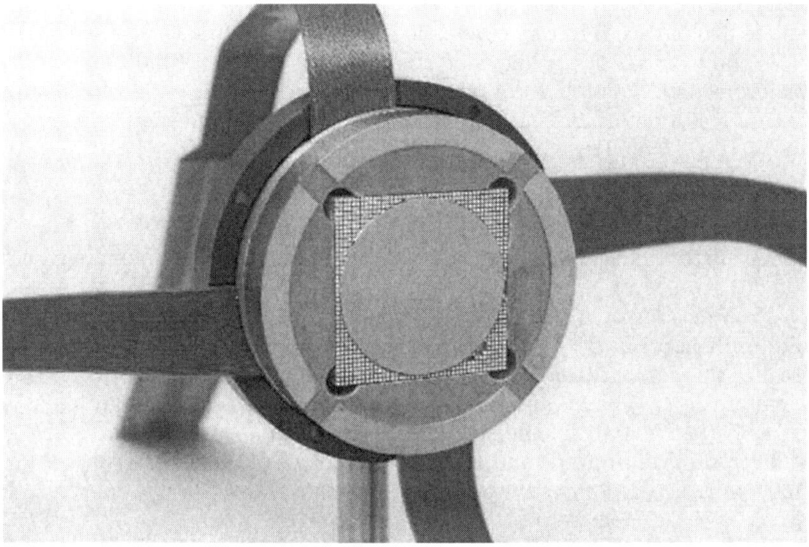

Figure 5. Xinetics deformable mirror.

can fall to a net 50-60% after 10 mirrors. Therefore, the present TPF coronagraph design uses aluminum for the primary and secondary mirror, so a pickoff mirror could feed accompanying UV instruments at fully adequate wavefront accuracy (but with no access to the DM). In the coronagraph instrument silver can then be used for the post-secondary mirrors to maximize visible-wavelength throughput. This enables a dual-use TPF planet-finding and astrophysics mission, as recommended by the National Research Council's decadal committee report "Astronomy and Astrophysics in the New Millennium" (Astronomy and Astrophysics Survey Committee, 2001). Technical challenges for the coatings include UV performance and the extreme uniformity of reflectivity required for coronagraph performance, as well as durability and stability in the space environment. System spectral measurement accuracy will depend on filter and detector performance and on calibration devices and techniques.

4. Effects of Vibration from Reaction Wheels

Modern techniques for mechanical isolation of reaction wheel vibrations can greatly reduce both high-frequency line-of-sight (LOS) jitter and dynamic distortions in the primary mirror surface. Using small and fast wheels produces less jitter than larger wheels, meeting the 1 milli-arcsecond rms maximum LOS jitter requirement over most speeds. Use of HST-class wheels with 1.5 Hz isolation will meet requirements except at a few resonances. These results have been modeled for an open-loop system: they don't engage any fine steering mirror pointing loop, but that could be added in to the system if necessary. The worst-case primary mirror deformation (at 41 Hz) gives a spatial/ temporal rms modeled result of 8.4 nm ($\sim$ 17 nm wavefront error), but adding in a 1.5 Hz isolator reduces the worst-case primary deformation (at 4.6 Hz) to 0.05 nm ($\sim$ 0.1 nm wavefront error). A further doubling of the wheel speed is expected to suppress even that small deformation by another order of magnitude.

5. Initial Thermal Control Design

Primary-mirror temperature variations can be controlled (including use of a sunshield) to a level of spatial variations (over a 24-hour cycle) below 0.03°C. Active heater control of the optical bench is required, because a passive design would result in primary mirror temperatures of −140°C; instead, primary mirror operating temperatures of 0°C are favored, to simplify manufacturing, testing, and calibration. (This can drive the bench heater power requirement to as much as 700 W.) At 0°C, the coefficients of thermal expansion of ULE^{TM} and Zerodur^{TM} are minimized, and the active heating provides precise control over primary mirror temperature gradients.

6. Observatory Ground Test Overview (Integrated System Test)

There are several potential large space-optics test facilities to choose from: the Mark I chamber (12.8 m × 25 m) at the Arnold Engineering Development Center in Tennessee, the Glenn Research Center Space Power Facility (Plum Brook) chamber (30.5 m × 37.2 m) in Ohio, and the Johnson A chamber (16.8 m × 27.4 m) in Texas. We have evaluated the test set-up for Plum Brook, the largest facility, which can offer a very low vibration level and is a NASA-controlled facility. There it would be possible to test the entire observatory, with only the sunshield and any extendable solar array removed for testing. In the chamber, vertical orientation of the telescope eliminates major distorting moments into the primary aperture. A vacuum test can be performed in a simulated on-orbit thermal environment.

7. Conclusions and Recommendations for UV/Optical Hubble 2

A large-mirror telescope with wavefront quality excellent for UV astronomy can be built and tested, and is feasible for launch in the 2015 timeframe. Technology readiness to meet this goal is shown in Table 2. Active-optics primary-mirror actuators could enable use of a very large (4 × 4 m or larger) high-quality segmented mirror. Incorporating a DM could control the wavefront to the accuracy needed for high-contrast coronagraph operation ($\lambda/10000$). Technology development, especially for the large mirror, should be adequately supported, to ensure that the system can achieve the high mechanical stability needed throughout an observation, to maintain the wavefront accuracy. Therefore thermal and vibrational control are key elements of the technology required. (Thermal control to milli-Kelvin levels may be the single key technology need, and should be attainable with a thermal shield plus computerized network of temperature sensors

Table 2. TPF Visible-light Coronagraph — Technology Readiness and Prospects (TRL 1=Principles Observed, TRL 9=Flight Proven)

NASA Technology Readiness Levels				
Technology	Current TRL	Future Work	Completion Date (Est.)	Resulting TRL
Large, lightweight optics	3	Lab demo of scale model	2005	6
Wavefront sensing with science camera	3	TPF technology testbed demos	2005	6
Deformable mirrors	5	Lifetime tests in lab	2003	6
Thermal control	4	NGST validation	2008	7
Binary pupil masks with edge adjustment	2	Lab demo of full scale masks	2003	6
Integrated model of full optical system	6	NGST validation	2008	8

and heaters — integrated modeling is needed to demonstrate the control level achievable.) For maximum collecting area and resolution, an elliptical monolith primary could be used and would fit in the launch vehicle fairing. This concept for the UV/Optical Hubble 2 will be expensive (up to $1.5B), but the observatory can host several instruments including the TPF coronagraph, provide extremely high spatial and spectral resolution, offer wavelength coverage in the UV/optical spectrum, and support a wide field-of-view.

References

Astronomy and Astrophysics Survey Committee, Astronomy and Astrophysics in the New Millennium, Board on Physics and Astronomy, Space Studies Board, National Research Council, 2001. Available on-line at http://www.nap.edu/books/0309070317/html

Shull, J.M., et al. 1999, The Emergence of the Modern Universe: Tracing the Cosmic Web, White Paper on ST2010, available at http://origins.colorado.edu/uvconf/white-final/UVOWG_final_jun20.html

Hubble's Science Legacy: Future Optical-Ultraviolet Astronomy from Space
ASP Conference Series, Vol. 291, 2003
K.R. Sembach, J.C. Blades, G.D. Illingworth, R.C. Kennicutt, Jr. (eds.)

Starting a UV Mission: Building from NGST Technology

J. Mather

Infrared Astronomy Branch, Code 685, NASA Goddard Space Flight Center, Greenbelt, MD 20771

Abstract. The Next Generation Space Telescope project provides many lessons for the the design of a new space ultraviolet telescope. I describe the NGST, discuss why it was not itself an ultraviolet telescope, and outline the needed technological advances and design studies to enable a UV telescope. The most important area for investment is in detectors, which govern the required choices in the entire mission design.

1. Introduction

The Next Generation Space Telescope (NGST)is a redshifted ultraviolet (UV) telescope. It is an excellent example of a mission conceived to serve a wide community of users, with a design driven by enormously challenging scientific goals. NGST is intended to see the first stars and galaxies, by using infrared detectors to measure the redshifted visible and ultraviolet light from the distant universe. It is intended to detect the light from the end of the cosmic dark ages, when the first objects may have formed around a redshift of 10–30. Infrared light can also penetrate dust clouds and reveal the formation processes for nearby stars and planets. The NGST will have about 6 m aperture and will orbit around the Sun - Earth Lagrange point L2, 1.5×10^6 km from Earth, where radiative cooling can lower its temperature to about 30 K. It will cover the wavelength range from 0.6 to 28 μm with cameras and spectrometers.

Why wasn't NGST a UV telescope? The first and obvious reason was that the "HST and Beyond" committee report (Dressler et al. 1996) recommended scientific objectives that could only be reached in the infrared. Part of this no doubt included the recognition of a huge infrared "discovery space" advantage over other equipment, which really just shows how poor our infrared capabilities have been in the past. When these goals were translated into detailed observing programs (The NGST Design Reference Mission, Stiavelli et al. 1997) it was clear that much was possible, providing a really large telescope could be built. There is also the obvious competition with present and future ground-based telescopes with adaptive optics, and the opinion has been expressed that since UV detectors are not near to unit quantum efficiency, we should work on those before deciding to build a new UV space telescope. It was also recognized that light-weight high quality mirrors much larger than the HST mirror are not now available, so it seems that an infrared mission might be the right thing to do next. Another aspect that worried our NGST engineering team was the difficulty of keeping UV optics clean enough, though we clearly have done this in the past.

The NGST scientific goals and initial design concept are very well summarized in the book by Stockman et al. (1997).

2. Hypothetical Design Objectives for a UV Space Telescope

Let me assume that there will be strong scientific reasons to seek the following performance specifications, bearing in mind that compromise is always feasible in the face of reality. I believe that in the time frame for a new space mission there is fortunately or unfortunately plenty of time to develop new technologies. I think we should seek diffraction limited imaging at 0.2 μm, with a 50 m^2 clear aperture (an 8 m circle, or a 4 x 16 m off-axis ellipse). Coronagraphic capability for planet searching and analysis, with extremely low stray light, would be an excellent objective, one which is already sought by another major NASA program with excellent public support. I think a general purpose observatory should also provide a very large field of view, to couple with very large detector arrays. Some proposed missions like the SNAP (Supernova Acceleration Probe) and the GEST (Galactic Exoplanet Survey Telescope) already require 10^9 pixels, and claim that they could be built with today's technology. I would expect that both medium and high spectral resolutions should be provided by the spectrometers, and that a multiplexed spectrometer based on the NGST microshutter arrays would be very powerful.

3. Mirrors

Mirrors meeting the NGST requirements were definitely not available when the mission was conceived and placed in the plan by NASA. In order to make such mirrors a reality, the NGST project (in collaboration with the DoD and NRO) invested $30M in several rounds of competitions among mirror vendors. More than 12 contracts were let, and at least three technologies are still serious candidates. A Kodak concept uses a pair of glass face sheets fused to a glass rib structure. A Goodrich concept uses a single face sheet with integral ribs cut by water jets, supported in numerous places by actuators. A Composite Optics concept uses a similar face sheet bonded to carbon fiber composite ribs. Remarkably enough it is possible to choose material combinations which have zero residual thermal stress at room temperature and again at a low temperature. A University of Arizona concept uses a thin, uniform glass face sheet supported at many places by actuators. A Ball Aerospace concept uses relatively small beryllium segments with integral ribs, each supported by position actuators and possessing a radius of curvature actuator as well. A less mature technology is silicon carbide, available in many forms. In principle it can be formed into nearly the right shape before grinding and polishing. All of these materials are brittle and fragile except for beryllium, and beryllium requires special safety procedures for machining and polishing. As a result, all the mirror technology programs have encountered technical difficulties that proved the wisdom of starting as soon as possible. Nevertheless, in my opinion there are many ways to reach the NGST goals and I have little doubt of eventual success.

How can larger mirrors be obtained? No hard boundaries have been encountered with the NGST mirrors, though it gets rapidly more difficult to support a mirror for polishing when it is supported on ever thinner webs and wires. It also gets rapidly more difficult to manage the sag of the mirror under the gravitational force, and to know how to test in a way that demonstrates that good performance will occur in a space environment. Larger mirrors than NGST's 6 m aperture are certainly feasible with NGST technology, but the NGST project had to reduce its requirement from 8 m to 6 m because of the cost and schedule associated with polishing the mirror. It seems likely based on the rate of progress that given a few more years of technology development, an 8 m mirror would be quite affordable (at least for infrared accuracies).

Can more accurate mirrors be made? The NGST experience shows two main problems to overcome. First, the NGST will be polished warm and operated cold, and despite efforts to use a homogeneous material, there have been differential contraction issues. If this is a stable phenomenon, then "cryo-figuring" can be used. Contrary to the implications of its name, this process is actually room-temperature figuring intended to polish in the negative of the measured errors of a cold mirror. It seems to be effective for the configurations so far explored for NGST, in the sense that the mirrors are stable and repeatable enough to be worth this effort. The second problem is that the mirror structures are too thin to withstand much polishing force, so various tricks must be used to protect them. These may in turn produce some distortions. One non-contacting final step is ion figuring, available from several sources.

If UV-quality mirrors cannot be made directly, then it is certainly feasible to build a deformable mirror to operate at an image of the imperfect mirror, usually the primary. This has been done many times for ground-based telescopes and has been explored for NGST. The problem is more difficult for NGST than for a warm telescope because of the change of piezo-electric coefficients of typical materials. Given enough actuators it is possible to correct for a very large number of bumps on the primary mirror and achieve coronagraphic performance. This was considered for NGST and studied in depth by John Trauger but was not part of the central NGST scientific program.

4. NGST Wavefront Sensing and Control

At the beginning of the NGST program it was thought to be highly risky to depend on wavefront sensing and control in space, although it had already been shown that some measurement algorithms gave perfect results when applied to the HST figure errors. Simulations have now been done which show that we can align the segments coarsely by an automated sequence using star images, and then determine relative piston errors using a "dispersed fringe sensor." This device is a small dispersive spectrometer, and a piston error between two adjacent mirror segments produces a stellar spectrum modulated by a cosine function of the piston error. The final step is to use phase retrieval based on the differences between star images taken at several different focus positions. This is the algorithm demonstrated on the HST mirror and now generalized to work well for a segmented system.

5. Off-axis Elliptical Monolithic Mirror

One concept considered for NGST was a telescope using an off-axis elliptical monolithic mirror. It would be launched with the mirror standing on end in the rocket shroud, and would have the obvious advantage that it is guaranteed to have a continuous surface, eliminating two of the three steps in wavefront sensing and control. It would also be possible to produce a much more perfect diffraction pattern, if the mirror were used in an off-axis configuration, so that there would be no shadowing or segment boundaries. This configuration was not popular with the NGST science team, because the strongest scientific driver was mirror area and not image quality. The science team also did not like an elliptical point spread function. There were no strong compensating engineering advantages, although one minor factor was that it might be easier to build a stiff secondary mirror tower. The mirror shape would be compatible with all of the main NGST mirror technologies, as all could allow for joining small segments to make a large piece before final polishing.

6. Contamination Control

In the early days of the NGST studies, many project engineers expressed strong desires that the wavelength range should not extend into the ultraviolet. It has taken a large effort and led to large costs to keep mirror surfaces clean in the past. The major hazard is the polymerization of hydrocarbons by solar UV radiation. A solution is to build it clean, bake it out, and bag it until launch, maybe even inside the rocket shroud. Then, in flight, keep the mirror warm compared with the condensation temperature of organic vapors, and keep it away from sunshine or the Earth's albedo. One good location is clearly the L2 point of NGST. If astronauts and the Shuttle are used to assemble the UV telescope, they should all be kept very clean as well, and they should do their work out of the sunlight.

It is worth considering on-orbit cleaning with ion beams or liquid CO_2. A test of a CO_2 cleaning system has been prepared for flight on the Shuttle. A more radical approach is to develop on-orbit recoating, but this might only work for one or two surfaces. If the reflective properties of truly bare metal are required for the needed wavelength range, this might be a compelling option.

7. Better Pointing

NGST and HST have about the same stability and angular resolution requirements, but a future UV telescope might be asked to do much better. The guide star acquisition problem is not all that simple, in that the potentially numerous guide stars are mostly red and faint. For NGST it is expected that a field of view of the order of 8 square arcmin is required to find enough 17th magnitude stars. This result depends critically on the assumed bandwidth of the pointing control system, and requires excellent suppression of vibrations by other means. As the sensor noise changes only slowly with brightness, inversely as the square root of the star photon count measured in a sampling time of about the inverse of the control bandwidth, there are no very clever solutions available except to find bright stars and suppress vibration.

8. Configuration Issues – Orbit

There are four basic orbit choices for a new UV telescope. First, the Low Earth orbit of HST offers the advantage of relatively easy access for servicing, but has penalties for the rapidly changing Earth - Sun geometry, and the Earth albedo which influences contamination control as well as temperature. Second, a polar sun-synchronous low Earth orbit like that of the Cosmic Background Explorer (COBE) is available, although we do not currently have a launch pad for Shuttles into polar orbit. This orbit offers a much more stable thermal environment, and avoids Earth eclipse for most of the year (depending on altitude). Third, a high Earth orbit, such as that used by the IUE and ISO, provides a thermally stable environment and avoids most albedo effects. Return for servicing in low Earth orbit is relatively easy. Finally, a deep space orbit, such as that used by NGST, MAP, Planck, Herschel, and SIRTF, is thermally stable, but it is difficult or impossible to retrieve the payload for servicing. Data rate capabilities are adequate for NGST with the L2 orbit, or even in solar orbit with a higher gain antenna for SIRTF, because typical exposure times are long.

9. Configuration Issues – Tube or Naked?

The NGST will be the first large optical space telescope to be launched without a protective tube. For NGST this is ideal because it allows radiative cooling to $\sim 20 - 30\,\mathrm{K}$. However, as a secondary result, the NGST mirror will have about 0.2% of its area covered with pits from micrometeoroids in 10 years. This is not thought to be a problem for the NGST, as little light is lost, and the scattered light is widely distributed so halos around bright stars will not be much affected. These pits may, however, increase the amount of stray light emitted by the sun shield that is scattered into the telescope line of sight.

For a UV telescope, there is no current requirement to be cold, so there is no reason not to have a shield or tube. The tube could help control the temperature of the optics for stability, by reducing the total heat radiated to outer space. For example, an area of 50 m^2 at 300 K with an emissivity of 0.05 radiates 1250 W. Therefore the main issue is probably thermal control. Heat is readily available but must be regulated carefully with a large number of local controllers and high passive conductance. As the system orientation with respect to the Sun will affect the boundary conditions, it may be best to suppress all direct heat input and provide it all electrically. Even the solar flux changes 7% peak-to-peak over the course of a year (it is brightest in winter). There are many ways to erect a baffle tube of plastic film on a frame or cable, as is used for NGST.

10. Servicing and Robotics

As noted before, the only currently available servicing capability is with the Space Shuttle at low altitude. However, many other organizations with different objectives and sources of funds are developing tele-robotics, including the defense establishment, medical technologists, and other NASA projects like the Space Station. There are also many applications in hazardous terrestrial locations such as underground pipes, oil wells, mines, and underwater. Even farming robots are under development. I predict that the difficulties preventing wide application of tele-robotics to space astronomy will finally be overcome. In consequence, even if it is not ready for NGST, I think it can be considered for maintaining a future UV observatory.

Many robotic capabilities were considered for NGST and its successors. A study was done by Ron Muller (1999), showing how an assembly robot could be built with a version of the robotic arms already made for the Space Shuttle and Space Station. Such a robot would have a central hub and two arms, each with manipulators, grabbers, and electrical connections. The robot could walk around the spacecraft, attaching itself to footholds and doing work by pre-programmed sequence. Experience suggests that humans in the control loop are not effective unless the round-trip time delay is a small fraction of a second, which nearly rules out humans in the loop at L2 for now. There could be a second robot for reliability.

Its advantages include allowing more efficient use of the rocket shroud volume. Robots can be clean and can operate in a cryogenic environment. There are then a much wider range of possible configurations for deployable systems, and there is the possibility of a flexible response to a failure. Spare parts can be carried along with the payload. For systems requiring complex adjustments (like interferometers), the flexibility of a robot might avoid the need for a very large number of specific actuators. Robots can operate near delicate optics, because they (presumably) do not get tired or impatient, and can follow pre-programmed sequences.

11. Better Detectors

Some planners believe that the optical technology must be attacked first before detectors are built, but I think the opposite. I think detector technology has an extremely long lead time and that huge improvements are possible given time for new ideas to be tried. Detector requirements drive the mission configuration - for instance, are the detectors cold? Do they need a cold telescope? Do they provide energy resolution? What kind of electronics do they need? What data rate do they produce? Do they inspire new scientific goals?

Detector investments can have a very high benefit per unit cost, much more than improvements in mirror area, e.g, but they need tens of $M to make progress. There is no military requirement for UV detectors of astronomical grade, so we only incidentally benefit from the huge investment in semiconductor technology. The NGST project successfully held competitions with substantial funds, and now has two good sources for near IR detectors. Moore's law works, but only if there is strong demand and reward for progress. In the IR, we have seen that the doubling time for detector speed over the past 50 years is only about a year. Let's try to do the same with the UV!

12. Instrument Technology

I don't know enough about this area to say much, but I do think that two types of instruments are worth pushing forwards. Clearly coronagraphs of several types exist, and were studied for NGST. The rotation shearing interferometer suppresses on-axis radiation, and is good for looking close to a star. A more standard coronagraph with multiple Lyot stops and a wavefront corrector was studied by J. Trauger for NGST. According to that study, a good coronagraph can cope with a segmented primary, although at some penalty in performance, using a matching segmented apodizing mask.

I also think that the multi-object spectroscopy developed so highly for ground-based telescopes will be essential. NGST has selected micro-shutter arrays to work with an imaging spectrograph and select around 1000 objects for simultaneous spectroscopy. The shutters will clearly work for UV studies as well.

13. Conclusions and Recommendations

Be ambitious: the time frame is long and the competition (for funds and scientific discoveries) is strong. Be poetic: the scientific goals must be as inspiring as the technical challenges are difficult. Start with the top payoff items: detector development needs to be supported well. Finally, don't give up!

Acknowledgments. The NGST project benefits strongly from the support of top NASA management, beginning with the statement made by Dan Goldin, the NASA Administrator, to the American Astronomical Society in 1996. He said the "HST and Beyond" report was much too cautious in requesting a 4 m NGST, and said NASA would build a much larger NGST. The NGST also benefits greatly from the work of Alan Dressler, who took the "HST and Beyond" report to Dan Goldin and established a relationship of trust. The GSFC ISIM study received scientific leadership from Matt Greenhouse, who is also the main point of contact for the instrument studies funded by NASA.

References

Dressler, A., et al. 1996, Exploration and the Search for Origins: A Vision for Ultraviolet-Optical-Infrared Space Astronomy, Report of the HST & Beyond Committee, (Washington, D.C.: AURA), on the web at http://ngst.gsfc.nasa.gov/project/bin/HST_Beyond.PDF

Muller, R. 1999, Robotic Assembly of a 20-Meter Space Telescope With Extension to 40-Meter Telescopes, on the web at http://www.ngst.nasa.gov/public/unconfigured/doc_0669/rev_01/robassm.pdf

Stiavelli, M., Stockman, H., & Burg, R. 1997, The Next Generation Space Telescope Design Reference Mission, ST-ECF Newsletter, 24 ,4, on the web at http://ecf.hq.eso.org/newsletter/stecf-nl-24/

Stockman, H.S., et al. 1997, The Next Generation Space Telescope: Visiting a Time When Galaxies Were Young, (Washington, D.C.: AURA), on the web at http://oposite.stsci.edu/ngst/initial-study/

Hubble's Science Legacy: Future Optical-Ultraviolet Astronomy from Space
ASP Conference Series, Vol. 291, 2003
K.R. Sembach, J.C. Blades, G.D. Illingworth, R.C. Kennicutt, Jr. (eds.)

HST++ Summary

Jeremy Mould

National Optical Astronomy Observatory, 950 N. Cherry Ave., Tucson, AZ 85726-6732

Abstract. At this meeting HST++ has earned its place on NASA's Origins theme 2003 roadmap as the large UV/optical mission. The strategic high ground HST++ should occupy is wide-field high-resolution imaging and ultraviolet spectroscopy. Key projects include the evolution of the intergalactic medium, the first generation of stars in the Local Group, and the expansion of the Universe.

1. The Discovery Space for HST++

The first speakers at this meeting stressed to us that the astrophysical landscape is sure to change by the time of the launch of the Hubble Space Telescope's successor. A clear vision of the discovery space of a new mission complements the quest for key projects in an important way. Dimensions of the discovery space include:

- ultraviolet spectroscopy 2.5 mag beyond the Cosmic Origins Spectrograph (COS)
- optical imaging 2.5 mag beyond the Advanced Camera for Surveys (ACS)
- spatial resolution of a few milli-arcseconds
- a field of view larger than ACS, sampled with at least a billion pixels

This description assumes an 8 meter effective diameter collecting area. Instrumental gains beyond this are possible, if science drivers are identified, and detector development (for example) is pursued. Discussion suggested that the extraordinary richness of the atomic spectrum in the ultraviolet was an additional feature of the discovery space.

2. The Astrophysics Facility Landscape in 2015

HST++ will complement facilities such as the Atacama Large Millimeter Array (ALMA), the Next Generation Space Telescope (NGST), the Giant Segmented Mirror Telescope (GSMT), and the Space Interferometry Mission (SIM). The scientific emphasis of the first three of these will be galaxy formation in the young Universe and star and planetary system formation. The SIM mission (and GAIA) will measure fundamental astronomical distances and motions, and lay down an accurate map of our Galaxy.

An ultraviolet/optical mission in the 2015 time frame would be very complementary with these facilities, and fill a major void in our understanding of the evolution of the intergalactic medium (IGM), including

- determining the equation of state of the baryonic Universe
- surveying hot gas (10^5– 10^7 K)
- measuring its large scale structure
- discovering how galaxies interact with this environment
- tracing the chemical evolution of both light elements, such as deuterium, and heavy elements in all their phases.

A wide field optical instrument could have a major impact on cosmology by studying

- weak lensing
- supernova cosmology

However, the 2000 decadal survey *Astronomy and Astrophysics in the New Millennium* expects these problems to be substantially addressed from the ground by means of the Large Synoptic Survey Telescope (LSST) , which is slated to furnish a digital survey of the sky each week. Its science drivers include:

- a weak lensing mass map of the Universe
- 100,000 supernovae per year in the redshift interval 1–2
- a survey of Earth crossing asteroids
- 10,000 primordial trans-Neptunian objects

3. "Killer Apps" for HST++

3.1. The Intergalactic Medium

We were encouraged on the first day to find some "killer applications" for the new mission. A good case was put to us in a series of presentations for identifying the large scale structure of the IGM at $z < 1.65$. The outcomes would include:

- complete knowledge of the evolution of Lyα forest and its large scale structure. This would require many QSO spectra to give many sight lines through the IGM. The new mission would fill the void in our understanding at redshifts less than the critical value (1.65) for Lyα to enter the window in which the UV penetrates to the ground

- tracing the evolution of damped Lyα systems to low redshifts. These systems, thought to evolve into normal galaxies, are the major baryon concentrations for all redshifts.

- understanding the chemical evolution of the IGM to the present day

- the correlation of this Lyα cloud structure to galaxies

The redshift survey required to accomplish the last of these goals could effectively be executed from the ground by the next generation multi object spectrograph (Barden 2002).

3.2. Understanding the Galaxy's Hot Corona

There is a strong suggestion from highly ionized clouds, and from the physics of high velocity clouds and the Magellanic Stream, that the Galaxy is bathed in a hot corona of low density gas at temperatures at $10^5 - 10^7$ K. Indeed, this phase of the IGM may be pervasive and amount to a global baryon reservoir.

To study the Galaxy in sufficient detail to establish this model requires that we maximize the number of sight lines for absorber spectroscopy and the spectroscopic resolution with which these sight lines are probed. This discovery of the Galaxy's environment and its ecology seems from the discussion at this meeting to be a second killer application for HST++. Its requirements on the mission are broadly compatible with the first, but spectroscopic resolution trades would need to be made carefully.

4. Stellar Populations "Killer Apps" for HST++

Stellar populations have proved a rewarding field with HST. There is a tendency, however, to try to win the war one island (or galaxy) at a time.

4.1. Galaxy Formation

A clear theme, such as galaxy formation needs to be developed and strategic outcomes targeted, such as

- the quest for the predicted dark halos which were unfilled or lightly filled with baryons

- the nature of the first generation of stars and its initial mass function (IMF)

- forming stellar halos by accretion

Each of these areas calls for wide field imaging with resolution several times that of HST. The frequency and characteristics of star streams like that discovered by Ibata et al. (2001) in M31 will teach us a great deal about stellar halo formation. The resolution of HST++ would open much of the Local Supercluster to such studies.

4.2. The Hubble Constant

Cepheids in Coma will be accessible to HST++. Coma's peculiar velocity has been estimated at ~1% of its redshift, thus cutting secondary distance indicators out of the classical extragalactic distance scale (see Kennicutt, Mould, & Freedman 1995). If HST++ can extend its Cepheid distance measurements to a few times the Virgo distance, the SN Ia zero-point will be verified within 5 years with a sample of 50 or more SN Ia hosts.

With the aid of SIM to accurately define the Cepheid zero-point, the error budget $\delta H/H_0$ can be reduced to approximately 2%. The principal outcome of this work will be a constraint, not so much on the age of the Universe, which will have been measured with Planck, as much as on the the dark energy. This results from measuring both Ωh^2 from the cosmic microwave background, and h separately. Together these should yield a determination of the equation of state of the Universe, $\delta w \approx 0.01$. The HST Key Project aimed to find how fast the Universe is expanding; HST++ can tell us *why* the Universe is expanding.

5. The Relation Between HST++ and GSMT

5.1. Where HST++ Will Complement GSMT

There are many instances to illustrate how the legacy of HST will be extended by the combined efforts of the next generation of telescopes.

- Star formation histories in Virgo can be studied in Main Sequence Population I (to 10^8 years) with HST++, and in its intermediate age and old population on the Asymptotic Giant Branch ($10^8 - 10^{10}$ years) with GSMT.

- The Universe at $z = 6$. The molecular clouds will be studied with ALMA; the star formation rate from Hα with NGST. The kinematics of galaxy formation and assembly will require GSMT. Finally, the Lyman continuum is a subject for HST++.

- The large scale structure of the IGM at $z < 1.65$. HST++ will probe the intergalactic medium densely with many lines of sight. The relationship between the structure in the Lyα forest and the galaxies will require a deep redshift survey at optical wavelengths. GSMT can provide that survey with its seeing-limited multi-object spectrograph.

5.2. Where GSMT Will Supersede HST

There are also instances where GSMT can complete work commenced with HST. An example is the evolution of the fundamental plane. The evolution of M/L from z = 0.5 to 1.5 has been clearly delineated by van Dokkum et al. (2001) and Kelson et al. (2000a,b). Structural parameters have been from HST and kinematics from Keck. Extension of this work to $z = 2$ to 3.5, where evolution is fastest can be accomplished observing both structure and kinematics from GSMT, using the conventional Mg lines at 517 nm redshifted to $1.6 - 2.3\mu$.

5.3. Extending the Stellar Populations Laboratory

Detailed star formation histories will be accessible with HST++ for M31 and for galaxies in the nearest groups. The prototype work is that of Holtzman et al. (1996) with HST/WFPC2 on the star formation history of the LMC. In all Local Group galaxies, the full star formation history is on display. Their color-magnitude diagrams (CMDs) show the number of stars formed each Gyr in the Hertzsprung gap. One simply has to count the stars as they cross. A basic assumption here is that the IMF is robust. This is verified by HST studies of local dwarfs, although one should be wary that it may be inapplicable to Population III.

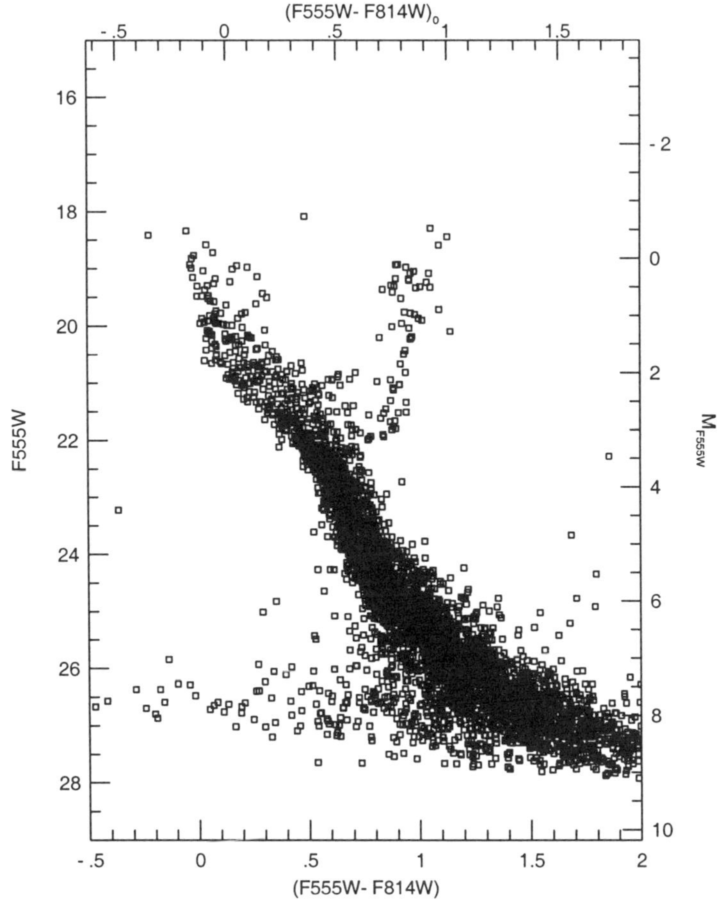

Figure 1. WFPC2 CMD of a field in the LMC from Holtzman et al. (1996). Information on the star formation history is provided by the stars crossing the Hertzsprung gap between the main sequence and giant branch.

6. The Importance of Resolution for HST++

An 8 meter HST++ does not cross any critical resolution thresholds for galactic nuclei. However, it is worth keeping in mind that the resolution standard for astronomical facilities will rise in the next decade.

6.1. Resolving Stars

There are a number of interesting applications of this resolution in stellar astrophysics.

- the hot O VI inner disk in CVs
- surface features of cool stars
- globular cluster cores
- star formation

Star formation applications are addressed in Beckwith's review (this volume).

6.2. Black Hole Environments

Probing the physics of black hole environments may call for comparable resolution from the ultraviolet to millimeter wavelengths.

Table 1. Resolution Comparison

Observatory	Resolution	λ
HST++ (4 m)	10 mas	200 nm
SIM	10 mas	600 nm
GSMT	10 mas	1.6 μm
NGST	300 mas	10 μm
ALMA	10 mas	1 mm

7. Organization

There is much to be gained from a coordinated ground-based science co-program. This should be designed at the outset. The recent Committee of Research in Astronomy & Astrophysics emphasized such coordination (Augustine et al. 2001). The new National Astronomy and Astrophysics Committee is intended to encourage the necessary NSF/NASA cooperation. Internationally, the Inter-Agency Consultative Group can assist by effecting NASA/ESA cooperation.

References

Augustine, N., et al. 2001, U.S. Astronomy & Astrophysics, Managing an Integrated Program, Washington: National Academy Press (in press), http://www.nas.edu/bpa/projects/brp

Barden, S. 2002, in ASP Conf. Ser., Wide Field Multi-Object Spectroscopy, eds. M. Brown & A. Dey, (San Francisco: ASP), in press

Holtzman, J., et al. 1996, ApJ, 466, 742

Ibata, R., Irwin, M., Lewis, G., Ferguson, A., & Tanvir, N. 2001, Nature, 412, 49

Kelson, D., Illingworth, G., van Dokkum, P., & Franx, M. 2000a, ApJ, 531, 137

Kelson, D., Illingworth, G., van Dokkum, P., & Franx, M. 2000b, ApJ, 531, 159

Kennicutt, R.C., Mould, J. & Freedman, W. 1995, AJ, 110, 1476

van Dokkum, P., Franx, M., Kelson, D., & Illingworth, G. 2001, ApJ, 553, L39

Hubble's Science Legacy: Future Optical-Ultraviolet Astronomy from Space
ASP Conference Series, Vol. 291, 2003
K.R. Sembach, J.C. Blades, G.D. Illingworth, R.C. Kennicutt, Jr. (eds.)

The Local Universe: The Emergence of Life

Steven V. W. Beckwith

Space Telescope Science Institute, 3700 San Martin Dr., Baltimore, MD 21218

Abstract. An observatory-class optical/UV telescope with a large aperture emphasizing better angular resolution for imaging and good contrast to study the environments of nearby stars is a natural successor for the Hubble Space Telescope (HST). The main scientific driver for this capability will be to understand the steps leading to the creation of other planets and the emergence of life in the Universe. The intellectual appeal of this new direction combined with the broad observational capability allowed by space observatories, make UV/optical successors to HST more valuable than special purpose missions targeted at a single scientific objective. Just as HST has proven to be a dominent resource for almost all fields of astronomy in the last 10 years, its successor will be an essential tool used to study phenomena that we cannot yet predict. By providing the best resolution at optical wavelengths, traditionally the most important in observing the Universe, it will be an essential comple ment to the other facilities we will have in the next decade: ALMA, NGST, and ground-based facilities such as GSMT and LSST.

1. Introduction

The Hubble Space Telescope (HST) was planned as a general purpose facility. It had a few obvious science drivers that motivated the construction of a space observatory, most notably the resolution of stars in nearby galaxies to measure the distance scale with Cepheid variables in the Virgo cluster. However, its real importance greatly transcends the few science areas for which it was built, and that was expected at the time it was planned (Spitzer 1990). Some of its most important uses include the study of the distant universe through the Hubble Deep Field and the use of SN Ia as standard candles to determine that the Universe is accelerating. It has had major impacts in almost every area of astronomy from observations of the planets in the Solar System to cosmology.

It is now inconceivable that we will not maintain a high-resolution optical/UV telescope in space. But justfication for any new space facility normally requires one or two "killer-applications," specific problems that are enabled by that facility, mainly because NASA understandably takes a mission-oriented approach to funding science. What will be the killer-aps for HST's successor, which we refer to as HST++?

HST was built when the thrust of astronomy was to expand the boundaries of the observable universe. Cosmology was the dominant intellectual force behind this exploration. Research aimed to reduce the complexity we see around us to a few simple rules – the physical laws – with a few parameters to explain the richness evident in any deep photograph of the heavens. Hubble's justification was to study known objects at very high angular resolution to test our theories of how these objects came about and behaved as part of the grand scheme. Although not primarily designed to observe the most distant stars and galaxies, HST became the principle tool to find them when we realized that they were compact and hence more amenable to observation with a medium-sized space telescope than a large, seeing-limited ground-based telescope.

This emphasis on cosmology may well change over the next decade. The Next Generation Space Telescope (NGST) is designed to bring this quest for more distant objects to an end, and the Giant Segemented Mirror Telescope (GSMT) will provide spectral follow-up on essentially all the distant targets discovered by NGST. Two satellites studying the cosmic microwave background, MAP and Planck, will observe the most distant radiation in very fine detail too. To be sure, we will not understand the big picture until we learn the nature of dark matter and dark energy, perhaps through laboratory experiments. However, if the standard big bang model of the Universe continues to gain support from new observations, we will improve our quantitative understanding of observational cosmology but not our conceptual view of the how it works. Conceptual mysteries of the Universe might then come from other areas.

The new mystery will be the laws that govern emergent phenomena, particularly the genesis of life and the evolution of sentient beings from inorganic matter. An emergent phenomenon is one whose character is so different from the characteristics of its parts that we cannot explain it easily in the standard reductionist terms of physics – at least it is not obvious how to do so now. Life is a good example of emergence. Intelligence and self-awareness are also good examples. We do not even agree on a conceptual framework for how life emerged on Earth. We cannot decide if it was inevitable, and hence common in the Universe, or unusual, perhaps a one-time occurrence. The nature of self-awareness is still a question better suited to philosophers than scientists, a situation much like that of cosmology a few hundred years ago. Some emergent phenomena, such as those arising in the study of cellular automata, are seen in many different areas of science, suggesting that the underlying rules do not depend on the laws governing elementary particles or the structure of space-time (Anderson 2002).

Finding other examples of life outside of the Solar System may be the most important advance in the understanding of its emergence. It is certainly one of the most interesting questions to lay persons. The advance of technology means that the search for extrasolar life is entering the realm of astronomy. This search and the associated research into the origins of planetary systems may make one of the most fundamental shifts in astronomical inquiry since the realization long ago that the understanding of the Universe's origins may also be within our grasp.

2. Scientific Questions Associated with Emergence of Life

If life is the ultimate example of an emergent phenomenon, it is not the only one. There are many steps leading up to the creation of planetary systems and their surface evolution that play a role in bringing about life. We will want to understand these processes in some detail to learn whether our situation is common or rare. Just as the observations of fully formed galaxies do not tell us how they came about, the observation of life on other planets will lead to similar questions about its origin, universality, and ubiquity.

There are at least two general lines of inquiry that an HST successor will need to follow. The first concerns gathering the scientific knowledge needed to identify life on other planets when they can be observed directly. Here, a good understanding of atmospheric physics gleaned from studying the atmospheres of Solar System objects will be essential to interpreting data on exo-planet atmospheres. That understanding will require resolutions of a few km at the distances to the outer planets, as stressed by McGrath (this volume) in her summary of the Solar System science to be done with HST++. Good resolution, not high contrast, is the important observational requirement, and modest spectral resolution will be important to allow studies of atmospheric constituents. Many of the important atomic and molecular lines used to diagnose atmospheric physics appear in the ultraviolet.

The second line of inquiry concerns the creation of planets and how they shape the environment for life. Does the richness of the interstellar medium automatically lead to the creation of life-bearing planets (Drake 1962)? Or are the events that led to the creation of the Earth so unusual, and the subsequent evolution of intelligence so unlikely, that we are rare if not unique in the Universe (Ward & Brownlee 2000)? It is not obvious that these questions are most easily addressed by hunting for life and simply looking at the statistics or by understanding enough about the steps along the way to make reasonable extrapolations to the number of life-bearing planets in the Galaxy.

We will want to study the environments around nearby star systems with enough resolution and dynamic range to see large planets, planetary debris such as our zodiacal light, and any additional matter generated by the stars. The kind of resolution and dynamic range needed to make images of terrestrial exoplanets orbiting other stars will certainly be adequate. More importantly, many of the important goals – studying the giant planets and the planetary debris – will be satisfied by telescopes with far less ambitious technical goals than the Terrestrial Planet Finder (TPF).

Studies of nascent planetary systems will be aided greatly by an HST++ working in concert with the Atacama Large Millimeter Array (ALMA). ALMA will be able to study the cold circumstellar matter with sub-second of arc resolution and sufficient spectral resolution to map velocity fields. HST++ will be more sensitive to large clumps of matter, planetesimals, and energetic phenomena such as collimated winds. High angular resolution and good (but not superb) dynamic range are prerequisites, but not high sensitivity. Full imaging capability will almost certainly be required especially to map extended light from circumstellar disks and long jets. Optical and ultraviolet lines for use as diagnostics will be important to these studies too.

Finally, the history of the Solar System will be a necessary part of the quest to understand other planetary systems. Understanding the Kuiper Belt, its objects and extent will be a focus of future Solar System research, and a successor to HST must have the capability to observe distant objects within the Solar System.

2.1. The Importance of Having an Optical Observatory in Space

Of the top ten most important findings from the HST program, about half of them could not have been predicted prior to its launch. Table 1 lists these topics. Traditionally, we have maintained observatories because some of the more important science requires general observational capability that cannot be foreseen at the time of construction. This science is often not discovery of new classes of objects, as outlined in Martin Harwit's book, *Cosmic Discovery.* Rather, it follows from new surveys of the sky or unexpected results from other observatories.

Table 1. Hubble's Top Ten

Rank	Topic	Program Size	Expected?
1	Galaxy formation (HDF)	Large (DD)	No
2	Distance Scale (H_0 KP)	Large	Yes
3	Giant Black Holes	Medium	Yes
4	Active Galaxy gas	Medium	Yes
5	Quasar Absorption Lines (KP)	Large	Yes
6	Accelerating Universe	Large	No
7	ISM Chemistry (KP)	Large	Yes
8	Gamma Ray Bursts	Medium/Large	No
9	Late stellar evolution	Medium?	Yes
10	Young star jets & disks	Small	No

Several recent topics highlight the need for continuous optical observational capability: measurements of universal acceleration using SN Ia as standard candles (Riess et al. 1998; Perlmutter et al. 1999), the study of gamma ray-burst sources, the imact of comet Shoemaker-Levy on Jupiter, size measurements of Kuiper belt objects, and the study of the most distant galaxies from the Hubble deep field.

The use of SN Ia as standard candles to measure the rate change of universal expansion relies on HST's resolution and sensitivity for the most distant objects. At redshifts greater than about 1.2, HST is uniquely capable of providing the infrared photometry needed to use SN Ia as standard candles. The discovery that the Universe is accelerating, and thus the support for a cosmological constant, may be one of the most important findings of the 20th century. Although not built with this science in mind, HST's capabilities enabled this field to leap forward quickly, once the effect was evident. In retrospect, HST's ability to measure distant supernovae alone could justify having a space telescope available at all times, and one possible HST successor would be a wide-field optical/IR telescope that could discover SN Ia at 10 to 100 times the present rate.

Gamma-ray bursters (GRBs) have been known since the late 1960s, but it was only in the late 1990s that gamma and X-ray satellites acquired the ability to determine the location of the bursts with reasonable accuracy. Once that occured, HST was needed to understand the location of the bursts in distant galaxies and the morphology of those galaxies. Simply having the observational capability available was essential to moving this field forward too.

The Hubble Deep Field turned up a large sample of high redshift galaxies that differ in appearance from local galaxies. Images of these galaxies with HST opened up the study of the early universe. Because these young galaxies are small, HST is still uniquely capable of providing useful images of the highest redshift objects we can detect. Not only was the telescope not designed to do this kind of science, early predictions actually stated that HST *could not* do this kind of science (Bahcall, Guhathakurta, & Schneider 1990). Despite this pessimism, the Hubble Deep Field has had more impact on astronomy than any other single observation from HST. Once again, it is the existence of HST's unique imaging capability that allowed this science to progress, but its importance was not foreseen.

There are several science areas enabled by a large space telescope that could be as important to our understanding of the Universe as the examples given above, although they are presently too speculative to provide the sole justification for a new mission. Potentially, the most interesting is a time-variable fine structure constant, $\alpha = 2\pi e^2/hc$. There are hints of a variation in α from quasar absorption line spectra at redshifts around 1 (Murphy et al. 2001; Webb et al. 2001), and very good limits on variations in terrestrial samples covering redshifts less than about 0.1 (e.g., Fujii et al. 2000). If α varies at high redshift, it will be essential to measure its value at low redshifts too, to understand the form of the variability, requiring ultraviolet observations with a telescope larger than HST (Williams 2002, personal communication).

A large space telescope would be able to study the distribution of baryons in the intergalactic medium (Shull et al. 1999); it would be the tool of choice to study microlensing by objects in the Galaxy, such as planets (Sahu et al. 2001; 2002), or in the distant universe; it could determine the stellar populations in nearby galaxies such as M31; and it could study the nature of close binary stars. Just as HST, it could provide excellent astrometric capability. These are a few of the scientific topics that are obvious, but just as with HST the most important are almost certainly beyond our imagination right now.

It is important for astronomy to have HST-like capability in the future to enable this kind of unforeseen science. A space telescope will provide the highest angular resolution at optical wavelengths that will be available to astronomers; it provides unique access to the ultraviolet portion of the spectrum; and it provides the best possible sensitivity to small objects at wavelengths crippled by high backgrounds on Earth at 0.7 - 2 μm. The wavelength range between the near ultraviolet and the near infrared has historically been the most important in the spectrum for gathering astronomical information – primarily because stars put out most of their energy in this band – and it is inconceivable that we would not have continuing coverage of this band with space telescopes in the future.

2.2. What is HST++?

A 10 m-class, filled-aperture optical coronagraph would enable all of the science discussed above, give us a huge increase in observational capability over HST, and satisfy the requirements of TPF if sufficient contrast could be achieved. This kind of facility represents the ideal next step after HST and NGST. Such a telescope is technically ambitious, however, and it is important to recognize that several intermediate steps would also enable much of the science while at the same time providing a series of technical challenges allowing us to build up to the ideal telescope. These are outlined in Table 2.

Table 2. Hubble Successors

Description	Size	Resolution	Science
UV Light Bucket	5 m	$\frac{\Delta\lambda}{\lambda} \geq 10^3$	$\alpha(t)$, IGM baryons disk chemistry
Optical Imager	5 m	$\Delta\theta \leq 20\,\mathrm{mas}$ contrast$\geq 10^6$	planetary atmospheres, circumstellar disks, exoplanet eclipses
Optical Coronagraph	5 m	$\Delta\theta \leq 20\,\mathrm{mas}$; contrast$\geq 10^8$	Image large exoplanets, young exoplanets, stellar populations, lensing
Optical Coronagraph	10 m	contrast $> 10^{10}$	TPF + other science

Each step brings in a new scientific area while providing a technical challenge along the road to a telescope that can image terrestrial-size exoplanets directly. As the contrast and resolution increase, we get an increasingly detailed look at the environments around nearby stars: disks, exo-zody light, giant planets, terrestrial-size planets. Each step also provides a general purpose observatory that enables astronomical observations of the sort that have made HST so successful. Each one will additionally bring in unique observational capabilities that are not foreseen from ground-based telescopes and thus preserve the realm for exploration of new problems just as new problems have contributed to HST's success, as Harwit stresses in his contribution to this conference.

References

Anderson, P.W. 2002, Physics Today, 55, 7, 56

Bahcall, J.N., Guhathakurta, P. & Schneider, D.P. 1990, Science, 248, 178

Drake, F.D. 1962, Intelligent Life in Space, (New York: Macmillan)

Fujii, Y, et al. 2000, Nucl. Phys., B573, 377

Harwit, M. 1981, Cosmic Discovery, (New York: Basic Books)

Murphy, M.T., et al. 2001, MNRAS, 327, 1208

Perlmutter, S., et al. 1999, ApJ, 517, 565

Riess, A., et al. 1998, AJ, 116, 1009

Sahu, K.C., et al. 2001, Nature, 411, 1022

Sahu, K.C., Anderson, J., & King, I. 2002, ApJ, 525, L21

Shull, J.M., et al. 1999, http://universe.gsfc.nasa.gov/roadmap/docs/SUVO.pdf

Spitzer, L., Jr. 1990, Astronomical Quarterly, 7, 129

Ward, P. & Brownlee, D. 2000, Rare Earth, (New York: Copernicus, Springer)
Webb, J.K., et al.. 2001, Phys. Rev. Lett., 87, 091301-1

Hubble's Science Legacy: Future Optical-Ultraviolet Astronomy from Space
ASP Conference Series, Vol. 291, 2003
K.R. Sembach, J.C. Blades, G.D. Illingworth, R.C. Kennicutt, Jr. (eds.)

Hubble's Science Legacy: A Technology Summary

J.H. Crocker

Lockheed Martin Space Systems Company, P.O. Box 179, Denver, CO 80201-0179

M. Menzel

Lockheed Martin Space Systems Company, 7474 Greenway Center Dr., Suite 200, Greenbelt, MD 20771

Abstract. This paper provides a summary of the technical discussions presented at the Hubble Science Legacy Workshop. As such, it is a compilation of presentations of the numerous authors on the development of large observatories for both ground and space applications. We have tried to include and expand on those issues, which will be the key drivers for a large optical-ultraviolet space observatory.

1. Introduction

The success of NASA's first generation of Great Observatories was based on the concept of using a multi-spectrum approach. Each of the Observatories was designed to cover a unique part of the electromagnetic spectrum. The Hubble Space Telescope (HST), the first Great Observatory, covered the visible and ultraviolet (UV); later the Compton Gamma Ray Observatory covered the γ-ray spectrum, followed by Chandra for the X-ray spectrum. In early 2003, the Space Infrared Telescope Facility (SIRTF), a cryogenic infrared (IR) telescope in an Earth-trailing orbit, will complete the first generation Great Observatory program.

NGST (the Next Generation Space Telescope) is, one hopes, the first of the next generation of NASA's Great Observatories. The scientific advances of the first Great Observatories are still exciting the public and scientific community alike. The new generation of space observatories will have at least an order of magnitude improvement in performance. A repeated theme has occurred every time the performance of telescopes has improved by a factor of ten. It has held true from Galileo's first telescope in the 1600s to Hubble in the 2000s. We rewrite the textbooks and, more importantly, we rewrite our understanding of the Universe and our place in it.

NGST is designed to study the ultraviolet light from the early Universe that has been redshifted to the infrared. This allows the visible and UV light from the first stars and galaxies to be detected in the infrared.

2. Astrophysical Problems and Future Space Missions

The presentations and discussions of the majority of this conference have highlighted the importance of the following on-going and future areas of astrophysical investigation:

- Cosmic microwave background
- Galaxy formation and assembly and evolution ($1 < z < 3$)
- Measurements of extragalactic dark matter and energy and their distributions
- Stellar populations of the local supercluster
- Active galactic nuclei
- Initial mass function
- Interstellar medium and molecular clouds
- Star and planet formation
- Planetary weather, magnetism, and evolution
- Exploration of Kuiper Belt objects
- Detection and investigation of the atmospheres of Jovian exo-planets
- Detection of terrestrial exo-planets

All of these areas can benefit from the unique advantages of space-based observatories. Both NASA and ESA have current space facilities, such as the HST and the Far Ultraviolet Spectroscopic Explorer (FUSE), actively engaged in this research, and both have formulated straw man concepts for future missions, such as the Next Generation Space Telescope (NGST), the Terrestrial Planet Finder (TPF), and the next generation UV-visible observatory.

Since this conference was concerned primarily with the future UV and visible missions to follow the HST, we shall discuss the requirements and technology needs for these. Several common "technology-driving" requirements become apparent, which can be summarized as follows:

- High sensitivity to attain the required signal-to-noise level of approximately 10 with reasonable observational efficiency
- Spectral coverage at least as far as FUSE is needed to include the important O VI diagnostic spectral line
- Diffraction limited optical quality over fields of view of approximately $10' \times 10'$ is needed to efficiently cover areas necessary for dark matter mapping and other imaging and spectroscopic survey programs

- High contrast imagery (the ability to detect a low luminosity object in proximity to a higher luminosity object) for detecting and analyzing exoplanets, circumstellar discs, or quasar host galaxies

These requirements drive technology specifically in the areas of instrument detectors, large lightweight telescope optics, space structures and deployment mechanisms, and pointing and vibration control. Furthermore, to maximize the science return of these observatories, technology should also address increasing their reliability and operational life and observational efficiency. Such technology development should be directed not only at the flight segment, but also at the ground segments of these systems. Technology should also consider ways to maintain these telescopes via servicing by manned or robotic means.

3. Development Process

Effective technology development has been and will continue to be critical to address the desires of the science community to solve key astrophysical problems with space telescopes. Monetary constraints for these future missions will increase the criticality of effective technology development planning. Identifying and planning the development of key enabling technologies is derived from classical systems engineering practices. Science needs are translated into technical engineering requirements to derive candidate straw man missions and designs that are analyzed for their technical and cost risks. Technology development programs are then formulated to reduce these risks to levels acceptable for full-scale system development. The success of this classical engineering approach relies to a large extent on the specificity to which science needs and their technical requirements can be defined. The science and the engineering communities must therefore strive for a healthy balance in which scientists push for maximum flexibility to accommodate the evolving science environment, and the engineers push for a high degree of requirements definition to enable performance optimization in the presence of constrained resources. These requirements must include not only science driven performance requirements, but also operational and programmatic requirements.

Although this "top down" approach has proven successful, Martin Harwitt points out scientists should be vigilant for opportunities from other development areas. Many key advances for astrophysics have occurred when scientists leveraged technology being developed for other applications without regard to these science problems. Such was the case with radio astronomy that had it roots in the developing radar field, and in the case of gamma ray bursters, which were first detected by military.

4. Space Telescope Uniqueness

The process of designing future space-based astronomical systems must trade these against future and current ground-based systems. Ground facilities are inherently less expensive, more reliable, and more maintainable. Space-based systems, however, can offer unique capabilities unattainable from the ground and yield higher performance. The state of this uniqueness space needs to be continually assessed as both ground and space technologies evolve. Currently space-based systems offer the following unique advantages that define the so-called "uniqueness space:"

- Spectral windows in the UV and near- and mid-infrared unavailable from ground-based observatories
- More efficient faint object detection systems by virtue of their freedom from atmospheric noise and attenuation
- Higher spatial resolution over wider fields of view due to their freedom from atmospheric distortions

The first of these is a fundamental uniqueness determined by atmospheric attenuation. The second of these needs to be considered in more detail. Ground-based telescopes can currently be made larger than space telescopes and compensate for speed loss due to atmospheric attenuation. However, even with smaller apertures, the lack of atmospheric attenuation and noise levels, and the less constrained modes of operations at locations like the Sun-Earth L2 point, enable space observatories to provide higher observational speed and efficiency. The third of these uniqueness areas also requires case-by-case consideration. Advances in adaptive optics (AO), particularly in multi-conjugate AO, have made ground-based observatory resolution very competitive to the resolution of space observatories at wavelengths lower than $0.6\,\mu$m over fields of view of up to a square arc minute. For wavelengths shorter than this, and for larger fields of view, space observatories provide currently unique capability that can leveraged for large survey programs, such as mapping dark matter distributions through gravitational image shear, locating supernovae or other serendipitous distance indicators, or performing spectroscopic surveys.

5. Detectors

Improvements in UV detectors in both performance and cost per pixel are perhaps the technology with the best leverage. Visible and IR detectors have benefited greatly from commercial and military investments. This has driven performance to near theoretical limits and cost per pixel to levels that have allowed the latest generation of Hubble instruments to be able to afford huge gains in detector array size with the corresponding increase in discovery space. UV detector performance improvement has a direct and large impact on observatory cost. Double the sensitivity of a detector and you can halve the collecting area of the telescope for the same sensitivity, dramatically reducing not only the cost of the mirrors but the mass of the observatory. Gains here in sensitivity will pay the highest dividends.

NASA and the science community must invest heavily in UV detectors. The money spent on IR and visible detectors by NASA for the resulting performance is very misleading. The funds were for the most part to package and test devices produced for commercial and military applications. To make progress in UV detectors, we may well need to spend an order of magnitude more than was spent on the visible and IR detectors. We will still gain more performance for a dollar spent on detectors than a dollar spent on better or more mirrors.

In addition to performance and cost, several other technical areas need to be addressed. UV detectors need to be solar blind; otherwise the UV photons are swamped by even small leakage of IR photons. This can be addressed directly in the physics of the detector or through the development of better filters, but addressed it must be. For CCD-based approaches, the radiation damage and resultant lifetime issue needs to be addressed.

New physical approaches for UV detectors are giving hope for better detector performance. These include two-dimensional photo-emissive and photoconductive approaches and three-dimensional energy-resolving detectors.

Detector performance improvements, while likely to provide the highest gain, will also take the longest to achieve. This is due in large part to the lack of drivers from military and civil applications. New semiconductor processes from the next generation of micro-electronics require UV lithography. This may also stimulate the need for better UV detectors, but not on the scope and scale that have driven the short doubling time for visible and IR detectors over the last several decades. Comparable sums of money are unlikely to be available in short time frames. Our approach then to UV detector is to increase funding as much as we can and to perform the development over a longer time. In my experience, usually the detectors drive the schedule and thus the cost of scientific space instruments. Getting new detectors developed, tested, and flown to support the next UV-optical space observatory needs to be started now if we are to hope for the required performance and cost enhancements.

6. Optics

When you build an instrument, it is the detectors that drive cost and schedule. When you build a telescope, it is the optics that drive the cost, schedule, and technical performance. For hundreds of years, the monolithic mirror has both driven and limited telescope aperture and performance. The Keck telescopes on the ground and NGST in space are breaking this paradigm. Small 1-2 meter optics are assembled and actively aligned to form the large collecting area. This allows a much larger aperture telescope to be realized than would ever be possible by scaling up monolithic designs. It also results in lower cost and weight for the final system. The segmentation of space optics allows optics larger than the launcher shroud, limited only by volume and the cleverness of the folding and deployment scheme.

For a UV observatory, we have an interesting dilemma. The next generation UV observatory may be able to achieve outstanding scientific results with a primary of 4 to 4.5 meters in diameter. UV telescopes are more sensitive to the effects of mirror segmentation, and they require higher accuracy of surface figure and smoothness. An important trade for the mirror will be the size and form. To achieve the smoothness required, a monolith is a less costly approach. A circular mirror is then limited to the size of the available launch shroud. We have two choices: bigger shrouds or different shaped mirrors. Both are possible. Early NGST studies demonstrated that shrouds up to about 6 meters in diameter could be developed for use on existing or planned launch vehicles for a cost of a few tens of millions of dollars. A second approach is the use of an elliptical primary, which could also incorporate an off-axis secondary. This would allow optics up to 4×16 meters to be launched on existing or planned systems and would dramatically reduce the issue of diffracted and scattered light from a segmented on-axis system. A 4×16 meter system of this nature would be able to achieve all of the science discussed in this conference.

I think the choice is clear. If a 4 meter circular aperture is acceptable for the science mission, it will require far less time and effort. The technology exists today to construct and polish such a mirror to the required figure and smoothness using existing facilities and techniques. Techniques developed as part of the AMSD program are scalable to the 4 meter class and should provide the mass and surface finish required. The 4×16 meter elliptical mirror appears feasible based on scaling some of these same technologies but will require new facilities for blank production and polishing. While this will be more expensive, it is by no means unprecedented. A new factor was built in Germany by Schott for the production of the four Very Large Telescope 8 meter blanks and extensive new facilities were constructed for the polishing at REOSC.

If the elliptical mirror is limited to about 4×8.5 meters, it is likely that existing facilities can be utilized. It is likely that the cost per unit area of the 4 meter monolith and the $\times 8.5$ meter elliptical mirrors will be comparable. This trade should be driven by the available mission funding and science. Going larger will likely be a step function of several tens of millions of dollars above about 8.5 meters.

A likely driver for the large diameter will be the need and ability to detect planets around other stars. A 4×16 meter telescope may be required to search hundreds of stars for earth-like planets. The small 8 meter class telescope equipped with a coronagraph could explore perhaps fewer than 100 nearby stars.

In all cases, it will be necessary to have the optic be extremely clean to preserve the UV performance. The efforts on HST to clean and keep clean not only the mirrors but also the observatory to prevent self-contamination during launch and operation were a considerable part of the cost drivers. While the techniques certainly exist and have been proven on HST, new technologies have been developed to clean optics late in the production and even on orbit. Ground-based telescopes now routinely use a spray-on peel-off coating to remove dust and some contaminates. Carbon dioxide snow, developed to clean and degrease semiconductors, is used at most major observatories and is very effective in removing even small particles. Several groups have proposed flight experiments to demonstrate cleaning of optics using this technique on orbit.

Using any of the currently available polishing approaches requires a trade-off between polishing time, optic mass, and surface smoothness. A light optic with thin face and support ribs will require a lighter polishing pressure and thus a longer polishing time for the same smoothness. To break this connection requires new approaches to polishing. One approach is to stress the optic and polish it as a sphere, spherical polishing providing smoother and faster results. Another approach is to stress the lap, under computer control, with the same effect. A third approach is called stress strip polishing, which is a combination of the two. In any event, a change in the polishing approach to speed up polishing and smoothness convergence could be a most effective way to reduce cost and schedule for this observatory.

The fundamental system trade for the UV optic for this next great observatory will be a value balance between science capability and cost driven by optic size, smoothness, mass, and cleanliness. A study of science discovery space for these key parameters would be most enlightening.

7. Not Just the Optics

Observatory performance is also limited by the second order effects. Perfect mirrors are quickly degraded by poor system performance. Precise and stable pointing in a low jitter environment will be critical to preserving the exquisite point-spread function of this observatory. For telescopes in the lower end of the range, new state of the art HST-like approaches are capable of achieving the required performance. For large optics with requirements to take advantage of the potential higher resolution, new techniques will be required. Sophisticated isolation systems, both active and passive, to decouple the payload from the spacecraft bus have been proposed and are currently in the laboratory demonstration phase. These systems will be able to eliminate concerns about jitter introduced by reaction wheels, solar array motion, and antenna slews. Combined with a fast steering system, they could enable the UV observatory to be placed in a more convenient Earth orbit. Because this observatory does not require the low temperature of NGST, distant L2-type orbits are not needed.

To minimize the effect of the thermal day/night cycle seen by HST, a Sun-sync or stable high Earth orbit could be employed. A key trade here will be the desirability of access if retrieval or service is needed. It will likely be the case that a 4 meter class observatory can be placed in low earth orbit retrievable by shuttle and have good performance. A larger, 8 meter class system with the stability to detect planets may be more suitable for a more temperature stable Sun-sync or high Earth orbit.

In earth orbit, the effect of stray light from the earth and moon will be more of an issue. Some type of light baffle will be required. This baffle will also serve to keep the temperature of the system more constant. Recent work on inflatable tubes as well as the work on deployable shades for NGST should serve as a starting point. For the smaller 4 meter system, it is likely that a fixed tube will be the least costly approach. Anything large will require some type of deployment system. Much of the NGST work in this area is directly applicable.

8. Wavefront Control

Wavefront control or at least management will be a significant system issue for a large UV telescope. Here again the size of the telescope and the need to detect planets will be a critical drive. Wavefront control consists of two parts, passive and active. The passive part is driven by misalignments and deformation of the optics due to primarily thermal changes. The system solution here is a product of material choices and insulation and the selection of the orbit. The active part is the trade off on bandwidth to sense and maintain the wavefront. For active wavefront control, an external source is usually monitored and adjustments are made to optical alignment and perhaps figure to maintain performance. Planet detection will be the most stressing driver for active wavefront control. This requirement will drive active alignment of the optics and figure correction. The correction can be done either on the primary mirror directly, or by the use of a small corrector deformable mirror located elsewhere. Something on the order of thousands of actuators either on the primary or deformable mirror (DM) will be required for the most stressing planet detection requirements proposed – perhaps a few thousand for the minimum case, a few tens of thousands for the latter. Much effort is being expended on DM technology. While not proposed for NGST due to the extreme cold temperatures, room temperature DMs have been demonstrated and good performance in the 100×100 actuator class achieved. One issue to watch with the DM is the second order effects of dimples introduced by "pinching" caused by non-perfect alignment of the actuators. These affect and quickly degrade performance if not controlled.

9. Summary

In summary, the key architecture and technical issues are dependent on the science requirements as they drive the telescope aperture size. The initial look at the science drivers seems to indicate that most of the science cases can be achieved by a telescope of modest aperture of approximately 4 meters. In this case, a fixed monolith with no deployment of the optics or stray light shield would be the best solution. It appears that the added requirement to detect and observe earth like planets is the scientific driver that would drive the system to larger apertures with more complex packaging and deployment approaches. The intermediate step of going to an elliptical mirror will scale in cost with the area up to about 4×8 meters. Above 8 meters, a significant price step may occur as new facilities are required to fabricate and polish the large optic. This step function in price needs to be considered in the light of the cost of a separate mission, such as TPF, to carry out the planet-finding mission alone. The combination of the two into a single mission could be most attractive.

In any event, the continuation of NASA's approach to a new series of Great Observatories that span the spectrum at increasing sensitivity and resolution will continue the advance of discovery. We truly live in a golden age of astronomy when revolutionary advances occur not once in a lifetime but repeatedly.

Hubble's Science Legacy: Future Optical-Ultraviolet Astronomy from Space
ASP Conference Series, Vol. 291, 2003
K.R. Sembach, J.C. Blades, G.D. Illingworth, R.C. Kennicutt, Jr. (eds.)

Hubble Science Legacy Workshop: Closing Remarks

Hashima Hasan

Office of Space Science, NASA Headquarters, Washington, DC 20546

Abstract. Objectives and expected outcomes of the Workshop are stated. The advisory structure of NASA's Office of Space Science (OSS) is outlined. The Workshop is placed in context of the OSS strategic planning process.

1. Introduction

The objective of the Hubble Science Legacy Workshop was to continue a community dialog started in Boulder, Colorado in 1998 to develop a consensus on scientific priorities in ultraviolet/optical astronomy after de-orbiting of HST in 2010. At this time there is no planned space mission after 2010 to address scientific questions that require astronomical observations in the ultraviolet/optical wavelength range. While the community has the opportunity to compete for small missions in NASA's Explorer and Discovery lines, these missions are designed for targeted science questions, and do not meet the needs of the larger community that are catered to by observatory class missions such at the Hubble Space Telescope (HST).

2. Workshop Outcome

A whole slew of exciting and important science problems that will likely remain after HST were presented at this conference, as were the technological challenges facing the community, as summarized in numerous papers in this Proceedings.

3. Where Do We Go From Here?

I have requested Drs. Garth Illingworth and Rob Kennicutt to set up a small Working Group, the charter of which will be to distill the ideas from this Workshop and other community documents into a coherent White Paper. A preliminary Executive Summary should be submitted to the roadmapping teams of NASA's Origins Subcommittee (OS) and Structure and Evolution of the Universe Subcommittee (SEUS) which are currently preparing inputs to the Office of Space Science (OSS) strategic plan.

The White Paper should address the following broad areas:

- Identify major science questions in ultraviolet/optical astronomy
  - What is the *compelling* science that would re-write textbooks the way HST science has done?
  - Does this science open new research windows not achieved by an approved mission in the OSS strategic plan? How do we go beyond Terrestrial Planet Finder (TPF) science?
  - Will the science excite the public?
- Identify the technology development required to achieve the above scientific goals
  - What is the current state-of-the-art technology and what technological breakthroughs are required (e.g., detector development, mirror technology, propulsion systems, etc.)?
  - Define a roadmap for technology development
- Identify near term research and analysis (R&A) needs
  - What basic research needs to be done today to prepare for the mission(s) of tomorrow?
  - What kind of mission concept studies are required?
- Define strategy for achieving science goals
  - Is a large ultraviolet/optical telescope required?
  - Would a series of competed Explorer class missions or Explorer + missions suffice?
  - Nature of facility/facilities
    - Spectroscopic?
    - Imaging?
    - Define wavelength range, resolution, sensitivity, etc. required

4. Context for White Paper

In order to place the White Paper in the context of NASA's strategic plan, I would once again like to summarize the process which was described in some detail by Dr. Philippe Crane at the start of the Workshop.

- OSS Strategic Planning Process
  - National Academy of Science defines fundamental science questions for Astronomy and Astrophysics every ten years
  - OSS prepares strategic plan every three years as input to congressionally mandated NASA strategic plan

    * Space Science Advisory Committee (SScAC) is responsible for OSS strategic plan
    * Each subcommittee of the SScAC receives community input (e.g., workshops, working group reports, professional society meetings, etc.) to prepare roadmap for its science theme within context of National Academy recommendations
    * Roadmaps submitted to SScAC for incorporation into OSS Strategic Plan
  - OSS Performance Plan and metrics prepared annually based on President's budget
  - OSS Performance Report prepared based on evaluation by SScAC and NASA Advisory Council

- OSS Advisory Structure (governed by Federal Advisory Committee Act)
  - Community advice to OSS is provided through its advisory committees. The top level committee, the SScAC, advises the Associate Administrator for Space Science. The SScAC has four subcommittees, each of which represents the four science themes within OSS. Each subcommittee provides advice to the OSS Science Director responsible for the theme. The two subcommittees relevant to Astronomy and Physics are the Origins Subcommittee, and the Structure and Evolution of the Universe Subcommittee.

References

The following is a list of resources containing more information of relevance to the goals of this Workshop.

Blades, J.C. 2002, Ultraviolet and Visible Detectors for Future Space Astrophysics Missions, (Washington: NASA, and Baltimore: Space Telescope Science Institute)

Far-IR, Sub-mm, MM Detector technology Workshop, Monterey, CA, April 1–3, 2002 (http://sofia-usra.arc.nasa.gov/det_workshop) (proceedings to be published)

Morse, J.A., Shull, M., & Kinney, A.L. 1999, Ultraviolet-Optical Space Astronomy Beyond HST, (San Francisco: Astronomical Society of the Pacific)

Origins 2002, Grand Tetons, WY, May 26–30, 2002 (proceedings to be published)

Salama, F., Leckrone, D., Mathis, J., McGrath, M., Miller, H.R., Phillips, T., Sanders, W., Smith, P., Snow, T., & Tielens, A. 2002, Laboratory Astrophysics White Paper: Summary of Laboratory Astrophysics Needs, http://www.astrochemistry.org/nasalaw.html

Second Workshop on New Concepts for Far-IR/Submillimeter Space Astronomy, College Park, MD, March 5–7, 2002 (proceedings to be published)

Shull, J.M., Savage, B.D., Morse, J.A., Neff, S.G., Clarke, J.T., Heckman, T., Kinney, A.L., Jenkins, E.J., Dupree, A.K., Baum, S. A., & Hasan, H.

1999, The Emergence of the Modern Universe: Tracing the Cosmic Web, (Boulder: University of Colorado)

Hubble's Science Legacy: Future Optical-Ultraviolet Astronomy from Space
ASP Conference Series, Vol. 291, 2003
K.R. Sembach, J.C. Blades, G.D. Illingworth, R.C. Kennicutt, Jr. (eds.)

Reflections on the Workshop

Garth D. Illingworth

University of California, Lick Observatory, Santa Cruz, CA 95064

Robert C. Kennicutt, Jr.

University of Arizona, Steward Observatory, Tucson, AZ 85721

Abstract. As co-chairs of the scientific organizing committee for the Workshop, we present some thoughts and reflections on a successful meeting. These comments will provide the starting point for the White Paper to be prepared for NASA.

The scientific discussions were very comprehensive, encompassing problems in cosmology, extragalactic astronomy, stellar populations, the ISM/IGM, star and planet formation, stellar astrophysics, and solar system astronomy. All of these subjects were considered in the context of other ground and space missions producing discoveries in the 2010–2020 time frame. It was noted that the observational capabilities for astronomy will be very rich in the coming decade, and will likely include facilities as diverse as SIRTF, GALEX, ALMA, NGST, CELT/GSMT/OWL, SIM, LSST, and SNAP.

The breadth of the science that could be addressed by space optical-UV missions is impressive. Focusing in, some key science objectives for the the next decade emerged. They included:

1. planet detection and characterization (distribution of planetary orbits and properties; inner planets and Earth-like planets);

2. star/planet formation (evolution and dynamics of protoplanetary disks; disk evolution; structural and spectral diagnostics; planetary signatures; synergy with observations at other wavelengths);

3. dispersed baryons in the Universe – missing mass and detection of the dominant baryon component (the "cosmic web"; the IGM and its interaction with star formation and the ISM in galaxies; structural, ionization, and chemical evolution of the IGM; star formation/galaxy/IGM feedback cycles);

4. galaxy formation and evolution (evolution of structure and stellar populations of distant galaxies as revealed in the redshifted UV, combined with observations of the redshifted optical and infrared with NGST and ALMA; complemented by "archeology" of nearby galaxies – the time history of galaxy development from stellar population studies);

5. central black hole formation and development – characterizing the role of black hole buildup and AGNs in galaxy formation and evolution (detailed

structure and dynamics of the centers of local galaxies, complemented by AGN activity and frequency in low- and high-redshift galaxies);

6. cosmological issues – distance scale, dark energy, and dark matter (improved measurement of the distance scale using Cepheids to much larger distances; the use of lensing to map cosmological geometry; supernovae as "standard candles").

While these items struck us as being likely to play a key role in the scientific case for NHST, we were encouraged that many imaginative science goals were also identified that ranged very broadly from planetary studies in our own solar system, through stellar properties and stellar evolution, to the ISM (in addition to the IGM emphasized above), to globular clusters and nearby galaxy properties. The combination of new discoveries with more incremental but fundamental advances that has made HST so powerful clearly remains as a framework for NHST.

While the breadth of the scientific potential was very encouraging, discussion also focused on what was called the "killer aps" (from software developers who look for the applications that make their products a "must buy" for consumers). We face the same problem in "selling" our very expensive missions. What are the science goals that capture the imagination, not just of our colleagues, but also the public, NASA, the agencies responsible for budgeting, and our representatives in government who will ultimately fund the mission.

The direct detection and imaging of planets can obviously be categorized as a "killer ap". Other potentially compelling applications that were suggested at the conference included the mapping of the "cosmic web", star/planet formation, and early galaxy formation/evolution (including the role of black holes). Which of these provides the strongest justification for NHST remains an open question at this stage, but the number of such applications in itself demonstrates the richness of fundamental science that would be produced by such a facility.

While the presentations mainly focused on the science rationale for an NHST, speakers were also invited to identify needed capabilities (e.g., spectral coverage and sensitivity, size of telescope, instrumental capabilities). The contributions in this area, and the followup discussion, was very informative and thought-provoking.

The thinking about the mission(s) led to a summary of what might be considered the core capabilities for NHST.

First, a factor 10, or greater, gain over HST is required. The capabilities that are unique to optical-UV astronomy in space, and thus are of the highest scientific importance could be seen as:

1. high-throughput (multiplexed) UV spectroscopy;

2. wide-field optical diffraction-limited imaging;

3. high dynamic range optical coronagraphy.

Such capabilities can be made more specific by noting that the discovery space for NHST could be characterized as being:

1. UV spectroscopy at least 2 magnitudes fainter than HST/COS;
2. imaging at least 2 magntudes fainter than HST/ACS and HST/WFC3;
3. resolution <10 mas (matching adaptive optics on CELT/GSMT at 2 μm);
4. larger field-of-view than AO systems (also larger than HST/ACS and HST/WFC3).

As an example, it was noted that the ISM/IGM program could be accomplished through a specific, modestly-priced mission that could be accommodated within an enlarged Explorer program of order \$400–500M using a ~2–3 m class telescope with a highly–efficient (multiplexing?) UV spectrograph with high-QE UV detectors.

Many of the other key science goals, however, required missions that were larger, and more challenging, especially for planet searches. For these the discussion identified a number of possible missions, including a ~4 m coronagraphic telescope that is a TPF precursor (called "TPF-lite" at the workshop), which could eventually lead to an ~8 m class telescope that is uniquely optical-UV in its science goals. Or, if TPF/Life Finder is ultimately an optical coronagraph and not an IR interferometer, then an ~8 m TPF could potentially have broader optical-UV science capabilities as well. It is these telescopes, as a phased program, that would be seen as "NHST". Under this scenario, with coupling to the TPF program, NHST could be a "program" of two or more missions, instead of a particular mission.

One approach might be as follows. All of these needs could be addressed by a phased series of missions, with the earlier mission(s) providing technology stepping stones toward an NHST on the 2015–2020 time scale. Specifically they could include the enhanced Explorer–class telescope, an ~4–8 m class UV-optical telescope with wide–field imaging and multiobject spectroscopy, and/or an ~8 m class optical coronagraphic telescope (for Earth-like planet detection and limited photometric/spectroscopic characterization of detected planets) with capability for wider field imaging and spectroscopy.

The technologies that will play a major role in any such optical-UV missions include optical systems, possibly with elliptical optics, deformable mirrors, and almost certainly light-weight optics, with particular attention being paid to optical throughput from coatings, high efficiency spectrograph gratings, multiobject capability for spectrographs, and new polishing technology (especially for coronagraphs). New concepts for coronagraphs is currently an area attracting a great deal of interest. Any coronagraphic telescope will require very careful systems design, paying particular attention to stability and thermal control. Extensive modeling and ground demonstrations will be needed in these areas.

Detectors continue to be a challenge, with improved UV performance, and particularly for many classes of device, attention to performance in a high radiation environment (since the likely location for all new major facilities is well beyond LEO).

A key area for consideration of the implementation of new missions is the role that manned space flight will play. This role could be much larger than is planned for more near-term missions (e.g., SIM, NGST). The trade studies have yet to be made, but the manned program could be involved through post-launch deployment and testing (and possibly subsequent servicing – though that would more likely be robotic given the likely orbits).

Presentations on the likely capabilities on the ground with large optical telescopes and the potential of adaptive optics (AO) in the near–infrared at wavelengths beyond $\sim$1 μm also provided a valuable framework. It is very clear that the optical regime presents huge challenges for AO on large ground-based telescopes, even well into the next decade, and will not be competitive with optical space telescopes, at least over significant fields of view.

The enthusiasm of the participants for NHST came through very clearly in the talks and the discussion. By building on the science goals and recommendations of the SUVO workshop in 1998, this workshop is preparing the path for more focused technical and scientific workshops on NHST. The planning for these is already underway. Together these will provide the underpinnings that NASA needs to build NHST into its long-range roadmap. This workshop was a key step in the process that will make NHST a real space observatory befitting its namesake.

Hubble's Science Legacy: Future Optical-Ultraviolet Astronomy from Space
ASP Conference Series, Vol. 291, 2003
K.R. Sembach, J.C. Blades, G.D. Illingworth, R.C. Kennicutt, Jr. (eds.)

Some Comments on the Need for Ultra-High Resolution Spectroscopy of the Cool ISM with a Future UV/Optical Space Telescope

B-G Andersson

Center for Astrophysics/Johns Hopkins University, 3400 N. Charles St., Baltimore, MD 21218

Abstract. In order to fully resolve both the kinematic structure and the intrinsic line-width of spectra from the cool interstellar medium (ISM) a resolving power of $R \gtrsim 600,000$ ($\delta v \sim 0.5$ km s^{-1}) is needed. Many unresolved issues of interstellar cloud formation, destruction, and star-formation can be uniquely addressed by spectrographs capable of such resolution. Given the large number of diagnostic transitions, both atomic and molecular, only available from space, we argue that an ultra-high resolution spectrometer should be included as a capability on future orbital ultraviolet (UV) telescopes.

1. Introduction

Typical line widths and component separations in the cool ISM are of the order 0.5-1 km s^{-1}, and b-values of <0.3 km s^{-1} have been measured (e.g., Welty & Hobbs, 2001; Andersson, Wannier, & Crawford 2002). Physical characteristics, such as: excitation, depletion, abundances, and ionization fraction vary between kinematic components, also on this scale. This is seen in ultra-high resolution spectroscopy (UHRS) observations, such as those by Barlow et al. (1995), where the ratio of Na I to Ca II towards ζ Oph varies between components. When such structure is not resolved, severe interpretation difficulties can arise. Unresolved multiple components, with differing excitation temperature, can for instance cause severe problems in the interpretation of H_2 spectra as demonstrated by Spitzer et al. (1974). This component-to-component variability holds information about mixing, UV-penetration, and chemistry:

- How far into a cloud do UV photons penetrate?
- Why is f(H_2) never ~1 (cf., Rachford et al. (2002))?
- On what scales do the chemistry and physics vary?
- What is the interaction between clump and inter-clump gas?

All of these are factors that influence the behavior and evolution of interstellar clouds and hence star formation. For instance, for the ζ Oph observations mentioned above, the Na I to Ca II ratio systematically decreases away from the velocity of the associated CO (mm-wave) emission, indicating a stratified, dynamical cloud.

The resolved line width of a component also provides important information about the state of the gas, notably about the temperature and level of turbu-

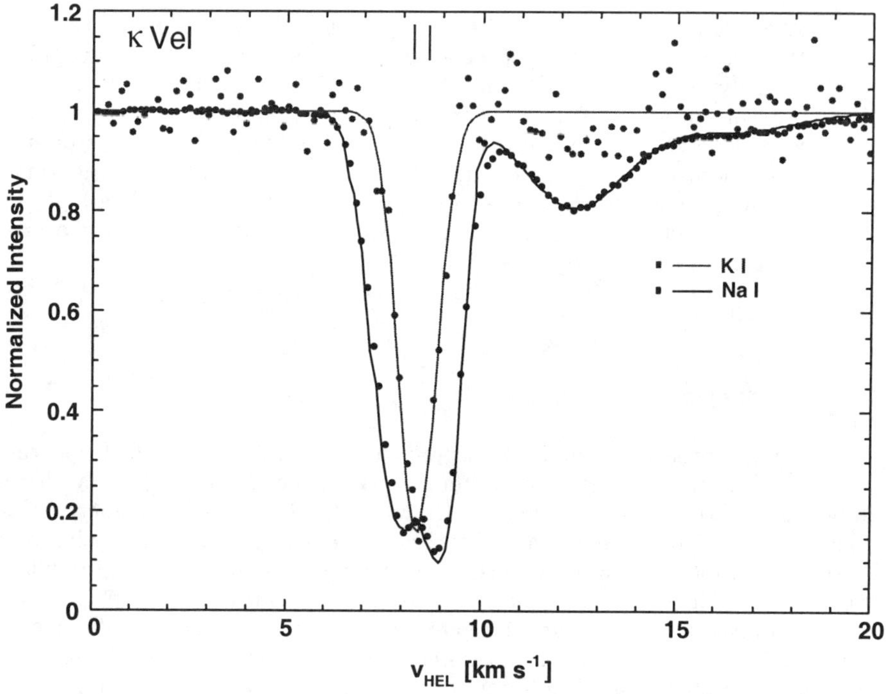

Figure 1. Dunkin and Crawford (1999) used UHRS data of K I and Na I to derive the thermal and turbulent broadening for the sight line towards κ Vel. They find that the turbulence is most likely sub-sonic in this case.

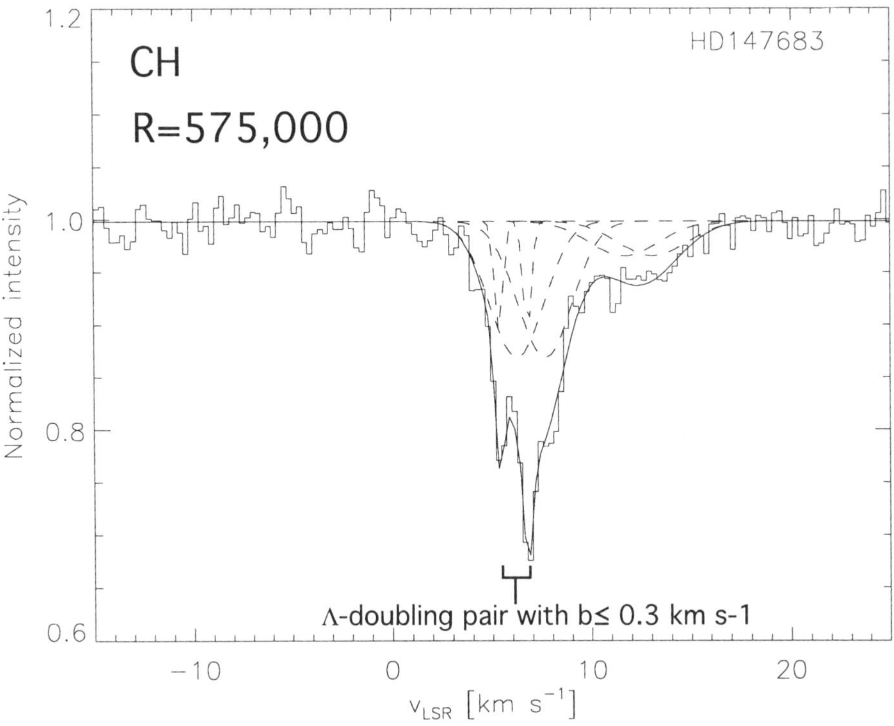

Figure 2. That the component structure in the cool ISM is complex and can show very narrow components is illustrated by this CH, UHRS spectrum of HD147683 (Andersson et al. 2002). Although acquired at a very high spectral resolution, the narrowest line in the spectrum is unresolved, yielding an upper limit on the b-value of <0.3 km s^{-1}. For the narrowest component, we find that, assuming sonic turbulence, a temperature of 16 K is derived. Assuming no turbulent broadening, a temperature of ~100 K is found.

lence. Turbulence must play a major role in the support of giant molecular clouds, but is poorly understood. Questions remain about what drives it and what the dissipation times are in different media. Mapping at radio/mm-wave wavelengths provides important data (Falgarone & Phillips 1991), but needs to be complemented by other tracers. By fully resolving lines and comparing line widths of species with different atomic masses we can get direct measurements of the turbulent velocities in interstellar clouds. By selecting species that are likely to be spatially co-existent and with as large a mass difference as possible, we can use the two observed line widths to solve for both the thermal and turbulent broadening as illustrated in Figure 1 (cf., Dunkin & Crawford 1999). Even with only single-species observations some interesting limits can be set (see Figure 2).

- Is the turbulence sub- or super-sonic?
- What is the scale of the end of the turbulent cascade?

While important questions about trace element abundances, isotope ratios and chemical evolution would also greatly benefit from or require the use of

UHRS data, this has been recently covered by Jenkins & Wallerstein (1999). The interested reader is referred to that excellent review.

2. Why UV/Optical UHRS from Space?

Radio/mm-wave observations routinely achieve 0.5 km s^{-1} resolution, but only a limited number of tracers are available. Emission line observations are also prone to ambiguities due to excitation and opacity uncertainties. We need UHRS optical/UV observations to expand the number of tracers, ionization states, and range of physical conditions explored, as well as to reliably compare to two-dimensional mapping provided by radio/mm-wave data.

As previously shown by Jenkins & Wallerstein (1999) and others, only a very small number of chemical species (atoms, ions, and molecules) have resonance lines accessible from the ground. By opening up the UV to UHRS observations, a much larger phase spacc and more varied physical conditions can be probed for physical and chemical structure.

3. Why Now?

Several recent detector developments will make UHRS from space less arduous. Photon counting CCDs are being developed by Jerram et al. (2001). Fast, photon counting microchannel plates with cross-stripped readouts allow for very small pixel sizes (Siegmund et al. 2001). These kinds of detectors will allow for high detector resolution and the ability to use anti-coincidence screening needed to make better spectrographs and dramatically reduce the background at the very high dispersions needed.

References

Andersson, B-G, Wannier, P.G., & Crawford, I.A. 2002, MNRAS, 333, 642

Barlow, M.J., et al. 1995, MNRAS, 272,

Dunkin, S.K., & Crawford, I.A. 1999, MNRAS, 302, 197

Falgarone, E., & Phillips, T.G. 1991, in IAU Symp. 147, Fragmentation of Molecular Clouds and Star Formation, 119

Jenkins, E.B., & Wallerstein, G. 1999, in ASP Conf. Ser. 164, Ultraviolet-Optical Space Astronomy Beyond HST, 118

Jerram, P., et al. 2001, Marconi Applied Technologies Preprint.

Rachford, B.L., et al. 2002, ApJ, in press

Siegmund, O.H., Tremsin, A.S., Abiad, R., & Vallerga, J.V. 2001, in Proc. SPIE Vol. 4498, UV/EUV and Visible Space Instrumentation for Astronomy and Solar Physics, eds. O.H. Siegmund, S. Fineschi, & M. A. Gummin, 131

Spitzer, L., Cochran, W.D., & Hirshfeld, A. 1974, ApJS, 28, 373

Welty, D.E., & Hobbs, L.M. 2001, ApJS, 133, 345

Hubble's Science Legacy: Future Optical-Ultraviolet Astronomy from Space
ASP Conference Series, Vol. 291, 2003
K.R. Sembach, J.C. Blades, G.D. Illingworth, R.C. Kennicutt, Jr. (eds.)

High Resolution Observations of Supernova Remnants

William P. Blair
Department of Physics and Astronomy, The Johns Hopkins University, 3400 N. Charles Street, Baltimore, MD 21218

Abstract. The Hubble legacy for observations of supernova remnants surely lies in its ability to provide high spatial resolution in both imaging and spectroscopy. I review work done over the last decade with various collaborators where we have used the high resolution of the Hubble Space Telescope (HST) to study spatial/spectral emission line variations on a wide range of physical scales. Recent work with high resolution far-ultraviolet spectroscopy from the Far Ultraviolet Spectroscopic Explorer (FUSE) satellite also demonstrates the importance of adequate spectral resolution to understand line profiles.

1. Introduction

Supernova remnants (SNRs) are an important means of enriching and energizing the interstellar medium (ISM). High spatial resolution for both imaging and spectroscopy of SNR shocks and their ejecta are made available by HST across the combined ultraviolet/optical range. This provides real advantages beyond those available previously. These advantages break down into several categories, depending largely on the distance to the objects in question. Recently, the advent of the FUSE satellite has demonstrated the importance of high resolution far-ultraviolet spectroscopy for a proper understanding of observed relative line intensities. I review some of these aspects briefly.

2. Galactic Remnants

Observations of Galactic SNRs with the HST have accessed interesting physical scales in the ionized and shock-heated gaseous filaments. In the Crab Nebula, where one sees primarily ejecta, HST images show a wide range of filament morphologies and resolved ionization structures (Hester et al. 1996; Blair et al. 1997; Sankrit et al. 1998). Statistical analysis shows most filaments are resolved ($0''.4$ – $0''.8$), corresponding to (1.2 – 2.4) $\times$ 10^{16} cm (at an assumed distance of 2 kpc). Continuum images show 'shadows' caused by foreground dense, dusty filament cores. Analysis of the extinction caused by these cores implies internal filament densities of ~40,000 cm^{-3}, about ten times higher than densities derived from emission line analyses. Some structures require magnetic confinement to remain stable.

The Cygnus Loop shows a SNR blast wave interacting with the inhomogeneous ISM instead of ejecta. The object is huge (3.5°) in angular size, so

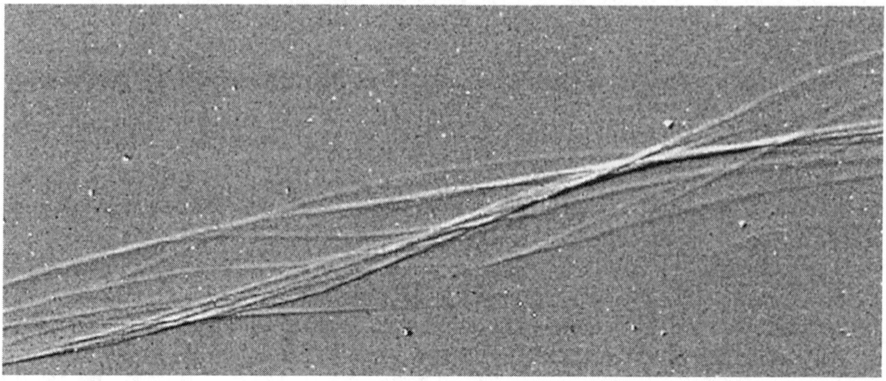

Figure 1. A difference image of HST/WFPC2 Hα images of the north-east Cygnus Loop filament taken four years apart. The motion of the filament is apparent.

HST can study only a small portion of the structure. One such study has concentrated on the far NE limb, where very faint, Balmer-dominated optical filaments demarcate the primary shock front as it first encounters the ISM (Blair et al. 1999). This filament shows remarkable structure. Long ribbons of faint Hα emission represent a nearly edge-on 'sheet' of gas at the shock front. This HST/WFPC2 image was used with a ground based image to refine the distance estimate downward from $\sim$770 pc to $\sim$440 pc, using the shock velocity of the filament as derived from HST/STIS and previous observations. The proper motion of the filament has now been confirmed with two epochs of HST data. Figure 1 shows a difference image made from data obtained in 1997 and 2001 where the $0\farcs3$ motion of the filament is evident.

We have used an HST/STIS-MAMA longslit far-ultraviolet (far-UV) spectrum to dissect this 'nonradiative' shock front at unprecedented spatial scales ($0\farcs1 = 6 \times 10^{14}$ cm). The far-UV lines should form over an observable scale as the gas is ionized behind the shock. In the STIS spectrum (Sankrit et al. 2000), the N V line is displaced behind the hydrogen emission by $\sim0\farcs3$. Models of this structure indicate the shock front has a velocity near 180 km s^{-1} and the preshock density is $2 - 4$ cm^{-3}.

We have also observed this filament using the FUSE satellite (Sankrit & Blair 2002). FUSE provides high resolution spectra in the 905 – 1187 Å range, which is a crucial new diagnostic ability. FUSE provides access to the O VI doublet lines near 1032,1038 Å, which arise in hotter portions of the shock. A small grid of FUSE spectra was obtained on $4''$ centers, stepping back from the shock front. Figure 2 shows the spectra corresponding to these aperture positions. Note how the central peak in λ1032 changes dramatically as the aperture is stepped back, but the 'wings' on this feature remain steady. These wings arise from other shock structures along the sight line that are not obvious in the image. High spectral resolution permits us to decouple these different components along the same sight line through the SNR.

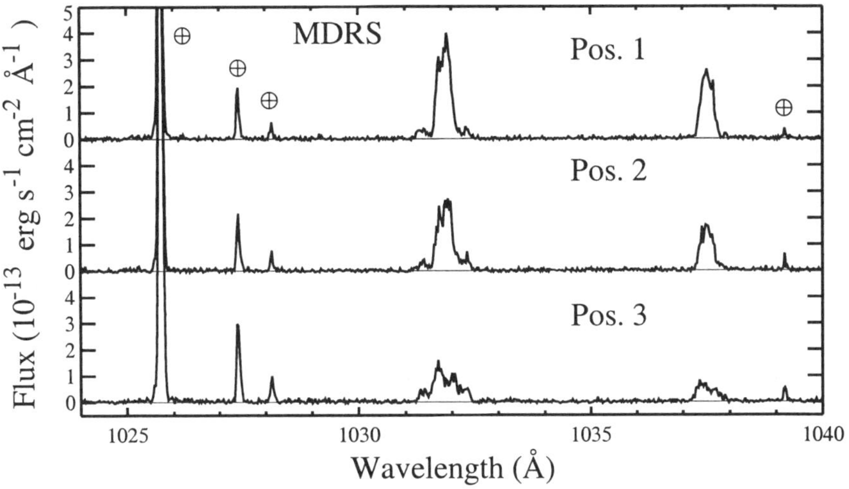

Figure 2. FUSE spectra from the north-east nonradiative shock in the Cygnus Loop. O VI λλ1032,1038 change in both shape and intensity as the aperture is stepped back into the post-shock region. Narrow lines due to airglow show the instrumental resolution.

3. Magellanic Cloud SNRs

Magellanic Cloud SNRs show a wealth of detail at HST resolution, and having the objects at a known distance is a great advantage. We have performed two detailed studies of the objects N132D (LMC) and E0102-7219 (SMC), two of the premiere remnants of core collapse SNe we have available for study (Morse et al. 1996; Blair et al. 2000). These objects have relatively little foreground extinction, making full optical/UV coverage possible. Ejecta and ISM components are clearly separated at HST resolution, and meaningful comparisons to other wavelengths become possible. The comparison of these two objects is like having a snapshot of two different ages of young SNR. E0102 is about three times smaller and younger than N132D. In the latter object, the O-rich ejecta are still clearly visible, but the blast wave has grown to $\sim$ 25 pc in diameter and is now interacting with the surrounding cavity wall, where a textbook example of a shock precursor is resolved. E0102 is dominated by the ejecta at optical and UV wavelengths, while the outer shock structures are visible at radio and X-ray wavelengths (Gaetz et al. 2000).

4. More Distant SNRs

Even at distances of $\sim$5 Mpc, HST has the power to resolve SNR structures and provide new insights. The SNR in NGC 4449 is known to be a young, rapidly expanding remnant of a core-collapse SN, embedded in an H II region (Blair, Kirshner, & Winkler 1983). The date of the explosion is not known, but the ~6000 km s^{-1} expansion velocity and optical spectrum dominated by oxygen lines are similar to Cas A in our Galaxy. I used HST/FOS and its 0."9 aperture to obtain a full UV/optical spectrum with almost no contamination from the surrounding H II region. This spectrum shows detailed line profiles and additional SNR lines not seen previously. The spectrum is dominated by O, Ne, S, and Mg, with weaker C emission than in the Magellanic Cloud SNRs.

There is an object in the face-on spiral NGC 6946 that was initially thought to be similar to the NGC 4449 object. It is very luminous at optical, X-ray, and radio wavelengths, but its optical spectrum did not show evidence of peculiar abundances or high velocities (Blair & Fesen 1994). HST/WFPC2 resolved the structure into a set of interlocking loops, one of which is much brighter at optical wavelengths and may represent a smaller, younger SNR interacting with an older adjacent object (Blair, Fesen, & Schlegel 2001). The great luminosity of the object arises partly from this interaction and partly from the inferred dense surrounding material into which the shock fronts are propagating (cf., Dunne et al. 2000).

The work summarized here is but a scratch on the surface of what could be done to better understand SNRs, their ejecta, and their interactions with the ISM. This work demonstrates the utility of high spatial resolution spectroscopy and narrow-band imaging and high spectral resolution spectroscopy of emission line objects at UV/optical wavelengths. For SNRs, using these data in conjunction with other multi-wavelength data sets provides a much broader understanding than that from any individual data set alone. I encourage the community to consider the need for these capabilities in future mission designs.

Acknowledgements – I would like to thank my collaborators on many of these projects. My attendance was supported by NASA contract NAS5-32985 to the Johns Hopkins University.

References

Blair, W.P., Davidson, K., Fesen, R.A., Uomoto, A., MacAlpine, G.M., & Henry, R.B.C. 1997, ApJS, 109, 473

Blair, W.P., & Fesen, R.A. 1994, ApJ, 424, L103

Blair, W.P., Fesen, R.A., & Schlegel, E.M. 2001, AJ, 121, 1497

Blair, W.P., Kirshner, R.P., & Winkler, P.F. 1983, ApJ, 272, 84

Blair, W.P., Sankrit, R., Raymond, J.C., & Long, K.S. 1999, AJ, 118, 942

Blair, W.P., et al. 2000, ApJ, 537, 667

Dunne, B.C., Gruendl, R.A., & Chu, Y.-H. 2000, AJ, 119, 1172

Gaetz, T.A., et al. 2000, ApJ, 534, L47

Hester, J.J., et al. 1996, ApJ, 456, 225

Morse, J.A., et al. 1996, AJ, 112, 509

Sankrit, R., Blair, W.P., Raymond, J. C., & Long, K.S. 2000, AJ, 120, 1925

Sankrit, R., et al. 1998, ApJ, 504, 344

Sankrit, R., & Blair, W.P. 2002, ApJ, 565, 297

Hubble's Science Legacy: Future Optical-Ultraviolet Astronomy from Space
ASP Conference Series, Vol. 291, 2003
K.R. Sembach, J.C. Blades, G.D. Illingworth, R.C. Kennicutt, Jr. (eds.)

Multiple Weak Lensing Deflections in the HDF–North

Tereasa G. Brainerd

Boston University, Department of Astronomy, 725 Commonwealth Ave., Boston, MA 02215

Abstract.
Using the ~600 galaxies in the Hubble Deep Field (North) and the flanking fields for which both spectroscopic redshifts and rest frame blue luminosities are known, we compute the theoretical galaxy–galaxy lensing shear field within the HDF–N. For this we adopt a fiducial model in which the halos of L^* galaxies have velocity dispersions of $\sigma_v^* = 156$ km s^{-1} and the characteristic extent of these halos is $s^* = 100h^{-1}$ kpc. We show that the probability of distant source galaxies being lensed by more than one galaxy in the HDF–N is large and that neglecting multiple weak deflections when computing the theoretical shear field results in a significant underestimate of the mean galaxy–galaxy lensing shear in the region of the HDF–N.

1. Introduction

The systematic weak gravitational lensing of distant galaxies by foreground galaxies has been detected at a statistically–significant level in a number of independent investigations (see, e.g., McKay et al. 2001 and references therein). This effect, known as galaxy–galaxy lensing, is now developing into a premier tool by which the nature of the dark matter halos around galaxies, as well as the galaxy–mass correlation function, can be constrained (see, e.g., Brainerd 2001).

Early work on the theory of galaxy–galaxy lensing (Brainerd, Blandford & Smail 1996) indicated that distant source galaxies which undergo galaxy–galaxy lensing are likely to have been lensed at a significant level by *two or more* foreground galaxies. That is, the light rays emanating from many source galaxies will have encountered multiple deflections due to the presence of several, yet comparable, weak galaxy lenses along the line of sight.

Although the existence of such multiple deflections in galaxy–galaxy lensing has been recognized, to date there has been no detailed investigation of the probability of multiple deflections occurring in deep data sets, nor of the degree to which the predicted shear field will be in error if multiple deflections are ignored in a calculation which infers fundamental halo parameters (e.g., velocity dispersions and cut–off radii) from observations of galaxy–galaxy lensing.

Here we use the well–studied Hubble Deep Field (North) and the flanking fields in order to make an accurate prediction for the weak lensing shear field within the HDF–N itself. The rest frame blue luminosities, L_B, and the spectroscopic redshifts of ~ 600 galaxies in this region of space are known (Cohen

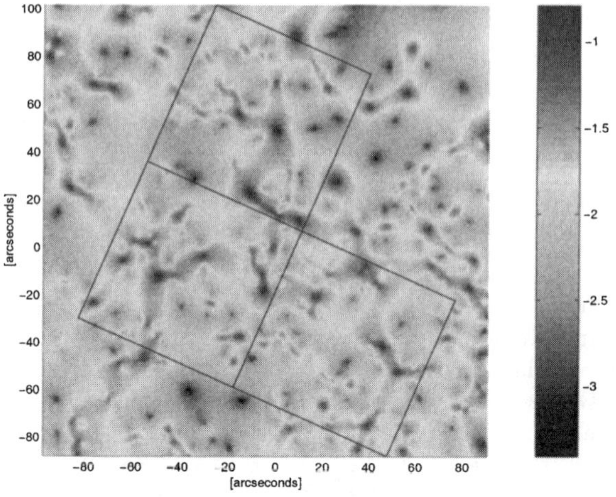

Figure 1. Theoretical galaxy–galaxy lensing shear field for the HDF–N.

2002) and through some simple scaling relations for the dark matter halos surrounding these galaxies, we can determine the theoretical galaxy–galaxy lensing shear field produced by these galaxies.

2. Lens Model

Following Brainerd et al. (1996), we model the density profiles of the dark matter halos of the galaxies in Cohen (2002) as:

$$\rho(r) = \frac{\sigma_v^2 s^2}{2\pi G r^2 \left(r^2 + s^2\right)}, \tag{1}$$

where G is Newton's constant and s is a characteristic halo radius. We let σ_v^* be the velocity dispersion of the halo of an L^* galaxy halo and we assume that a Tully–Fisher or Faber–Jackson type of relation will hold for each galaxy. We therefore have $\sigma_v/\sigma_v^* = (L_B/L_B^*)^{1/4}$, where σ_v is the velocity dispersion of a halo in which a galaxy with luminosity L_B resides. For simplicity, we also assume that the total mass–to–light ratio of a galaxy is constant independent of its luminosity, in which case we have that the radii of the halos of galaxies with $L_B \neq L_B^*$ scale with the radii of the halos of L^* galaxies as $s/s^* = (L_B/L_B^*)^{1/2}$. Throughout we adopt fiducial values of $\sigma_v^* = 156$ km s^{-1} and $s^* = 100h^{-1}$ kpc for the halos of L_B^* galaxies, which are comparable to the characteristic halo parameters that have been inferred from previous studies of galaxy–galaxy lensing.

Shown in Figure 1 is the theoretical weak lensing shear field produced by the galaxies in Cohen (2002). Here we have specifically included the effects of multiple weak lensing deflections on the final shapes of the source galaxies. We have adopted an open cosmology with $\Omega_0 = 0.3$, $\Lambda_0 = 0$, and $H_0 = 60$ km s^{-1} Mpc^{-1}, but the results are essentially independent of this choice (see Wright & Brainerd

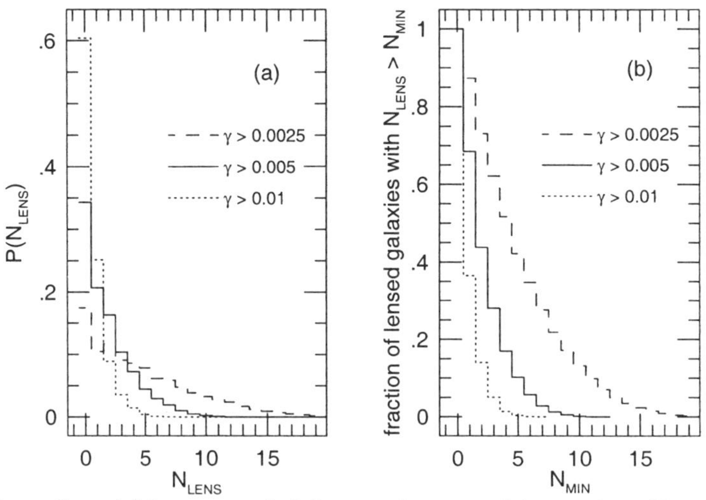

Figure 2. (a) Lensing probability as a function of the number of lenses and the minimum shear due to a single deflection. (b) Fraction of lensed sources which have been lensed by more than $N_{\rm MIN}$ lenses.

2002). Our source galaxies are chosen to have $19 < I < 23$, and we have adopted a redshift distribution of the form:

$$z_0 = k_z[z_m + z'_m(I - I_m)], \tag{2}$$

where z_m is the median redshift, I_m is the median I-band magnitude, and z'_m is the derivative of the median redshift with respect to I. This is in good agreement with both the no–evolution model and with the redshift distribution of galaxies with $19 \le I \le 22$ obtained from the Canada-France Redshift Survey (LeFèvre et al. 1996).

3. Results

Shown in Figure 2a is the probability that a given source in the HDF–N will have been lensed by $N_{\rm LENS}$ objects, with a minimum shear value per single deflection of γ (dashed line: $\gamma = 0.0025$; solid line: $\gamma = 0.005$; dotted line: $\gamma = 0.010$). Shown in Figure 2b is the fraction of lensed sources which have been lensed by *more than* $N_{\rm min}$ foreground lenses with a minimum shear value per single deflection of γ. The typical shear due to galaxy–galaxy lensing is $\gamma \sim 0.005$ and, so, we see from Figure 2 that there is a substantial probability (of order 45%) that a given source galaxy has been lensed by *two or more* galaxies with a shear equal to or greater than the expected mean value.

Shown in Figure 3 is the mean galaxy–galaxy lensing shear, centered on each of the galaxies from Cohen (2002) which are within the HDF–N. Solid squares show the correct shear as a function of lens–source separation, $\gamma(\theta)$, when all multiple weak lensing deflections are included in the computation of the final image shapes of the sources (i.e., this is the mean, circularly–averaged shear which is obtained from the shear field shown in Figure 1). Open squares show

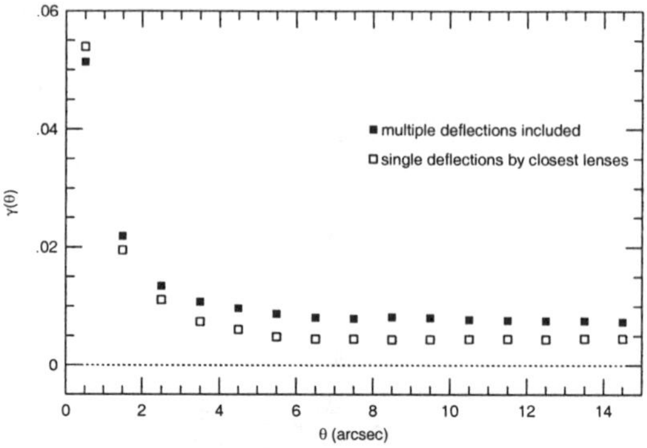

Figure 3. Mean, circularly–averaged shear due to the lens galaxies in the HDF–N. Solid squares: all weak deflections are included. Open squares: only single deflections due to the closest lens are included.

the mean shear that one obtains if the multiple weak deflections are *ignored* in the calculation and, instead, each source galaxy is assumed to have been lensed by only one foreground galaxy (taken to be the galaxy which is nearest to it in projection on the sky). From Figure 3, then, neglecting the multiple weak deflections leads to a substantial under prediction of the mean shear due to galaxy–galaxy lensing. Therefore, if one were to use a single–deflection model to constrain the parameters of the dark matter halos of the HDF–N galaxies, an overestimate of the halo mass would occur. When compared to our fiducial shear field (i.e., Figure 1), the best–fitting single–deflection calculation yields a halo model with $\sigma_v^* = 198$ km s^{-1} and $s^* = 435h^{-1}$ kpc, which corresponds to a mass for the halo of an L^* galaxy which is a factor of ~ 2 too large within a radius of $100h^{-1}$ kpc (see also Wright & Brainerd 2002).

Acknowledgments. Support under NSF contract AST–0098572 and STScI contract HST–AR–08746.01–A is gratefully acknowledged. This work is based in part on archival observations with the NASA/ESA Hubble Space Telescope obtained at the Space Telescope Science Institute, which is operated by AURA, Inc. under NASA contract NAS–26555.

References

Brainerd, T.G., Blandford, R.D. & Smail, I. 1996, ApJ, 466, 623

Brainerd, T.G. 2001, in ASP Conf. Ser. 237, Gravitational Lensing: Recent Progress & Future Goals, eds. T. Brainerd & C. Kochanek, (San Francisco: ASP), 379

Cohen, J. 2002, ApJ, 567, 672

LeFèvre, O., Hudon, D., Lilly, S.J., et al. 1996, ApJ, 461, 534

McKay, T.A., Sheldon, E.S., et al. 2002, ApJ, submitted (astro-ph/0108013)

Wright, C.O. & Brainerd, T.G. 2002, ApJ, submitted (astro-ph/0205297)

Hubble's Science Legacy: Future Optical-Ultraviolet Astronomy from Space
ASP Conference Series, Vol. 291, 2003
K.R. Sembach, J.C. Blades, G.D. Illingworth, R.C. Kennicutt, Jr. (eds.)

100 Times Faster and 3 Times Sharper: Background-Dominated Observations of Stellar Populations with an 8-meter Optical-UV Space Telescope

Thomas M. Brown

Space Telescope Science Institute, 3700 San Martin Drive, Baltimore, MD 21218

Abstract. An 8 m successor to the Hubble Space Telescope (HST) would make incredible gains in the study of stellar populations, especially in the Local Group. If diffraction-limited at 0.5 μm, the "Next HST" could produce high-resolution imaging of faint sources over a wide field in 1 percent of the time needed with the HST. With these capabilities, photometry of the ancient main sequence could be obtained for many sight lines through Local Group galaxies, thus determining directly the ages of their structures and providing a formation history for the Local Group populations.

1. Introduction

One of the primary quests of observational astronomy is measuring the formation history of giant galaxies. Recently, renewed interest in formation via accretion of dwarf galaxies has been sparked by the discoveries of the Sagittarius dwarf galaxy falling into the Milky Way (Ibata et al. 1994) and of a tidal stream in the Andromeda halo (Ibata et al. 2001). With large investments of Hubble Space Telescope (HST) time, color-magnitude diagrams that reach the ancient main sequence can be constructed for selected fields in galaxies of the Local Group, thus providing accurate ages for their structures via the same techniques traditionally used to date the populations in Galactic globular clusters. Unfortunately, these studies are limited by the long integration times needed to reach the main sequence at the distance of Andromeda (the nearest spiral to our own), and by the stellar crowding that can be addressed at the HST resolution. However, an 8 m optical-UV space telescope, diffraction-limited at 0.5 μm, would crack this field wide open, because of a simple concept that is often overlooked in discussions of an HST successor: The time to reach a given signal-to-noise for background-dominated photometry of point sources scales as aperture to the fourth power, for a telescope that is diffraction-limited at a given wavelength.

2. Background-Dominated Photometry

At a given wavelength, the width of a resolution element decreases linearly with the aperture diameter (D), and thus the area of a resolution element on the sky decreases as D^2. Because the collecting area increases as D^2, the sky counts in a resolution element remain constant as aperture increases. The signal-to-noise ratio (S/N) for background-dominated photometry thus scales as:

$$S/N \;\propto\; \frac{S\times t\times(\frac{D}{d})^2}{\sqrt{B\times t + S\times t\times(\frac{D}{D_\odot}\times\frac{d_\odot}{d})^2}} \;\propto\; \frac{S\times t\times(\frac{D}{d})^2}{\sqrt{B\times t}} \;\propto\; S\times\sqrt{t}\times(\frac{D}{d})^2$$

where the source rate is S, the sky background rate is B ($B \gg S$), the exposure time is t, and the source distance is d. Thus, with other parameters fixed, the distance at which you can see a source scales as D, the flux you can measure scales as D^{-2}, and the exposure needed scales as D^{-4}. Current HST observing programs studying faint stars would be 100 times faster with an 8 m optical-UV space telescope, without assuming any advances in optical coatings, detector efficiencies, etc. (see Figure 1). These dramatic gains will be even larger than those realized by the Next Generation Space Telescope (a 6 m telescope, diffraction limited at 1 – 2 μm, with higher background in its near-IR channels).

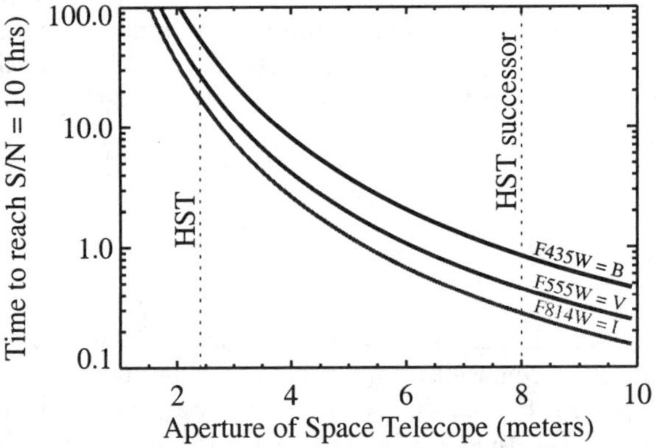

Figure 1. The exposure time needed to obtain photometry of the Sun at the distance of Andromeda, assuming a telescope that is diffraction-limited at 0.5 μm. The Sun makes a good fiducial because of its familiarity and its proximity to the MSTO in the color-magnitude diagram of an old stellar population.

Many difficult problems, requiring large investments of HST observing time (more than 100 orbits), become trivial with an 8 m version of the HST. The shorter integrations make it possible to study many more sight lines through Local Group galaxies, and the superior resolution allows investigation into structures that are much more crowded. Because the exposure time for a given S/N is so much faster, one can also study variability in regimes previously unreachable. Examples of programs that would benefit greatly from a larger optical-UV space telescope include age-dating Local Group populations using the main sequence turnoff (MSTO), studying the white dwarf cooling curve in Galactic globular clusters, and microlensing searches for dark matter.

3. Simulations

To dramatize the gains made in imaging of stellar populations, I have created two simulations of imaging in the bulge of Andromeda, of which only small subsections are shown in Figure 2. The first simulation shows a combination of three bands from a hypothetical HST imaging program using the Advance Camera for Surveys (ACS) and its Wide Field Camera (WFC), assuming the native pixel size (50 mas). This gray-scale image combines F435W data (120 hrs) for the blue channel, F606W data (42 hrs) for the green channel, and F814W data (49 hrs) for the red channel. Although such deep exposures could detect isolated stars on the ancient main sequence at this distance, the crowding and surface brightness of the unresolved stars in the bulge and disk would preclude detection of the MSTO. The second simulation shows the same field as imaged with the same bandpasses using the "Next HST" (NHST; an 8 m version of the HST), assuming $1/10^{\text{th}}$ the exposure time; note that only $1/100^{\text{th}}$ of the HST time is needed to match the HST S/N. The NHST simulation resolves the ancient MSTO. The NHST would likely have two imaging cameras: a wide-field camera that undersamples the point spread function (PSF) with an approximately 1 square degree field, and a high-resolution camera that critically samples the PSF (7.5 mas pixels) with a field-of-view comparable to the current ACS/WFC on HST ($204'' \times 204''$). The NHST simulation uses the high-resolution camera, which provides the same field, but with a PSF that is both critically sampled and 3.3 times narrower than the PSF in the HST simulation.

The simulations reproduce the surface brightness in a field about 8 arcmin from the center of Andromeda, where the bulge and disk populations have approximately the same surface brightness (Baggett, Baggett, & Anderson 1998). The full simulated images include a globular cluster, young open cluster, and tidal stream from a disrupted galaxy (10% overdensity). All of the components (disk, bulge, stream, and clusters) were constructed using the isochrones of Bertelli et al. (1994). Figure 2 shows $7.1'' \times 5.4''$ subsections in the vicinity of the Andromeda globular cluster.

4. Summary

For background-dominated photometry of stellar populations, an 8 m version of the HST would be 100 times faster than the current HST, and provide 3 times the resolution. Although the increased resolution would allow photometry in fields that are currently impossible with the HST, the reduction in exposure time is not an advance that should be treated lightly. Obviously, for a given field that can be resolved with the HST, one can always argue that requesting one hundred orbits of HST time would be easier than building an 8 m HST and observing for one orbit. However, a factor of 100 in exposure time allows one to resolve the main sequence in dozens of sight lines throughout the Local Group, a program that is simply impossible with the HST, no matter what fraction of its remaining lifetime is devoted to the problem. Such a program would allow the direct age determination of structures (disks, bulges, halos) in Local Group galaxies, providing a formation history for these galaxies and their components.

References

Baggett, K.S.J., Baggett, S.M., & Anderson, W.E. 1998, AJ, 116, 1626

Bertelli, G., et al. 1994, A&AS, 106, 275

Ibata, R.A., Gilmore, G., & Irwin, M.J. 1994, Nature, 370, 194

Ibata, R., et al. 2001, Nature, 412, 491

Figure 2. Simulated images in the bulge of Andromeda, near a globular cluster (see text for details). *Top panel:* HST/ACS/WFC; *bottom panel:* NHST.

Hubble's Science Legacy: Future Optical-Ultraviolet Astronomy from Space
ASP Conference Series, Vol. 291, 2003
K.R. Sembach, J.C. Blades, G.D. Illingworth, R.C. Kennicutt, Jr. (eds.)

The Stellar Imager (SI): An Ultra-High Angular Resolution Ultraviolet/Optical Observatory

Kenneth G. Carpenter and Richard G. Lyon

NASA-Goddard Space Flight Center, Greenbelt, MD 20771

Carolus J. Schrijver

Lockheed-Martin Advanced Technologies Center, Palo-Alto, CA 94304

Lee Mundy

University of Maryland, College Park, MD 20742

Ronald J. Allen and Jay Rajagopal

Space Telescope Science Institute, 3700 San Martin Dr., Baltimore, MD 21218

Abstract. The Stellar Imager (SI) is envisioned as a space-based, ultraviolet-optical interferometer composed of 10 or more one-meter class elements distributed with a maximum baseline of 0.5 km and providing a resolution of 60 micro-arcseconds at 1550 Å. It will image stars and binaries with sufficient resolution to enable long-term studies of stellar magnetic activity patterns and their evolution with time, for comparison with those on the Sun. It will also sound their interiors through astroseismology to image internal structure, differential rotation, and large-scale circulations. SI will enable us to understand the various effects of magnetic fields of stars, the dynamos that generate these fields, and the internal structure and dynamics of the stars in which these dynamos operate. The ultimate goal of the mission is to achieve the best-possible forecasting of solar activity as a driver of climate and space weather on time scales ranging from months up to decades, and an understanding of the impact of stellar magnetic activity on life in the Universe. The road to that goal will revolutionize our understanding of stars and stellar systems, the building blocks of the Universe. Fitting naturally within the NASA and ESA long-term time lines, SI complements defined missions, and with them will show us entire other solar systems, from the central star to their orbiting planets. In this paper we describe the scientific goals of the mission, the performance requirements needed to address those goals, and the design concepts now under study. We also note other fields of astrophysics that would benefit from a long-baseline interferometer in space and report on the current status of the mission concept development efforts.

1. Introduction

The primary science goals of the SI mission are to study spatio-temporal stellar magnetic activity patterns in a sample of stars covering a broad range of activity, in order to: 1) understand the underlying dynamo process(es) and thereby enable improved forecasting of solar activity on time scales of days to centuries, 2) understand the impact of stellar magnetic activity on astrobiology, 3) enable astroseismology (acoustic imaging) to measure internal stellar structure and rotation and their relationship to the dynamo, and 4) complete the assessment of external solar systems by imaging the central stars of systems in which the IR-interferometry missions [Terrestrial Planet Finder (TPF), IRSI/Darwin, Planet Imager (PI)] find and image planets.

2. Science Requirements and Performance Goals

There are two broad science requirements for the mission, both of which require a population study of cool stars representing a broad range of magnetic activity: surface imaging to detect and monitor the evolution of active regions, and astroseismology (acoustic imaging) to obtain information on the sub-surface layers. To understand the dynamo, we need to know how magnetic fields are generated and behave in different circumstances - the Sun is only one example and provides insufficient constraints on theories of dynamos, turbulence, structure, and internal mixing. We must observe other stars to establish how mass, rotation, brightness and age affect the patterns of activity and thereby provide robust observational constraints on dynamo theories.

The design requirements for imaging of stellar surface activity include the acquisition of ultraviolet (UV) images (with ~1000 pixels over the stellar surface) to ensure visibility of surface manifestations of the dynamo. Dark starspots in the visible-light photosphere are small in most stars and have low contrast with the surrounding bright stellar surface. High-contrast bright spots are seen in UV (chromospheric, transition-layer) emission (e.g., Mg II h&k 2800 Å, C IV 1550 Å) from plages above the surface wherever it is penetrated by strong magnetic fields, making them ideal activity diagnostics. Modest integration times (~hours for dwarfs to days for giants) are required to avoid smearing of images due to rotation, proper motions, and activity evolution. Requirements for imaging of stellar interiors by seismology include short integration times (minutes for dwarf stars to hours for giant stars) and thus broadband optical wavelengths to get sufficiently high fluxes, and low-resolution imaging to measure non-radial resonant waves (~30-100 total resolution elements over the stellar surface). Flexible interferometer configurations are required in both modes.

3. Strawman Mission Concept

The current leading architecture concept for Stellar Imager is that of a 0.5 km diameter, space-based, UV-optical Fizeau Interferometer composed of a reconfigurable array of 10 - 30 one-meter-class spherical array elements on "mirrorsats". Those elements direct light to an image-plane beam combination facility in a hub at the prime focus, as shown in Figure 1. This design will provide: an angular

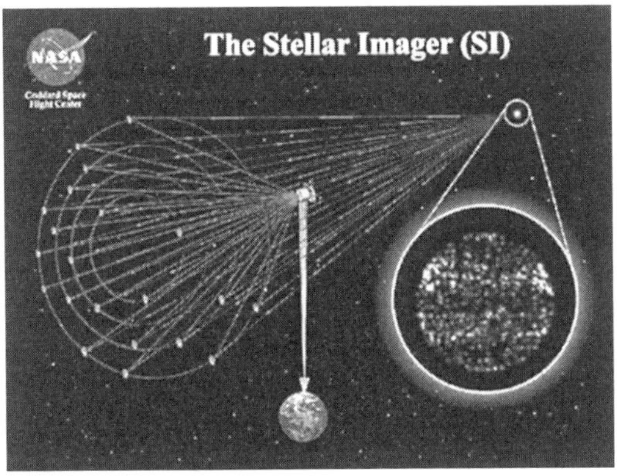

Figure 1. An artist's concept of a 21-element Stellar Imager design option.

resolution of 60 and 120 micro-arcseconds at 1550 Å and 2800 Å, ~1000 pixels of resolution over the surface of nearby dwarf stars, observations in ~10 Å UV pass-bands around C IV (100,000 K) and Mg II h&k (10,000 K), and broadband observations in the near-UV or optical continuum (formed at 3,000-10,000 K). SI is designed as a long-term mission with a requirement of a 10 year lifetime and a goal of 20 years, to allow the study of significant portions of stellar magnetic activity cycles. Individual telescopes and the central hub can be refurbished or replaced as needed. SI will be located in a Lissajous orbit around the Sun-Earth Lagrange point L2, which has both a small and very well characterized gravity gradient that permits precision formation flying and which should be accessible in the 2020 time frame for servicing and upgrade by robotic and/or manned missions.

3.1. Place in NASA/ESA Strategic Roadmaps

SI is on the strategic path of NASA Origins and ESA interferometry missions. It is a stepping stone towards crucial technology: SI is comparable in complexity to the TPF and Darwin nulling-IR-interferometers, and it may serve as a useful pathfinder for the Planet Imager. SI addresses science goals of three NASA Office of Space Science Themes: *understand why the Sun varies* (Sun Earth Connection), *understand the origin of stars, planetary systems, and life* (Origins), *understand the structure and evolution of stars* (Structure and Evolution of the Universe). It is complementary to the planetary imaging interferometers. TPF, IRSI/Darwin, and PI null the stellar light to find and image planets. SI images the central star to study the effects of that star on the habitability of those planets. TPF, SI, IRSI/Darwin, and PI together provide complete views of other solar systems.

3.2. SI and General Astrophysics

A long-baseline interferometer in space could benefit many fields of astrophysics (cf., Bely et al. 1996). Potential targets include: *Active galactic nuclei* – the transition zone between broad-line regions and narrow-line regions, origin and orientation of jets; *Quasi-stellar objects and black holes* – close-in structure, radiation from accretion processes; *Supernovae* – close-in structure; *Stellar interiors* – internal structure of stars outside solar parameters; *Hot stars* – hot polar winds, non-radial photospheric pulsations; *Envelopes and shells of Be-stars*; *Spectroscopic binary stars* – observe companions and orbits, determine stellar properties, perform key tests of stellar evolution; *Interacting binary stars* – resolve mass-exchange, dynamical evolution and accretion, study more efficient dynamos; *Cool, evolved giant and supergiant stars and long-period and semi-regular variable stars* – spatio-temporal structure of extended atmospheres/winds and shocks; and *Extrasolar planet detection* – via transits.

3.3. Enabling Technologies

Stellar Imager will rely on a number of critical technologies, including: precision formation flying, lightweight mirror technology, coarse ranging and array alignment, on-board autonomous computing and control systems, and closed loop optical control to maintain array alignment based on the science data.

Study of these technologies is ongoing at GSFC, JPL, various universities, and in industry, and significant leveraging and cross-fertilization will occur across projects [e.g., with the Next Generation Space Telescope (NGST) and TPF]. Series of testbeds are in operation or are under development at GSFC, including the: Wavefront Control Testbed to study image-based optical control methods for NGST, Phase Diverse Testbed to study extended scene phase diversity optical control with moving array elements, Wide-Field Imaging Interferometry Testbed to study extending the field of Michelson imaging interferometers, and the Fizeau Interferometry Testbed to study closed-loop control of an array of elements, as well as assess and refine technical requirements on hardware, control, and imaging algorithms. Studies of the full SI mission as well as pathfinder concepts continue in GSFC's Integrated Design Center.

4. Summary/Status

SI is currently included in the far-horizon NASA Sun-Earth Connection Roadmap. The mission concept continues to be developed by GSFC in collaboration with LMATC, NRL/NPOI, STScI, and the University of Maryland. Further information on the mission can be found on the internet at the following URL: http://hires.gsfc.nasa.gov/~si. We plan to gather and utilize additional community input and produce a book summarizing the science and societal motivations for the mission, the technology roadmap, and the most promising architecture options. We invite you to join us in the definition and realization of this mission. Please contact K. Carpenter (kgc@stargate.gsfc.nasa.gov) or C. Schrijver (schryver@lmsal.com) with your comments and suggestions.

References

Bely, P. and the Space Interferometry Study Team, June 1996, ESA SCI(96)7.

Hubble's Science Legacy: Future Optical-Ultraviolet Astronomy from Space
ASP Conference Series, Vol. 291, 2003
K.R. Sembach, J.C. Blades, G.D. Illingworth, R.C. Kennicutt, Jr. (eds.)

High Resolution Ultraviolet Quasar Absorption Line Spectroscopy of a z ~ 1 Galaxy Group

Jie Ding

Department of Astronomy & Astrophysics, The Pennsylvania State University, 525 Davey Laboratory, University Park, PA 16802

Abstract. We used ultraviolet spectra from HST/STIS ($R = 30,000$), together with optical spectra from Keck/HIRES ($R = 45,000$), to study the three Mg II-selected absorption systems at $z = 0.9254$, 0.9276, and 0.9342 toward the quasar PG 1206 + 459. A multi-phase gaseous structure, with low-ionization components produced in small condensations and high-ionization ones in diffuse clouds, is indicated in all three systems. Each system is likely to represent a different galaxy with absorption due to some combination of interstellar medium, coronal gas, halo gas, and high-velocity clouds. Even with the improved sensitivity of HST/COS, we will only be able to obtain high-resolution ultraviolet spectra of the brightest quasars in the sky. A larger telescope with ultraviolet coverage will enable quasar absorption line studies of hundreds of galaxies, including a wide range of galaxy types and environments at low and intermediate redshifts.

1. Introduction

The light beams from distant quasars pass through various galaxies and probe physical properties of gaseous structures in the galaxies. Therefore, studying absorption features in the background quasar spectrum is a unique method for tracing the cosmic evolution of the Universe. Strong Mg II absorbers are almost always associated with luminous galaxies ($> 0.05L^*$). Thus, using this tool we can study the predecessors of the giant spiral and elliptical galaxies that we see in the nearby universe.

Three Mg II systems (A, B, and C), at redshifts $z = 0.9254$, 0.9276, and 0.9342, are found along the line of sight toward the quasar PG 1206 + 459. In a previous study, Churchill & Charlton (1999) found the three systems to be multi-phase absorbers with Mg II clouds embedded in extended, high-ionization gas that gives rise to C IV, N V, and O VI. Their analysis was based upon the combination of low-resolution data from HST/FOS and high-resolution data from Keck/HIRES. However, many important issues remained unresolved, such as (1) whether C IV, N V and O VI arise in the same layer of gas and whether their profiles are smooth or have sub-structure; (2) whether the high-ionization phase envelops the low-ionization phase or whether it is offset in velocity; (3) whether the majority of Si IV in the $z = 0.9276$ system arises in a single Mg II cloud that is similar to a Milky Way high-velocity cloud.

In May 2001, we obtained a stunning high-resolution ($R = 30,000$) HST/STIS spectrum, covering Lyα and high-ionization transitions Si IV, C IV, and N V, for the three systems. Through photoionization modeling of the various chemical transitions in this spectrum and in the earlier Keck/HIRES spectrum, the metallicities, abundance patterns, and ionization states of absorbing gas clouds have been constrained (for details on the modeling technique, see Ding et al. 2002). The results are presented in the following sections along with the physical interpretations of individual phases of gas.

2. System A at z = 0.9254

Six clouds, spread over ~ 200 km s^{-1} in velocity space, are needed to fit the Mg II absorption. Ionization parameters are constrained to be $-2.8 \lesssim \log U \lesssim -2.5$ and the metallicity is required to be super-solar, if a solar abundance pattern is assumed. These clouds, all having $\log \mathrm{N(H\,I)} < 16$, contribute negligibly to the partial Lyman limit break. The dotted lines in Figure 1 represent the contribution from the Mg II phase, for our best model.

A diffuse phase is required to fit the residuals in C IV, N V, and O VI, unaccounted for by the Mg II phase. Collisional ionization is ruled out as a mechanism to produce the N V absorption, due to the narrow features in the high-resolution spectrum. The ionization parameters of the seven high-ionization clouds are constrained to be $-1.5 \lesssim \log U \lesssim -0.6$. A super-solar metallicity is also required in this phase, unless N is enhanced. In Figure 1, solid lines represent the contribution from this diffuse phase.

Neither the low-ionization phase nor the diffuse phase could fully account for the residuals to the red of the C IV profiles. An additional component, with $b(C) \sim 6$ km s^{-1}, seems to be necessary. This component could be photoionized or collisionally ionized. A near-solar metallicity is required in either case. The photoionized component, with an intermediate ionization parameter $\log U \simeq -2$, is represented by the dashed-dotted curves in Figure 1.

The large kinematic spread of the Mg II, with no dominant component, the lack of a Lyman limit break, and the multiple-component structure in N V suggest that this system does not represent a traditional disk/corona structure.

3. System B at z = 0.9276

The strong, blended Mg II profiles could be fit with five narrow components, spread over ~ 200 km s^{-1} in velocity space. The ionization parameters of the low-ionization phase clouds are $-3.2 \lesssim \log U \lesssim -2.5$ and the metallicity is constrained to be $\log Z \simeq -0.1$, by the partial Lyman limit break detected in the FOS spectrum. The dotted lines in Figure 1 represent the contribution from this phase.

The residuals in C IV, N V, and O VI indicate a highly ionized, diffuse phase. A broad component, with $b(N) \sim 50$ km s^{-1}, is needed to fit the smooth N V absorption. An offset, narrower one, with $b(C) \sim 14$ km s^{-1}, is also needed to account for the blue part of C IV. An ionization parameter of $\log U \simeq -1.6$ and a metallicity of $\log Z \simeq -0.6$ are required for both clouds for a photoionization model. The solid lines in Figure 1 show the two diffuse components. Collisional ionization is ruled out in this phase due to the observed ratio of C III to C IV absorption.

An additional component, with $b(Si) \sim 10$ km s^{-1}, is required to produce the observed Si IV absorption unaccounted for by either the Mg II phase or the diffuse phase, in the cloud furthest to the red. This component could be photoionized or collisionally ionized. The collisional component, with a temperature $\log T \sim 4.8$, is represented by the dashed-dotted curves in Figure 1.

The Mg II phase is likely to arise in the disk of a galaxy. The broad component in the diffuse phase may represent a corona structure similar to that of the Milky Way. Two candidate absorbing galaxies appear in a WIYN image of the quasar field at ~ 25 kpc and ~ 35 kpc.

4. System C at z = 0.9342

A weak, unresolved Mg II cloud, aligned with the bulk of the C IV absorption in velocity space, is separated from System B, by $\sim 1,000$ km s^{-1} (see Figure 1). If the Mg II cloud does not give rise to C IV, the ionization parameter of this phase is constrained to be $\log U \lesssim -3$ and the metallicity is $\log Z \simeq -0.6$. If, instead, the majority of C IV is produced by the Mg II clouds, the ionization parameter is constrained to be $-2.2 \lesssim \log U \lesssim -2$ and the metallicity is close to the solar value.

Regardless of the ionization state of the Mg II phase, the strong absorption in O VI (detected in the low-resolution HST/FOS spectrum) requires an additional, highly ionized component. If this component is photoionized, the ionization parameter would either be $\log U \simeq -1.5$ or $\log U \gtrsim -0.7$, depending upon whether C IV arises in the same phase or not. Alternatively, if the diffuse component is collisionally ionized, the temperature of this phase is constrained to be $\log T \gtrsim 5.4$. Figure 1 shows a model in which C IV is produced in the Mg II phase and O VI (not shown) is produced in a more highly photoionized phase. The dotted lines represent the model contributions from the Mg II cloud, and the solid lines show those from the diffuse component.

In addition, a cloud offset ~ -40 km s^{-1} from the center of the Mg II absorption is required by the Lyα profile. This cloud also gives rise to the blue part of the C IV profiles. The dashed-dotted curves in Figure 1 represent this offset cloud.

The simple, weak absorption feature suggests that this system may be a high-velocity cloud or a dwarf galaxy. Two candidate associated galaxies are found in the WIYN image. They would be dwarfs and have impact parameters of 20–40 kpc, if at $z \sim 1$.

References

Churchill, C.W., & Charlton, J.C. 1999, AJ, 118, 59

Ding, J., Charlton, J.C., Churchill, C.W., & Palma, C. 2002, in preparation

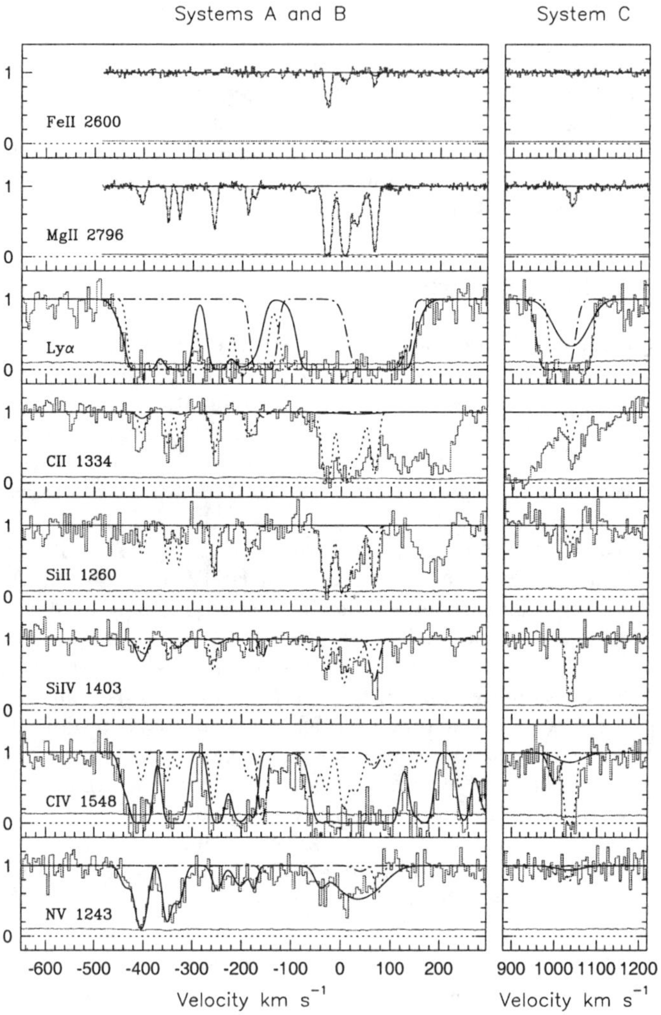

Figure 1. Various key transitions are displayed in velocity space, with velocity the zero-point at $z = 0.9276$. The solid histograms represent the normalized HST/STIS and Keck/HIRES data. The dotted lines show the contribution from the low-ionization phase in each system. The solid curves represent the photoionized, diffuse phase. The dashed-dotted lines show the intermediate phase.

Hubble's Science Legacy: Future Optical-Ultraviolet Astronomy from Space
ASP Conference Series, Vol. 291, 2003
K.R. Sembach, J.C. Blades, G.D. Illingworth, R.C. Kennicutt, Jr. (eds.)

Constraining the Heliosphere: The Need for High-Resolution Observations of Nearby Interstellar Matter

Priscilla C. Frisch

The University of Chicago, Department of Astronomy and Astrophysics, 5640 South Ellis Ave., Chicago, IL 60637

Abstract. High-resolution ultraviolet observations of nearby bright and faint stars are required to evaluate changes in the past and future Galactic environments of the Sun, and the possible impact of these changes on the interplanetary environment at 1 AU (around the Earth). The boundary conditions of the heliosphere and interplanetary environments are constrained by the characteristics of the surrounding interstellar material (ISM), which changes on time scales of 10^3–10^5 years. An increase in the density of the interstellar cloud surrounding the Solar System to 10 cm^{-3} decreases the heliosphere radius by about an order of magnitude. Ultraviolet observations of nearby stars at high spectral resolution (>300,000) and high signal-to-noise are required to evaluate future modifications to heliosphere properties by the ISM.

1. Introduction

High-resolution ultraviolet (UV) observations of nearby bright and faint stars are required to evaluate changes in the past and future Galactic environments of the Sun, and the possibly impact of these changes on the interplanetary environment at 1 AU (around the Earth) and in the outer Solar System.

2. The Heliosphere, Interstellar, and Interplanetary Matter

The interplanetary environment around the Earth and the boundary conditions of the heliosphere depend on the physical properties of the interstellar cloud surrounding the Solar System (known as the Local Interstellar Cloud, or LIC). These boundary conditions vary with the space motions of the Sun[1] (~13.5 km s^{-1}) and interstellar clouds (0 – 30 km s^{-1}) through the Local Standard of Rest (LSR). The primary visitors from beyond the heliopause are neutral atoms, large dust grains (radii >0.20 μm, Frisch et al. 1999), and cosmic rays with energies >1 GeV. Presently ~98% of the diffuse material in the Solar System is the neutral component of interstellar gas which flows relatively freely

[1]The LSR velocity used here is the Hipparcos value, corresponding to a solar motion towards the Galactic coordinates l = 27.7°, b = 32.4° at the velocity V = 13.4 km s^{-1} (Dehnen & Binney 1998).

through the heliopause region. The solar wind plasma and interstellar neutral densities are equal at ~6 AU because of the $1/R^2$ falloff in solar wind density. The solar wind shields Earth from the low pressure and low density interstellar cloud around and within the Solar System ($P_{\rm thermal} \sim 2500$ cm^{-3} K). Variations in the properties of the cloud surrounding the Solar System yield variations in the heliosphere configuration and the amount of ISM in the inner Solar System.

The heliopause location and heliosphere configuration depend on the relative ram pressures of the solar wind and ISM (Holzer 1989). The current boundary conditions of the heliosphere are a relative Sun-cloud velocity of ~26 km s^{-1}, n(H I) ~ 0.24 cm^{-3}, n(e) ~ 0.1 cm^{-3}, T $\sim 7,000$ K, X(H) = 0.31 and X(He) = 0.48, based on radiative transfer models of ISM within 3 pc, and pickup ion and ISM absorption data (Slavin & Frisch 2002). Today, the heliosphere radius is large (~150 AU) and varies with the solar cycle.

Increasing the ISM neutral density to 10 atoms cm^{-3} has been shown to contract the heliopause radius by an order of magnitude and the heliopause becomes Rayleigh-Taylor unstable from solar wind mass-loading (Zank & Frisch 1999). This change in the heliosphere size would modify the 1 AU interplanetary environment of the Earth. The flux of interstellar neutrals at the Earth's orbit increases to ~2 cm^{-3}, the heliopause radius contracts to $R \sim 10 - 14$ AU, and outer planets are exposed to the raw ISM, dramatically altering the interplanetary environment of the outer Solar System.

3. Variations in Properties of ISM Surrounding the Solar System

The Sun is embedded in a cluster of interstellar cloudlets flowing outwards from the region of the Loop I superbubble. High-resolution ground-based observations of Ca II (resolution 0.3–1.0 km s^{-1}) show that the ISM flow past the Solar System contains at least six distinct velocity component groups suggestive of separate clouds. About 96 components are observed towards 60 nearby stars. Several of the nearest stars show ~1 component per 1.4–1.6 pc. Ca II component velocities towards nearby stars exhibit random velocities (r.m.s. dispersion σ ~5 km s^{-1}) distributed about a bulk flow velocity with upstream direction towards Loop I. In the LSR, the bulk flow velocity corresponds to a motion of –17.0 km s^{-1} with an upstream direction of l = 2.3°, b = –5.2° (Frisch, Grodnicki, & Welty 2002).

Cloud velocity is a proxy for cloud structure. Typical relative Sun-ISM velocities of 0 to 40 km s^{-1} (or more) are to be expected. An encounter with a 'new' cloud is thus possible every ~25,000 years, and the potential for variations in the Galactic environment of the Sun exists for even shorter time scales.

Variable boundary conditions yield variable heliosphere properties. One hundred thousand years ago the Sun was immersed in the interior of the Local Bubble (T $\sim 10^6$ K, $n_p \sim 0.005$ cm^{-3}), and the heliosphere radius was similar to today but with no interstellar neutrals inside of the Solar System (e.g., Mueller et al. 2002). The Sun must have emerged from the Local Bubble interior and entered the cloud now surrounding the Sun $2 \times 10^3 - 1 \times 10^5$ years ago (Frisch 1994). As the Sun continues to travel through the outflow of cloudlets from the Loop I region, new cloudlets will be encountered.

The nearest star, α Cen (1.3 pc), is located ~30° from the LIC LSR upstream direction (Frisch et al. 2002); however the velocity of the ISM towards

α Cen differs by several km s^{-1} from the projected LIC velocity suggesting the Sun may encounter the α Cen cloudlet within 10^4 years. Another cloudlet with d $<$ 5 pc and a heliocentric velocity of $\sim$30 km s^{-1} is located in the solar apex direction. If this cloud has 0 km s^{-1} tangential velocity, the Sun will encounter this cloud within $\sim$1–2 x 10^5 years (Frisch et al. 2002).

4. Advantage of High-Resolution UV data

Evaluating possible changes in the boundary conditions of the heliosphere requires UV observations of the individual cloudlets near the Sun, since the required resonance transitions of dominate ions are not accessible from the ground. The tag which identifies individual cloudlets is component velocity, and high-resolution optical Ca II data show the component structure of interstellar clouds is under-sampled in the UV at resolutions of 3 km s^{-1}. Welty, Morton, & Hobbs (1996) have shown a crowding of optical components in velocity space, such that the number of detected components increases exponentially with improved instrumental resolution. As a result, observations at the resolution of HST/GHRS and HST/STIS appear to miss over 60% of the absorption components. High-resolution and high signal-to-noise observations in low column density sight lines ($<10^{18}$ cm^{-2}) towards nearby stars provide the best opportunity to resolve individual cloudlets free from the well-known but difficult-to-specify uncertainties resulting from blended components.

Optical observations of Ca II towards the star α Oph (14 pc) provide an example of the gains expected from improved UV resolution. At a resolution $\lambda/\delta\lambda \sim 10^6$, data from the Ultra-High Resolution Facility at the Anglo-Australian Observatory show four absorption components, compared to three components found at the lower resolution of 250,000 (e.g., Crawford & Dunkin 1995; Welty et al. 1996; Crawford 2001). Doppler b-value varies from 1.06 to 4.04 for the four components (or T = 2700 K to 39000 K, if no turbulence, Crawford 2001). The weak Ca II feature towards α Oph at –31.8 km s^{-1} is formed in a cloud within 5 pc of the Sun (Frisch et al. 2002), and the b-value is alternately reported as 1.5 km s^{-1} and 4.0 km s^{-1} in the analysis of high-resolution data. This difference emphasizes that high-resolution data must also be high signal-to-noise data to remove uncertainties in interpreting observations of very low column density cloudlets.

Separating turbulence from thermal broadening requires UV observations with resolutions $\sim$1 km s^{-1}, and understanding the ionization and densities of the individual components requires resolving the component structure for a range of ions (e.g., Fe II, Mg II, Mg I). Resolving out possible cold (100 K) cloudlets embedded in the flow of ISM past the Sun requires resolutions on the order of 10^6. Such high resolution will permit determining the thermal width of Mg I lines in cool neutral (100 K, b = 0.26 km s^{-1}) versus warm partially ionized (T = 7000 K, b = 2.2 km s^{-1}) cloudlets, so that the physics of Mg I in the local ISM (i.e., radiative versus dielectronic recombination) can be resolved.

Factor of two or greater variations in Fe II abundances are seen in cloudlets observed towards nearby stars, showing that small-scale structure and pressure variations remain to be discovered in the immediate Galactic environment of the Sun with high-resolution UV data.

5. Conclusions

The morphology of the nearby ISM (< 30 pc), and interstellar pressure variations affecting the heliosphere and interplanetary medium surrounding the Earth, *can only be found* from high-resolution ($\lambda/\delta\lambda > 300,000$) observations of interstellar absorption lines in the ultraviolet wavelength band (912 — 3000 Å). The relative simplicity of sight lines to nearby stars, combined with high-resolution high signal-to-noise UV data, provides the best opportunity for obtaining accurate abundances and cloud properties for interstellar clouds.

References

Crawford, I.A., & Dunkin, S.K. 1995, MNRAS, 273, 219

Crawford, I.A. 2001, MNRAS, 327, 841

Dehnen, W., & Binney, J.J. 1998, MNRAS, 298, 387

Frisch, P.C. 1994, Science, 265, 1423

Frisch, P.C., et al. 1999, ApJ, 525, 492

Frisch, P.C., Grodnicki, L., & Welty, D.E. 2002, ApJ, August issue

Holzer, T.E. 1989, ARA&A, 27, 199

Mueller, H.R., Frisch, P.C., & Zank, G.P. 2002, in preparation

Slavin, J.D., & Frisch, P.C. 2002, ApJ, 565, 364

Welty, D.E., Morton, D.C., & Hobbs, L.M. 1996, ApJS, 106, 533

Zank, G.P., & Frisch, P.C. 1999, ApJ, 518, 965

Hubble's Science Legacy: Future Optical-Ultraviolet Astronomy from Space
ASP Conference Series, Vol. 291, 2003
K.R. Sembach, J.C. Blades, G.D. Illingworth, R.C. Kennicutt, Jr. (eds.)

The Exquisite Spectrum of RX J1230.8+0115

Rajib Ganguly, Joseph Masiero, and Jane C. Charlton

Department of Astronomy & Astrophysics, The Pennsylvania State University, 525 Davey Laboratory, University Park, PA 16802

Kenneth R. Sembach

The Space Telescope Science Institute, 3700 San Martin Dr., Baltimore, MD 21218

Abstract. The past twelve years have filled the HST data archive with many gems waiting to be mined. We present one such gem - an exquisite STIS-E140M spectrum of the nearby quasar RX J1230.8+0115. The line of sight toward this quasar passes by the Virgo cluster, one degree away from 3C 273. The spectrum is filled with an eclectic variety of absorption systems - the Galaxy, high velocity clouds, several Lyman-alpha absorbers possibly associated with Virgo, two intervening metal-line systems, and a complex of four associated absorbers. In this spectrum, the associated absorption complex is detected primarily in the high ionization N V species. There are many indications that this complex arises from gas truly intrinsic to the QSO central engine. One system has a smooth wind-like profile, indicative of an outflow - a mini-BAL. Another system is line-locked in N V with the mini-BAL. The highest redshift system lies redward of the N V emission line peak. Preliminary profile fits indicate that the gas causing the absorption in the associated systems only partly occults the QSO central engine.

1. Introduction

RX J1230.8+0115 ($V = 14.42$, $z_{\rm em} = 0.117$) was first detected by the ROSAT All-Sky Survey (Read, Miller, & Hasinger 1998, A&A, 335, 121) and is one of the brightest QSOs in the optical bandpass. The line-of-sight toward this quasar lies 54′ from 3C 273 and 11°.5 from the center of the Virgo Cluster. In the following figures, we highlight the rich HST/STIS-E140M spectrum which shows metal-line absorption from the Galaxy and high velocity clouds (Figure 1), the Virgo Cluster (Figure 2), and an intrinsic absorption complex (Figure 3). The intrinsic absorption complex toward RX J1230.8+0115 provides an excellent opportunity to examine the physical conditions and geometry of a variety of intrinsic absorbers. Far ultraviolet spectra (e.g., FUSE) can provide O VI and Lyman series information necessary to constrain the ionization conditions. Multiple observations to look for changes in the N V, Ly α, and C IV profiles can provide important constraints on the geometry and density of the gas. Thanks to Michael Rauch who proposed to obtain this rich spectrum with the HST.

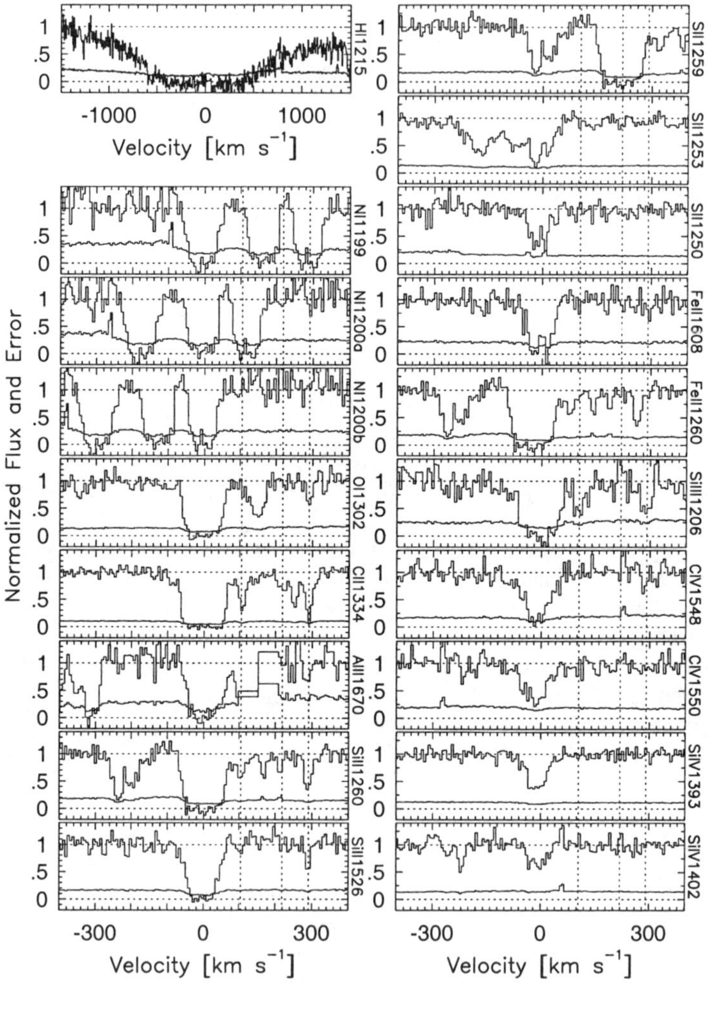

Figure 1. Absorption by the Galaxy is detected in a wide range of neutral and ionized species: H I, N I, O I, C II, Al II, Si II, S II, Fe II, Si III, C IV, and Si IV. High velocity clouds are also detected in neutral (H I, O I) and low ionization (C II, Si II) species over the redshift range $0.00035 < z_{\rm abs} < 0.00097$ ($105\ {\rm kms}^{-1} < \Delta {\rm v}_{\rm abs} < 290\ {\rm kms}^{-1}$), but not in higher ionization species (Si IV and C IV). Detected transitions are shown aligned in velocity relative to $z_{\rm abs} = 0.0$. The HVCs are marked with vertical dashed lines.

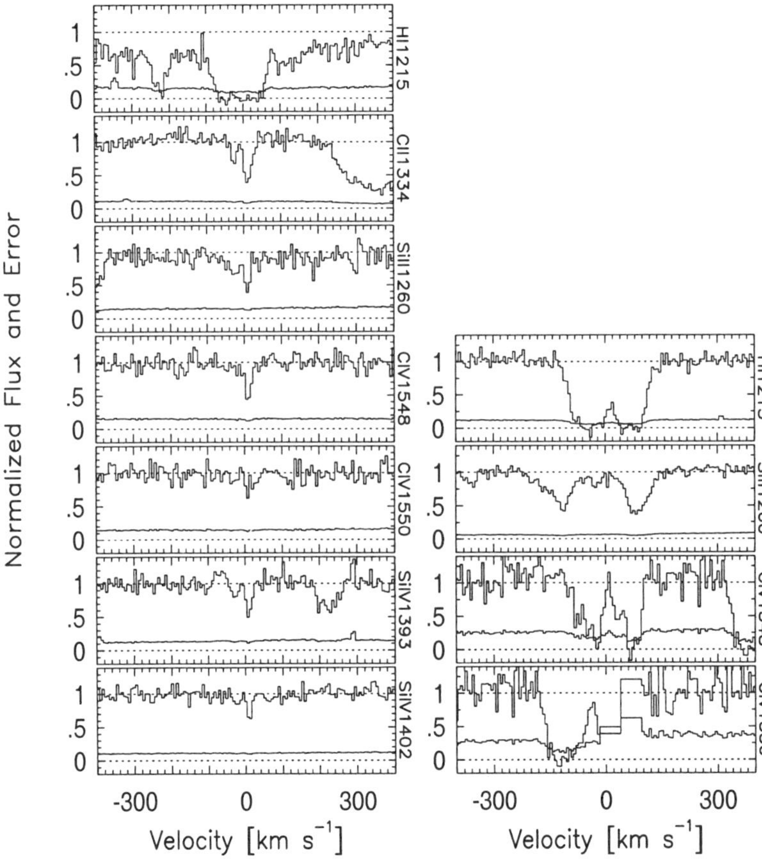

Figure 2. Absorption from the Virgo Cluster is detected at $z_{\text{abs}} = 0.0057$ in H I, C II, Si II, C IV, and Si IV. In the left-side plot, detected transitions are shown in velocity relative to $z_{\text{abs}} = 0.0057$. In the right-side plot, detected transitions of H I, Si II, and C IV are shown in velocity relative to $z_{\text{abs}} = 0.0778$. This metal line system is clearly beyond the Virgo Cluster.

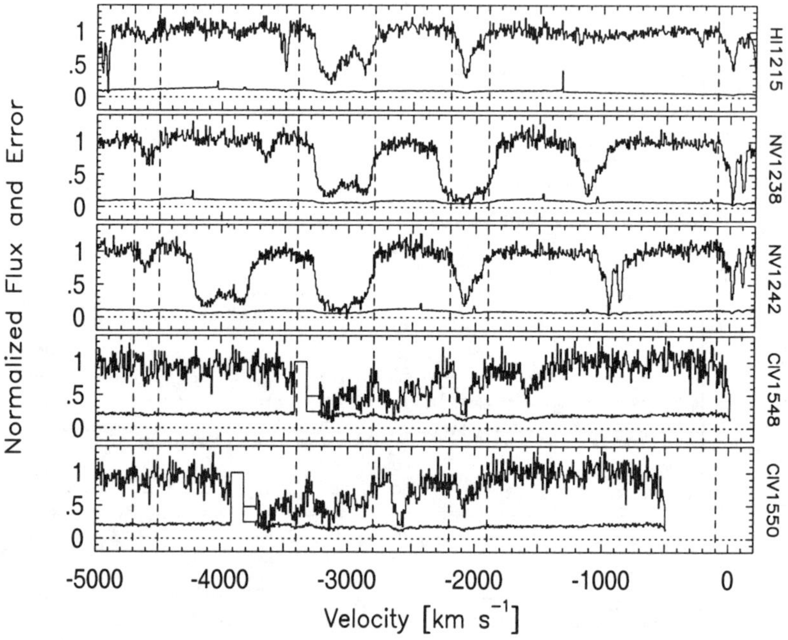

Figure 3. A complex of four associated absorbers ($z_{\rm abs}$ = 0.1000, 0.1058, 0.1093, 0.1162) is detected in H I, C IV, and N V. The spectra are shown (fluxes and errors) with $z_{\rm em} = 0.117$ as the velocity zero-point. The four systems are marked with vertical dashes. The first system ($\Delta v_{\rm abs} = -4600$ km s^{-1}) is a weak system. The second system ($\Delta v_{\rm abs} = -3020$ km s^{-1}) has a broad, smooth trough indicative of a wind - a mini-BAL. The N Vλ1242 transition is blended with the N Vλ1238 transition from the third system ($\Delta v_{\rm abs} = -2070$ km s^{-1}), a classic case of line-locking. The fourth system ($\Delta v_{\rm abs} = -215$ km s^{-1}) is, in fact, redward of the N V emission line peak, despite having a velocity that is blueshifted relative to $z_{\rm em}$. There are many indications that the entire complex of lines is due to intrinsic gas - strong high ionization lines (relative to Ly α), wind-like kinematics, and line-locking. Preliminary profile fits confirm this - all systems show components that only partly occult the QSO.

Hubble's Science Legacy: Future Optical-Ultraviolet Astronomy from Space
ASP Conference Series, Vol. 291, 2003
K.R. Sembach, J.C. Blades, G.D. Illingworth, R.C. Kennicutt, Jr. (eds.)

The HST Legacy and Future for High Spatial Resolution Spectroscopy of Winds and Outflows from Young Stars

C.A. Grady

NOAO, Eureka Scientific, and GSFC Code 681, NASA/GSFC, Greenbelt, MD 20771

B. Woodgate, R. Kimble, S. Heap, and T. Gull

Laboratory for Astronomy & Solar Physics, Code 681, NASA/GSFC, Greenbelt, MD 20771

Abstract. Accretion and conspicuous mass loss are some of the defining features of pre-Main Sequence evolution of young stars and their protoplanetary disks, with implications for the stellar properties, and the chemistry of the disk. HST/STIS observations have demonstrated the utility of spectral imaging at H I Lyα as a means of probing the winds from young stars into the equivalent of Jupiter's orbit. We discuss the potential of this technique with HST and potential ultraviolet-optical successor missions to distinguish between wind mechanisms, and to probe the collimation region for bipolar outflows.

1. Introduction

Mass loss is one of the most conspicuous, observational features of pre-Main Sequence evolution of young stars (Bally & Reipurth 2001). However, it also plays a key role in determining the subsequent angular momentum evolution of the star. The geometry of the mass loss is also diagnostic of the presence of stellar magnetic fields, especially when collimated outflows are detected. Moreover, material entrained in uncollimated winds is also important, since it can potentially seed the outer regions of the protoplanetary disk with chemically processed grains, potentially accounting for the compositional diversity of comets and meteorites (Shu et al. 2001; Nuth et al. 2001; Hill et al. 2002). Historically, such studies have been carried out for objects where the presence of an outflow was detected via direct imaging resulting in a bias toward objects which are moderately to heavily embedded in the optical. Recent HST coronagraphic imaging of nearby pre-Main Sequence stars has revealed that Herbig-Haro knots associated with collimated outflows are more common, and are associated with stars that are older, more massive, and much less embedded in their natal molecular clouds than previously anticipated (Devine et al. 2000). Mapping these outflows typically requires use of techniques to enhance the contrast of the wind or other nebulosity to the stellar direct, scattered, and diffracted light. In the optical, such studies have relied on narrow-band filters, Fabry-Perot interferometers, or coronagraphic techniques.

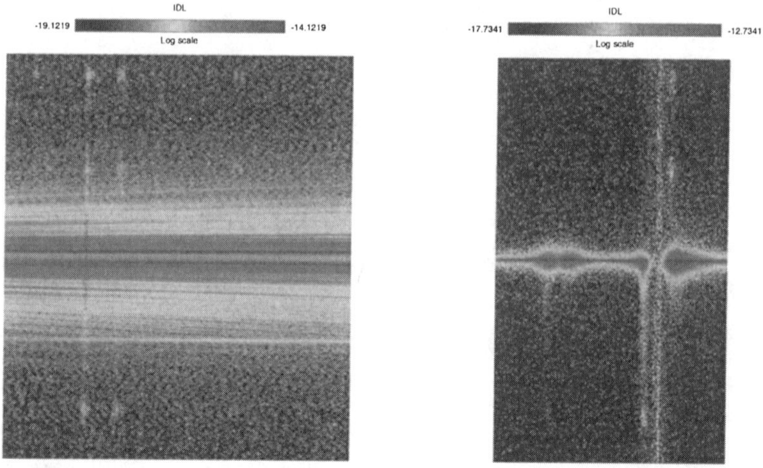

Figure 1. 52″ × 0.2″ spectra of HD 163296 in the optical and at Lyα. *Left*: Optical spectrum showing emission in [O I], Hα+[N II], [S II] and other lines. *Right*: Ultraviolet spectrum showing emission in Si III (1) and Lyα. Both single-orbit images span ±10″.

2. Why Study Winds in the UV?

High contrast spectroscopic studies of mass loss from young stars need not be limited to the optical. In fact, the ultraviolet (UV) offers many of the same advantages, without the need to reject the bulk of the stellar light, or to obscure the region immediately around the star. For the nearest pre-Main Sequence stars, the stellar photosphere drops in flux by a factor of 100–1000 between Hα and Lyα. For case A recombination, the wind imaged in Lyα will be a factor of 10 brighter than at Hα. UV observations, particularly at Lyα, are also subject to interstellar extinction by dust, and more importantly, absorption by atomic hydrogen in the line of sight. However, there are a number of pre-Main sequence stars which have sufficiently low line-of-sight extinction to permit us to take advantage of the increasing contrast provided by observations at Lyα, and many of these stars have spatially unresolved Lyα detections dating back to the International Ultraviolet Explorer (Blondel et al. 1993). The majority of these stars are interior to, or on the edge of, the Local Bubble at d≤200 pc.

3. Results to Date

As of June 2002, three Herbig Ae stars have been spectrally imaged with moderate resolution at Lyα and all show extended emission. STIS coronagraphic imaging of HD 163296 revealed the presence of a circumstellar disk, and a chain of nebulous knots (Devine et al. 2000; Grady et al. 2000), enabling us to optimally place a 52″ × 0.2″ slit along the chain of knots, both in the optical and at Lyα. While the optical spectrum (see Figure 1) demonstrated that the knots were low-excitation Herbig-Haro objects, the region within 4″ of the star was swamped by the residual stellar scattered and diffracted light, even after occulting the star with a 0.5″ bar. In contrast, at Lyα no occulting bar is required, and contamination of the spectrum by light from shorter wavelengths which is diffracted and scattered into the spectral image is negligible. As a result, we are able to trace outflow material in to within 0.025″ from the star. The Lyα spectral image demonstrates that the knots seen in the optical are part of a more extensive outflow, which exits the unresolved point source at -500 km s^{-1} and gradually decelerates. Follow-on long slit observations have demonstrated that the opening angle for the outflow must be under 15° within 2″ (240 AU) of the star, while the opening angle 4–10″ from the star is $\leq 7°$.

The long-slit data can be valuable even without optical detection of a chain of Herbig-Haro objects. While coronagraphic imagery of HD 104237 did not resolve the disk, low resolution (R = 1000) long slit spectra obtained in 2001 October demonstrated that Lyα was spatially extended, with an apparent opening angle of 20° within 1″ of the star. Follow-on R = 10,000 imaging spectroscopy in 2002 April demonstrate that the jet extends 1″ from the star. The counterjet could not be detected in the short (700 s) G140L spectra, but can be traced 2.6″ from the star in the deeper G140M observation (Danks et al. 2002). Similarly, a long slit spectrum of HD 100546 with the slit placed along the disk's semi-major axis (e.g., orthogonal to the orientation expected for a bipolar outflow) revealed Lyα emission extending 200 AU in radius from the star, with 70% of the emission originating outside of the stellar point source. This distance is a factor of 2 beyond the "snow line" for a B9 star and may indicate the presence of a disk wind.

4. NHST

HST and STIS have demonstrated the power of spatially resolved spectroscopy, particularly in the far-UV for probing the geometry and structure of mass loss from nearby, young stars which are accessible to coronagraphic imaging, and to detailed chemical studies in the mid-IR. However, the HST studies are limited to spatially mapping the disks at distances equivalent to Jupiter's orbit and beyond, with no ability to probe the equivalent of the terrestrial planet region. To explore the interaction between stellar winds and the protoplanetary disk, and to assess the relative mass loading of collimated wind (X-wind) and disk wind components we need to probe $\approx$ a factor of 10 closer to the star. Understanding the wind geometry has implications for the formation of chondrules (Shu 2001), annealed silicates and seed grains for Fischer-Tropsch catalysis of organics in the disk (Nuth et al. 2001; Hill et al. 2002), and may ultimately have implications for the organic chemistry of the forming Earth and the origin of life.

Given the priority that the astronomical community and NASA have given to studies of the origin of planetary systems, we would argue that any successor to HST with a UV capability must achieve *at least* a factor of 5 improvement over HST to be able to separately sample collimated wind components (originating at the inner edge of the accretion disk) and uncollimated (e.g., disk winds which may arise in the inner 5 AU of the disk) winds. Improvements in our ability to simultaneous map an outflow with spectral resolution comparable to that achieved by HST, such as can be provided by an UV integral field spectrometer is also highly desirable, in that such a capability would enable us to

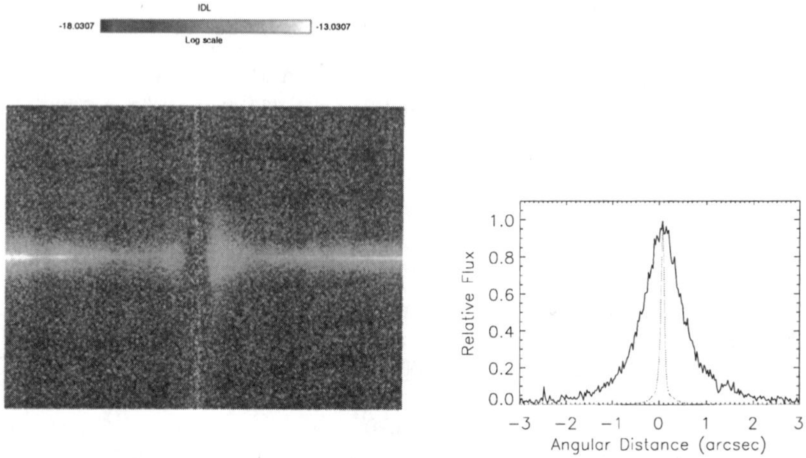

Figure 2. Lyα emission extends ±200 AU from HD 100546 along the semi-major axis of the r = 515 AU disk (Grady et al. 2001). *Left*: The spectral image. *Right*: A spatial cross-cut through the brightest portion of the Lyα emission is shown together with a point source.

fully map a sufficiently large sample to separate system-to-sytem variations from evolutionary effects depending upon the stellar mass or system age.

Acknowledgments. Support for the STIS IDT was provided by NASA Guaranteed Time Observer (GTO) funding to the STIS Science Team in response to NASA A/O OSSA-4-84 through the Hubble Space Telescope Project at Goddard Space Flight Center. CAG was supported through transfer of funds to The National Optical Astronomy Observatories. NOAO is operated by the Association of Universities for Research in Astronomy (AURA), Inc., under cooperative agreement with the National Science Foundation.

References

Bally, J., & Reipurth, B. 2001, ARA&A 39, 403

Blondel, P.F.C., Talavera, A., & Tjin A Djie, H.R.E. 1993, A&A 268, 624

Danks, A. et al. 2002, in preparation

Devine, D. et al. 2000, ApJ 542, L115

Grady, C.A. et al. 2000, ApJ 544, 895

Grady, C.A. et al. 2001, AJ 122, 3396

Grady, C.A. et al. 2002, (in preparation).

Hill, H.G.M., Grady, C.A., Nuth, J.A. III, Heap, S.R. 2002, LPI, 33, 1247

Nuth, J.A. III, 2002, LPI, 33, 1814

Shu, F.H., Shang, H., Gounelle, M., Glassgold, A.E., & Lee, T. 2001 ApJ 548, 1029

Hubble's Science Legacy: Future Optical-Ultraviolet Astronomy from Space
ASP Conference Series, Vol. 291, 2003
K.R. Sembach, J.C. Blades, G.D. Illingworth, R.C. Kennicutt, Jr. (eds.)

The HST Legacy and Future for High Contrast Imaging of of Disks and Outflows from Young Stars

C.A. Grady

NOAO, Eureka Scientific, & GSFC Code 681, NASA/GSFC, Greenbelt, MD 20771

B. Woodgate and C. Bowers

Laboratory for Astronomy & Solar Physics, Code 681 NASA/GSFC, Greenbelt, MD 20771

G. Schneider

Steward Observatory, University of Arizona, Tucson, AZ 85721

Abstract. Understanding the formation of planetary systems is one of the key problems for 21st century astronomy. High spatial resolution, high contrast imaging with HST has played a central role in demonstrating that the dissipation of infrared excesses between 1.5 and 5 Myr coincides with a transition from the protoplanetary disk to a centrally cleared debris disk. We discuss the contribution of HST coronagraphic imaging to our understanding of planetary system formation, and the potential for ultraviolet-optical successors to HST.

1. Introduction

Imaging protoplanetary and planetary systems requires studying material within a few hundred AU of the star. Direct imaging either with high Strehl ratio from the ground or with the Hubble Space Telescope (HST) is sufficient to image bright nebulosity around heavily obscured or intrinsically faint stars (Stapelfeldt et al. 1999; Krist et al. 2000). However, detecting material around optically bright stars, which are typically orders of magnitude brighter than the nebulosity, requires techniques for rejecting the direct light, such as using the polarization of dust-scattered light (Kuhn et al. 2001), or narrow band imaging to restrict the bandpass to selected emission lines. An alternative approach, which can sample both reflection and emission line nebulosity, is to occult the star and suppress the residual diffracted and scattered light by means of an apodizing stop (Lyot 1939).

2. The STIS White-Light Coronagraph

The Space Telescope Imaging Spectrograph (STIS) is equipped with a focal plane wedge structure to occult the star and, depending upon the position along

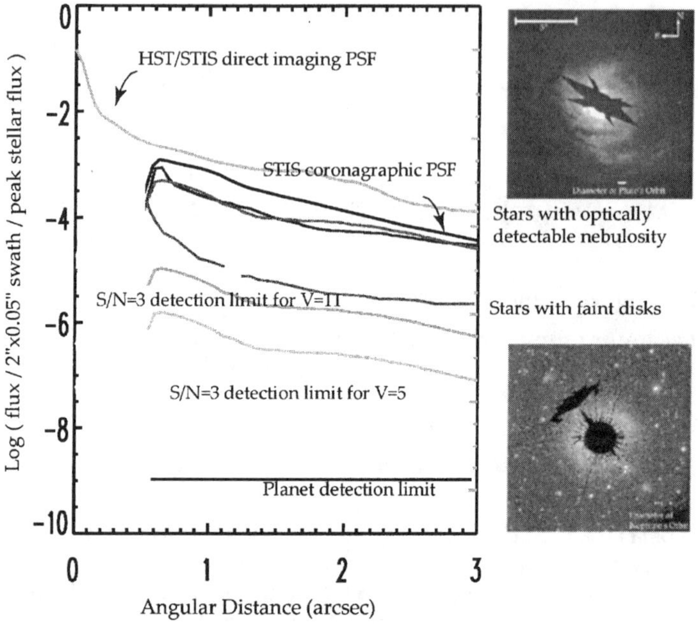

Figure 1. The performance of the STIS white-light coronagraph.

the wedge, a region 0.3–1.5″ from the star, pupil plane outer edge blocking, and an unfiltered CCD to create a simple white-light coronagraph. The coronagraph has a bandpass of 0.2-1.0μm and is sensitive to reflection and emission line nebulosity. To identify circumstellar nebulosity we make use of observations made at two independent spacecraft roll angles, and remove the point spread function (PSF) contribution using a library of color-matched PSF stars.

STIS has been used to coronagraphically image stars spanning $15 \leq V \leq 0$. As shown in Figure 1, our nebulosity detections span a factor of 40 in surface brightness ranging from objects which have surface brightness comparable to the residual coronagraphic PSF, down to disks which are a factor of 40 fainter. Our limiting surface brightness for V=6.5-8 Herbig Ae stars where color-matched but non-contemporary PSF data are used in the coronagraphic data reduction is $\approx 10^{-6}$ of the peak stellar flux at 2″ from the star. Experimentation with higher S/N, contemporary calibration observations suggests that at V=4-5 the limiting contrast is closer to 10^7. Thermally-driven focus variations between the science and calibration observations limit our ability to detect circumstellar nebulosity at $r \leq 1''$ (Figure 1).

3. Results to Date

Coronagraphic imaging of pre-Main Sequence (PMS) stars has resulted in a wealth of disk, outflow, companion, and large-scale nebulosity detections. The stars with the largest-scale nebulosity ($800 \leq r \leq 1300$ AU) in our survey appear to be single stars. The slow decline in surface brightness with distance from these stars is consistent with detection of circumstellar envelopes. The presence of such envelopes around 2 to 10 Myr old stars indicates that, for single stars, the envelope can persist through most of the PMS lifetime of the star. Our data suggest that envelopes are fully dissipated around closer binary and multiple systems prior to 0.5 Myr. Similarly, large ($r \geq 200$ AU) disks are found only for the widely separated ($r \geq 3''$) or single stars. Our survey has resulted in the detection of microjets around both T Tauri stars with $t \leq 3$ Myr, and Herbig Ae stars with $t \leq 4 - 6$ Myr.

Our sample includes many stars with large disks which have previously been imaged in the millimeter. The survey data point to a change in the optical detection frequency of large disks around 5 Myr, with fewer stars in this age range detected than in the younger systems (e.g., MWC 480, Grady et al. 2001b; Augereau et al. 2001). This suggests that either grain agglomeration has occurred or that the grains have begun to settle to the disk mid-plane, resulting in steeper disk surface brightness power laws, or both. Interestingly, this change in the large-scale disk properties appears to coincide with the age range where near-stellar IR excesses become less common (Haisch et al. 2001), and when structure in the disk, in the form of central clearing, or partial arcs begins to appear, potentially marking the presence of planets in the disks. The most important result here is that development of large-scale structure is not limited to the terrestrial planet region, may affect much of the disk, and occurs by 5-8 Myr, much earlier than had been predicted by minimum-mass Solar nebular models, while developing more slowly than expected for disk instabilities to produce giant gaseous protoplanets (see review by Wuchterl et al. 2000).

4. NHST

The available data for mature exo-planetary systems and the coronagraphic surveys indicate that planetary systems are more diverse in size and structure than has historically been considered, and that this diversity begins to be apparent before 1 Myr, while the star is still surrounded by nebulosity. Observations of the final protoplanetary phases of the disk evolution are therefore likely to be a critical component in understanding both the typical evolution of planetary systems, and the origin of our own Solar System. The first such observations have been taken with the current generation of high-contrast imaging surveys on HST. While useful, these observations have been limited to the closest and highest surface brightness protoplanetary disks (Grady et al. 1999, 2000, 2001a,b; Weinberger et al. 2002), and the brightest debris disks (Weinberger et al. 1999; Mouillet et al. 2001; Heap et al. 2000). The optical non-detections of systems known to have large disks from millimeter observations points to the need to a new generation of observations with contrasts relative to the total stellar flux $\geq 10^6$. HST/ACS, as a result of improved apodization, is expected to provide up to an order of magnitude improvement in contrast at $r \geq 1.5''$ over what HST/STIS and HST/NICMOS have achieved, but this may be insufficient to detect some of these systems, let alone the fainter debris disks. The HST coronagraphs, also do not provide the requisite high contrast imaging within 1″ of the star.

Thus, any successor mission to HST should be capable of providing high contrast imaging ($\geq 10^8$) with an angular scale which is well-matched to the sizes of protoplanetary disks at the distances of the nearest and best-studied star formation regions and young stellar associations. Since the initial signatures of the presence of planets appear to occur in systems with appreciable nebulosity, any high contrast imaging capability should be sufficiently flexible that it can detect point sources (e.g., brown dwarf or planetary companions) in the presence of nebulosity.

Acknowledgments. Support for the STIS IDT was provided by NASA Guaranteed Time Observer (GTO) funding to the STIS Science Team in response to NASA A/O OSSA-4-84 through the Hubble Space Telescope Project at Goddard Space Flight Center. CAG was supported through transfer of funds to The National Optical Astronomy Observatories. NOAO is operated by the Association of Universities for Research in Astronomy (AURA), Inc., under cooperative agreement with the National Science Foundation.

References

Augereau, J.C., et al. 1999, A&A 350, L51

Augereau, J.C., et al. 2001, A&A 365, 78

Grady, C.A., et al. 1999, ApJ 523, L151

Grady, C.A., et al. 2000, ApJ 544, 895

Grady, C.A., et al. 2001a, AJ 122, 3396

Grady, C.A., et al. 2001b, in ASP Conf. Ser. 244, Young Stars Near Earth: Progress and Prospects, eds. R. Jayawardhana & T.P. Greene, (San Francisco: ASP), 209

Heap, S.R., et al. 2000, ApJ 539, 435

Krist, J.E., et al. 2000, ApJ 538, 793

Krist, J.E., Stapelfeldt, K.R., & Watson, A.M. 2002, ApJ 570, 785

Kuhn, J.R., Potter, D., & Parise, B. 2001, ApJ 553, L189

Haisch, K.E., Jr., Lada, E.A., & Lada, C.J. 2001, ApJ 553, L153

Lyot, B. 1939 MNRAS, 99, 580

Mouillet, D., Lagrange, A.M., Augereau, J.C., & Ménard, F. 2001 A&A 372, L61

Stapelfeldt, K.R., et al. 1999, ApJ 516. L95

Weinberger, A., et al. 1999 ApJ 525, L53

Weinberger, A., et al. 2002, ApJ 566, 409

Wuchterl, G., Guillot, T., & Lissaur, J.J. 2000, in Protostars & Planets IV, eds. V. Mannings, A.P. Boss, & S.S. Russell, (Tucson: University of Arizona Press), 1081

Hubble's Science Legacy: Future Optical-Ultraviolet Astronomy from Space
ASP Conference Series, Vol. 291, 2003
K.R. Sembach, J.C. Blades, G.D. Illingworth, R.C. Kennicutt, Jr. (eds.)

High Angular Resolution Combined with Excellent Spectral Performance: A Necessity for the Next Ultraviolet/Visible Space Telescope

Theodore R. Gull

Code 681, NASA Goddard Space Flight Center, Greenbelt, MD 20771

Abstract. For complex astronomical objects with fine detail, excellent angular resolution combined with appropriately-matched spectral resolution can provide much useful information and insight on physical conditions within the structures. The performance of future large ultraviolet/optical space telescopes must include instruments that simultaneously provide both excellent spectral performance and excellent angular performance. Two examples of spatially resolved spectroscopy, accomplished with the Space Telescope Imaging Spectrograph, are described.

1. Introduction

Many desired attributes of future ultraviolet/optical space telescopes were described at this conference. We all agree that more efficient light-gathering capability is necessary. The equivalent of a four to eight meter aperture "light bucket" with very efficient instruments may satisfy many observers, especially those who are interested in spectroscopy ranging from spectra of very faint objects to studies of the interstellar medium (ISM) and the intergalactic medium (IGM). However, angular resolution and optical stability must be folded into the specifications as many "point-like" sources will prove to have spatial structure resolvable near the diffraction-limited capability of future telescopes. Each quasar is at the core of a faint galaxy; an active galactic nucleus resides in a faint galaxy; gamma ray bursters are associable with very faint, distant galaxies. A massive star with large mass loss rate, a binary star system, mass-change system, or a protoplanetary system often can be resolved into a stellar component(s) and an extended structure. Appropriate imaging and spectral resolution can bring out much insight on the physical conditions within these objects.

The criterion for excellent spatial response must include more than just the resolution capability of the primary aperture. The finish of the primary and secondary mirrors, the choice of a monolithic primary versus segmented components and the mechanical stability of the optical system all contribute to the overall telescope performance. In addition, very careful specifications must be included for the instruments in the focal area. Combining light scatter of the instrument optics, the diffraction and scatter of the dispersing element (usually gratings), and the imaging performance of the detector must be addressed.

This discussion does not provide specifications for these additional criteria. Instead a few observations successfully obtained with the Space Telescope

Imaging Spectrograph (STIS) are provided as examples. While the STIS Instrument Definition Team pushed technology to the limits of the mid-1990s, we found some instrumental parameters limit the total system performance. We point these out, not to disparage STIS (which is indeed a superbly performing, well-calibrated imaging spectrograph), but to sensitize users of future space telescopes with focal plane instruments as to what issues should be addressed up front and during the facility development. The STIS CCD has measurable internal scatter and fringing effects in the near-red. Grating scatter in the blue and the ultraviolet (plus telescope scatter) leads to limits in detectability of faint objects in the vicinity of a bright point source. Fiducials in the focal plane for the spectrograph apertures can be used to cut down on the scattering properties.

2. Why Space?

Ultraviolet studies can only be done in space due to atmospheric absorption. Science goals can drive optical performance to the angular diffraction limit in the ultraviolet. In the local rest frame ($z \leq 1.5$) the case is easily made when considering the relationship of quasars to host galaxies, gamma ray bursters within their environments, and the structure of Seyfert galaxies. In the local cluster of galaxies, individual stars and their immediate environment, especially when it comes to massive winds and ejecta, become important. Within our galaxy, the case is best made that highly evolved stars and young stars have very energetic ejecta seen in the form of bipolar structures with disks and even jets. High spatial resolution with good, appropriately-matched spectroscopy can only lead to improved knowledge of these systems.

Often the statement is made that better angular resolution can be achieved on the ground and that the resolution is limited only by the active optics and the primary aperture. If the goal is to image low contrast structures against a relatively bright background, then indeed such a goal can be achieved on the ground. However, active, rapid correction can compensate only for major aberrations of the atmosphere. Contrast ratios of 100-fold can be achieved in the near-red for relatively large isoplanatic patches. Detecting very faint structures near a bright point-like source proves to be much more difficult. Nulling interferometers are in their infancy, but promise to improve this situation greatly both on the ground and in space. Diffraction effects of multi-element primaries contribute high frequency structure to the telescope point spread function (PSF). Coupled with thermal changes, this structure can cause a continuously changing PSF. All of these effects lead to a less than optimal PSF that may provide a good full width at half maximum measurement, but will not provide sufficiently high contrast to measure faint structures in the vicinity of a bright point source.

2.1. Eta Carinae

The STIS study of Eta Carinae and its ejecta, commonly known as the Homunculus, is an excellent example of what an imaging spectrograph behind near-diffraction-limited optics can contribute to understanding the physical conditions. Eta Carinae is a massive stellar system - quite likely a binary - that ejected several solar masses in the 1840s. Today the HST/WFPC2 images reveal a bipolar reflection nebula with an extended intervening skirt. But that is

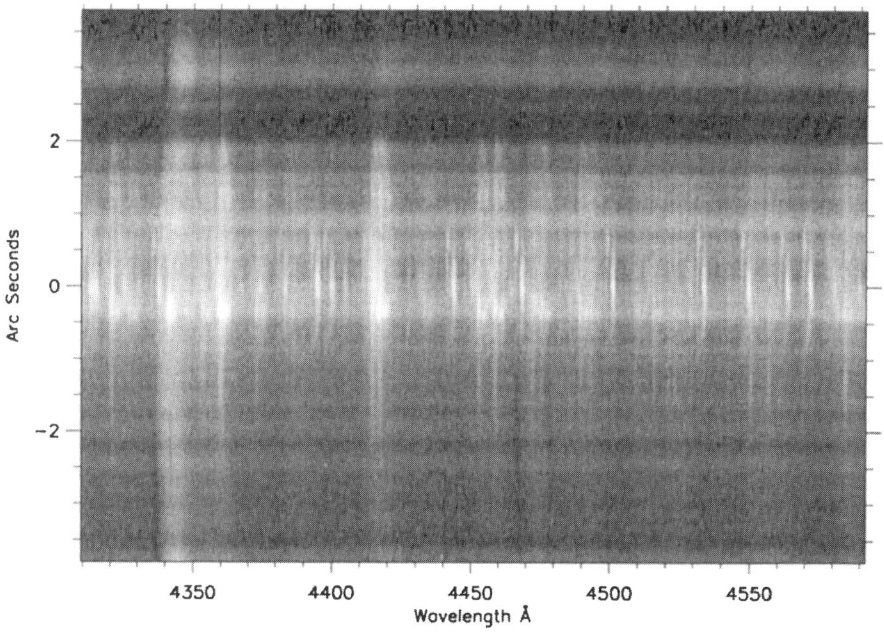

Figure 1. A STIS nebular spectrum of the Strontium patch in Eta Carina. Velocity and spatial shape help differentiate this region from other Homunculus structures.

not all of the story! Within the Homunculus ($10'' \times 18''$) lies an ionized bipolar structure called the Little Homunculus, extending $2''$ above and below Eta Carinae. With the high angular resolution of HST, the STIS visible spectra can be recorded with $R \approx 7000$ and $0.05''$/CCD pixel; this suppresses the bright, scattered continuum and allows us to see the thousands of emission lines that define the structure of the Little Homunculus and other emission structures. Within the skirt of the Homunculus are several "neutral" regions that exhibit no hydrogen emission, but are seen in the light of hundreds of emission lines from many heavy ions such as Fe II, Ni II, Ti II, and Sr II. Based upon the spatial structure and the velocity changes, we have identified nearly 1000 emission lines associated with this "Strontium patch" (Figure 1). Within a few light days of the central source is an ionized sphere, likely the extended wind, and the Weigelt blobs. We have identified over two thousand emission lines in the Weigelt blobs and are building models to determine the temperature, density, and heavy element abundances. Immediately around the central source is a reflection shell, seen in continuum and several emission lines extending a fraction of a lightday in diameter (Figure 2). Within the central source spectrum are the thousands of absorption lines - over 500 unique lines identified in the spectral region from 2400-3150A - some with up to twenty velocity components! Thousands more absorption lines below 2400 Å are now being identified.

Observations with IUE, with FUSE, with HST/GHRS and, in the near future, with HST/COS would not separate these components due to limited an-

Figure 2. Partial STIS echellogram of Eta Carinae near 2000 Å. The tilted absorptions are due to forground nebular ejecta. The vertical absorptions are due to the ISM. The nearly-point source (0.05″) is surrounded by a 0.14″ diameter shell.

gular resolution (COS does not replace STIS, but performs excellently for point sources in the 1150–3200 Å region; it will not provide spatial/spectral information at the diffraction limit of HST.) Here, STIS uniquely provides considerable spatial information on the location of many emission features. Until the next generation large ultraviolet/visible telescope is working in space, STIS is the only promise of providing this information. We note that the complex IUE and FUSE spectra obtained with larger apertures have been very difficult to study as 90 to 95% of the ultraviolet flux comes from scattered light in the Eta Carina region (i.e., the ratio of FUSE/STIS flux measures at 1175–1185 Å is 20 and the ratio of IUE/STIS at 1700 Å is 10). The absorption lines detected by IUE and FUSE are smeared in velocity due to the rapidly-expanding, scattering ejecta. By contrast the STIS spectra with 0.05″ angular resolution show very sharp absorption lines, most originating from intervening ejecta in twenty velocity systems.

3. Conclusions

Excellent imaging and spectroscopic performance is necessary for future ultraviolet/optical telescopes to provide unique insight into the structure of complex sources ranging from distant host galaxies to galactic stellar systems.

The Hubble Space Telescope Project provided resources for the STIS GTO Team. Observations and GO funding were provided through the STScI. We sincerely thank the support of Don Lindler and Keith Feggans in the data reduction and preparation of this paper.

Hubble's Science Legacy: Future Optical-Ultraviolet Astronomy from Space
ASP Conference Series, Vol. 291, 2003
K.R. Sembach, J.C. Blades, G.D. Illingworth, R.C. Kennicutt, Jr. (eds.)

A Detailed View of the Photosphere of the Hot White Dwarf G191-B2B from STIS

J.B. Holberg

Lunar and Planetary Lab., University of Arizona, Tucson, AZ 85721

M.A. Barstow

Dept. of Physics and Astronomy, University of Leicester, Leicester, LE1 7RH UK

I. Hubeny

National Optical Astronomy Observatory, P.O. Box 26732, 950 N. Cherry Ave., Tucson, AZ 85726-6732

M.S. Sahu and F.C. Bruhweiler

Goddard Space Flight Center, Greenbelt, MD 20771

W.B. Landsman

Science Systems and Applications Inc., Code 681, Goddard Space Flight Center, Greenbelt, MD 20771

Abstract. G191-B2B, in addition to being a well known standard star, is also the archetype for the class of hot, heavy-element-rich DA white dwarfs. We have assembled a unique, high signal-to-noise spectrum of G191-B2B from a co-addition of 22 E140H and 40 E230H Space Telescope Imaging Spectrograph (STIS) spectra. This superb dataset fully covers the wavelength range from 1155 Å to 3169 Å and presents an unprecedented, detailed view of the ultraviolet (UV) stellar absorption lines in the spectrum of a white dwarf star. We use this spectrum to search for a number of previously undetected heavy elements in the G191-B2B photosphere and to estimate elemental abundances and place upper limits on the abundances of a number of key species. An accurate photospheric inventory of heavy elements in hot degenerate stars is basic to our understanding of the process of chemical diffusion in these stars. These observations highlight the wealth of information present in echelle-resolution spectra of white dwarfs such as G191-B2B and the need for a succeeding generation of UV instrumentation capable of providing data of this type for other degenerate stars.

1. Introduction

The hot hydrogen-rich (DA) white dwarf G191-B2B is well-known as a photometric and spectroscopic standard star and serves as one of the primary calibration standards for Hubble Space Telescope (HST) spectroscopy. G191-B2B is also the prototype of an intensively studied class of metal-rich white dwarfs that exhibit numerous narrow interstellar-like UV absorption lines due to trace levels of elements heavier than hydrogen. The abundance of these heavy elements is primarily controlled by a diffusive equilibrium between gravitational settling and outward radiative forces. The study of these lines in the UV spectra of white dwarfs such as G191-B2B provides significant information on the chemical structure and evolution of white dwarf photospheres. The presence of these UV photospheric features was first discovered by Bruhweiler & Kondo (1981, 1983) in IUE echelle spectra of G191-B2B and other hot white dwarfs. Since that time, observations of G191-B2B by numerous UV, far-UV, and extreme-UV instruments have revealed an increasingly detailed and complex array of photospheric lines. Therefore, it is appropriate from a scientific, as well as a calibration, standpoint that G191-B2B was the subject of an extensive series of HST/STIS calibration observations. No white dwarf has ever been observed with this unique combination of high spectral resolution (R $\sim$ 120,000), high signal-to-noise (S/N), and extensive wavelength coverage. As a result, many new details of the photosphere of G191-B2B are evident in these data.

2. Data Reduction and Models

Between 2000 and 2001, G191 B2B was extensively observed with HST/STIS at the highest echelle resolution. A number of standard and non-standard grating settings using were used with both the E140H and E230H gratings to cover the entire range of vacuum UV wavelengths observable with the STIS MAMA detectors. In all, there are 22 E140H and 42 E230H spectra totaling 85,276 s obtained in the $0\farcs2 \times 0\farcs2$ aperture. We have co-added these spectra and assembled them into two overlapping spectra: one covering 1155.0 Å to 1687.9 Å for the E140H grating, and one spanning 1625.8 Å to 3189 Å for the E230H grating. In Figure 1 we plot these two spectra.

The data were processed using standard `CALSTIS` procedures. However it is important to note that improved treatments of the scattered light correction were employed as well a new procedure, developed by D. Lindler, which automatically determines echelle blaze in the application of grating ripple corrections. These procedures have been uniformly applied by one of us (WBL) to all spectra in our dataset. The resulting spectra were sampled at a sufficient spectral resolution to insure that the data points remain statistically independent. Signal-to-noise as a function of wavelength for these spectra range from $\sim$100 in the best exposed regions to $\sim$30 in less-well-covered regions. In wavelength regions having the highest S/N, absorption features with equivalent widths of less than 3mÅ are routinely observed.

A series of models of G191-B2B were calculated with the model atmosphere code `TLUSTY` (Hubeny 1988; Hubeny & Lanz 1995). Specifically, a sequence of non-LTE models was investigated and adjustments made to T_{eff}, $\log g$, and

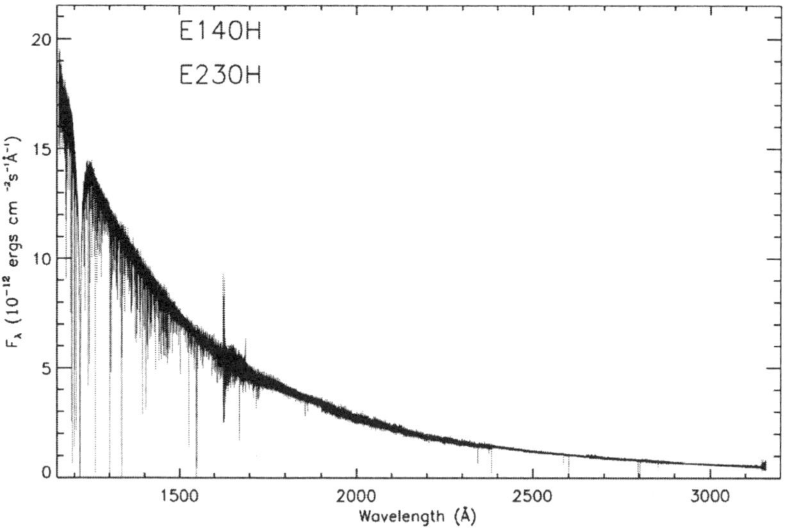

Figure 1. Co-added HST/STIS spectra of G191-B2B.

the Fe and Ni abundances that produced a good overall match to the Fe V and Fe IV lines as well as the Ni V and Ni IV lines. In the final model, abundances of other species such as C, N, and O were adjusted to improve fits for these features. The specific model discussed here used T_{eff} = 54,000 and $\log g$ = 7.50. The heavy elements were assumed to have a homogeneous distribution in the photospheric levels where the UV lines are formed. The He abundance was set at log (He/H) = –5. The opacity sampling treatment of line blanketing was used as opposed the previous approach of the opacity distribution function. This allowed an improved treatment of collisional excitation rates, resulting in Fe and Ni abundances a factor of two lower than previous estimates using the opacity distribution function.

3. Photospheric Features

The spectrum of G191-B2B is particularly rich in lines due to Fe V and Fe IV as well as Ni V and Ni IV (see Holberg et al. 1998). Fe and Ni abundances of log (Fe/H) = –5.50 and log (Ni/H) = –6.15, respectively, reproduce the observed lines very well. No other iron group elements are observed. In particular, it is possible to set limits of log (X/H) = –8 for the relative abundances of Cr, Mn, and Co.

Aluminum was previously identified in International Ultraviolet Explorer echelle data by Holberg et al.; however, no useful abundance estimates were possible due to the low *S/N* of those spectra. The present data clearly show the strong Al III resonance lines. Weaker excited Al III features at 1605.766,

1611.814, and 1611.873 Å are also seen. All imply an aluminum abundance of log (Al/H) = –6.52 . In contrast to Al, no Mg II or Mg III lines are seen in these data, in spite of a number of potentially strong transitions in the observed wavelength region; an upper limit of log (Mg/H) = –7.0 is implied.

Relative abundances of log (C/H) = –6.70, log (N/H) = –7.0, and log (O/H) = –5.82 were found to reproduce the observed features for these elements. A major discrepancy exists, however, for the C IV resonance doublets. As previously noted by Bruhweiler et al. (1999) and discussed by Vennes & Lanz (2001), these lines contain strong non-photospheric blue-shifted components, possibly related to mass loss. A number of lines due the second and third ionization stages of Si are evident. The observed Si IV resonance lines are reproduced well by the model for an abundance of log (Si/H) = –6.4. However, weak blue-shifted components are seen on the wings of both doublet lines. These have the same velocity as the non-photospheric C IV features and may be related.

4. Conclusions

STIS E140H and E230H calibration observations have been combined to produce a full high-resolution, high-S/N view of the spectrum of G191-B2B over the entire 1155 Å – 3190 Å region. Many new lines and several ionization states of key elements have been seen for the first time. No previously unidentified elements were detected, however. Derived abundances agree well with the results of earlier STIS spectra of G191-B2B obtained by Vennes & Lanz (2001). These new data yield the most detailed and accurate inventory of heavy element abundances for any hot white dwarf. Future UV instrumentation with high dispersion capability can produce similar observations for many other hot white dwarfs.

Acknowledgments. This work was funded by Space Telescope Science Institute Archive Grant AR9202.

References

Bruhweiler, F.C., & Kondo, Y. 1981, ApJ, 248, L123

Bruhweiler, F.C., & Kondo, Y. 1983, ApJ, 269, 657

Bruhweiler, F.C., Barstow, M., Holberg, J., & Sahu, M. 1999, BAAS, 31, 1419

Holberg, J.B., Barstow, M.A., & Sion, E.M. 1998, ApJS, 119, 207

Hubeny, I. 1988, Comput. Phys. Comun., 52, 103

Hubeny, I., & Lanz, T. 1995, ApJ, 439, 875

Vennes, S., & Lanz, T. 2001, ApJ, 553, 339

Hubble's Science Legacy: Future Optical-Ultraviolet Astronomy from Space
ASP Conference Series, Vol. 291, 2003
K.R. Sembach, J.C. Blades, G.D. Illingworth, R.C. Kennicutt, Jr. (eds.)

Resolving Star Forming Regions in Distant Galaxies: A Case Study with Blue Spheroid Candidates out to z =1

Myungshin Im

SIRTF Science Center, California Institute of Technology/JPL, Mail Stop 220-6, Pasadena, CA 91125

Abstract. Blue Spheroid Candidates (BSCs) are galaxies which morphologically resemble early-type galaxies, but have blue colors. Theories predict the existence of massive early-type galaxies with active star formation induced by recent merger events, and some speculate that BSCs might be such objects. Alternatively, they could simply be starbursting low-mass galaxies. In order to distinguish between these two possibilities, we study the kinematics, structural properties, and masses of ten BSCs using Keck and Hubble Space Telescope (HST) data. Our study reveals that BSCs are likely to be low mass systems ($M < 10^{10} M_{\odot}$) rather than direct progenitors of the present-day massive ellipticals. Although our study reveals important clues on the nature of BSCs, precise measurement of their dynamical mass is very challenging due to the small size of the objects and the lack of spatially-resolved spectra. Getting such information requires a large, optical space telescope, or future adaptive optics observations with large ground-based telescopes.

1. Introduction

Currently popular hierarchical models of galaxy formation predict increased star-forming activities within high redshift galaxies due to merger events (e.g., Cole et al. 2000). According to the theory, massive early-type galaxies form via major merger events and grow via accretion events, both of which happen frequently at redshifts between $z = 0$ and $z = 1$ (e.g., Benson et al. 2002). If so, we should be able to find distant spheroidal galaxies with blue star-forming regions possibly on top of red underlying stellar populations.

2. Blue Spheroid Candidates

Previous studies have discovered blue objects with spheroidal shape (BSCs). Some of these objects appear to have blue nucleated regions on top of a redder underlying stellar population (Menanteau et al. 2001; Schade et al. 1999; Im et al. 2001). We identify 10 BSCs from HST/WFPC2 images of the Deep Extragalactic Evolutionary Probe (DEEP) Groth Strip Survey. Figure 1 shows the rest frame U-B color gradients of BSCs along their major axis, as well as a two-dimensional color map. From Figure 1, we find that five, possibly six, of the 10 BSCs show a tendency to have blue inner regions and red outer regions. Could

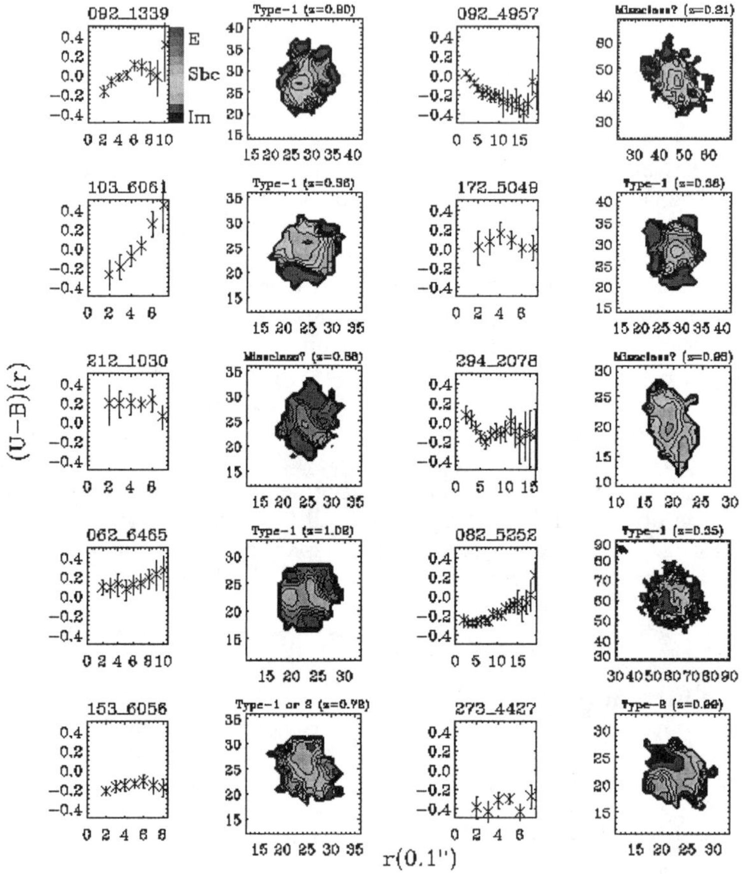

Figure 1. Azimuthally averaged rest-frame $U - B$ color profiles plotted along their major axes, and 2-dimensional $U - B$ color maps of BSCs. For the 1-D color profile, the x-axis shows the distance from the center of the object, r, in 0.1 arcsec. For the 2-D color map, the x and y axes are also in units of pixels. Type-1 BSCs have a central region significantly bluer than the outer region, which in many cases is almost as red as typical field E/S0 galaxies. Note that the expected $U - B$ colors of passively evolving stellar populations after 0.1 Gyr bursts range from 0.1 to 0.5 depending on the age and metallicity at the redshift of interest.

they be the massive elliptical galaxies undergoing merger events as predicted in the semi-analytical models? Figure 2 shows the half-light radius (r_{hl}) versus the velocity dispersion (σ) measured from our Keck spectra for these BSCs. Also plotted is the same relation for various types of local galaxies, and for normal, red E/S0s out to $z = 1$. Figure 2 shows that no BSCs resemble present-day, massive, early-type galaxies or distant red E/S0 galaxies. They are more likely to be low-mass spheroids undergoing starburst, or simple misclassifications.

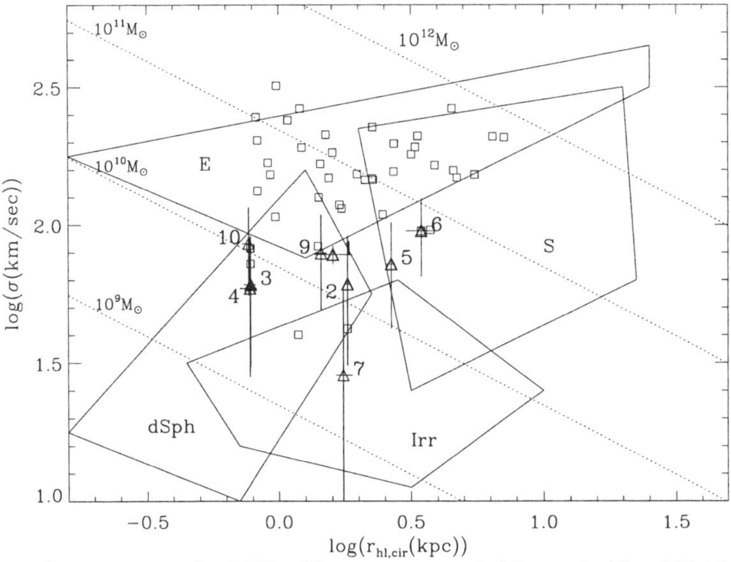

Figure 2. σ vs. r_{hl} for BSCs. The spaces occupied by typical local Hubble types are taken from Phillips et al. (1997). BSCs (triangles) are generally low mass objects that lie in or near the space occupied by dEs. Red E/S0 galaxies (Im et al. 2002) are plotted with squares. The numbers in the figure correspond to the object number in Table 2 of Im et al. (2001).

3. Future Prospects

Our study of BSCs demonstrates that spatially-resolved color, structural, and spectral information of distant galaxies can be used to test galaxy formation and evolution models. While it is true that the HST images combined with the Keck spectra provides such information to some degree, it is still not good enough for more detailed study of many distant galaxies. The typical sizes of BSCs are small ($r_{hl} \sim 0\farcs3$), and the main uncertainty on the mass estimate comes from the insufficient resolution of the HST images and the Keck spectra – the blue regions are not resolved, thus it is impossible to give accurate estimates on the mass of the star-forming regions. With the future 8 m-class optical-ultraviolet space telescope, we will be able to study the complex star formation history of a large sample of distant, faint galaxies in detail. Figure 3 shows simulated images of faint galaxy with $r_{hl} \sim 0\farcs3$ for the Hubble Space Telescope Advanced Camera for Surveys versus a future diffraction-limited 8 m-class optical-ultraviolet space telescope. Star-forming regions and morphological details are clearly visible with the future 8 m-class telescope, while the HST image provides only a limited view of the galaxy.

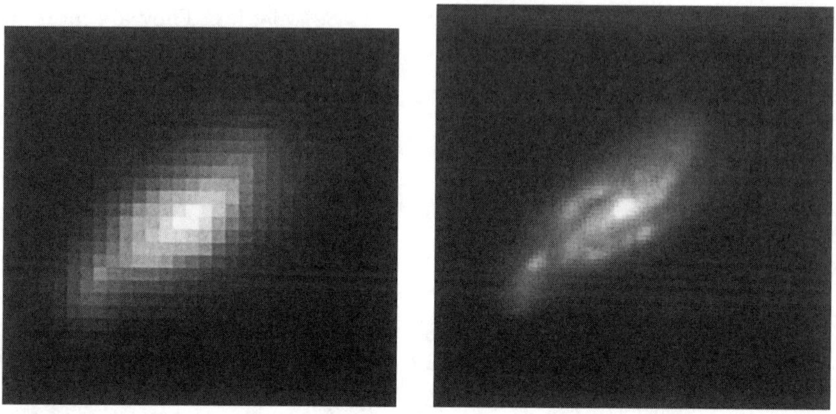

Figure 3. Simulated images of a distant galaxy with $r_{hl} = 0\farcs3$. The left figure is the simulation for HST/ACS, and the right figure is the simulation for a future 8 m space telescope. The image size is $1\farcs5$ by $1\farcs5$.

Acknowledgments. This work is supported by NSF grant AST-9529098, and STScI grant AR-8767.01. We also acknowledge the support of the Jet Propulsion Laboratory, California Institute of Technology, which is operated under contract with NASA.

References

Benson, A.J., Ellis, R.S., & Menanteau, F. 2002, MNRAS, submitted

Cole, S., et al. 2000, MNRAS, 319, 168

Im, M., et al. 2001, AJ, 122, 750

Im, M., et al. 2002, ApJ, 571, 136

Menanteau, F., et al. 2001, MNRAS, 322, 1

Phillips, A.C., et al. 1997, ApJ, 489, 543

Schade, D., et al. 1999, ApJ, 525, 31

Hubble's Science Legacy: Future Optical-Ultraviolet Astronomy from Space
ASP Conference Series, Vol. 291, 2003
K.R. Sembach, J.C. Blades, G.D. Illingworth, R.C. Kennicutt, Jr. (eds.)

Sensing Absorption Lines from the Interstellar and Intergalactic Media at High Signal-to-Noise and Resolution: How Far Can We Go?

Edward B. Jenkins

Princeton University Observatory, Princeton, NJ 08544-1001

Abstract. From 1960 to the present, the products of effective area, multiplexing capability, and wavelength resolving power of individual ultraviolet spectrographs and their telescopes have doubled about every 16 months. Assuming this trend will continue, we consider not only what new scientific questions can be addressed, but also how many new elements and ionization stages can be detected before the spectra become confusion limited. These issues are explored for investigations of gas complexes inside our Galaxy and in the low-z intergalactic medium.

1. Evolution in Capability

For UV spectrographs and telescopes operating in space, there are many properties that govern their capabilities to meet new scientific challenges. (1) *Effective area* $A_{\rm eff}$: the product of the unobscured area of the telescope multiplied by all of the efficiencies of the instrument subsystems (including the detector's quantum efficiency). (2) *Multiplexing power* M: the number of independent spectral elements that can be recorded simultaneously. (3) *Wavelength resolving power* $R \equiv \lambda/\Delta\lambda$: a quantity that governs the ability to distinguish features with small wavelength separations and, for absorption lines, to reduce the amount of unnecessary additional noise from the surrounding continuum. (4) *Total wavelength coverage.* (5) *Background flux* $F_{\rm bkg}$: the background count level whose statistical fluctuations can account for a reduction in S/N by a factor $(1+F_{\rm bkg}/F_{\rm source})^{1/2}$ [should include an equivalent $F_{\rm bkg}$ contribution created by any extra noise from a detector or its signal processing system]. (6) *Maximum S/N*: highest level of accuracy that can be achieved in compensating for variations in the detector's sensitivity versus wavelength. (7) *Maximum allowable flux*: the largest flux that can be recorded without saturating the signal or damaging the detector. (8) *Scattered light uncertainties*: for deep absorption lines, poorly known scattered light levels can compromise the accuracy in determining optical depths. If the scattered light level is large, $F_{\rm bkg}$ is increased.

Improvements in the product of the first three properties, $A_{\rm eff}$, M, and R for various space observatories from 1960 to the present seem to have exhibited a doubling time of about 16 months (Jenkins 1999). An extrapolation of this trend to the years 2015–2020 could be represented by a telescope and a high resolution spectrograph with $A_{\rm eff} = 15,000\,{\rm cm}^2$ (perhaps an 8 m diameter telescope feeding an echelle spectrograph and detector with a combined efficiency of 3% – a modest advancement over the currently flying STIS spectrograph), a resolving power $R = 5 \times 10^5$ ($0.6\,\mathrm{km\,s^{-1}}$), and a single-exposure coverage of 300Å ($M = 10^5$). This combination is suitable for high-precision recordings of interstellar absorption features in the spectra of early-type stars. For probing gas systems in the low-z intergalactic medium (or in other galaxies), we might envision a larger effective area $A_{\rm eff} = 47,000\,\mathrm{cm}^2$, obtained once again with an 8 m diameter telescope aperture, but with a spectrograph having an efficiency equivalent to the Cosmic Origins Spectrograph (COS) to be installed soon on HST. A reasonable evolution of capability might call for $R = 50,000$ ($6\,\mathrm{km\,s^{-1}}$) and a coverage of 900Å in a single exposure [again following the trend found by Jenkins (1999)].

2. Greater Sensitivity for Registering Less Abundant Species

It is generally recognized that important new insights can arise from observations of constituents with very low abundances. A range of interesting problems that can be addressed by observing weak absorption features at high resolution has been discussed by Jenkins & Wallerstein (1999) and will not be repeated here. They derived a formula for the sensitivity of detecting (at the 1σ level in an observing time t) minor constituents with abundances X/H relative to hydrogen,

$$\begin{aligned}\log(X/\mathrm{H}) &= 0.5(\log \Delta v - \log F_\lambda - \log A_{\rm eff} - \log t) \\ &\quad - \log \lambda - \log(f\lambda) - \log[N(\mathrm{H\,I})] + 13.614\,, \qquad (1)\end{aligned}$$

where Δv is the larger of either the velocity resolving power c/R or the line's velocity dispersion parameter b divided by 0.6 (FWHM of a Gaussian profile), F_λ is the target's flux (expressed in erg $\mathrm{cm^{-2}s^{-1}\AA^{-1}}$), $f\lambda$ is the product of the transition's strength and wavelength (in Å), and $N(\mathrm{H\,I})$ is the hydrogen column density (in cm^{-2}).

3. Some Concrete Examples

Figure 1 shows the brightnesses of a selection of suitable target stars as a function of their foreground hydrogen column densities. For the straight line running along the lower edge of the points and an integration time of 40,000 s to achieve a 2σ detection of a line, Equation 1 reduces to

$$\log(X/\mathrm{H}) + \log f\lambda = -(7.17 + 0.5 \log A_{\rm eff}) \qquad (2)$$

if b is set to a representative value of $5\,\mathrm{km\,s^{-1}}$. For the high resolution spectrograph described earlier, $\log A_{\rm eff} = 3.70$ if we assume that 3 exposures are needed to cover a respectable range in wavelength, say, from 1150 to 2050Å.

For a high sensitivity spectrograph viewing a Lyman limit system with $N(\mathrm{H\,I}) = 10^{18.5}\mathrm{cm}^{-2}$ against a background quasar with $F_\lambda = 10^{-15}$ erg cm^{-2} s^{-1} $\mathrm{\AA}^{-1}$, again one could achieve a 2σ detection in 40,000 seconds for lines with $b = 5\,\mathrm{km\,s^{-1}}$ with

$$\log(X/\mathrm{H}) + \log f\lambda = -(2.08 + 0.5 \log A_{\rm eff}) \qquad (3)$$

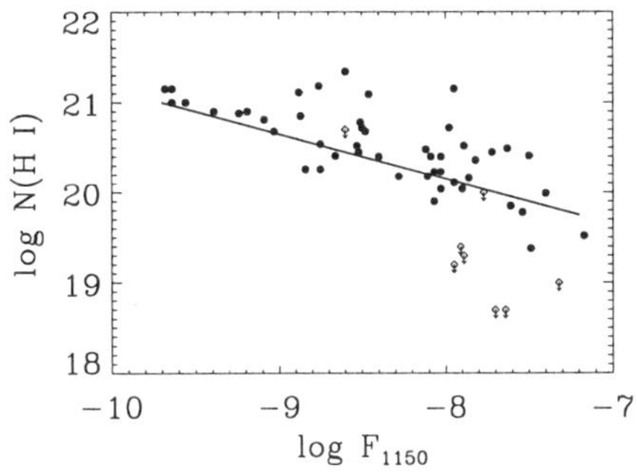

Figure 1. Logarithmic plots of hydrogen column densities $N(\mathrm{H\,I})$ *vs.* the ultraviolet fluxes F_λ at 1150Å for some representative, early-type, broad-lined ($v \sin i > 100\,\mathrm{km\ s^{-1}}$) bright stars that could be used for high sensitivity research on interstellar lines. This figure is adapted from Jenkins & Wallerstein (1999).

The spectrograph described earlier for this purpose has $\log A_{\mathrm{eff}} = 4.67$.

In both cases, it is important to remember that the calculations apply only to situations where the S/N is not limited by any of the secondary instrument parameters, such as the background flux level, maximum allowable flux, or maximum S/N.

4. Emergence of New Species or a Confusion Limit?

It is clear that using ever more powerful telescopes and spectrographs to study absorption features has two important benefits: (1) the availability of better quality information on lines of moderate strength and (2) the ability to detect features that were too weak to see with less powerful facilities. The gains associated with the former are self evident, but the latter must be examined in more detail. We now address the question: Do new species in fact come into view, or do we sink into a useless tangle of weak features belonging to species already appearing elsewhere in the spectrum?

Figure 2 shows for two different circumstances how rapidly new species can be expected to come into view as the sensitivity to weak lines increases. The strengths of atomic lines were computed on the assumption that the element abundances followed their respective cosmic ratios relative to hydrogen, but with an allowance for depletion onto dust grains following known empirical trends with the elements' condensation temperatures (Wasson 1985). Column densities for ionization stages that differed from the most abundant forms in H I regions were reduced by appropriate factors. Any absorption feature shifted by less than $10\,\mathrm{km\ s^{-1}}$ relative to a stronger line was deemed to be unobservable. For panel (a) the cumulative distribution shown in the figure starts at a value for $\log(X/\mathrm{H}) + \log f\lambda = -4.5$, which corresponds to lines with a central optical depth $\tau_0 = 3.0$ if $\log N(\mathrm{H}) = 20.5$ and $b = 5\,\mathrm{km\ s^{-1}}$. Lines stronger than this

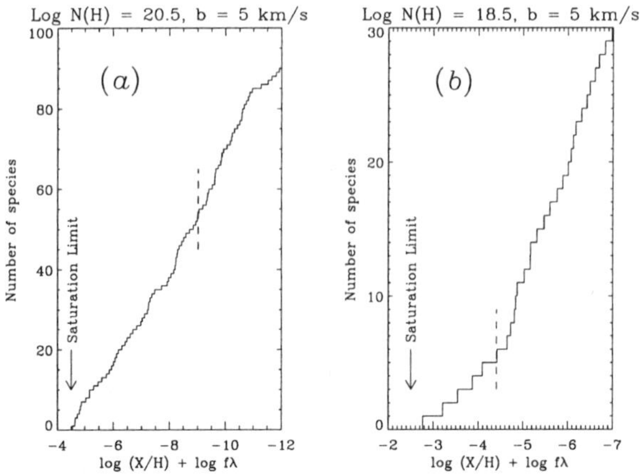

Figure 2. Emergence of new species (elements and their different stages of ionization) as a function of $\log(X/\mathrm{H}) + \log f\lambda$ (the scale for this quantity is inverted: increasing sensitivity for weaker lines goes toward the right). In panel (a), lines start to be counted when they become unsaturated, assuming $\log N(\mathrm{H}) = 20.5$ and $b = 5\,\mathrm{km\ s}^{-1}$. Panel (b) is similar, except that $\log N(\mathrm{H}) = 18.5$, a value that represents a typical strong Lyman limit system. For extragalactic systems with sub-solar abundances, $N(\mathrm{H})$ must be increased accordingly. The vertical dashed lines show the thresholds reached by the high-resolution and high-sensitivity spectrographs discussed in the text, after applying Eqs. 2 and 3, respectively.

limit are probably too saturated to be useful. For panel (b) the saturation limit moves to −2.5.

The outcomes shown in Figure 2 apply to a line of sight with a single velocity component. To varying degrees, extragalactic systems will suffer from interference by unrelated systems, including the profuse Ly-α forest lines. The figure shows that new species appear approximately in direct proportion to the logarithm of the spectrum's S/N. This result is intermediate between the pessimistic view that confusion produces diminishing returns and, at the opposite extreme, a prediction that "a thousand flowers will bloom" as we cross a threshold where a vast sea of new, low-abundance elements starts to appear.

References

Jenkins, E.B. 1999, in Chemical Evolution from Zero to High Redshift, eds. J.R. Walsh & M.R. Rosa, (Berlin: Springer), 280

Jenkins, E.B., & Wallerstein, G. 1999, in ASP Conf. Ser. 164, Ultraviolet-Optical Space Astronomy Beyond HST, eds. J.A. Morse, J.M. Shull, & A.L. Kinney, (San Francisco: ASP), 118

Wasson, J.T. 1985, Meteorites – Their Record of Early Solar-System History, (New York: Freeman)

Hubble's Science Legacy: Future Optical-Ultraviolet Astronomy from Space
ASP Conference Series, Vol. 291, 2003
K.R. Sembach, J.C. Blades, G.D. Illingworth, R.C. Kennicutt, Jr. (eds.)

Spatial Heterodyne Spectroscopy: An Emerging Technology for Interference Spectroscopy

Fred L. Roesler

University of Wisconsin, Madison, WI 53706

J.M. Harlander

St. Cloud State University, St. Cloud, MN 56301

J.G. Cardon and C.R. Englert

U. S. Naval Research Laboratory, Code 7641, Washington, DC 20375

R.J. Reynolds, K. Jaehnig, S. Watchorn, E.J. Mierkiewicz, and J. Corliss

University of Wisconsin, Madison, WI 53706

Abstract. Spatial Heterodyne Spectroscopy (SHS) stands poised to play a significant role in astrophysics from space and ground-based platforms. Compared with conventional instruments of similar resolving power and sensitivity, the field-widened spatial heterodyne spectrometer is highly compact and robust, has a throughput roughly 10^4 times higher, and is very forgiving of optical flatness defects. This paper will briefly describe the technique, its advantages, and applications in progress.

1. The SHS Principle

The principles of spatial heterodyne spectroscopy (SHS) are here described only briefly. More complete descriptions are in Harlander et al. (1992), Harlander et al. (2002), and Watchorn et al. (2001). SHS is a Fourier transform spectroscopy (FTS) method based on a Michelson interferometer modified by replacing the mirrors in each arm with diffraction gratings. In Figure 1, for each wavelength in the wavefront entering the interferometer, two wavefronts exit the interferometer with a wavelength-dependent crossing angle between them. This produces a superposition of Fizeau fringes with wavelength dependent spatial frequencies localized near the gratings and imaged on the detector. The image is a constant plus the Fourier transform of the input spectrum about the heterodyne wavelength (the wavelength producing parallel output wavefronts). The Fourier transform of the image produces the spectrum within a limited spectral range about the heterodyne wavelength. The diffraction-limited resolving power, R, of the grating system (the number of grooves, including both gratings, imaged on the detector) is achieved. If the pattern is imaged by N pixels, the total number of resolvable frequencies is $N/2$, giving a bandwidth of $N\lambda/2R$; a filter blocks light outside this range. No mechanical part is moved in this process.

The high throughput of interference spectrometers is achieved, resulting in a 100-fold gain compared to conventional grating instruments of the same size and resolution. Field-widening by inserting fixed prisms in each arm yields an additional 100-fold throughput gain. At $R = 50,000$, angles up to $\pm 5°$ degrees can be used within the interferometer. Since R is not limited by pixel size, long focal length cameras are not needed. These characteristics lead to SHS instruments much smaller and lighter than conventional instruments of comparable R and sensitivity, and thus more compatible with space flight.

For ultraviolet (UV) and far-ultraviolet (FUV) applications, conventional interference spectroscopy techniques are usually ineffective because of their high optical component flatness requirements ($\lambda/10$ for FTS and $\lambda/100$ for Fabry-Perot spectroscopy). In a conventional Fourier transform spectrometer, the detector is placed where defects over the entire wavefront reduce fringe visibility, requiring $\lambda/4$ or less of accumulated path difference variation over the entire wavefront. In SHS, each point of the fringe pattern samples only a small portion of the wavefront, and fringe contrast remains high if the flatness variation over the small area is within $\lambda/4$. For large optics, variations of several wavelengths can be tolerated if flatness is good over a few mm. Fringe waviness caused by non-flatness can be corrected during data reduction (Harlander et al. 1994).

2. Applications: SHIMMER

The Spatial Heterodyne Imager for Mesospheric Radicals (SHIMMER) (Harlander et al. 2002) was motivated by the successful Middle Atmosphere High Resolution Spectrograph Investigation (MAHRSI) experiment in which OH fluorescence between 307 nm and 310 nm in the sunlit atmosphere was used to measure global mesospheric OH abundance. Photochemical models fail to predict the observed results. MAHRSI demonstrated a successful approach and the need for further measurements from space, but it is too large and heavy for current space flight opportunities.

In laboratory tests, SHIMMER outperforms MAHRSI in a much smaller and lighter package (Harlander et al. 2002). SHIMMER has met performance requirements, and is scheduled as a mid-deck experiment on STS-112. Figure 2a shows SHIMMER in its mid-deck configuration with the cover removed. Figure 2b shows a monolithic version of SHIMMER. Since no adjustment is required once all optics have been set, complex levers and the heavy cage holding the separate optical elements are not needed after adjustment. Optics and spacers were manufactured with high accuracy and assembled by optical contact. The stable, robust module is well-suited for space flight.

3. Applications: O II 372.7 nm Emission from the ISM

A northern hemisphere survey of the component of the interstellar medium (ISM) known as the warm (10,000 K) ionized medium (WIM) was recently completed using the WHαM (Wisconsin Hα Mapper) Fabry-Perot instrument. Evidence has emerged for previously unrecognized energy sources heating the ISM;

[O II] 372.7 nm emission is believed to be the key to verifying their existence. SHS was chosen for this problem because WHαM does not work at 372.7nm, while SHS works extremely well in the near ultraviolet, is much more tolerant of flatness defects than F-Ps, and can potentially exceed WHαM's sensitivity.

A prototype field-widened 372.7 nm spatial heterodyne spectrometer was tested at the McMath-Pierce telescope on Kitt Peak. An interferogram from the Orion nebula emission is shown in Figure 2c. Its Fourier transform is the spectrum shown in Figure 2d. The spectrometer was set so that the the heterodyne wavelength was between the two lines of the doublet, and the gratings were tilted perpendicular to the dispersion direction to produce a 2-D interference pattern. This adjustment is described in Harlander et al. (1992). A scaled-up version designed to achieve the required sensitivity is under construction.

4. Applications: C IV 155.0 nm Emission from the ISM

Interstellar gas plays a fundamental role in the life cycle of stars, yet its nature is not fully known. The very faint emission lines that serve as tracers of the 10^5 K gas, such as the C IV doublet, are found in the FUV, and are observable only from space. Conventional instruments of sufficient resolving power and sensitivity to map the distribution and kinematics of the hot ISM are far too large for economical space application. The field-widened spatial heterodyne spectrometer provides the required performance in a very small package. The Fabry-Perot technique is not practical because its optical tolerances cannot be met in the FUV.

A sounding rocket flight in August 1999 (Watchorn et al. 2001) aimed at measuring C IV emission from the Cygnus loop demonstrated successful mechanical and optical design. Narrow-band reflecting multilayers were used for filtering out-of-band light, and calcium fluoride was used for the transmitting elements. Adequate FUV interferometric performance for SHS was achieved. Post-recovery measurements confirmed that the overall optical efficiency was too low to recover useful scientific data. Recent progress in FUV optics and handling driven by 157 nm lithography should solve the optical efficiency problem in the future.

Acknowledgments. Parts of this work were supported by the NASA Office of Space Science, the U. S. Airforce Space Test Program, and NSF grants ATM-9612228 and AST-0138228.

References

Harlander, J.M., Reynolds, R.J., & Roesler, F.L. 1992, ApJ, 396, 730

Harlander, J.M. Tran, H.T., Roesler, F.L., Jaehnig, K.P., Seo, S.M., Sanders, W.T., & Reynolds, R.J. 1994, Proc. SPIE, 2280, 310

Harlander, J.M., Roesler, F.L., Cardon, J.G., Englert, C.R., & Conway, R.R. 2002, Appl. Optics, 41, 1343

Watchorn, S., Roesler, F.L., Harlander, J.M., Jaehnig, K.P., Reynolds, R.J., & Sanders, W.T. 2001, Proc. SPIE, 4498, 284

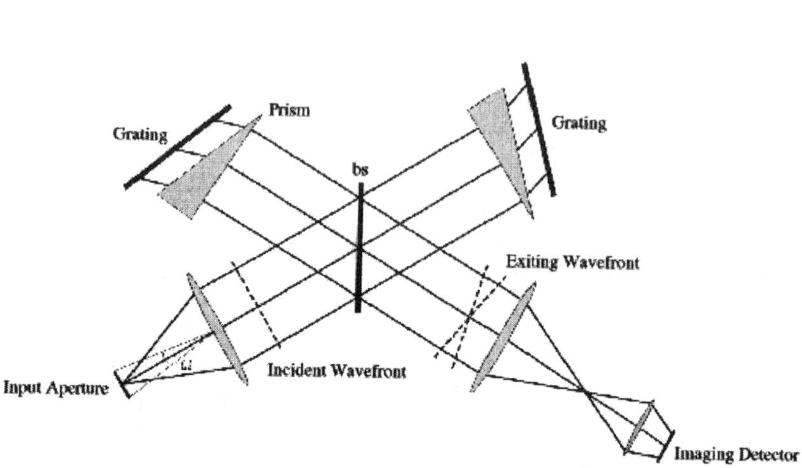

Figure 1. Field-widened spatial heterodyne spectrometer optical layout. Conventional blazed reflection gratings are used. Field-widening is provided by the fixed prisms in each arm. Element bs is the beamsplitter-compensating plate combination. The angle $\Omega \approx 10°$ for $R = 50,000$ with field-widening.

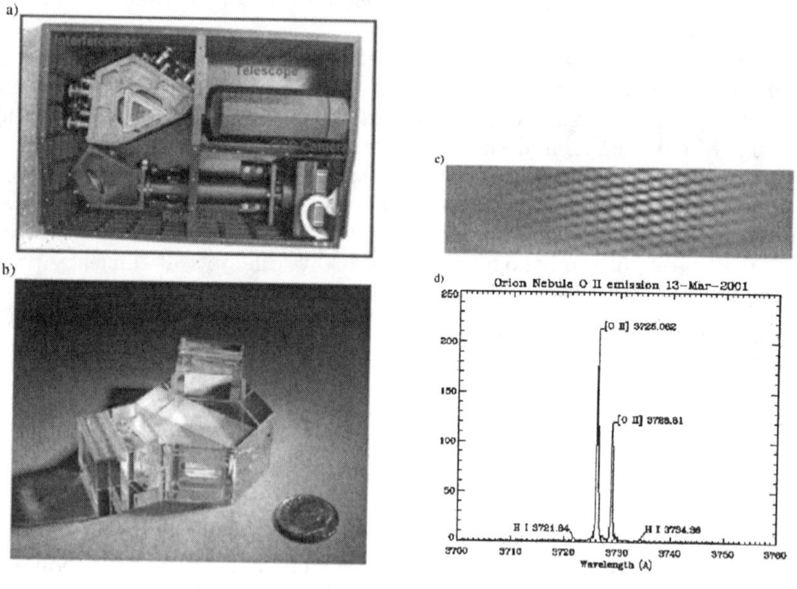

Figure 2. (a) SHIMMER configured for the Shuttle middeck experiment. It is the size of an airline carry-on bag. (b) An optically contacted, fused silica spatial heterodyne spectrometer for 308 nm. Note the dime for size comparison. (c) SHS interferogram obtained from 373 nm emission from the Orion nebula. (d) Spectrum derived from the interferogram in (c).

Hubble's Science Legacy: Future Optical-Ultraviolet Astronomy from Space
ASP Conference Series, Vol. 291, 2003
K.R. Sembach, J.C. Blades, G.D. Illingworth, R.C. Kennicutt, Jr. (eds.)

UV-IR Science Prospects with TES Imaging Arrays

R.W. Romani, J. Burney, P. Brink, B. Cabrera, and P. Castle

Dept. of Physics, Stanford University, Stanford, CA 94305-4060

T. Kenny and E. Wang

Dept. of Mech. Eng., Stanford University, Stanford, CA 94305-4021

B. Young

Dept. of Physics, Santa Clara University, Santa Clara, CA 95053

A.J. Miller and S.W. Nam

NIST, M.C. 814, 325 Broadway, Boulder, CO 80305

Abstract. We are developing photon-counting cameras employing cryogenic arrays of energy-resolving TES (transition edge sensor) pixels. These are being tested in ground-based instruments, but will have their greatest impact when employed on space platforms, where they can cover the 100 nm–10 μm range with high time- and moderate energy- resolution. Here we summarize briefly existing device performance, current directions in array camera development, and anticipated capabilities.

1. Introduction and Present UV-IR TES Performance

Cryogenic energy-resolving photon detectors show considerable promise for advanced instrumentation in several wavebands. In the near-infrared (near-IR) through ultraviolet (UV), these devices provide noise-free photon counting with $\leq 1\mu$s time resolution at $\delta E \approx 0.15$ eV energy resolution. Early work focused on superconducting tunnel junction devices (Perryman, Foden, & Peacock 1993), but alternative cryogenic technologies show great promise. Our group has pursued the application of superconducting TES devices to this problem (Cabrera et al. 1998). Our detectors are pixel arrays of 40 nm thick tungsten (W) films patterned on Si. When cooled below their ~ 100 mK transition temperature and voltage biased to produce negative electro-thermal feedback (Irwin 1995), an absorbed photon decreases the Joule-heating current. This pulse signal is read out with a DC SQUID array (Welty & Martinis 1993). Bi-layer thin films, a variety of substrates, and multiplex schemes to read out larger arrays are also under development and TES array technology is finding interesting application from the sub-mm (SCUBA-2) to the X-ray (Con-X) regimes.

Miller et al. (2000) summarize performance of present W TES devices. We routinely achieve $\delta E \sim 0.15$ eV at ~ 3 eV, absolute GPS photon times to

300 ns and single-pixel count rates of ~ 30 kHz. Energy discrimination rejects all cosmic rays and allows no dark current above the noise floor. The intrinsic bare W quantum efficiency is $\sim 50\%$ and in astronomical systems we have achieved efficiencies on the sky as high as $\sim 20\%$.

TES resolution is given by $\Delta E_{FWHM} = 2.355[4k_BT^2C(n/2)^{1/2}/\alpha]^{1/2}$, where T and C are the temperature and heat capacity of the W e^- system, $n = 5$ for electron-phonon conduction and $\alpha = (d\ln R/d\ln T)_{V=const}$. The pixel size sets C; the maximum (saturation) energy is $E_{max} \approx CT_c/\alpha$ and therefore the optimum detector design gives $\Delta E_{FWHM,min} = 2.355[4k_BT(n/2)^{1/2}E_{max}]^{1/2}$. Thus, for our typical 20μm pixel and $E_{max} = 10$ eV we obtain $\Delta E_{FWHM,min} \approx 0.05$ eV. Somewhat higher resolution can be obtained with lower E_{max} by operating in the saturation regime, measuring pulse shape rather than peak height. In the present devices on Si, 58% of the photon energy is lost from the W e^- system to substrate phonons, giving an expected $\Delta E = 0.088$ eV; for comparison the measured resolution is 0.12 eV FHWM. Interestingly, although the physics is different, these sums predict $\Delta E_{FWHM,min}$ quite similar to theoretical limits and goal resolutions for existing superconducting tunnel junction (STJ) devices. The eventual choice of detector is probably best decided on the basis of achieved performance and ease of manufacture. At low E the thermodynamic fluctuation noise floor allows detection to $\sim 10\mu$m. Of course, this full sensitivity range cannot be exploited from the ground, and we have employed a variety of filtering schemes to suppress unacceptably high thermal count rates beyond 2μm.

To illustrate the astronomical utility of these detectors, we have packaged TES arrays into simple demonstration systems, based on a He dilution refrigerator used for device development, and performed some basic observations. To filter the IR background, we employ a ~ 3 m length of high-OH optical fiber spooled at 4K in the dewar. The OH bands of this cold filter provided effective blocking at $\sim 1.7\mu$m, but allow transmission in the atmospheric windows. At the telescope, the atmosphere and optics limit us to ~ 3.5 eV, but in the lab we detected $E > 7$ eV photons. With such systems we demonstrated the first astronomical spectra taken with a cryogenic optical detector and using small telescopes have made unique, scientifically valuable, measurements of the Crab pulsar and other astronomical objects (Romani et al. 1999; Romani et al. 2001), including photon counting spectro-photopolarimetry in the optical/IR.

On small telescopes these test set-ups have achieved good efficiency, although system efficiencies of $\leq 10\%$ are typical. As we are fiber-coupled we have to date only employed a small subset of the TES array (e.g., 2×3 out of 6×6 for the McDonald 2.7 m Crab observations). Also, for ease of fabrication, the present TES arrays only employ a single lithographic layer of wiring; these wiring leads are responsive to photons, which complicates the energy PSF and spectral calibration. Finally, fiber coupled schemes, while attractive for test purposes, do not match well to the eventual (space platform) imaging applications for which these TES devices should have their largest impact.

2. Current Efforts

All of these factors can be mitigated with some simple design changes, and so we are assembling a first dedicated TES system for ground-based optical/UV astronomy. We are currently assembling an array in a compact, portable adiabatic demagnetization refrigerator (ADR) with dewar windows allowing direct imaging on the focal plane array. This solution is suitable for Cassegrain or Nasmyth applications at moderate to large aperture ground-based telescopes. The system accepts a slow f/10–f/20 beam and focal reduces with a cryogenic all-reflective system to allow good coupling to the devices. At the focal plane we read out a 32 pixel array. The existing design may be run at up to 8×8 format by adding more read channels; with modest changes to the device fabrication, arrays of 4–6 times this area could be made. With a window solution, we employ heat-absorbing filters passing primarily in the optical/UV, so the system efficiency can be increased by appropriate anti-reflectance coatings. Finally, to increase the fill factor and suppress energy PSF features from the wires and substrate, we will employ a focusing mask that collects the light to the $20 \times 20\mu$m W TES pixels.

3. Future Potential from Space Platforms

As ground-based tests demonstrate, these cryogenic technologies most immediately impact faint source observations when broad features in the spectral energy distribution and rapid time variability encode the key physics. Compact object studies, especially spin-powered pulsars and accretion-powered CVs and black holes are thus the 'killer-app' for first generation TES systems. Some of this science can be done from the ground, but the broad (0.1–10 eV) energy window accessed from space presents opportunities for uniquely powerful investigations.

The good quantum efficiency (QE) and noise properties of these devices encourages application to a wider range of problems, especially in the UV. Clearly, the main driver here is array format. Imaging applications providing a low-resolution ultra-broad band spectrum of extragalactic sources (see Romani et al. 1999; Mazin & Brunner 2000) over a modest field are very appealing. However, these arrays may also be useful in different formats such as order-sorting detectors for echelle spectroscopy in the optical/UV. The ability to read out more than the modest $\sim 10^2$ simultaneous channels covered in existing cryogenic system designs is therefore quite important. One fundamental limit is the complexity and heat load of the wiring delivering the signals to room temperature. In the case of TES devices there is a demonstrated (Chervenak et al. 1999) scheme for multiplex readout at moderate count rates of a number of TES pixels by a single SQUID channel. Multiplex technology development is also being driven by other TES applications and should certainly increase the useful arrays sizes by factors of $\sim 30 - 10^3$. There are other schemes for allowing large cryogenic pixel count without a prohibitive number of wires to low temperature (e.g, "kinetic inductance" coupling; Zmuidzinas et al. 2002).

What is needed to give TES imaging arrays (or comparable cryogenic technology) an exciting science role on a space-based platform? We envision three levels of capability. The first stage is a small $\sim 8 \times 8 - 16 \times 16$ pixel imaging array with low (~ 0.1 eV) energy resolution detecting the near IR-UV band with good QE. This already provides dramatic advance on time resolved studies of Galactic compact objects, with a roughly PSF-matched format and the option for local background subtraction, polarization and similar modes. This array could also perform area spectroscopy of small extragalactic objects reaching remarkably faint magnitudes. It may be built as a fairly straightforward rescaling of present systems using today's techniques. Such a (special purpose) system would enable very exciting science from a small aperture dedicated facility or a sub-orbital platform and in our opinion it should be pursued both for the science and for the demonstration of these detectors in a flight program.

A more capable facility would have 1–10 kilopixel format and would require some form of multiplex read-out. These read channels could be used both as an imaging array and in an elongated (order sorting) spectrographic format. It should be possible to have the same (expensive) read channels address several (cheap) array formats in a single dewar. This array appears to be a natural, near term ($\sim$ 5 yr) extension of present programs. If such an array can be coupled with a reliable closed-cycle, space-qualified cryocooler+ADR system, a strong science case can be made for a dedicated small aperture system. Grid filtering and passive optics cooling in space could allow simultaneous access to $\sim 0.2 - 10$ eV and imaging and spectroscopic modes, as above.

The ultimate instrument would have $\geq 10^5$ pixels, which with the high QE, ultra-broad sensitivity range, noiseless detection and intrinsic energy resolution would exceed the capabilities of CCD/MAMA arrays for a variety of mainstream astrophysics and cosmology applications. Such an array might be dynamically addressed, reading out only an interesting pixel subset. Light travel time ensures that sub-μs time stamps are unlikely to be of interest over large areas, so cryogenic preprocessing to accumulate simple spectral images would also be desirable. Schemes to build such an array and to accomplish the required read-out and processing are presently at the cartoon stage; hence prognostication of a development time is rather pointless. However, intermediate steps are quite compelling so with sufficient resources, progress toward such ultimate astronomical detectors is likely to occur.

Acknowledgments. This work was supported in part by NASA grant NAG5-3775 and made use of the Stanford Nanofabrication Users Network funded by NSF award ECS-9731294.

References

Cabrera, B., et al. 1998, ApPhL, 73, 735

Chervenak, J.A., et al. 1999, ApPhL, 74, 4043

Irwin, K. 1995, ApPhL 66, 1998

Mazin, B.A., & Brunner, R.J. 2000, AJ, 120, 2721

Miller, A.J., et al. 2000, NucIm, A444, 445

Perryman, M.A.C, Foden, C.L., & Peacock, A. 1993, NucIm, A325, 319

Romani, R.W., et al. 1999, ApJ, 521, L153

Romani, R.W., et al. 2001, ApJ, 563, 221

Welty, R., & Martinis, J. 1993, ITAS, 3, 2605

Zmuidzinas, J., Mazin, B., Vayonakis, A., Day, P. & Leduc, H.G. 2002, in Monterey Workshop Proceedings, in press

Hubble's Science Legacy: Future Optical-Ultraviolet Astronomy from Space
ASP Conference Series, Vol. 291, 2003
K.R. Sembach, J.C. Blades, G.D. Illingworth, R.C. Kennicutt, Jr. (eds.)

Advances in MCP Sensors for UV/Visible Astronomy

Oswald H.W. Siegmund, John V. Vallerga, and Anton S. Tremsin

Space Sciences Laboratory, University of California Berkeley, CA 94720

Abstract. Advances in photocathodes such as diamond (40% QE in UV), GaN (>30% QE in NUV), and GaAs (50% QE in visible/NIR), silicon microchannel plates (low background and distortion), and photon counting readouts (cross strip, <5 μm spatial resolution at 10^6 gain) are poised to make a significant impact on future ultraviolet (UV) and optical instruments. We describe these advances and discuss their significance.

1. Introduction

Recently some significant improvements in the most important performance parameters for UV/optical microchannel plate (MCP) detector systems, such as quantum detection efficiency (QDE), background rate, and spatial resolution have been achieved. These technologies will have a significant effect on the next generation of space-borne spectroscopy and imaging instruments.

2. Photocathodes

In the UV/Visible regime alkali halide, CsTe, and multi-alkali photocathode materials as semi-transparent, and opaque layers have been the standard for many years (Siegmund 1998). To obtain improvements in the QDE, bandpass coverage, and stability of cathodes, we have been investigating other materials such as diamond, GaN and GaAs. Diamond has been used as a cathode in the far-UV (Tremsin & Siegmund 2000) with good results, and its bandgap of 5.47 eV (227nm) and ability to have negative electron affinity promise further improvements. We have fabricated boron-doped polycrystalline diamond films on Si substrates and measured their QDE for different activation states. Figure 1 shows the QDE for the diamond film in its natural state compared with the same film after plasma hydrogenation. Under air exposure the activated film loses only 10% of its efficiency over 18 hours, but this can be restored by re-hydrogenation. Even after cleaning in water and alcohol (Tremsin & Siegmund 2000), the QDE can be regained this way. We are investigating Cs activation to increase the QDE further, but this is likely not to be as stable as produced by hydrogenation. GaN is also a very interesting photocathode with application to wavelengths below ~400 nm. Opaque cathodes with efficiencies of up to 30% at 200 nm have been made (see Ulmer et al., this volume), and we are currently working to produce GaN cathodes with even higher efficiencies and tunable cutoffs. GaAs photocathodes have been widely used for night vision devices,

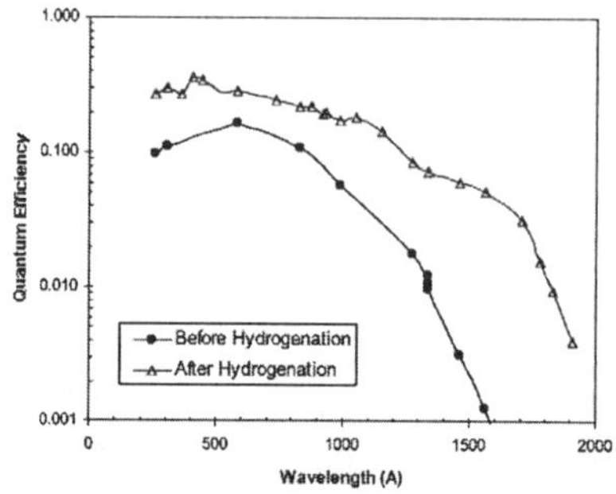

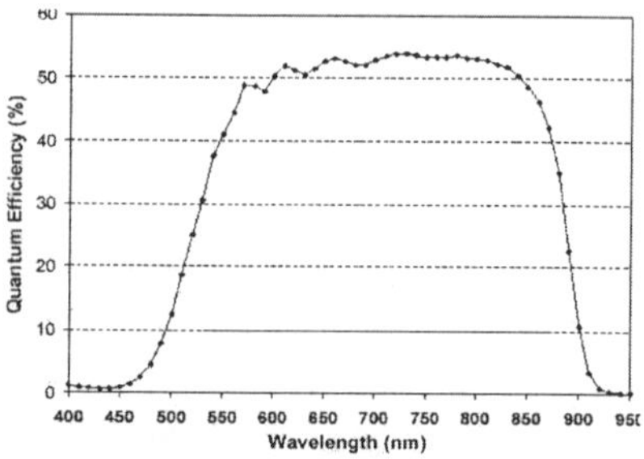

Figure 1. *Left*: Polycrystalline diamond QDE versus wavelength before and after hydrogenation.

Figure 2. *Right*: GaAs QDE in the optical/NIR (Courtesy, R. Benz, ITT).

but not in Astronomical instrumentation. Recent advances have raised the QDE to better than 50% over the visible range (Fig.2). Since the GaAs background rate can be reduced to <10 events s^{-1} at -20 °C, higher sensitivity than CCDs is possible for low S/N observations. We are currently incorporating GaAs into our MCP/readout devices to produce high-efficiency visible/NIR devices with high spatial and time resolution, along with high immunity to cosmic rays events.

3. Microchannel Plates

Silicon MCPs are produced by photolithographic processing of Si wafers. The manufacture of Si MCPs has undergone significant changes over a relatively short period of time. The samples we have tested are 25 mm format with $\sim$7 μm pores, with both square and hexagonal pore shapes (Figure 3). We have measured the gain of single Si MCPs and the gain, pulse height, gain uniformity, background, and quantum detection efficiency of Si MCPs in stacks. Hexagonal-pore MCPs with high gain and open area ratios of >75% have been successfully fabricated (Siegmund et al. 2001a). Gains of nearly 10^4 (Figure 4) for a single Si MCP have been achieved, and the quantum detection efficiency for Si MCPs has been shown to be the same as glass MCPs. The Si MCP background is as low as $\sim$ 0.02 events sec^{-1} cm^{-2} without shielding, giving significant improvement over even low-noise glass MCPs. The image flat fields are free of any patterned modulation, and the gain uniformity is relatively good. Si MCPs offer low stopping power for X-rays, γ-rays, and cosmic rays, stability to very high temperatures (>800°C), and lack of reactivity with photocathodes.

4. Cross Strip Anodes

The cross strip anode is a coarse (0.5 mm) multi-layer metal and ceramic cross strip pattern. The cross strip anode scheme uses a ceramic (alumina) substrate coated with insulating and conductor layers. On one surface of the substrate the conductors are fabricated as a set of fingers approximately 0.5 mm wide. Then

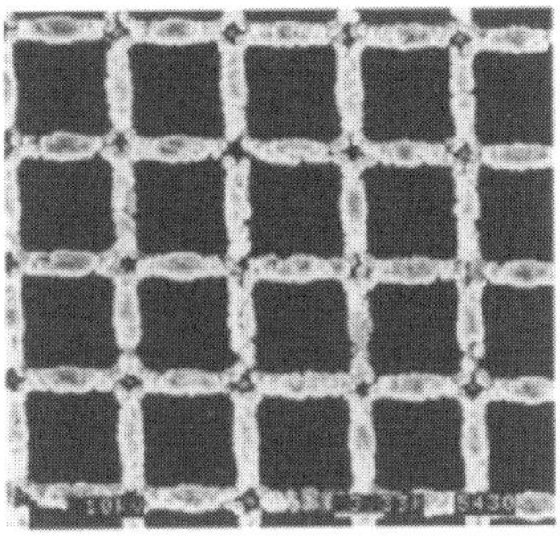

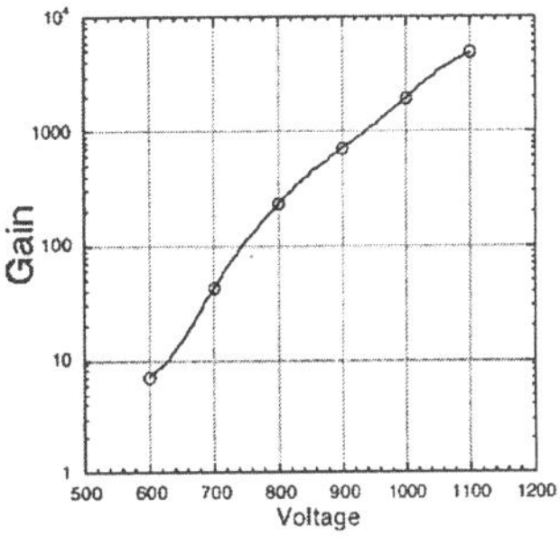

Figure 3. *Left*: Si MCP with 7μm pores, coated with diamond.

Figure 4. *Right*: Gain of a 7μm pore Si MCP versus voltage.

sets of insulating and conducting fingers are applied in the orthogonal direction such that 50% of the bottom layer is left exposed. The top layer and bottom layer are used to collect the charge from the MCPs. The MCP charge cloud is matched to the anode period (Figure. 5) so that charge is collected on several neighboring fingers to ensure an accurate event centroid can be determined. Event positions are encoded by direct sensing of the charge on each strip and subsequent determination of the charge cloud centroid for each event. This event position encoding is accomplished with preamplifiers on the anode, subsequent analog to digital conversion of individual strip charge values, and a software or hardware centroid determination.

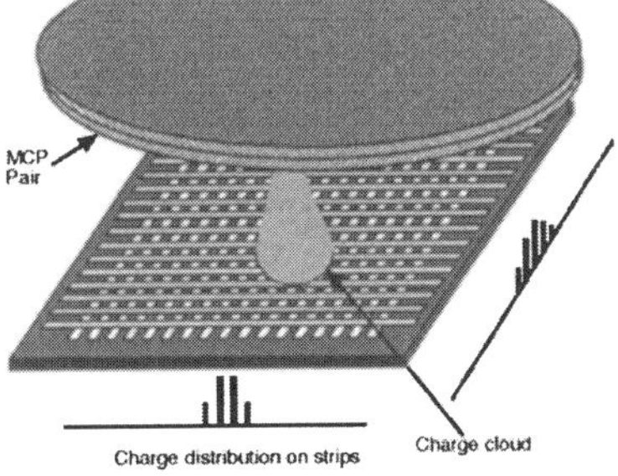

Figure 5. *Left*: Schematic of a XS anode and MCP pair, with the charge cloud depicted.

Figure 6. *Right*: Photo of a 32 × 32 mm XS anode with hermetic signal bias to the backside.

The strip that corresponds to the center peak of the charge cloud gives the coarse position. The calculation of the charge centroid to a small fraction of the strip width ($\sum nQ_n / \sum Q_n$) [Q_n is charge on $\mathrm{n}^{t}h$ strip] may be accomplished by a hardware or software sum and division. If the signal to noise ratio and the number of digitized bits are sufficient, centroids of better than 1/100 of a strip are possible. Since the strip capacitances can be small (few pF), the preamplifier noise can also be small and the time constants kept short (<10 ns). The high speed of digital logic and fast analog to digital converters, combined with an anode with custom chip electronics, may accomplish encoding at photon counting rates $\gg$1 MHz and with low power consumption ($\sim$2 W) suitable for applications in space astrophysics.

For an initial 8×8 mm anode (Siegmund et al. 2001b) we have resolved 12 μm MCP pores (Figure 7) while using low gain ($\sim 5 \times 10^6$) in an MCP pair configuration. The image linearity is good enough to enable distortions in the MCP pore pattern to be seen as well as dead and distorted pores. The spatial resolution is high enough (<3 μm) that the pores of even a 7 μm pore MCP can be resolved at low gain ($\sim 2 \times 10^6$). The low MCP gain requirement will enhance the local counting rate capacity and overall lifetime of a microchannel plate detector system with cross strip readout. We have also made a 32×32 mm anode format (Figure 6) where the signals are routed to the anode backside, providing a compact, electronically upgradeable configuration.

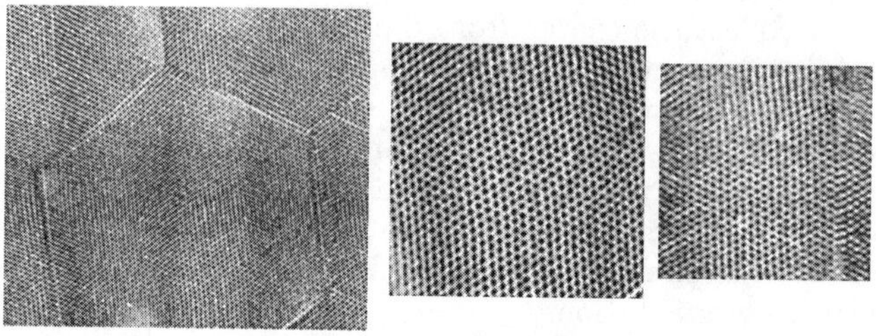

Figure 7. *Left*: Image detail of a 12μm pore MCP pair obtained at 5x10^6 gain with the 8×8 mm XS anode. *Middle*: Small zoomed area of image at left. *Right*: Image of a 7μm MCP pair obtained at 2×10^6 gain.

Acknowledgments. This work was supported by NASA grants NAG5-8667, and NAG-9149. We would like to thank Dr. J. Millaud and R. Abiad for work on the electronics, J. Hull and J. Malloy for systems design and engineering, and Nanosciences Inc. for Si MCPs.

References

Siegmund, O.H.W. 1998, in Methods of Vacuum Ultraviolet Physics, Ch. 2, 2nd edition, eds. J.A.R. Samson & D.L. Ederer (Burlington, MA: Academic Press)

Tremsin, A.S., & Siegmund, O.H.W. 2000, Proc. SPIE, 4139, 16

Siegmund, O.H.W., et al. 2001a, Proc. SPIE, 4497, 139

Siegmund, O.H.W., et al. 2001b, Proc. SPIE, 4498, 131

Hubble's Science Legacy: Future Optical-Ultraviolet Astronomy from Space
ASP Conference Series, Vol. 291, 2003
K.R. Sembach, J.C. Blades, G.D. Illingworth, R.C. Kennicutt, Jr. (eds.)

The Physical Conditions and Metal Enrichment of Low-Redshift Interstellar and Intergalactic Media: The Benefits of High-Resolution Ultraviolet Spectra

Todd M. Tripp and David V. Bowen

Department of Astrophysical Sciences, Princeton University, Princeton, NJ 08544

Abstract. To underscore the value of high spectral resolution for the study of low-z QSO absorption lines, we briefly present Space Telescope Imaging Spectrograph echelle data that demonstrate how increased resolution leads to dramatic improvement in line measurements. We show that even $R = \lambda/\Delta\lambda \approx 20{,}000$ is insufficient for some measurements. The higher the resolution the better, but $R \approx 50{,}000$ is adequate for many outstanding questions about the intergalactic medium and the interstellar media of galaxies that can be probed using QSO absorption lines.

1. Introduction

QSO absorption lines provide a sensitive probe of the physical conditions and enrichment of the diffuse intergalactic medium (IGM)as well as the interstellar media (ISM) of galaxies. It is well know that at low redshifts, the critical absorption lines are all in the ultraviolet and must be observed from space. However, the need for high spectral resolution is not always fully appreciated. This poster presented several examples, from real data obtained with the Space Telescope Imaging Spectrograph (STIS), of the benefits of high-resolution spectroscopy for the study of QSO absorbers. Some of these examples are reproduced in this short contribution to illustrate the importance of high spectral resolution as we consider options for more powerful telescope/instrument combinations to follow in the footsteps of the Hubble Space Telescope (HST). The data used for these demonstrations were obtained with the echelle modes of STIS, which provide the following resolutions: E140M – R = 46,000 (FWHM = 7 km s^{-1}); E140H with the "Jenkins Slit" – R = 200,000 (FWHM = 1.5 km s^{-1}).

Of course, the spectral resolution required depends on the nature of the measurement. For example, due to the lower mass of hydrogen, H I absorption lines are often much broader than their corresponding metal lines ($b^2 = 2kT/m$ for thermal broadening), and consequently the resolution requirements are more stringent for metals than for H I studies. Using E140M STIS data, Davé & Tripp (2001) showed that the majority of low$-z$ H I Lyα lines have $b >$ 15 km s^{-1}, i.e., well-resolved at E140M resolution. Therefore, a resolving power $R = \lambda/\Delta\lambda \approx$ 20,000 is probably adequate for most H I measurements. Similarly, highly-ionized species such as O VI are often quite broad (e.g., Tripp et al. 2000; Tripp 2002) and may be sufficiently resolved at this resolution. However, the lower-ionization

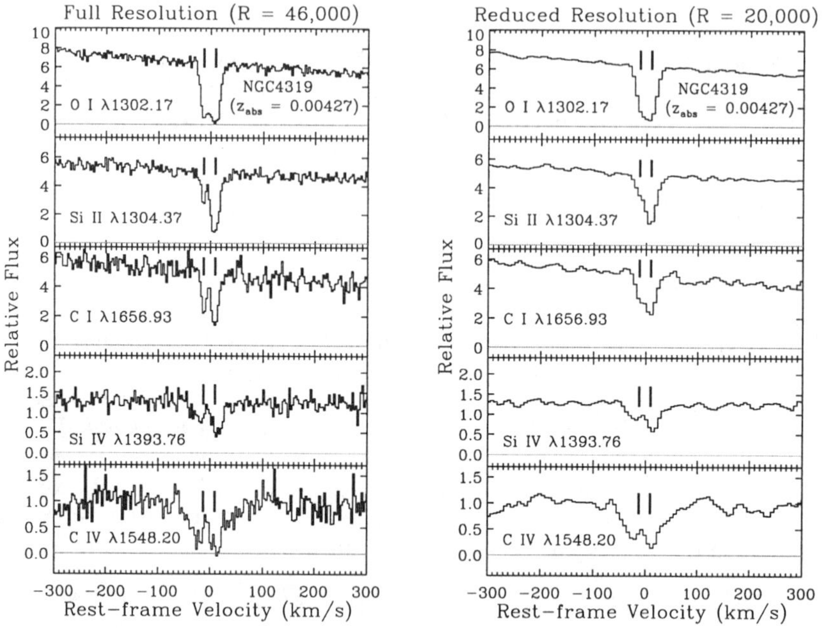

Figure 1. Absorption lines of the spiral galaxy NGC 4319 ($z_{\rm abs} = 0.00427$) in the spectrum of Mrk 205 at $R = 46{,}000$ (left panels), and $R = 20{,}000$ (right panels).

lines associated with O VI systems are often narrow and intricate (e.g., Tripp et al. 2000; Simcoe et al. 2002). To study the detailed component structure of metal lines (e.g., column densities and line widths of *individual components*), resolving power substantially better than 20,000 is often required. It is crucial to examine individual components to properly constrain the conditions of the gas and to study abundance patterns, which sometimes show remarkable variation from component to component (e.g., Ganguly et al. 1998).

2. The ISM of NGC 4319

The sight line to the QSO Mrk 205 provides a useful example of the benefits of high resolution. This sight line passes through the interarm region of the foreground spiral galaxy NGC 4319 at a projected distance of 2.8 h_{100}^{-1} kpc from the galaxy center (see Bowen et al. 1995 and references therein). Selected absorption lines due to the ISM of this spiral, recorded at the full 7 km s^{-1} E140M resolution, are shown in the left panels of Figure 1. At 7 km s^{-1} resolution, two components are readily apparent in the profiles of C I, O I, and Si II. The high ion (Si IV and C IV) profile components are remarkably similar, but not identical, to those of the low ions. $R = 46{,}000$ is sufficient to constrain the component parameters for all of the species in Figure 1 except O I, where the two components are severely blended. Unfortunately, at $R = 20{,}000$, much of this

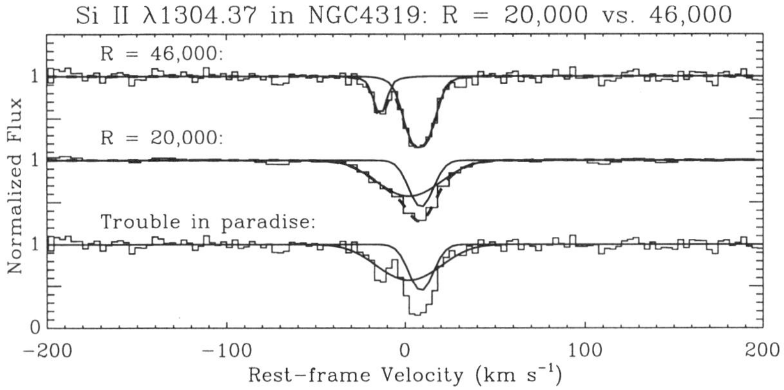

Figure 2. Comparison of profile-fitting results for the Si II λ1304.37 transition in the ISM of NGC 4319 at (top) the full $R = 46{,}000$ resolving power of the original STIS data, and (middle) the same data reduced to $R = 20{,}000$. In each case, the data are plotted with a histogram, the two components that were fitted to the data are shown with solid lines, and the combined fit is shown with a dashed line. The two components from the $R = 20{,}000$ fit are overplotted on the full resolution data at the bottom of the figure; clearly the reduction in resolution led to seriously erroneous results.

information is lost, as shown in the right panels of Figure 1. At this resolution, the O I line appears to be a single component, and moreover the various species appear to have rather different component structure. Two components are still apparent (albeit badly blended) in Si II and C I. However, attempting to fit two components to the $R = 20{,}000$ Si II data leads to serious errors, as illustrated in Figure 2. The lower resolution is not sufficient to back out the correct parameters of the individual components. It is worth noting that the equivalent widths and total (i.e., integrated) column densities are correctly measured from the lower-resolution data, but the individual components are poorly constrained.

3. C I in Damped Lyα Absorbers

The component structure in Figures 1 and 2 is relatively simple. Figure 3 shows a more daunting example: Milky Way C I lines in the sight line to HD 210839 (λ Cep). Here $R = 20{,}000$ is certainly insufficient, but with $R = 200{,}000$ and coverage of many C I lines (i.e., broad wavelength coverage), the C I profiles can be effectively analyzed (Jenkins & Tripp 2001). The $E(B-V)$ and $N(\mathrm{H\,I})$ are substantially higher toward λ Cep than in NGC 4319, but C I lines have been detected in several damped Lyα QSO absorbers (e.g., Quast et al. 2002; Petitjean et al. 2002) that do have high H I column densities. For analysis of these systems, very high spectral resolution is paramount.

High resolution is expensive, and it is certain that compromises between resolution and throughput will be required in the design of the HST successor, based on the science drivers. In our experience, the resolution provided by the E140M echelle mode of STIS ($R \approx 50{,}000$) is adequate for a large fraction of

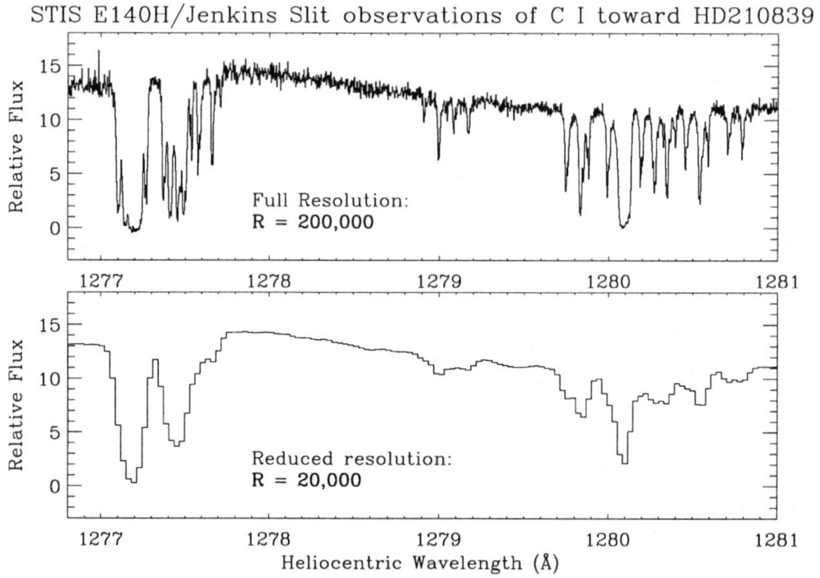

Figure 3. *Top*: STIS E140H echelle spectra of HD 210839 (λ Cep) at R = 200,000 (from Jenkins & Tripp 2001). *Bottom*: Same data reduced to R = 20,000. The estimated distance to this star is 880 pc in the direction $l = 103.8°, b = 2.6°$; E(B-V) = 0.57.

the current questions about the IGM and ISM, but STIS can only observe the brightest targets. An ability to go deeper at this resolution would be highly valuable.

References

Bowen, D.V., Blades, J.C., & Pettini, M. 1995, ApJ, 448, 634

Davé, R., & Tripp, T.M. 2001, ApJ, 553, 528

Ganguly, R., Churchill, C. W., & Charlton, J. C. 1998, ApJ, 498, L103

Jenkins, E.B., & Tripp, T.M. 2001, ApJS, 137, 297

Petitjean, P., Srianand, R., & Ledoux, C. 2002, MNRAS, 332, 383

Quast, R., Baade, R., & Reimers, D. 2002, A&A, 386, 796

Simcoe, R.A., Sargent, W.L.W., & Rauch, M., 2002, ApJ, in press [astro-ph/0206497]

Tripp, T.M. 2002 in ASP Conf. Ser. 254, Extragalactic Gas at Low Redshift, eds. J.S. Mulchaey & J.T. Stocke, (San Francisco: ASP), 323

Tripp, T.M., Savage, B.D., & Jenkins, E.B. 2000, 534, L1

Hubble's Science Legacy: Future Optical-Ultraviolet Astronomy from Space
ASP Conference Series, Vol. 291, 2003
K.R. Sembach, J.C. Blades, G.D. Illingworth, R.C. Kennicutt, Jr. (eds.)

Progress in the Fabrication of GaN Photocathodes

Melville P. Ulmer

Dept. of Physics & Astronomy, Northwestern University, Evanston, IL 60208-2900

Bruce W. Wessels

Dept. of Materials Science & Engineering, Northwestern University, Evanston, IL 60208-3108

Oswald H.W. Siegmund

Space Science Laboratory, University of California, Berkeley, CA 94720

Abstract. Greatly improved ultraviolet-visible-blind imaging system are necessary for future space missions. GaN photocathode based devices are a likely choice for these missions. We demonstrate this by describing our current work on making GaN phototubes, and we suggest the path to follow to make further improvements.

1. Introduction

For astrophysics applications we have the following major requirements for ultraviolet (UV) detectors: a high (at least 0.4) quantum efficiency (QE) below the long wavelength cutoff, a low (no better than 10^{-6}) QE above the long wavelength cutoff, a tunable band-gap, and a low dark current ($\simeq 10^{-14}$ nA/μm^2) for single photon counting. Large formats (our goal is $15,000 \times 15,000$) and high dynamic range (both local and global) are important for most astronomical applications as well (see Ulmer, Razeghi, & Bigan 1995 and references therein). Microchannel plates (MCPs) currently win over CCD-like solid state p-i-n devices made from III-V semi-conductors. The dark current in the p-i-n devices is at least 10^5 times too high (e.g., Parish et al. 1999), and there is no clear path to reducing the dark current in these p-i-n devices. The development of photocathodes is the approach most likely to lead to devices for UV space missions over the next 10 years. We have, therefore, embarked on making GaN-based photocathodes.

Figure 1 shows a compilation of the QE of several devices for comparison with Figure 2. These figures demonstrate the potential for GaN based photocathodes. A single data point (QE of 50% $\pm$ 6%) at 312.6 nm from Machuca et al. (2000) is not shown. The result by an independent group plus our work demonstrates that GaN photocathodes are indeed an excellent choice for future UV space missions. Also, for comparison, in Figure 2 we have plotted the QE of diamond and CsTe photocathodes. For a narrow-band response these may prove adequate, but for a broad band with a tunable (via alloying with Al or

In) long wavelength cutoff, GaN-based (or other wide band-gap materials) films are necessary. We also remind the reader that the QE of opaque photocathodes can exceed 50%.

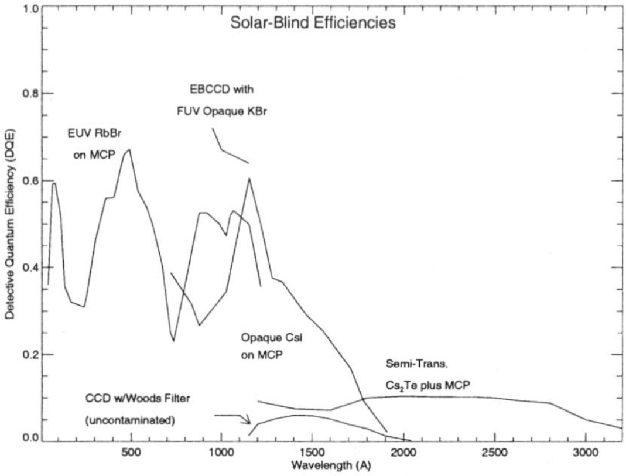

Figure 1. The detective efficiency versus energy for various devices (from Ulmer et al. 1995).

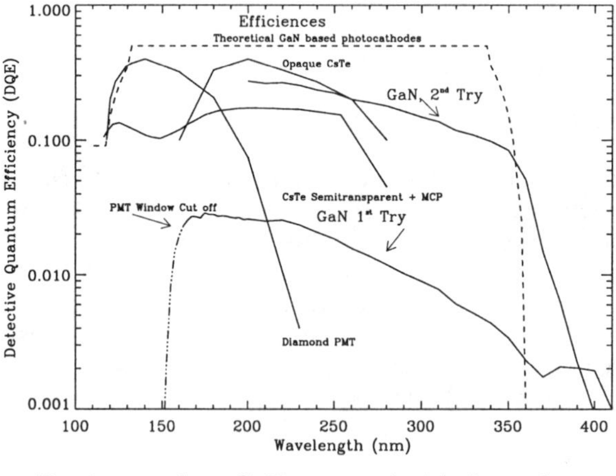

Figure 2. Previous work on GaN compared with the performance of other photocathodes. The CsTe MCP data are from Joseph (1995) and Kimble et al. (1999). The diamond response is from a Hamamatsu phototube specification document R6800U-26, but see also Tremsin & Siegmund (2000). The opaque CsTe data are from a Hamamatsu photomultiplier tube catalog: R1220 PMT.

2. Main Points About Using GaN

To use GaN as a photocathode, the GaN must be p-type with an n-type ~ 1 nm activation surface layer to produce "band-gap bending". This produces the desired negative electron affinity (see Shahedipour et al. 2002, and references therein). The preferred dopant is Mg. The relatively high ionization energy of Mg (about 200 meV) results in only about a 1% contribution to the charge carrier concentration, and hence 10^{20} atoms cm^{-3} are needed to produce a carrier concentration of 10^{18} cm^{-3}. It is this number of 10^{18} cm^{-3} that we hypothesize is the level above which only marginal increases in QE will be gained. However, our own work (Shahedipour et al. 2002; Ulmer et al. 2001) suggests that high conductivity is the key factor in the bulk film property. This is demonstrated in Figure 3. We have continued our research by having photocathodes made both by O. Siegmund (presented separately, this conference) and by ITT night vision (K. Passmore & A. Smith, private communication). Their resulting photocathodes had QEs within a few % of the "second try" by Hamamatsu shown in Figure 2. However, the Figure 2 "second try" is still the best performance with our GaN. Furthermore, the film used by Siegmund had a conductivity of 0.36 (Ohm cm)$^{-1}$ and carrier concentration of 2×10^{17} cm^{-3}, and the film used by Passmore and Smith had a conductivity of 0.55 (Ohm cm)$^{-1}$ and carrier concentration of 3×10^{17} cm^{-3}. These characteristics are markedly higher than those of the films used by Hamamatsu (see Figure 3). Perhaps increasing the conductivity above ~ 0.13 (see Figure 2) will not improve the performance of GaN-based photocathodes. It is more likely that changes in the surface treatment by our new collaborators (Siegmund, Passmore, and Smith) will lead to improved QEs. Since the results achieved to date by Siegmund, Passmore, and Smith were based on their very first attempts at processing this GaN material, it is quite plausible that the limit of QE performance has not yet been reached. More work is needed with films that have conductivities exceeding ~ 0.13 (Ohm cm)$^{-1}$ to determine the importance of surface preparation versus bulk film properties.

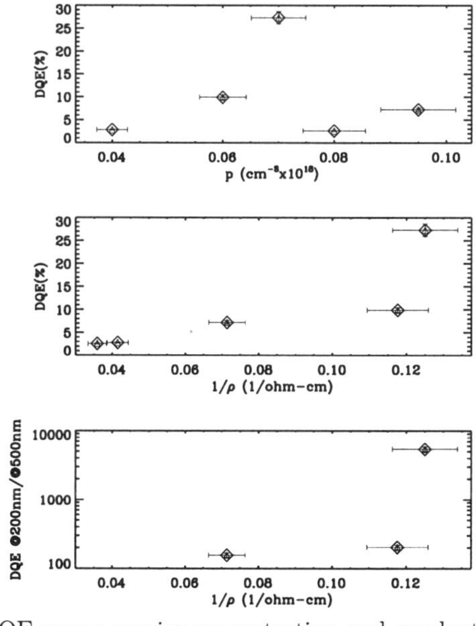

Figure 3. QE versus carrier concentration and conductivity are shown in the top two rows. In the third row we show a "figure of merit" (the ratio of the QE at 200 nm to the QE at 500 nm) versus conductivity.

There are several issues yet to be addressed:

1. What is the threshold above which increased conductivity and/or carrier concentration has no measurable affect on the optimal possible photocathode performance?

2. Can the QE as function of wavelength be improved by achieving carrier concentrations of 10^{18} and/or conductivity as high as 5 (Ohm cm)$^{-1}$ via co-doping (see Korotkov, Gregie, & Wessels 2000)? However, the increased number of lattice defects caused by co-doping (Han et al. 2002) may cause a degradation in the long wavelength performance below the band-gap rather than an improvement.

3. What other avenues besides the ones discussed above are there to improve QE? One likely answer is the placement of a wider band-gap material below the active layer (e.g., AlN/AlGaN and then the active GaN surface – see, for example, Machuca et al. 2000). The wider band-gap material acts as an "electron mirror" and can, in principle, produce QEs greater than 50%.

4. Can coating GaN onto Si MCPs be accomplished to produce large format detectors made with opaque GaN-based photocathodes?

Acknowledgments. This work was supported in part by grants from NSF (DMR-9705134) and NASA (NAG5-6730, NAG5-9149, NAG5-8667, & NAG5-11547). We thank Anton Tremsin, Joel Gregie, Fatemeh Shahedipour, T. Nahashi, Bing Han, and Charles Joseph for useful comments and supplying some of the data used here. We also thank Keith Passmore, Arlynn Smith, Bruce Woodgate, and Tim Norton for useful discussions.

References

Han, B., Gregie, J.M, Ulmer, M.P., & Wessels, B.W. 2002, in preparation

Joseph, C.L., Argabright, V.S., Abraham, J., Dieball, D., Franka, S., Styonavich, M., van Houten, C., Danks, A., & Woodgate, B.E. 1995, Proc. SPIE, 2551, 248

Kimble, R.A., et al. 1999, Proc. SPIE, 3764, 209

Korotkov, R.Y., Gregie, J.M., & Wessels, B.W. 2001, ApPhL, 78, 222

Machuca, F., Sun, Y., Liu, Z., Ioakeimidi, K., Pianetta, P., & Pease, R.F.W. 2000, J. Vac Sci. Technol. B, 18, 3042.

Parish, G., Keller, S., Kozodoy, P., Ibbetson, J.P., Marchand, H., Fini, P.T., Fleischer, S.B., DenBaars, S.P., Mishra, U.K., Tarsa, E.J. 1999, ApPhL, 75, 247

Shahedipour, F.S., Ulmer, M.P., Wessels, B.W., Joseph, C.L., & Nihashi, T. 2002, IEEE Journal of Quantum Electronics, 38, 333

Tremsin, A.S., & Siegmund, O.H. 2000, Proc. SPIE, 4139, 16

Ulmer, M.P., Razeghi, M., & Bigan, E. 1995, Proc. SPIE, 2397, 210

Ulmer, M.P., Wessels, B.M., Shahedipour, F., Korotokov, R.Y., Joseph, C.L., & Nihashi, T. 2001, Proc. SPIE, 4288, 246

Hubble's Science Legacy: Future Optical-Ultraviolet Astronomy from Space
ASP Conference Series, Vol. 291, 2003
K.R. Sembach, J.C. Blades, G.D. Illingworth, R.C. Kennicutt, Jr. (eds.)

Modeling the Breakup of Comet Shoemaker-Levy 9

Kevin J. Walsh and Derek C. Richardson

Dept. of Astronomy, University of Maryland, College Park, MD 20742-2421

Terrence W. Rettig

Dept. of Physics, University of Notre Dame, Notre Dame, IN 46556

Abstract. We present preliminary simulations of the tidal breakup of Comet Shoemaker-Levy 9. The simulations investigate the manner in which bulk density, diameter, spin period, and orientation of the spin axis affect the breakup of the comet. The diagnostics used to explore these parameters are the position angle, length, and mass distribution of the fragment train. The simulation results are compared to a large and detailed observational record of the comet's encounter, including HST observations. The analysis could not have been accomplished without data provided by space-based instrumentation. Solar system studies like these that require high spatial resolution will benefit from the Next Generation Space Telescope.

1. Introduction

In 1992 Comet Shoemaker-Levy 9 (SL9) passed within 1.3 Jovian radii of Jupiter and was pulled apart into a fragment train that coalesced into $\sim$ 21 cometary clumps. These clumps, discovered by Shoemaker et al. (1993), were later observed with HST (Weaver et al. 1994, 1995) and were tracked until they impacted Jupiter over a span of one week in 1994 (Chodas & Yeomans 1996).

The SL9 breakup required a very fragile progenitor, possibly a "rubble pile" (Asphaug & Benz 1994; Richardson et al. 2002). Observations of the SL9 clumps (their number and relative size) and of the fragment train as a whole (its length and position angle on the sky) constrain the properties of the progenitor. Previous studies (Scotti & Melosh 1993; Boss 1994; Solem 1994; Asphaug & Benz 1994, 1996; Rettig, Sobczak, & Hahn 1996) constrained the progenitor bulk density to be between 0.5 and 1.0 $\mathrm{g\,cm^{-3}}$ and the diameter to be between 1.5 and 2 km. The progenitor spin is degenerate with bulk density, since both determine the effective surface gravity, but reasonable values of the bulk density imply spin periods no faster than 6 hours with obliquity less than 90° (Asphaug & Benz 1996).

Observations of the dynamics of the SL9 clumps also provide clues as to the nature of the progenitor. Evidence for secondary fragmentation was observed (e.g., Fragment Q1 was inferred to have separated from Fragment Q2 in April

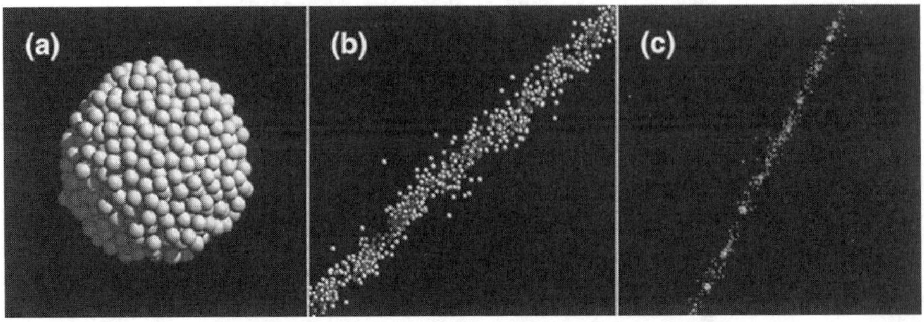

Figure 1. Snapshots of the tidal disruption of a rubble-pile comet. Clumps form via gravitational instability along the fragment train.

1993; see Sekanina 1996), which is a strong test for the rubble pile hypothesis. From the HST observations, the dust production and sizes as well as the outflow velocities suggested the dust was collisionally produced and the fragments were at least partially in the form of "swarms" of particles (Rettig et al. 1996; Hahn, Rettig, & Mumma 1996; Hahn & Rettig 2000). There has been a debate as to whether the fragments were solid objects or swarms of particles. The answer is critical to understanding the nature of cometary nuclei.

2. Simulations

Adopting the rubble pile scenario that the SL9 progenitor consisted of tiny fragments held together by gravity, we seek to constrain more precisely the SL9 progenitor properties using a sophisticated numerical code called `pkdgrav` (Richardson et al. 2000) that can handle both particle collisions and gravitation. In this scenario, Jovian tides stretch the fragments into a linear assemblage that collapses into distinct clumps under their own self gravity once outside the Roche zone of the planet (Figure 1).

To date we have tested over 100 configurations varying progenitor bulk density (0.3, 0.5, and 0.7 $\mathrm{g\,cm^{-3}}$), diameter (0.8, 1.0, and 1.2 km), spin (6, 9, 12 hours, and no rotation), and obliquity (spin vector pointing along the orbital momentum vector, pointing directly toward Jupiter, pointing along the direction of motion, and one case in between these three). The initial conditions for each run were based on a numerical integration of Fragment K from 6.0 July 1994 to a point before perijove when the comet was outside the Roche limit at 6 Jovian radii. The integrations included Saturn and the Earth, spanned over 720 days of real time, and were performed using $\sim$1000 particles each.

For each run the position angle and length of the fragment train was measured as a function of time. These were measured as they would have been observed from Earth, so as to compare with observation. Uniformity of the fragment train was measured simply by comparing the relative masses of each clump at a given instant. Identification of clumps was performed using an automated procedure that ignores groupings of fewer than 3 particles.

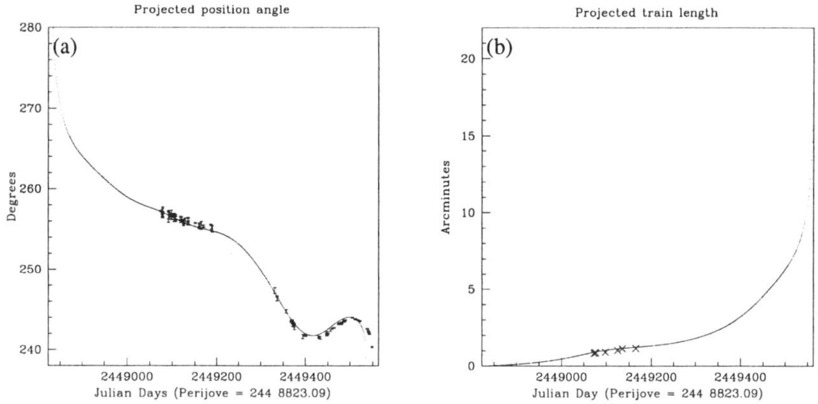

Figure 2. Position angle (a) and train length (b) as seen from Earth. Crosses or points with error bars are actual observations.

3. Results

The position angles for all the simulations showed good qualitative agreement with the observations, apart from systematic deviations at late times which we ascribe to errors in the initial conditions (which did not include observations taken just before impact). However, the simulations did show significant mutual differences at early to mid times, with separations up to 3–5°, which rules out some cases based on observations; e.g., runs with high density (0.7 $\mathrm{g\,cm^{-3}}$) and no spin, and runs with low density (0.3 $\mathrm{g\,cm^{-3}}$) and fast spin (6 hours). An example of a reasonable fit is shown in Figure 2a.

Measurements of the observed train length were published only over a short interval, between March and April 1993 (Scotti & Melosh 1993), and are dependent on instrument sensitivity. The measurements serve as a lower limit: if the simulated train is shorter than the observed length, then that case can be ruled out. Conversely, a simulated train may be much longer than that of the observations if some fragments were too faint to observe. So far in our simulations no set of parameters can be eliminated on this basis alone (see Figure 2b) for an example). We also measured a near constant rate of growth of the fragment train, suggesting that the rate at which the fragment train grows is almost entirely independent of density, diameter, rotation rate, or rotation direction.

Constraints provided by published observations of fragment sizes based on relative intensities (Weaver et al. 1995) are in agreement with the position angle results (i.e., that high density / low spin and low density / high spin cases can be ruled out). This gives us confidence that we have bracketed the relevant regime so that we can focus on a narrower set of parameters for determining the overall best fits to the data in future work. We also note that during our simulations some variability in clump mass was detected, suggesting each individual fragment was actually a swarm of particles that was constantly changing. To explore this in more detail a series of very high resolution simulations will provide more resolution in particle size to investigate cases of secondary fragmentation.

4. Future Work

The current work has provided insight into constraints on tidal disruption outcome (e.g., how rotation affects the position angle of the fragment train). Within any given subset of parameters the density was a dominant factor in determining the quality of fits to observations. Though the observational bias involved with train length prohibits the combination of all the constraints to determine a single set of best-fit parameters at this time, the elimination of some provides insight into the range of possible parameters as a whole. The mass fraction analysis ruled out most of the high density / slow spin and low density / fast spin cases, and the position angle measurements supported this finding.

Much of the data set was the result of space-based observations that provided the spatial resolution required to analyze the dust comae and secondary fragmentation events. It is only with such capabilities that a better understanding of the formation and destruction of cometary nuclei can be produced. Future observations, with new-technology space-based telescopes, will provide the community with many innovative ideas and directions not conceived of previously.

Acknowledgments. The authors would like to thank L. Benner, P. Chodas, and Z. Sekanina for their assistance with this project.

References

Asphaug, E., & Benz, W. 1994, Nature, 370, 120

Asphaug, E., & Benz. W. 1996, Icarus, 121, 225

Boss, A.P., 1994, Icarus, 107, 422

Chodas, P.W., & Yeomans, D. K. 1996, in Proc. IAU Colloq. 156, The Collision of Comet Shoemaker-Levy 9 and Jupiter, eds. K.S. Noll, H.A. Weaver & P.D. Feldman, (Cambridge: STScI), 1

Hahn, J.M., & Rettig, T.W. 2000, Icarus, 146, 501

Hahn, J.M., Rettig, T.W., & Mumma, M.J. 1996, Icarus, 121, 291

Rettig, T., Sobczak, G., & Hahn, J. 1996, Icarus, 121, 281

Richardson, D.C., Quinn, T., Stadel, J., & Lake, G. 2000, Icarus, 143, 45

Richardson, D.C., Leinhardt, Z.M., Melosh, H.J., Bottke, W.F., Jr., & Asphaug, E. 2002, in Asteroids III, eds. W.F. Bottke, Jr., A. Cellino, P. Paolicchi & R.P. Binzel (Tucson: Univ. of Arizona Press), in press

Scotti, J.V., & Melosh, H.J. 1993, Nature, 365, 733

Sekanina, Z. 1996, in Proc. IAU Colloq. 156, The Collision of Comet Shoemaker-Levy 9 and Jupiter, ed. K.S. Noll, H.A. Weaver & P.D. Feldman, (Cambridge: STScI), 55

Shoemaker, C.S., Shoemaker, E.M., & Levy, D. 1993, IAU Circ. 5725

Solem, J.C. 1994, Nature, 370, 349

Weaver, H.A., et al. 1994, Science, 263, 787

Weaver, H.A., et al. 1995, Science, 267, 1282

Hubble's Science Legacy: Future Optical-Ultraviolet Astronomy from Space
ASP Conference Series, Vol. 291, 2003
K.R. Sembach, J.C. Blades, G.D. Illingworth, R.C. Kennicutt, Jr. (eds.)

The HST Legacy at Midlife: What Is Needed in the UV and Can It Be Done?

Willem Wamsteker, for WIC (WSO Implementation Committee)[1]

ESA VILSPA, P.O. Box 507272, 28080 Madrid, Spain

Abstract. The superb capabilities of the Hubble Space Telescope (HST) as a multi-wavelength observatory have supplied a new recognition of observational astrophysics, using both imaging and spectroscopic capabilities. We will try to evaluate what the HST legacy will need. As HST will in the near future be equipped with a new set of instruments (COS, WFC3) one cannot avoid being somewhat speculative. The absence of future ultraviolet (UV) capabilities presents a serious drawback in our ability to address the problems of today and those of the future.

1. The Space Astrophysics Picture 1990–2015

The Hubble science legacy (HSL) cannot be seen as a stand-alone issue. We need to be aware of the space missions planned for the time when the HSL will happen. The HSL should be seen in the overall needs for astrophysics in the programmatic context of the space agencies; therefore we need to understand the complete space astrophysics future. We have limited ourselves to space since we will essentially deal with the UV domain. Ground-based telescopes are not really a variable for the issues addressed here (the primary lifetime for ground-based telescopes is $\geq$ 30 years). The critical issue is shown in Figure 1, which gives a complete overview of the space astrophysics capabilities that are currently planned to be available. Since such planning schedules are under continuous review, this table is kept as a dynamic item at the Web site: `http://wso.esa.vilspa.es` for future reference.

To define the HSL outlook will require placing the Hubble Space Telescope in the context of space astrophysics according to current space agency planning (Weiler 2001; Southwood 2002). Progress in pure science is generally driven more by human intellectual activities than by other sociologically better-organized human endeavors, such as important budgetary plans. This presents a fundamental problem for policy makers, in that thc purely intellectual activities are usually rather unpredictable and their influence is often uncertain. Generally, the expenditures associated with the prestigious and stimulating projects in physics

[1]WIC: W. Wamsteker, *ESA-VILSPA, Spain*; B. Shustov, *INASAN, Russia*; M. Barstow, *Univ. Leicester, United Kingdom*; N. Brosch, *TAU, Israel*; Fu-Zhen Cheng, *USTC-CfA, China*; M. Dennefeld; *IAP, France*; M. Dopita, *ANU, Australia*; A.I. Gomez de Castro, *UCM, Spain*; H. Haubold, *UN/OOSA, Austria*; N. Kappelmann, *IAAT, Germany*; I. Pagano, *Obs. Astroph. di Catania, Italy*; J. Sahade, *CONAE, Argentina*

or astronomy are publicly funded. This is accomplished by an open political process in which conveying scientific knowledge to the public is considered as important as the need for a strong diversity of science.

It was shown (Wamsteker 1999) that in an environment where a lot of technology needed to be developed for non-science purposes, resources were applied to the science, because the knowledge base was firmly based in academia. In exchange for the required needs, regular funding was available for academic research, since the required technology could only be developed with the support of academia. This created a very effective scheme where, between the different space agencies, a continuous access to space was assured for the astrophysics community for all photon energies requiring space.

In Figure 1 we show the space astrophysics picture confined to what could be considered to be remote sensing of the Universe. The choice of the current interval is driven by the fact that every mission launched or in operation between 1995 and 2005 will have a significant influence on the HSL. The first important thing which should be recognized (see also Wamsteker 1999) is that before 1990, space astrophysics was clearly dominated and limited to the X-ray and UV domain, with only small participation in the infrared (IR) and the γ-ray domains.

It is quite obvious from Figure 1 that the wavelength domains (and therefore the associated technologies) have matured, so that now, instead of two main domains, we are facing the needs of astrophysical detection and observation capabilities from space extending from $\sim$ 500 GeV to 1 GHz, including also basic physics (non-photon astrophysics). This has the important consequence that the resources of the space agencies have to be distributed over three times as many industrial and academic communities, and more technology developments will have to be carried by basic science funding.

Another consequence is that, since space missions are generally very efficient data gatherers, the analysis and interpretative capabilities of the scientific communities are becoming very stretched. This has lead to the current emphasis on the development of "virtual observatories" which, although very commendable, are no replacement for an active observing capability. The absence of capabilities in a wavelength range will strongly affect important aspects of observational astrophysics, which is the direct and relatively fast interplay between results and new questions requiring new observations.

Space science missions are made to validate old theories and to develop new theories. They suffer from budgetary constraints, and although this aspect cannot be excluded from the discussion, this is not the right place to tackle that aspect in detail. On the other hand, it is important to point out that in the conditions outlined above, serious new pressures will be brought on anything that one wants to do. The very careful evaluation of the overall costing of projects cannot be seen any longer as a somewhat isolated activity in a specific community. The comparative constraints from different communities and science interests enter as previously unknown parameters in the selection process.

2. How to See HSL Now

The UV has a special and unique importance which cannot be addressed by any of the missions currently planned. Most of this is highlighted in more

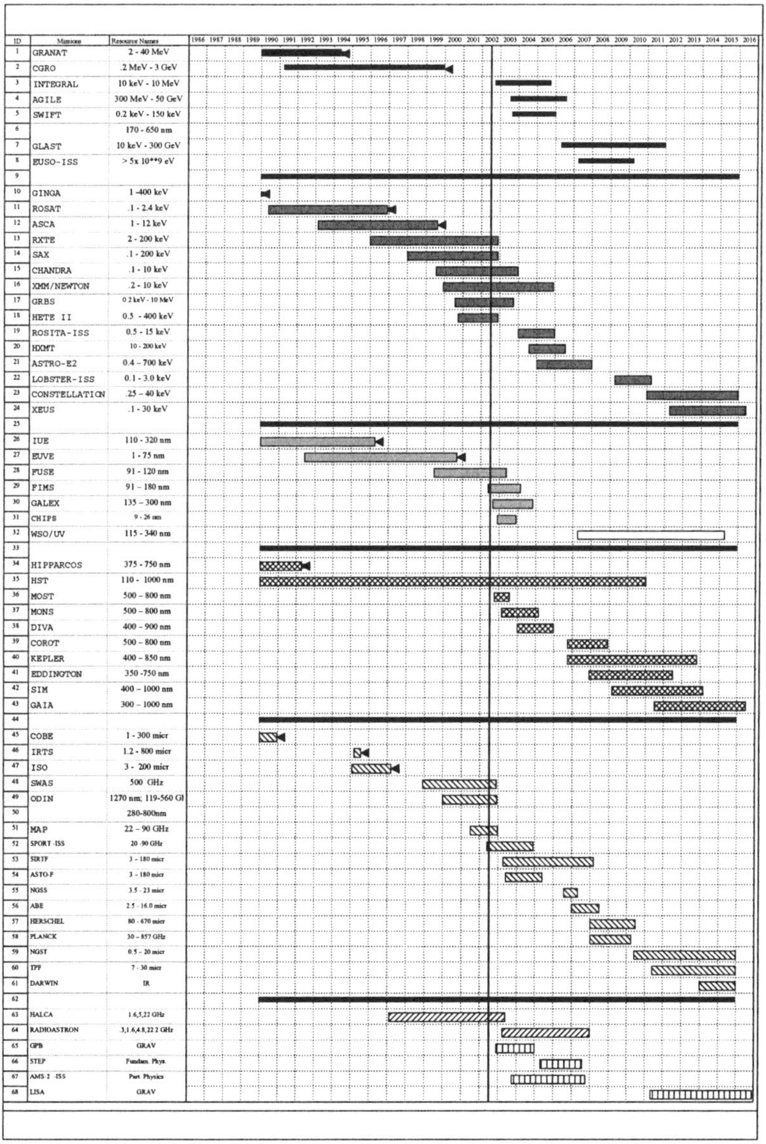

Figure 1. This figure shows the timeline of astrophysics from space between 1990–2015. The start of the filled bars indicates the launch of a mission and the length of the bars corresponds to the nominal mission duration. The ◁ symbol to the right side of the horizontal duration bar indicates a mission which has terminated operations. In this chart, the current epoch is indicated by thin vertical black line and the missions are coded according to energy, wavelength or frequency domain. Missions associated with the International Space Station are indicated with *–ISS* after the name.

detail in the topical talks in this meeting. Evaluating the issues indicated in §1, one can question the association of HSL with the UV/optical domain. This domain combination is historically correct. HST was launched originally as a UV/optical observatory, but its current life and the foreseen future instruments suggest that a better description of HST after midlife will be an optical/near-IR observatory, with a UV capability. This is already clear from the statistics for the demand and the allocation of time for the 11th observing cycle of HST. This shows that all UV activities of HST comprise 16% of the demand and 17% of the allocation, while the optical/near-IR accounts for $\sim$ 84% of the demand and 72% of the allocation. (Machetto et al. 2002). Although one could say that this is simply a demonstration of the science interest of the community, such simplistic arguments cannot be maintained. The scientific community is intelligent enough to apply only for those studies for which an instrument is suitable.

Consider the following constraining factors: (1) The general economical environment has changed, and draconian cost control is a very strong requirement. (2) Telescope size is major cost driver, as a consequence of the requirements on the launcher. (3) Multi-function observatories have serious cost drivers and risk parameters. (4) The current broad science planning timelines of the Space Agencies supply the environment in which scientists should look at the future. Taking these environmental constraints on the definition of future UV missions, one can try to evaluate, in global way, the requirements of a UV mission which would build upon the HST Science Legacy and supply a unique capability.

We give below the requirements that can be brought together in an efficient UV mission, are feasible on the timescales needed, and are under cost restrictions of the environment indicated by the four points above (Wamsteker et al. 2000; Oberto et al. 2001). A mission of the following characteristics will present an essential contribution to the HSL and will be able to address the many questions in the astrophysics of today as outlined throughout this Workshop: wavelength range: 105–320 nm; high-Earth orbit or L2; large continuous viewing zone; high instrumental stability; large FOV; large dynamic range; UV sensitivity matched to 8 m-class ground-based telescopes; spectral resolution $R \approx 1000$ and $R \approx 10^5$–10^6; near diffraction limited imaging; long duration integration; rapid response capability; bright source suppression; repeat visit capability; flexible scheduling and operations; availability 1–2 years after the GALEX sky survey; and early involvement of educational institutes on a global scale.

References

Machetto, D. et al. 2002, STScI Newsletter, vol. 19, (01) 7

Oberto, R., et al. 2001, JPL/ADP Assessment of WSO Proposal, CL# 01- 1168

Southwood, D. 2002, Cosmic Vision 2020, ESA-PR 36-2002

Wamsteker, W. 1999, in ASP Conf. Ser. Vol. 164, UV-Optical Space Astronomy Beyond HST, eds. J.A. Morse, J.M. Shull, & A.L. Kinney, (San Francisco: ASP), 261

Wamsteker, W., et al. 2000, WSO/UV Assessment Study ESA/CDF-05 (A)

Weiler, E.J. 2000, NASA Space Science Strategic Plan: The Space Science Enterprise

Hubble's Science Legacy: Future Optical-Ultraviolet Astronomy from Space
ASP Conference Series, Vol. 291, 2003
K.R. Sembach, J.C. Blades, G.D. Illingworth, R.C. Kennicutt, Jr. (eds.)

On the Ionization of Heavy Elements in Diffuse Interstellar Clouds

Daniel E. Welty

University of Chicago, Astronomy & Astrophysics Center, 5640 S. Ellis Ave., Chicago, IL 60637

Abstract. We briefly discuss several puzzles related to the ionization of heavy elements in diffuse interstellar clouds — in our Galaxy and in the SMC.

1. Introduction

It is commonly assumed that photoionization and radiative recombination dominate the ionization equilibrium of heavy elements in diffuse interstellar clouds: $\Gamma\ n(\mathrm{X\,I}) = \alpha\ n_e\ n(\mathrm{X\,II})$, where Γ is the photoionization rate, α is the recombination rate, and n_e is the electron density. Recent observational and theoretical studies have suggested, however, that additional processes may often play a significant role. For example, the electron densities inferred from the ratios $N(\mathrm{X\,I})/N(\mathrm{X\,II})$, for different elements X, can differ by more than an order of magnitude along a given line of sight, if photoionization equilibrium is assumed. Moreover, the patterns of relative n_e observed in different lines of sight show both similarities and differences — suggesting that multiple additional processes besides photoionization and radiative recombination significantly affect the ionization balance of heavy elements — to different degrees in different lines of sight. Where independent estimates can be made for both n_e and the local hydrogen density (n_{H}), the resulting fractional ionization is often much greater than the value 2×10^{-4} that would be expected if most of the electrons come from ionization of carbon and other heavy elements.

2. Relative Electron Densities from X I/X II

Several recent studies have suggested that the ratios $N(\mathrm{X\,I})/N(\mathrm{X\,II})$, for different elements X but for the same assumed T and radiation field, can yield markedly different values for n_e (HD 215733, Fitzpatrick & Spitzer 1997; 23 Ori, Welty et al. 1999). For the highest column density clouds toward 23 Ori, the average n_e derived independently from nine pairs of neutral and singly ionized species range over a factor ~ 20. We have estimated n_e for the main interstellar clouds toward 23 Ori and four other stars, based on high-resolution optical and ultraviolet (HST/GHRS) spectra (Welty, Hobbs, & Morton 2002). Figure 1 shows those n_e, relative to the respective averages (n_0) of the values for C, Na, and K; n_e/n_0 should be fairly insensitive to the actual values of T and the ambient radiation fields characterizing the clouds in the different lines of sight.

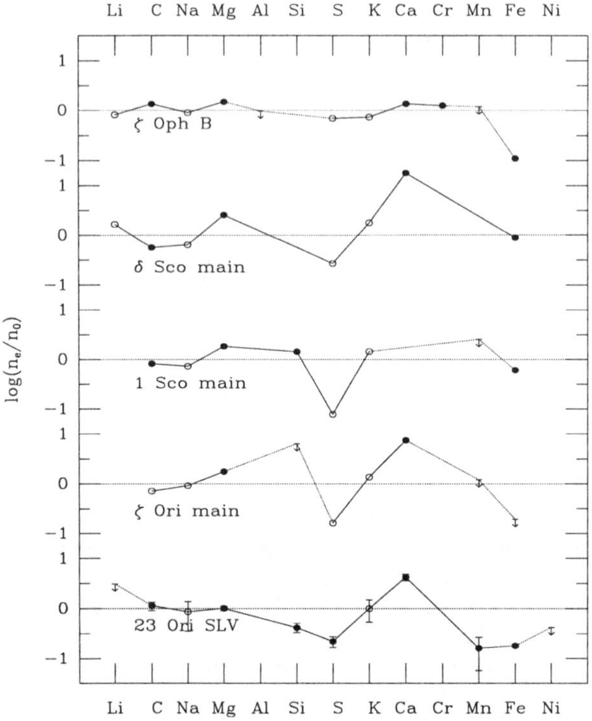

Figure 1. Logarithm of relative electron density (n_e/n_0) for the main components toward five stars. Open circles denote cases where N(X II) was estimated based on N(Zn II) and typical depletions; representative 1 σ error bars are shown for 23 Ori. Note the similarities between 23 Ori and ζ Ori, and between 1 Sco and δ Sco — and the differences between those two pairs and ζ Oph. The points for S, Mn, Fe, and Ni are often low; Ca is often high.

Figure 1 shows that the differences in relative n_e found toward 23 Ori are neither unique nor universal. The n_e inferred for the main clouds toward ζ Ori, for example, are significantly lower than those toward 23 Ori, but show a very similar pattern — with roughly similar values from C, Na, Mg, and K, low values from S and Fe, and a high value from Ca. The patterns seen toward 1 Sco and δ Sco are also similar, except that the n_e derived from Si and Fe are comparable to the values derived from C, Na, Mg, and K — rather than much lower, as seen toward 23 Ori. In the main clouds toward ζ Oph, however, most elements — even S and Ca — give very similar n_e, and only Fe gives a low value.

The differences in the element-to-element pattern of relative n_e for these five lines of sight suggest that the differences in n_e inferred for any given line of sight are not just due to a single process or factor — but instead result from multiple additional processes that selectively affect the ionization balance of the various elements, to different degrees in the different lines of sight. Lepp et al. (1988), Fitzpatrick & Spitzer (1997), Welty et al. (1999), Weingartner & Draine (2001), and Welty et al. (2002) all have discussed some possible factors and/or processes: incorrect f-values for individual transitions, errors in the photoionization and/or recombination rates, differences in the shape of the radiation field, dielectronic recombination (increasing Ca I and Mg II in warm gas), dependence of depletion on local hydrogen density, dissociative recombination of CH^+ (increasing C I), charge exchange between singly ionized species and large molecules / small grains (enhancing the corresponding neutral species), and charge exchange with protons (reducing some of the trace neutral species). While some of these processes may have some effect, none appears able to account for all the apparent differences in inferred n_e.

3. Fractional Ionization

The other puzzling aspect of the ionization balance in the main clouds toward 23 Ori is the apparently high fractional ionization: $n_e/n_{\rm H} \sim 0.01$. This is a factor $\sim$ 50 higher than the value $\sim 2\times10^{-4}$ expected from photoionization of carbon and other abundant heavy elements, and a factor $\sim$ 10 higher than the values obtained by including ionization of H and He due to cosmic rays and x-rays (e.g., Wolfire et al. 1995; Weingartner & Draine 2001). Similar estimates for $n_e/n_{\rm H}$ in the main clouds in eight other sight lines (in Orion, Sco-Oph, and Cepheus) reveal several other cases where $n_e/n_{\rm H}$ is clearly higher than the value expected from photoionization of heavy elements alone (Welty et al. 2002). After accounting for (probably) elevated radiation fields in the Sco-Oph region, it is possible that most of the clouds in this small sample have $n_e/n_{\rm H}$ significantly in excess of 2×10^{-4}; there may be a range in $n_e/n_{\rm H}$ of more than a factor of 10. The apparently high fractional ionization could imply either that n_e has been overestimated (e.g., if charge exchange processes are important) or that hydrogen is slightly ionized even in predominantly neutral gas — or both.

4. Main SMC Clouds Toward Sk 155

Studies of the interstellar medium (ISM) in the Magellanic Clouds (MC) explore somewhat different environmental conditions (metallicity, dust-to-gas ratio, ambient radiation field, ultraviolet extinction) than those typically probed in the Galactic ISM. Determination of elemental abundances, depletions, and physical conditions in the MC ISM therefore provides interesting tests for theoretical models of both interstellar clouds and interstellar dust grains.

Using high-resolution (FWHM $\sim$ 1.4 km s^{-1}) Na I spectra, STIS echelle spectra (FWHM $\sim$ 2.7 km s^{-1}), and Far Ultraviolet Spectroscopic Explorer (FUSE) spectra (FWHM $\sim$ 25 km s^{-1}), we have begun to address these issues in the various interstellar clouds toward the SMC star Sk 155 (Welty et al. 2001 and in prep.). The STIS spectra have revealed unusual relative abundances/depletions in several of the main SMC components — severe depletions for Cr, Fe, and Ni but negligible depletions for Mg and Si — a pattern which has not been seen in the Galactic ISM. Analysis of the C I fine-structure excitation, for several of the main SMC components, indicates thermal pressures $n_{\rm H}T$ of about $10^{3.9-4.1}$ cm^{-3}K. Such pressures are higher than those typically found in our Galaxy (Jenkins & Tripp 2001), but are consistent with predictions for lower metallicity gas (e.g., Wolfire et al. 1995). Using $T_{01}(H_2) \sim 85$ K, the n_H range from about 80–150 cm^{-3} in the two main SMC components. If we assume typical depletions for C and Na, and that the radiation field is 5 times the Galactic WJ1 field (e.g., Lequeux 1989), then we estimate $n_e \sim$ 0.02–0.10 cm^{-3} in several of the main SMC components. The corresponding values estimated from N(Mg II)/N(Mg II) are factors of 2–6 higher — comparable to some of the discrepancies seen in our Galaxy (Figure 1). The fractional ionization in those SMC components thus ranges from about 0.00020 to 0.00125 — factors of $\sim$ 5–30 higher than the value $\sim 4{\times}10^{-5}$ "expected" (for the lower metallicity SMC) if the electrons come from ionization of carbon and other heavy elements. Again, such differences are comparable to those found locally.

5. Summary / Conclusions

Estimates of n_e and n_H based on the assumptions of photoionization equilibrium and $n_e/n_H \sim 2{\times}10^{-4}$ must be considered to be very uncertain — particularly if they are based on only a single ratio N(X I)/N(X II). Observations of lines from many different neutral/first ion pairs, for additional sight lines sampling different environmental conditions, may help us to understand the various processes affecting the ionization balance in diffuse clouds. High-resolution ultraviolet spectra of faint targets (e.g., heavily reddened Galactic stars and stars in the Magellanic Clouds) will be very useful.

Acknowledgments. Support for this work has been provided by NASA LTSA grant NAG5-3228; by NASA grants GO-2251.01-87A and GO-08145.01-A from the Space Telescope Science Institute, which is operated by AURA, Inc., under NASA contract NAS5-26555; and by NASA grant NAG5-10370 (FUSE).

References

Fitzpatrick, E.L., & Spitzer, L. 1997, ApJ, 475, 623

Jenkins, E.B., & Tripp, T.M. 2001, ApJS, 137, 297

Lepp, S., Dalgarno, A., van Dishoeck, E.F., & Black, J.H. 1988, ApJ, 329, 418

Lequeux, J. 1989, in Recent Developments of Magellanic Cloud Research, eds. K.S. de Boer, F. Spite, & G. Stasinska (Paris: Paris Obs.)

Weingartner, J.C., & Draine, B.T. 2001, ApJ, 563, 842

Welty, D.E., & Hobbs, L.M. 2001, ApJS, 133, 345

Welty, D.E., Hobbs, L.M., Lauroesch, J.T., Morton, D.C., Spitzer, L., & York, D.G. 1999, ApJS, 124, 465

Welty, D.E., Hobbs, L.M., & Morton, D.C. 2002, ApJ, submitted

Welty, D.E., Lauroesch, J.T., Blades, J.C., Hobbs, L.M., & York, D.G. 2001, ApJ, 554, L75

Wolfire, M.G., Hollenbach, D., McKee, C.F., Tielens, A.G.G.M., & Bakes, E.L.O. 1995, ApJ, 443, 152

Author Index

A LIST OF THE VOLUMES

Published
by

THE ASTRONOMICAL SOCIETY OF THE PACIFIC
(ASP)

An international, nonprofit, scientific and educational organization founded in 1889

All book orders or inquiries concerning

THE ASTRONOMICAL SOCIETY OF THE PACIFIC
CONFERENCE SERIES
(ASP - CS)

and

INTERNATIONAL ASTRONOMICAL UNION VOLUMES
(IAU)

should be directed to the:

The Astronomical Society of the Pacific Conference Series
390 Ashton Avenue
San Francisco CA 94112-1722 USA

Phone: 800-335-2624 (Within USA)
Phone: 415-337-2126
Fax: 415-337-5205

E-mail: service@astrosociety.org
Web Site: http://www.astrosociety.org

Complete lists of proceedings of past IAU Meetings are maintained at the IAU Web site at the URL: http://www.iau.org/publicat.html

Volumes 32 - 189 in the IAU Symposia Series may be ordered from:

Kluwer Academic Publishers
P. O. Box 117
NL 3300 AA Dordrecht
The Netherlands

Kluwer@wKap.com

ASP CONFERENCE SERIES VOLUMES

Published by the Astronomical Society of the Pacific

PUBLISHED: 1988 (* asterisk means OUT OF STOCK)

Vol. CS -1 PROGRESS AND OPPORTUNITIES IN SOUTHERN HEMISPHERE OPTICAL ASTRONOMY: CTIO 25TH Anniversary Symposium
eds. V. M. Blanco and M. M. Phillips
ISBN 0-937707-18-X

Vol. CS-2 PROCEEDINGS OF A WORKSHOP ON OPTICAL SURVEYS FOR QUASARS
eds. Patrick S. Osmer, Alain C. Porter, Richard F. Green, and Craig B. Foltz
ISBN 0-937707-19-8

Vol. CS-3 FIBER OPTICS IN ASTRONOMY
ed. Samuel C. Barden
ISBN 0-937707-20-1

Vol. CS-4 THE EXTRAGALACTIC DISTANCE SCALE: Proceedings of the ASP 100th Anniversary Symposium
eds. Sidney van den Bergh and Christopher J. Pritchet
ISBN 0-937707-21-X

Vol. CS-5 THE MINNESOTA LECTURES ON CLUSTERS OF GALAXIES AND LARGE-SCALE STRUCTURE
ed. John M. Dickey
ISBN 0-937707-22-8

PUBLISHED: 1989

Vol. CS-6 * SYNTHESIS IMAGING IN RADIO ASTRONOMY: A Collection of Lectures from the Third NRAO Synthesis Imaging Summer School
eds. Richard A. Perley, Frederic R. Schwab, and Alan H. Bridle
ISBN 0-937707-23-6

PUBLISHED: 1990

Vol. CS-7 PROPERTIES OF HOT LUMINOUS STARS: Boulder-Munich Workshop
ed. Catharine D. Garmany
ISBN 0-937707-24-4

Vol. CS-8 * CCDs IN ASTRONOMY
ed. George H. Jacoby
ISBN 0-937707-25-2

Vol. CS-9 COOL STARS, STELLAR SYSTEMS, AND THE SUN: Sixth Cambridge Workshop
ed. George Wallerstein
ISBN 0-937707-27-9

Vol. CS-10 * EVOLUTION OF THE UNIVERSE OF GALAXIES: Edwin Hubble Centennial Symposium
ed. Richard G. Kron
ISBN 0-937707-28-7

Vol. CS-11 CONFRONTATION BETWEEN STELLAR PULSATION AND EVOLUTION
eds. Carla Cacciari and Gisella Clementini
ISBN 0-937707-30-9

Vol. CS-12 THE EVOLUTION OF THE INTERSTELLAR MEDIUM
ed. Leo Blitz
ISBN 0-937707-31-7

PUBLISHED: 1991

Vol. CS-13 THE FORMATION AND EVOLUTION OF STAR CLUSTERS
ed. Kenneth Janes
ISBN 0-937707-32-5

ASP CONFERENCE SERIES VOLUMES

Published by the Astronomical Society of the Pacific

PUBLISHED: 1991 (* asterisk means OUT OF STOCK)

Vol. CS-14 ASTROPHYSICS WITH INFRARED ARRAYS
ed. Richard Elston
ISBN 0-937707-33-3

Vol. CS-15 LARGE-SCALE STRUCTURES AND PECULIAR MOTIONS IN THE UNIVERSE
eds. David W. Latham and L. A. Nicolaci da Costa
ISBN 0-937707-34-1

Vol. CS-16 Proceedings of the 3rd Haystack Observatory Conference on ATOMS, IONS, AND MOLECULES: NEW RESULTS IN SPECTRAL LINE ASTROPHYSICS
eds. Aubrey D. Haschick and Paul T. P. Ho
ISBN 0-937707-35-X

Vol. CS-17 LIGHT POLLUTION, RADIO INTERFERENCE, AND SPACE DEBRIS
ed. David L. Crawford
ISBN 0-937707-36-8

Vol. CS-18 THE INTERPRETATION OF MODERN SYNTHESIS OBSERVATIONS OF SPIRAL GALAXIES
eds. Nebojsa Duric and Patrick C. Crane
ISBN 0-937707-37-6

Vol. CS-19 RADIO INTERFEROMETRY: THEORY, TECHNIQUES, AND APPLICATIONS, IAU Colloquium 131
eds. T. J. Cornwell and R. A. Perley
ISBN 0-937707-38-4

Vol. CS-20 FRONTIERS OF STELLAR EVOLUTION: 50th Anniversary McDonald Observatory (1939-1989)
ed. David L. Lambert
ISBN 0-937707-39-2

Vol. CS-21 THE SPACE DISTRIBUTION OF QUASARS
ed . David Crampton
ISBN 0-937707-40-6

PUBLISHED: 1992

Vol. CS-22 NONISOTROPIC AND VARIABLE OUTFLOWS FROM STARS
eds. Laurent Drissen, Claus Leitherer, and Antonella Nota
ISBN 0-937707-41-4

Vol CS-23 * ASTRONOMICAL CCD OBSERVING AND REDUCTION TECHNIQUES
ed. Steve B. Howell
ISBN 0-937707-42-4

Vol. CS-24 COSMOLOGY AND LARGE-SCALE STRUCTURE IN THE UNIVERSE
ed. Reinaldo R. de Carvalho
ISBN 0-937707-43-0

Vol. CS-25 ASTRONOMICAL DATA ANALYSIS, SOFTWARE AND SYSTEMS I - (ADASS I)
eds. Diana M. Worrall, Chris Biemesderfer, and Jeannette Barnes
ISBN 0-937707-44-9

Vol. CS-26 COOL STARS, STELLAR SYSTEMS, AND THE SUN: Seventh Cambridge Workshop
eds. Mark S. Giampapa and Jay A. Bookbinder
ISBN 0-937707-45-7

Vol. CS-27 THE SOLAR CYCLE: Proceedings of the National Solar Observatory/Sacramento Peak 12th Summer Workshop
ed. Karen L. Harvey
ISBN 0-937707-46-5

ASP CONFERENCE SERIES VOLUMES

Published by the Astronomical Society of the Pacific

PUBLISHED: 1992 (asterisk means OUT OF STOCK)

Vol. CS-28 AUTOMATED TELESCOPES FOR PHOTOMETRY AND IMAGING
eds. Saul J. Adelman, Robert J. Dukes, Jr., and Carol J. Adelman
ISBN 0-937707-47-3

Vol. CS-29 Viña del Mar Workshop on CATACLYSMIC VARIABLE STARS
ed. Nikolaus Vogt
ISBN 0-937707-48-1

Vol. CS-30 VARIABLE STARS AND GALAXIES
ed. Brian Warner
ISBN 0-937707-49-X

Vol. CS-31 RELATIONSHIPS BETWEEN ACTIVE GALACTIC NUCLEI AND STARBURST GALAXIES
ed. Alexei V. Filippenko
ISBN 0-937707-50-3

Vol. CS-32 COMPLEMENTARY APPROACHES TO DOUBLE AND MULTIPLE STAR RESEARCH, IAU Colloquium 135
eds. Harold A. McAlister and William I. Hartkopf
ISBN 0-937707-51-1

Vol. CS-33 * RESEARCH AMATEUR ASTRONOMY
ed. Stephen J. Edberg
ISBN 0-937707-52-X

Vol. CS-34 ROBOTIC TELESCOPES IN THE 1990's
ed. Alexei V. Filippenko
ISBN 0-937707-53-8

PUBLISHED: 1993

Vol. CS-35 * MASSIVE STARS: THEIR LIVES IN THE INTERSTELLAR MEDIUM
eds. Joseph P. Cassinelli and Edward B. Churchwell
ISBN 0-937707-54-6

Vol. CS-36 PLANETS AROUND PULSARS
ed. J. A. Phillips, S. E. Thorsett, and S. R. Kulkarni
ISBN 0-937707-55-4

Vol. CS-37 FIBER OPTICS IN ASTRONOMY II
ed. Peter M. Gray
ISBN 0-937707-56-2

Vol. CS-38 NEW FRONTIERS IN BINARY STAR RESEARCH: Pacific Rim Colloquium
eds. K. C. Leung and I.-S. Nha
ISBN 0-937707-57-0

Vol. CS-39 THE MINNESOTA LECTURES ON THE STRUCTURE AND DYNAMICS OF THE MILKY WAY
ed. Roberta M. Humphreys
ISBN 0-937707-58-9

Vol. CS-40 INSIDE THE STARS, IAU Colloquium 137
eds. Werner W. Weiss and Annie Baglin
ISBN 0-937707-59-7

Vol. CS-41 ASTRONOMICAL INFRARED SPECTROSCOPY: FUTURE OBSERVATIONAL DIRECTIONS
ed. Sun Kwok
ISBN 0-937707-60-0

ASP CONFERENCE SERIES VOLUMES

Published by the Astronomical Society of the Pacific

PUBLISHED: 1993 (* asterisk means OUT OF STOCK)

Vol. CS-42 GONG 1992: SEISMIC INVESTIGATION OF THE SUN AND STARS
ed. Timothy M. Brown
ISBN 0-937707-61-9

Vol. CS-43 SKY SURVEYS: PROTOSTARS TO PROTOGALAXIES
ed. B. T. Soifer
ISBN 0-937707-62-7

Vol. CS-44 PECULIAR VERSUS NORMAL PHENOMENA IN A-TYPE AND RELATED STARS, IAU Colloquium 138
eds. M. M. Dworetsky, F. Castelli, and R. Faraggiana
ISBN 0-937707-63-5

Vol. CS-45 LUMINOUS HIGH-LATITUDE STARS
ed. Dimitar D. Sasselov
ISBN 0-937707-64-3

Vol. CS-46 THE MAGNETIC AND VELOCITY FIELDS OF SOLAR ACTIVE REGIONS, IAU Colloquium 141
eds. Harold Zirin, Guoxiang Ai, and Haimin Wang
ISBN 0-937707-65-1

Vol. CS-47 THIRD DECENNIAL US-USSR CONFERENCE ON SETI -- Santa Cruz, California, USA
ed. G. Seth Shostak
ISBN 0-937707-66-X

Vol. CS-48 THE GLOBULAR CLUSTER-GALAXY CONNECTION
eds. Graeme H. Smith and Jean P. Brodie
ISBN 0-937707-67-8

Vol. CS-49 GALAXY EVOLUTION: THE MILKY WAY PERSPECTIVE
ed. Steven R. Majewski
ISBN 0-937707-68-6

Vol. CS-50 STRUCTURE AND DYNAMICS OF GLOBULAR CLUSTERS
eds. S. G. Djorgovski and G. Meylan
ISBN 0-937707-69-4

Vol. CS-51 OBSERVATIONAL COSMOLOGY
eds. Guido Chincarini, Angela Iovino, Tommaso Maccacaro, and Dario Maccagni
ISBN 0-937707-70-8

Vol. CS-52 ASTRONOMICAL DATA ANALYSIS SOFTWARE AND SYSTEMS II - (ADASS II)
eds. R. J. Hanisch, R. J. V. Brissenden, and Jeannette Barnes
ISBN 0-937707-71-6

Vol. CS-53 BLUE STRAGGLERS
ed. Rex A. Saffer
ISBN 0-937707-72-4

PUBLISHED: 1994

Vol. CS-54 * THE FIRST STROMLO SYMPOSIUM: THE PHYSICS OF ACTIVE GALAXIES
eds. Geoffrey V. Bicknell, Michael A. Dopita, and Peter J. Quinn
ISBN 0-937707-73-2

Vol. CS-55 OPTICAL ASTRONOMY FROM THE EARTH AND MOON
eds. Diane M. Pyper and Ronald J. Angione
ISBN 0-937707-74-0

Vol. CS-56 INTERACTING BINARY STARS
ed. Allen W. Shafter
ISBN 0-937707-75-9

ASP CONFERENCE SERIES VOLUMES

Published by the Astronomical Society of the Pacific

PUBLISHED: 1994 (* asterisk means OUT OF STOCK)

Vol. CS-57 STELLAR AND CIRCUMSTELLAR ASTROPHYSICS
eds. George Wallerstein and Alberto Noriega-Crespo
ISBN 0-937707-76-7

Vol. CS-58 * THE FIRST SYMPOSIUM ON THE INFRARED CIRRUS AND DIFFUSE INTERSTELLAR CLOUDS
eds. Roc M. Cutri and William B. Latter
ISBN 0-937707-77-5

Vol. CS-59 ASTRONOMY WITH MILLIMETER AND SUBMILLIMETER WAVE INTERFEROMETRY,
IAU Colloquium 140
eds. M. Ishiguro and Wm. J. Welch
ISBN 0-937707-78-3

Vol. CS-60 THE MK PROCESS AT 50 YEARS: A POWERFUL TOOL FOR ASTROPHYSICAL INSIGHT, A Workshop of the Vatican Observatory --Tucson, Arizona, USA
eds. C. J. Corbally, R. O. Gray, and R. F. Garrison
ISBN 0-937707-79-1

Vol. CS-61 ASTRONOMICAL DATA ANALYSIS SOFTWARE AND SYSTEMS III - (ADASS III)
eds. Dennis R. Crabtree, R. J. Hanisch, and Jeannette Barnes
ISBN 0-937707-80-5

Vol. CS-62 THE NATURE AND EVOLUTIONARY STATUS OF HERBIG Ae/Be STARS
eds. Pik Sin Thé, Mario R. Pérez, and Ed P. J. van den Heuvel
ISBN 0-9837707-81-3

Vol. CS-63 SEVENTY-FIVE YEARS OF HIRAYAMA ASTEROID FAMILIES: THE ROLE OF COLLISIONS IN THE SOLAR SYSTEM HISTORY
eds. Yoshihide Kozai, Richard P. Binzel, and Tomohiro Hirayama
ISBN 0-937707-82-1

Vol. CS-64 * COOL STARS, STELLAR SYSTEMS, AND THE SUN:
Eighth Cambridge Workshop
ed. Jean-Pierre Caillault
ISBN 0-937707-83-X

Vol. CS-65 * CLOUDS, CORES, AND LOW MASS STARS:
The Fourth Haystack Observatory Conference
eds. Dan P. Clemens and Richard Barvainis
ISBN 0-937707-84-8

Vol. CS-66 * PHYSICS OF THE GASEOUS AND STELLAR DISKS OF THE GALAXY
ed. Ivan R. King
ISBN 0-937707-85-6

Vol. CS-67 UNVEILING LARGE-SCALE STRUCTURES BEHIND THE MILKY WAY
eds. C. Balkowski and R. C. Kraan-Korteweg
ISBN 0-937707-86-4

Vol. CS-68 * SOLAR ACTIVE REGION EVOLUTION: COMPARING MODELS WITH OBSERVATIONS
eds. K. S. Balasubramaniam and George W. Simon
ISBN 0-937707-87-2

Vol. CS-69 REVERBERATION MAPPING OF THE BROAD-LINE REGION IN ACTIVE GALACTIC NUCLEI
eds. P. M. Gondhalekar, K. Horne, and B. M. Peterson
ISBN 0-937707-88-0

Vol. CS-70 * GROUPS OF GALAXIES
eds. Otto-G. Richter and Kirk Borne
ISBN 0-937707-89-9

ASP CONFERENCE SERIES VOLUMES

Published by the Astronomical Society of the Pacific

PUBLISHED: 1995 (* asterisk means OUT OF STOCK)

Vol. CS-71	TRIDIMENSIONAL OPTICAL SPECTROSCOPIC METHODS IN ASTROPHYSICS, IAU Colloquium 149 eds. Georges Comte and Michel Marcelin ISBN 0-937707-90-2
Vol. CS-72	MILLISECOND PULSARS: A DECADE OF SURPRISE eds. A. S Fruchter, M. Tavani, and D. C. Backer ISBN 0-937707-91-0
Vol. CS-73	AIRBORNE ASTRONOMY SYMPOSIUM ON THE GALACTIC ECOSYSTEM: FROM GAS TO STARS TO DUST eds. Michael R. Haas, Jacqueline A. Davidson, and Edwin F. Erickson ISBN 0-937707-92-9
Vol. CS-74	PROGRESS IN THE SEARCH FOR EXTRATERRESTRIAL LIFE: 1993 Bioastronomy Symposium ed. G. Seth Shostak ISBN 0-937707-93-7
Vol. CS-75	MULTI-FEED SYSTEMS FOR RADIO TELESCOPES eds. Darrel T. Emerson and John M. Payne ISBN 0-937707-94-5
Vol. CS-76	GONG '94: HELIO- AND ASTERO-SEISMOLOGY FROM THE EARTH AND SPACE eds. Roger K. Ulrich, Edward J. Rhodes, Jr., and Werner Däppen ISBN 0-937707-95-3
Vol. CS-77	ASTRONOMICAL DATA ANALYSIS SOFTWARE AND SYSTEMS IV - (ADASS IV) eds. R. A. Shaw, H. E. Payne, and J. J. E. Hayes ISBN 0-937707-96-1
Vol. CS-78	ASTROPHYSICAL APPLICATIONS OF POWERFUL NEW DATABASES: Joint Discussion No. 16 of the 22nd General Assembly of the IAU eds. S. J. Adelman and W. L. Wiese ISBN 0-937707-97-X
Vol. CS-79 *	ROBOTIC TELESCOPES: CURRENT CAPABILITIES, PRESENT DEVELOPMENTS, AND FUTURE PROSPECTS FOR AUTOMATED ASTRONOMY eds. Gregory W. Henry and Joel A. Eaton ISBN 0-937707-98-8
Vol. CS-80 *	THE PHYSICS OF THE INTERSTELLAR MEDIUM AND INTERGALACTIC MEDIUM eds. A. Ferrara, C. F. McKee, C. Heiles, and P. R. Shapiro ISBN 0-937707-99-6
Vol. CS-81	LABORATORY AND ASTRONOMICAL HIGH RESOLUTION SPECTRA eds. A. J. Sauval, R. Blomme, and N. Grevesse ISBN 1-886733-01-5
Vol. CS-82 *	VERY LONG BASELINE INTERFEROMETRY AND THE VLBA eds. J. A. Zensus, P. J. Diamond, and P. J. Napier ISBN 1-886733-02-3
Vol. CS-83 *	ASTROPHYSICAL APPLICATIONS OF STELLAR PULSATION, IAU Colloquium 155 eds. R. S. Stobie and P. A. Whitelock ISBN 1-886733-03-1
ATLAS	INFRARED ATLAS OF THE ARCTURUS SPECTRUM, 0.9 - 5.3 μm eds. Kenneth Hinkle, Lloyd Wallace, and William Livingston ISBN: 1-886733-04-X

ASP CONFERENCE SERIES VOLUMES

Published by the Astronomical Society of the Pacific

PUBLISHED: 1995 (* asterisk means OUT OF STOCK)

Vol. CS-84 THE FUTURE UTILIZATION OF SCHMIDT TELESCOPES, IAU Colloquium 148
eds. Jessica Chapman, Russell Cannon, Sandra Harrison, and Bambang Hidayat
ISBN 1-886733-05-8

Vol. CS-85 * CAPE WORKSHOP ON MAGNETIC CATACLYSMIC VARIABLES
eds. D. A. H. Buckley and B. Warner
ISBN 1-886733-06-6

Vol. CS-86 FRESH VIEWS OF ELLIPTICAL GALAXIES
eds. Alberto Buzzoni, Alvio Renzini, and Alfonso Serrano
ISBN 1-886733-07-4

PUBLISHED: 1996

Vol. CS-87 NEW OBSERVING MODES FOR THE NEXT CENTURY
eds. Todd Boroson, John Davies, and Ian Robson
ISBN 1-886733-08-2

Vol. CS-88 * CLUSTERS, LENSING, AND THE FUTURE OF THE UNIVERSE
eds. Virginia Trimble and Andreas Reisenegger
ISBN 1-886733-09-0

Vol. CS-89 ASTRONOMY EDUCATION: CURRENT DEVELOPMENTS, FUTURE COORDINATION
ed. John R. Percy
ISBN 1-886733-10-4

Vol. CS-90 THE ORIGINS, EVOLUTION, AND DESTINIES OF BINARY STARS IN CLUSTERS
eds. E. F. Milone and J. -C. Mermilliod
ISBN 1-886733-11-2

Vol. CS-91 BARRED GALAXIES, IAU Colloquium 157
eds. R. Buta, D. A. Crocker, and B. G. Elmegreen
ISBN 1-886733-12-0

Vol. CS-92 * FORMATION OF THE GALACTIC HALO .... INSIDE AND OUT
eds. Heather L. Morrison and Ata Sarajedini
ISBN 1-886733-13-9

Vol. CS-93 RADIO EMISSION FROM THE STARS AND THE SUN
eds. A. R. Taylor and J. M. Paredes
ISBN 1-886733-14-7

Vol. CS-94 MAPPING, MEASURING, AND MODELING THE UNIVERSE
eds. Peter Coles, Vicent J. Martinez, and Maria-Jesus Pons-Borderia
ISBN 1-886733-15-5

Vol. CS-95 SOLAR DRIVERS OF INTERPLANETARY AND TERRESTRIAL DISTURBANCES: Proceedings of 16th International Workshop National Solar Observatory/Sacramento Peak
eds. K. S. Balasubramaniam, Stephen L. Keil, and Raymond N. Smartt
ISBN 1-886733-16-3

Vol. CS-96 HYDROGEN-DEFICIENT STARS
eds. C. S. Jeffery and U. Heber
ISBN 1-886733-17-1

Vol. CS-97 POLARIMETRY OF THE INTERSTELLAR MEDIUM
eds. W. G. Roberge and D. C. B. Whittet
ISBN 1-886733-18-X

ASP CONFERENCE SERIES VOLUMES

Published by the Astronomical Society of the Pacific

PUBLISHED: 1996 (* asterisk means OUT OF STOCK)

Vol. CS-98 FROM STARS TO GALAXIES: THE IMPACT OF STELLAR PHYSICS ON GALAXY EVOLUTION
eds. Claus Leitherer, Uta Fritze-von Alvensleben, and John Huchra
ISBN 1-886733-19-8

Vol. CS-99 COSMIC ABUNDANCES:
Proceedings of the 6th Annual October Astrophysics Conference
eds. Stephen S. Holt and George Sonneborn
ISBN 1-886733-20-1

Vol. CS-100 ENERGY TRANSPORT IN RADIO GALAXIES AND QUASARS
eds. P. E. Hardee, A. H. Bridle, and J. A. Zensus
ISBN 1-886733-21-X

Vol. CS-101 ASTRONOMICAL DATA ANALYSIS SOFTWARE AND SYSTEMS V – (ADASS V)
eds. George H. Jacoby and Jeannette Barnes
ISBN 1080-7926

Vol. CS-102 THE GALACTIC CENTER, 4th ESO/CTIO Workshop
ed. Roland Gredel
ISBN 1-886733-22-8

Vol. CS-103 THE PHYSICS OF LINERS IN VIEW OF RECENT OBSERVATIONS
eds. M. Eracleous, A. Koratkar, C. Leitherer, and L. Ho
ISBN 1-886733-23-6

Vol. CS-104 PHYSICS, CHEMISTRY, AND DYNAMICS OF INTERPLANETARY DUST, IAU Colloquium 150
eds. Bo Å. S. Gustafson and Martha S. Hanner
ISBN 1-886733-24-4

Vol. CS-105 PULSARS: PROBLEMS AND PROGRESS, IAU Colloquium 160
ed. S. Johnston, M. A. Walker, and M. Bailes
ISBN 1-886733-25-2

Vol. CS-106 THE MINNESOTA LECTURES ON EXTRAGALACTIC NEUTRAL HYDROGEN
ed. Evan D. Skillman
ISBN 1-886733-26-0

Vol. CS-107 COMPLETING THE INVENTORY OF THE SOLAR SYSTEM:
A Symposium held in conjunction with the 106th Annual Meeting of the ASP
eds. Terrence W. Rettig and Joseph M. Hahn
ISBN 1-886733-27-9

Vol. CS-108 M.A.S.S. -- MODEL ATMOSPHERES AND SPECTRUM SYNTHESIS:
5th Vienna - Workshop
eds. Saul J. Adelman, Friedrich Kupka, and Werner W. Weiss
ISBN 1-886733-28-7

Vol. CS-109 COOL STARS, STELLAR SYSTEMS, AND THE SUN: Ninth Cambridge Workshop
eds. Roberto Pallavicini and Andrea K. Dupree
ISBN 1-886733-29-5

Vol. CS-110 BLAZAR CONTINUUM VARIABILITY
eds. H. R. Miller, J. R. Webb, and J. C. Noble
ISBN 1-886733-30-9

Vol. CS-111 MAGNETIC RECONNECTION IN THE SOLAR ATMOSPHERE:
Proceedings of a Yohkoh Conference
eds. R. D. Bentley and J. T. Mariska
ISBN 1-886733-31-7

ASP CONFERENCE SERIES VOLUMES

Published by the Astronomical Society of the Pacific

PUBLISHED: 1996 (* asterisk means OUT OF STOCK)

Vol. CS-112 THE HISTORY OF THE MILKY WAY AND ITS SATELLITE SYSTEM
eds. Andreas Burkert, Dieter H. Hartmann, and Steven R. Majewski
ISBN 1-886733-32-5

PUBLISHED: 1997

Vol. CS-113 EMISSION LINES IN ACTIVE GALAXIES: NEW METHODS AND TECHNIQUES, IAU Colloquium 159
eds. B. M. Peterson, F.-Z. Cheng, and A. S. Wilson
ISBN 1-886733-33-3

Vol. CS-114 YOUNG GALAXIES AND QSO ABSORPTION-LINE SYSTEMS
eds. Sueli M. Viegas, Ruth Gruenwald, and Reinaldo R. de Carvalho
ISBN 1-886733-34-1

Vol. CS-115 GALACTIC CLUSTER COOLING FLOWS
ed. Noam Soker
ISBN 1-886733-35-X

Vol. CS-116 THE SECOND STROMLO SYMPOSIUM:
THE NATURE OF ELLIPTICAL GALAXIES
eds. M. Arnaboldi, G. S. Da Costa, and P. Saha
ISBN 1-886733-36-8

Vol. CS-117 DARK AND VISIBLE MATTER IN GALAXIES
eds. Massimo Persic and Paolo Salucci
ISBN-1-886733-37-6

Vol. CS-118 FIRST ADVANCES IN SOLAR PHYSICS EUROCONFERENCE:
ADVANCES IN THE PHYSICS OF SUNSPOTS
eds. B. Schmieder. J. C. del Toro Iniesta, and M. Vázquez
ISBN 1-886733-38-4

Vol. CS-119 PLANETS BEYOND THE SOLAR SYSTEM
AND THE NEXT GENERATION OF SPACE MISSIONS
ed. David R. Soderblom
ISBN 1-886733-39-2

Vol. CS-120 LUMINOUS BLUE VARIABLES: MASSIVE STARS IN TRANSITION
eds. Antonella Nota and Henny J. G. L. M. Lamers
ISBN 1-886733-40-6

Vol. CS-121 ACCRETION PHENOMENA AND RELATED OUTFLOWS, IAU Colloquium 163
eds. D. T. Wickramasinghe, G. V. Bicknell, and L. Ferrario
ISBN 1-886733-41-4

Vol. CS-122 FROM STARDUST TO PLANETESIMALS:
Symposium held as part of the 108th Annual Meeting of the ASP
eds. Yvonne J. Pendleton and A. G. G. M. Tielens
ISBN 1-886733-42-2

Vol. CS-123 THE 12th 'KINGSTON MEETING': COMPUTATIONAL ASTROPHYSICS
eds. David A. Clarke and Michael J. West
ISBN 1-886733-43-0

Vol. CS-124 DIFFUSE INFRARED RADIATION AND THE IRTS
eds. Haruyuki Okuda, Toshio Matsumoto, and Thomas Roellig
ISBN 1-886733-44-9

Vol. CS-125 ASTRONOMICAL DATA ANALYSIS SOFTWARE AND SYSTEMS VI
eds. Gareth Hunt and H. E. Payne
ISBN 1-886733-45-7

ASP CONFERENCE SERIES VOLUMES

Published by the Astronomical Society of the Pacific

PUBLISHED: 1997 (* asterisk means OUT OF STOCK)

Vol. CS-126 FROM QUANTUM FLUCTUATIONS TO COSMOLOGICAL STRUCTURES
eds. David Valls-Gabaud, Martin A. Hendry, Paolo Molaro, and Khalil Chamcham
ISBN 1-886733-46-5

Vol. CS-127 PROPER MOTIONS AND GALACTIC ASTRONOMY
ed. Roberta M. Humphreys
ISBN 1-886733-47-3

Vol. CS-128 MASS EJECTION FROM AGN (Active Galactic Nuclei)
eds. N. Arav, I. Shlosman, and R. J. Weymann
ISBN 1-886733-48-1

Vol. CS-129 THE GEORGE GAMOW SYMPOSIUM
eds. E. Harper, W. C. Parke, and G. D. Anderson
ISBN 1-886733-49-X

Vol. CS-130 THE THIRD PACIFIC RIM CONFERENCE ON
RECENT DEVELOPMENT ON BINARY STAR RESEARCH
eds. Kam-Ching Leung
ISBN 1-886733-50-3

PUBLISHED: 1998

Vol. CS-131 BOULDER-MUNICH II: PROPERTIES OF HOT, LUMINOUS STARS
ed. Ian D. Howarth
ISBN 1-886733-51-1

Vol. CS-132 STAR FORMATION WITH THE INFRARED SPACE OBSERVATORY (ISO)
eds. João L. Yun and René Liseau
ISBN 1-886733-52-X

Vol. CS-133 SCIENCE WITH THE NGST (Next Generation Space Telescope)
eds. Eric P. Smith and Anuradha Koratkar
ISBN 1-886733-53-8

Vol. CS-134 BROWN DWARFS AND EXTRASOLAR PLANETS
eds. Rafael Rebolo, Eduardo L. Martin, and Maria Rosa Zapatero Osorio
ISBN 1-886733-54-6

Vol. CS-135 A HALF CENTURY OF STELLAR PULSATION INTERPRETATIONS:
A TRIBUTE TO ARTHUR N. COX
eds. P. A. Bradley and J. A. Guzik
ISBN 1-886733-55-4

Vol. CS-136 GALACTIC HALOS: A UC SANTA CRUZ WORKSHOP
ed. Dennis Zaritsky
ISBN 1-886733-56-2

Vol. CS-137 WILD STARS IN THE OLD WEST: PROCEEDINGS OF THE 13th NORTH
AMERICAN WORKSHOP ON CATACLYSMIC VARIABLES
AND RELATED OBJECTS
eds. S. Howell, E. Kuulkers, and C. Woodward
ISBN 1-886733-57-0

Vol. CS-138 1997 PACIFIC RIM CONFERENCE ON STELLAR ASTROPHYSICS
eds. Kwing Lam Chan, K. S. Cheng, and H. P. Singh
ISBN 1-886733-58-9

Vol. CS-139 PRESERVING THE ASTRONOMICAL WINDOWS:
Proceedings of Joint Discussion No. 5 of the 23rd General Assembly of the IAU
eds. Syuzo Isobe and Tomohiro Hirayama
ISBN 1-886733-59-7

ASP CONFERENCE SERIES VOLUMES

Published by the Astronomical Society of the Pacific

PUBLISHED: 1998 (* asterisk means OUT OF STOCK)

Vol. CS-140 SYNOPTIC SOLAR PHYSICS --18th NSO/Sacramento Peak Summer Workshop
eds. K. S. Balasubramaniam, J. W. Harvey, and D. M. Rabin
ISBN 1-886733-60-0

Vol. CS-141 ASTROPHYSICS FROM ANTARCTICA:
A Symposium held as a part of the 109th Annual Meeting of the ASP
eds. Giles Novak and Randall H. Landsberg
ISBN 1-886733-61-9

Vol. CS-142 THE STELLAR INITIAL MASS FUNCTION: 38th Herstmonceux Conference
eds. Gerry Gilmore and Debbie Howell
ISBN 1-886733-62-7

Vol. CS-143 * THE SCIENTIFIC IMPACT OF THE GODDARD HIGH RESOLUTION SPECTROGRAPH (GHRS)
eds. John C. Brandt, Thomas B. Ake III, and Carolyn Collins Petersen
ISBN 1-886733-63-5

Vol. CS-144 RADIO EMISSION FROM GALACTIC AND EXTRAGALACTIC COMPACT SOURCES, IAU Colloquium 164
eds. J. Anton Zensus, G. B. Taylor, and J. M. Wrobel
ISBN 1-886733-64-3

Vol. CS-145 ASTRONOMICAL DATA ANALYSIS SOFTWARE AND SYSTEMS VII – (ADASS VII)
eds. Rudolf Albrecht, Richard N. Hook, and Howard A. Bushouse
ISBN 1-886733-65-1

Vol. CS-146 THE YOUNG UNIVERSE GALAXY FORMATION AND EVOLUTION AT INTERMEDIATE AND HIGH REDSHIFT
eds. S. D'Odorico, A. Fontana, and E. Giallongo
ISBN 1-886733-66-X

Vol. CS-147 ABUNDANCE PROFILES: DIAGNOSTIC TOOLS FOR GALAXY HISTORY
eds. Daniel Friedli, Mike Edmunds, Carmelle Robert, and Laurent Drissen
ISBN 1-886733-67-8

Vol. CS-148 ORIGINS
eds. Charles E. Woodward, J. Michael Shull, and Harley A. Thronson, Jr.
ISBN 1-886733-68-6

Vol. CS-149 SOLAR SYSTEM FORMATION AND EVOLUTION
eds. D. Lazzaro, R. Vieira Martins, S. Ferraz-Mello, J. Fernández, and C. Beaugé
ISBN 1-886733-69-4

Vol. CS-150 NEW PERSPECTIVES ON SOLAR PROMINENCES, IAU Colloquium 167
eds. David Webb, David Rust, and Brigitte Schmieder
ISBN 1-886733-70-8

Vol. CS-151 COSMIC MICROWAVE BACKGROUND AND LARGE SCALE STRUCTURES OF THE UNIVERSE
eds. Yong-Ik Byun and Kin-Wang Ng
ISBN 1-886733-71-6

Vol. CS-152 FIBER OPTICS IN ASTRONOMY III
eds. S. Arribas, E. Mediavilla, and F. Watson
ISBN 1-886733-72-4

Vol. CS-153 LIBRARY AND INFORMATION SERVICES IN ASTRONOMY III -- (LISA III)
eds. Uta Grothkopf, Heinz Andernach, Sarah Stevens-Rayburn, and Monique Gomez
ISBN 1-886733-73-2

ASP CONFERENCE SERIES VOLUMES

Published by the Astronomical Society of the Pacific

PUBLISHED: 1998 (* asterisk means OUT OF STOCK)

Vol. CS-154 COOL STARS, STELLAR SYSTEMS AND THE SUN: Tenth Cambridge Workshop
eds. Robert A. Donahue and Jay A. Bookbinder
ISBN 1-886733-74-0

Vol. CS-155 SECOND ADVANCES IN SOLAR PHYSICS EUROCONFERENCE: THREE-DIMENSIONAL STRUCTURE OF SOLAR ACTIVE REGIONS
eds. Costas E. Alissandrakis and Brigitte Schmieder
ISBN 1-886733-75-9

PUBLISHED: 1999

Vol. CS-156 HIGHLY REDSHIFTED RADIO LINES
eds. C. L. Carilli, S. J. E. Radford, K. M. Menten, and G. I. Langston
ISBN 1-886733-76-7

Vol. CS-157 ANNAPOLIS WORKSHOP ON MAGNETIC CATACLYSMIC VARIABLES
eds. Coel Hellier and Koji Mukai
ISBN 1-886733-77-5

Vol. CS-158 SOLAR AND STELLAR ACTIVITY: SIMILARITIES AND DIFFERENCES
eds. C. J. Butler and J. G. Doyle
ISBN 1-886733-78-3

Vol. CS-159 BL LAC PHENOMENON
eds. Leo O. Takalo and Aimo Sillanpää
ISBN 1-886733-79-1

Vol. CS-160 ASTROPHYSICAL DISCS: An EC Summer School
eds. J. A. Sellwood and Jeremy Goodman
ISBN 1-886733-80-5

Vol. CS-161 HIGH ENERGY PROCESSES IN ACCRETING BLACK HOLES
eds. Juri Poutanen and Roland Svensson
ISBN 1-886733-81-3

Vol. CS-162 QUASARS AND COSMOLOGY
eds. Gary Ferland and Jack Baldwin
ISBN 1-886733-83-X

Vol. CS-163 STAR FORMATION IN EARLY-TYPE GALAXIES
eds. Jordi Cepa and Patricia Carral
ISBN 1-886733-84-8

Vol. CS-164 ULTRAVIOLET–OPTICAL SPACE ASTRONOMY BEYOND HST
eds. Jon A. Morse, J. Michael Shull, and Anne L. Kinney
ISBN 1-886733-85-6

Vol. CS-165 THE THIRD STROMLO SYMPOSIUM: THE GALACTIC HALO
eds. Brad K. Gibson, Tim S. Axelrod, and Mary E. Putman
ISBN 1-886733-86-4

Vol. CS-166 STROMLO WORKSHOP ON HIGH-VELOCITY CLOUDS
eds. Brad K. Gibson and Mary E. Putman
ISBN 1-886733-87-2

Vol. CS-167 HARMONIZING COSMIC DISTANCE SCALES IN A POST-HIPPARCOS ERA
eds. Daniel Egret and André Heck
ISBN 1-886733-88-0

Vol. CS-168 NEW PERSPECTIVES ON THE INTERSTELLAR MEDIUM
eds. A. R. Taylor, T. L. Landecker, and G. Joncas
ISBN 1-886733-89-9

ASP CONFERENCE SERIES VOLUMES

Published by the Astronomical Society of the Pacific

PUBLISHED: 1999 (* asterisk means OUT OF STOCK)

Vol. CS-169 11th EUROPEAN WORKSHOP ON WHITE DWARFS
eds. J.-E. Solheim and E. G. Meištas
ISBN 1-886733-91-0

Vol. CS-170 THE LOW SURFACE BRIGHTNESS UNIVERSE, IAU Colloquium 171
eds. J. I. Davies, C. Impey, and S. Phillipps
ISBN 1-886733-92-9

Vol. CS-171 LiBeB, COSMIC RAYS, AND RELATED X- AND GAMMA-RAYS
eds. Reuven Ramaty, Elisabeth Vangioni-Flam, Michel Cassé, and Keith Olive
ISBN 1-886733-93-7

Vol. CS-172 ASTRONOMICAL DATA ANALYSIS SOFTWARE AND SYSTEMS VIII
eds. David M. Mehringer, Raymond L. Plante, and Douglas A. Roberts
ISBN 1-886733-94-5

Vol. CS-173 THEORY AND TESTS OF CONVECTION IN STELLAR STRUCTURE: First Granada Workshop
ed. Álvaro Giménez, Edward F. Guinan, and Benjamín Montesinos
ISBN 1-886733-95-3

Vol. CS-174 CATCHING THE PERFECT WAVE: ADAPTIVE OPTICS AND INTERFEROMETRY IN THE 21st CENTURY,
A Symposium held as a part of the 110th Annual Meeting of the ASP
eds. Sergio R. Restaino, William Junor, and Nebojsa Duric
ISBN 1-886733-96-1

Vol. CS-175 STRUCTURE AND KINEMATICS OF QUASAR BROAD LINE REGIONS
eds. C. M. Gaskell, W. N. Brandt, M. Dietrich, D. Dultzin-Hacyan, and M. Eracleous
ISBN 1-886733-97-X

Vol. CS-176 OBSERVATIONAL COSMOLOGY: THE DEVELOPMENT OF GALAXY SYSTEMS
eds. Giuliano Giuricin, Marino Mezzetti, and Paolo Salucci
ISBN 1-58381-000-5

Vol. CS-177 ASTROPHYSICS WITH INFRARED SURVEYS: A Prelude to SIRTF
eds. Michael D. Bicay, Chas A. Beichman, Roc M. Cutri, and Barry F. Madore
ISBN 1-58381-001-3

Vol. CS-178 STELLAR DYNAMOS: NONLINEARITY AND CHAOTIC FLOWS
eds. Manuel Núñez and Antonio Ferriz-Mas
ISBN 1-58381-002-1

Vol. CS-179 ETA CARINAE AT THE MILLENNIUM
eds. Jon A. Morse, Roberta M. Humphreys, and Augusto Damineli
ISBN 1-58381-003-X

Vol. CS-180 SYNTHESIS IMAGING IN RADIO ASTRONOMY II
eds. G. B. Taylor, C. L. Carilli, and R. A. Perley
ISBN 1-58381-005-6

Vol. CS-181 MICROWAVE FOREGROUNDS
eds. Angelica de Oliveira-Costa and Max Tegmark
ISBN 1-58381-006-4

Vol. CS-182 GALAXY DYNAMICS: A Rutgers Symposium
eds. David Merritt, J. A. Sellwood, and Monica Valluri
ISBN 1-58381-007-2

Vol. CS-183 HIGH RESOLUTION SOLAR PHYSICS: THEORY, OBSERVATIONS, AND TECHNIQUES
eds. T. R. Rimmele, K. S. Balasubramaniam, and R. R. Radick
ISBN 1-58381-009-9

ASP CONFERENCE SERIES VOLUMES

Published by the Astronomical Society of the Pacific

PUBLISHED: 1999 (* asterisk means OUT OF STOCK)

Vol. CS-184 THIRD ADVANCES IN SOLAR PHYSICS EUROCONFERENCE: MAGNETIC FIELDS AND OSCILLATIONS
eds. B. Schmieder, A. Hofmann, and J. Staude
ISBN 1-58381-010-2

Vol. CS-185 PRECISE STELLAR RADIAL VELOCITIES, IAU Colloquium 170
eds. J. B. Hearnshaw and C. D. Scarfe
ISBN 1-58381-011-0

Vol. CS-186 THE CENTRAL PARSECS OF THE GALAXY
eds. Heino Falcke, Angela Cotera, Wolfgang J. Duschl, Fulvio Melia, and Marcia J. Rieke
ISBN 1-58381-012-9

Vol. CS-187 THE EVOLUTION OF GALAXIES ON COSMOLOGICAL TIMESCALES
eds. J. E. Beckman and T. J. Mahoney
ISBN 1-58381-013-7

Vol. CS-188 OPTICAL AND INFRARED SPECTROSCOPY OF CIRCUMSTELLAR MATTER
eds. Eike W. Guenther, Bringfried Stecklum, and Sylvio Klose
ISBN 1-58381-014-5

Vol. CS-189 CCD PRECISION PHOTOMETRY WORKSHOP
eds. Eric R. Craine, Roy A. Tucker, and Jeannette Barnes
ISBN 1-58381-015-3

Vol. CS-190 GAMMA-RAY BURSTS: THE FIRST THREE MINUTES
eds. Juri Poutanen and Roland Svensson
ISBN 1-58381-016-1

Vol. CS-191 PHOTOMETRIC REDSHIFTS AND HIGH REDSHIFT GALAXIES
eds. Ray J. Weymann, Lisa J. Storrie-Lombardi, Marcin Sawicki, and Robert J. Brunner
ISBN 1-58381-017-X

Vol. CS-192 SPECTROPHOTOMETRIC DATING OF STARS AND GALAXIES
ed. I. Hubeny, S. R. Heap, and R. H. Cornett
ISBN 1-58381-018-8

Vol. CS-193 THE HY-REDSHIFT UNIVERSE: GALAXY FORMATION AND EVOLUTION AT HIGH REDSHIFT
eds. Andrew J. Bunker and Wil J. M. van Breugel
ISBN 1-58381-019-6

Vol. CS-194 WORKING ON THE FRINGE: OPTICAL AND IR INTERFEROMETRY FROM GROUND AND SPACE
eds. Stephen Unwin and Robert Stachnik
ISBN 1-58381-020-X

PUBLISHED: 2000

Vol. CS-195 IMAGING THE UNIVERSE IN THREE DIMENSIONS: Astrophysics with Advanced Multi-Wavelength Imaging Devices
eds. W. van Breugel and J. Bland-Hawthorn
ISBN 1-58381-022-6

Vol. CS-196 THERMAL EMISSION SPECTROSCOPY AND ANALYSIS OF DUST, DISKS, AND REGOLITHS
eds. Michael L. Sitko, Ann L. Sprague, and David K. Lynch
ISBN: 1-58381-023-4

Vol. CS-197 XVth IAP MEETING DYNAMICS OF GALAXIES: FROM THE EARLY UNIVERSE TO THE PRESENT
eds. F. Combes, G. A. Mamon, and V. Charmandaris
ISBN: 1-58381-24-2

ASP CONFERENCE SERIES VOLUMES

Published by the Astronomical Society of the Pacific

PUBLISHED: 2000 (* asterisk means OUT OF STOCK)

Vol. CS-198 EUROCONFERENCE ON "STELLAR CLUSTERS AND ASSOCIATIONS: CONVECTION, ROTATION, AND DYNAMOS"
eds. R. Pallavicini, G. Micela, and S. Sciortino
ISBN: 1-58381-25-0

Vol. CS-199 ASYMMETRICAL PLANETARY NEBULAE II: FROM ORIGINS TO MICROSTRUCTURES
eds. J. H. Kastner, N. Soker, and S. Rappaport
ISBN: 1-58381-026-9

Vol. CS-200 CLUSTERING AT HIGH REDSHIFT
eds. A. Mazure, O. Le Fèvre, and V. Le Brun
ISBN: 1-58381-027-7

Vol. CS-201 COSMIC FLOWS 1999: TOWARDS AN UNDERSTANDING OF LARGE-SCALE STRUCTURES
eds. Stéphane Courteau, Michael A. Strauss, and Jeffrey A. Willick
ISBN: 1-58381-028-5

Vol. CS-202 * PULSAR ASTRONOMY – 2000 AND BEYOND, IAU Colloquium 177
eds. M. Kramer, N. Wex, and R. Wielebinski
ISBN: 1-58381-029-3

Vol. CS-203 THE IMPACT OF LARGE-SCALE SURVEYS ON PULSATING STAR RESEARCH, IAU Colloquium 176
eds. L. Szabados and D. W. Kurtz
ISBN: 1-58381-030-7

Vol. CS-204 THERMAL AND IONIZATION ASPECTS OF FLOWS FROM HOT STARS: OBSERVATIONS AND THEORY
eds. Henny J. G. L. M. Lamers and Arved Sapar
ISBN: 1-58381-031-5

Vol. CS-205 THE LAST TOTAL SOLAR ECLIPSE OF THE MILLENNIUM IN TURKEY
eds. W. C. Livingston and A. Özgüç
ISBN: 1-58381-032-3

Vol. CS-206 HIGH ENERGY SOLAR PHYSICS – *ANTICIPATING HESSI*
eds. Reuven Ramaty and Natalie Mandzhavidze
ISBN: 1-58381-033-1

Vol. CS-207 NGST SCIENCE AND TECHNOLOGY EXPOSITION
eds. Eric P. Smith and Knox S. Long
ISBN: 1-58381-036-6

ATLAS VISIBLE AND NEAR INFRARED ATLAS OF THE ARCTURUS SPECTRUM 3727-9300 Å
eds. Kenneth Hinkle, Lloyd Wallace, Jeff Valenti, and Dianne Harmer
ISBN: 1-58381-037-4

Vol. CS-208 POLAR MOTION: HISTORICAL AND SCIENTIFIC PROBLEMS, IAU Colloquium 178
eds. Steven Dick, Dennis McCarthy, and Brian Luzum
ISBN: 1-58381-039-0

Vol. CS-209 SMALL GALAXY GROUPS, IAU Colloquium 174
eds. Mauri J. Valtonen and Chris Flynn
ISBN: 1-58381-040-4

Vol. CS-210 DELTA SCUTI AND RELATED STARS: Reference Handbook and Proceedings of the 6th Vienna Workshop in Astrophysics
eds. Michel Breger and Michael Houston Montgomery
ISBN: 1-58381-043-9

ASP CONFERENCE SERIES VOLUMES

Published by the Astronomical Society of the Pacific

PUBLISHED: 2000 (* asterisk means OUT OF STOCK)

Vol. CS-211 MASSIVE STELLAR CLUSTERS
eds. Ariane Lançon and Christian M. Boily
ISBN: 1-58381-042-0

Vol. CS-212 FROM GIANT PLANETS TO COOL STARS
eds. Caitlin A. Griffith and Mark S. Marley
ISBN: 1-58381-041-2

Vol. CS-213 BIOASTRONOMY `99: A NEW ERA IN BIOASTRONOMY
eds. Guillermo A. Lemarchand and Karen J. Meech
ISBN: 1-58381-044-7

Vol. CS-214 THE Be PHENOMENON IN EARLY-TYPE STARS, IAU Colloquium 175
eds. Myron A. Smith, Huib F. Henrichs and Juan Fabregat
ISBN: 1-58381-045-5

Vol. CS-215 COSMIC EVOLUTION AND GALAXY FORMATION: STRUCTURE, INTERACTIONS AND FEEDBACK
The 3rd Guillermo Haro Astrophysics Conference
eds. José Franco, Elena Terlevich, Omar López-Cruz, and Itziar Aretxaga
ISBN: 1-58381-046-3

Vol. CS-216 ASTRONOMICAL DATA ANALYSIS SOFTWARE AND SYSTEMS IX
eds. Nadine Manset, Christian Veillet, and Dennis Crabtree
ISBN: 1-58381-047-1 ISSN: 1080-7926

Vol. CS-217 IMAGING AT RADIO THROUGH SUBMILLIMETER WAVELENGTHS
eds. Jeffrey G. Mangum and Simon J. E. Radford
ISBN: 1-58381-049-8

Vol. CS-218 MAPPING THE HIDDEN UNIVERSE: THE UNIVERSE BEHIND THE MILKY WAY THE UNIVERSE IN HI
eds. Renée C. Kraan-Korteweg, Patricia A. Henning, and Heinz Andernach
ISBN: 1-58381-050-1

Vol. CS-219 DISKS, PLANETESIMALS, AND PLANETS
eds. F. Garzón, C. Eiroa, D. de Winter, and T. J. Mahoney
ISBN: 1-58381-051-X

Vol. CS-220 AMATEUR - PROFESSIONAL PARTNERSHIPS IN ASTRONOMY: The 111th Annual Meeting of the ASP
eds. John R. Percy and Joseph B. Wilson
ISBN: 1-58381-052-8

Vol. CS-221 STARS, GAS AND DUST IN GALAXIES: EXPLORING THE LINKS
eds. Danielle Alloin, Knut Olsen, and Gaspar Galaz
ISBN: 1-58381-053-6

PUBLISHED: 2001

Vol. CS-222 THE PHYSICS OF GALAXY FORMATION
eds. M. Umemura and H. Susa
ISBN: 1-58381-054-4

Vol. CS-223 COOL STARS, STELLAR SYSTEMS AND THE SUN: Eleventh Cambridge Workshop
eds. Ramón J. García López, Rafael Rebolo, and María Zapatero Osorio
ISBN: 1-58381-056-0

Vol. CS-224 PROBING THE PHYSICS OF ACTIVE GALACTIC NUCLEI BY MULTIWAVELENGTH MONITORING
eds. Bradley M. Peterson, Ronald S. Polidan, and Richard W. Pogge
ISBN: 1-58381-055-2

ASP CONFERENCE SERIES VOLUMES

Published by the Astronomical Society of the Pacific

PUBLISHED: 2001 (* asterisk means OUT OF STOCK)

Vol. CS-225 VIRTUAL OBSERVATORIES OF THE FUTURE
eds. Robert J. Brunner, S. George Djorgovski, and Alex S. Szalay
ISBN: 1-58381-057-9

Vol. CS-226 12th EUROPEAN CONFERENCE ON WHITE DWARFS
eds. J. L. Provencal, H. L. Shipman, J. MacDonald, and S. Goodchild
ISBN: 1-58381-058-7

Vol. CS-227 BLAZAR DEMOGRAPHICS AND PHYSICS
eds. Paolo Padovani and C. Megan Urry
ISBN: 1-58381-059-5

Vol. CS-228 DYNAMICS OF STAR CLUSTERS AND THE MILKY WAY
eds. S. Deiters, B. Fuchs, A. Just, R. Spurzem, and R. Wielen
ISBN: 1-58381-060-9

Vol. CS-229 EVOLUTION OF BINARY AND MULTIPLE STAR SYSTEMS
A Meeting in Celebration of Peter Eggleton's 60th Birthday
eds. Ph. Podsiadlowski, S. Rappaport, A. R. King, F. D'Antona, and L. Burderi
IBSN: 1-58381-061-7

Vol. CS-230 GALAXY DISKS AND DISK GALAXIES
eds. Jose G. Funes, S. J. and Enrico Maria Corsini
ISBN: 1-58381-063-3

Vol. CS-231 TETONS 4: GALACTIC STRUCTURE, STARS, AND THE INTERSTELLAR MEDIUM
eds. Charles E. Woodward, Michael D. Bicay, and J. Michael Shull
ISBN: 1-58381-064-1

Vol. CS-232 THE NEW ERA OF WIDE FIELD ASTRONOMY
eds. Roger Clowes, Andrew Adamson, and Gordon Bromage
ISBN: 1-58381-065-X

Vol. CS-233 P CYGNI 2000: 400 YEARS OF PROGRESS
eds. Mart de Groot and Christiaan Sterken
ISBN: 1-58381-070-6

Vol. CS-234 X-RAY ASTRONOMY 2000
eds. R. Giacconi, S. Serio, and L. Stella
ISBN: 1-58381-071-4

Vol. CS-235 SCIENCE WITH THE ATACAMA LARGE MILLIMETER ARRAY (ALMA)
ed. Alwyn Wootten
ISBN: 1-58381-072-2

Vol. CS-236 ADVANCED SOLAR POLARIMETRY: THEORY, OBSERVATION, AND INSTRUMENTATION, The 20th Sacramento Peak Summer Workshop
ed. M. Sigwarth
ISBN: 1-58381-073-0

Vol. CS-237 GRAVITATIONAL LENSING: RECENT PROGRESS AND FUTURE GOALS
eds. Tereasa G. Brainerd and Christopher S. Kochanek
ISBN: 1-58381-074-9

Vol. CS-238 ASTRONOMICAL DATA ANALYSIS SOFTWARE AND SYSTEMS X
eds. F. R. Harnden, Jr., Francis A. Primini, and Harry E. Payne
ISBN: 1-58381-075-7

Vol. CS-239 MICROLENSING 2000: A NEW ERA OF MICROLENSING ASTROPHYSICS
ed. John Menzies and Penny D. Sackett
ISBN: 1-58381-076-5

ASP CONFERENCE SERIES VOLUMES

Published by the Astronomical Society of the Pacific

PUBLISHED: 2001 (* asterisk means OUT OF STOCK)

Vol. CS-240 GAS AND GALAXY EVOLUTION,
A Conference in Honor of the 20th Anniversary of the VLA
eds. J. E. Hibbard, M. P. Rupen, and J. H. van Gorkom
ISBN: 1-58381-077-3

Vol. CS-241 CS-241 THE 7TH TAIPEI ASTROPHYSICS WORKSHOP ON COSMIC RAYS IN THE UNIVERSE
ed. Chung-Ming Ko
ISBN: 1-58381-079-X

Vol. CS-242 ETA CARINAE AND OTHER MYSTERIOUS STARS: THE HIDDEN OPPORTUNITIES OF EMISSION SPECTROSCOPY
eds. Theodore R. Gull, Sveneric Johannson, and Kris Davidson
ISBN: 1-58381-080-3

Vol. CS-243 FROM DARKNESS TO LIGHT: ORIGIN AND EVOLUTION OF YOUNG STELLAR CLUSTERS
eds. Thierry Montmerle and Philippe André
ISBN: 1-58381-081-1

Vol. CS-244 YOUNG STARS NEAR EARTH: PROGRESS AND PROSPECTS
eds. Ray Jayawardhana and Thomas P. Greene
ISBN: 1-58381-082-X

Vol. CS-245 ASTROPHYSICAL AGES AND TIME SCALES
eds. Ted von Hippel, Chris Simpson, and Nadine Manset
ISBN: 1-58381-083-8

Vol. CS-246 SMALL TELESCOPE ASTRONOMY ON GLOBAL SCALES, IAU Colloquium 183
eds. Wen-Ping Chen, Claudia Lemme, and Bohdan Paczyński
ISBN: 1-58381-084-6

Vol. CS-247 SPECTROSCOPIC CHALLENGES OF PHOTOIONIZED PLASMAS
eds. Gary Ferland and Daniel Wolf Savin
ISBN: 1-58381-085-4

Vol. CS-248 MAGNETIC FIELDS ACROSS THE HERTZSPRUNG-RUSSELL DIAGRAM
eds. G. Mathys, S. K. Solanki, and D. T. Wickramasinghe
ISBN: 1-58381-088-9

Vol. CS-249 THE CENTRAL KILOPARSEC OF STARBURSTS AND AGN: THE LA PALMA CONNECTION
eds. J. H. Knapen, J. E. Beckman, I. Shlosman, and T. J. Mahoney
ISBN: 1-58381-089-7

Vol. CS-250 PARTICLES AND FIELDS IN RADIO GALAXIES CONFERENCE
eds. Robert A. Laing and Katherine M. Blundell
ISBN: 1-58381-090-0

Vol. CS-251 NEW CENTURY OF X-RAY ASTRONOMY
eds. H. Inoue and H. Kunieda
ISBN: 1-58381-091-9

Vol. CS-252 HISTORICAL DEVELOPMENT OF MODERN COSMOLOGY
eds. Vicent J. Martínez, Virginia Trimble, and María Jesús Pons-Bordería
ISBN: 1-58381-092-7

PUBLISHED: 2002

Vol. CS-253 CHEMICAL ENRICHMENT OF INTRACLUSTER AND INTERGALACTIC MEDIUM
eds. Roberto Fusco-Femiano and Francesca Matteucci
ISBN: 1-58381-093-5

ASP CONFERENCE SERIES VOLUMES

Published by the Astronomical Society of the Pacific

PUBLISHED: 2002 (* asterisk means OUT OF STOCK)

Vol. CS-254 EXTRAGALACTIC GAS AT LOW REDSHIFT
eds. John S. Mulchaey and John T. Stocke
ISBN: 1-58381-094-3

Vol. CS-255 MASS OUTFLOW IN ACTIVE GALACTIC NUCLEI: NEW PERSPECTIVES
eds. D. M. Crenshaw, S. B. Kraemer, and I. M. George
ISBN: 1-58381-095-1

Vol. CS-256 OBSERVATIONAL ASPECTS OF PULSATING B AND A STARS
eds. Christiaan Sterken and Donald W. Kurtz
ISBN: 1-58381-096-X

Vol. CS-257 AMiBA 2001: HIGH-*Z* CLUSTERS, MISSING BARYONS, AND CMB POLARIZATION
eds. Lin-Wen Chen, Chung-Pei Ma, Kin-Wang Ng, and Ue-Li Pen
ISBN: 1-58381-097-8

Vol. CS-258 ISSUES IN UNIFICATION OF ACTIVE GALACTIC NUCLEI
eds. Roberto Maiolino, Alessandro Marconi, and Neil Nagar
ISBN: 1-58381-098-6

Vol. CS-259 RADIAL AND NONRADIAL PULSATIONS AS PROBES OF STELLAR PHYSICS, IAU Colloquium 185
eds. Conny Aerts, Timothy R. Bedding, and Jørgen Christensen-Dalsgaard
ISBN: 1-58381-099-4

Vol. CS-260 INTERACTING WINDS FROM MASSIVE STARS
eds. Anthony F. J. Moffat and Nicole St-Louis
ISBN: 1-58381-100-1

Vol. CS-261 THE PHYSICS OF CATACLYSMIC VARIABLES AND RELATED OBJECTS
eds. B. T. Gänsicke, K. Beuermann, and K. Reinsch
ISBN: 1-58381-101-X

Vol. CS-262 THE HIGH ENERGY UNIVERSE AT SHARP FOCUS: CHANDRA SCIENCE, held in conjunction with the 113th Annual Meeting of the ASP
eds. Eric M. Schlegel and Saeqa Dil Vrtilek
ISBN: 1-58381-102-8

Vol. CS-263 STELLAR COLLISIONS, MERGERS AND THEIR CONSEQUENCES
ed. Michael M. Shara
ISBN: 1-58381-103-6

Vol. CS-264 CONTINUING THE CHALLENGE OF EUV ASTRONOMY: CURRENT ANALYSIS AND PROSPECTS FOR THE FUTURE
eds. Steve B. Howell, Jean Dupuis, Daniel Golombek, Frederick M. Walter, and Jennifer Cullison
ISBN: 1-58381-104-4

Vol. CS-265 ω CENTAURI, A UNIQUE WINDOW INTO ASTROPHYSICS
eds. Floor van Leeuwen, Joanne D. Hughes, and Giampaolo Piotto
ISBN: 1-58381-105-2

Vol. CS-266 ASTRONOMICAL SITE EVALUATION IN THE VISIBLE AND RADIO RANGE, IAU Technical Workshop
eds. J. Vernin, Z. Benkhaldoun, and C. Muñoz-Tuñón
ISBN: 1-58381-106-0

Vol. CS-267 HOT STAR WORKSHOP III: THE EARLIEST STAGES OF MASSIVE STAR BIRTH
ed. Paul A. Crowther
ISBN: 1-58381-107-9

Vol. CS-268 TRACING COSMIC EVOLUTION WITH GALAXY CLUSTERS
eds. Stefano Borgani, Marino Mezzetti, and Riccardo Valdarnini
ISBN: 1-58381-108-7

ASP CONFERENCE SERIES VOLUMES

Published by the Astronomical Society of the Pacific

PUBLISHED: 2002 (* asterisk means OUT OF STOCK)

Vol. CS-269 THE EVOLVING SUN AND ITS INFLUENCE ON PLANETARY ENVIRONMENTS
eds. Benjamín Montesinos, Álvaro Giménez, and Edward F. Guinan
ISBN: 1-58381-109-5

Vol. CS-270 ASTRONOMICAL INSTRUMENTATION AND THE BIRTH AND GROWTH OF ASTROPHYSICS: A Symposium held in honor of Robert G. Tull
eds. Frank N. Bash and Christopher Sneden
ISBN: 1-58381-110-9

Vol. CS-271 NEUTRON STARS IN SUPERNOVA REMNANTS
eds. Patrick O. Slane and Bryan M. Gaensler
ISBN: 1-58381-111-7

Vol. CS-272 THE FUTURE OF SOLAR SYSTEM EXPLORATION, 2003-2013
Community Contributions to the NRC Solar System Exploration Decadal Survey
ed. Mark V. Sykes
ISBN: 1-58381-113-3

Vol. CS-273 THE DYNAMICS, STRUCTURE AND HISTORY OF GALAXIES
eds. G. S. Da Costa and H. Jerjen
ISBN: 1-58381-114-1

Vol. CS-274 OBSERVED HR DIAGRAMS AND STELLAR EVOLUTION
eds. Thibault Lejeune and João Fernandes
ISBN: 1-58381-116-8

Vol. CS-275 DISKS OF GALAXIES: KINEMATICS, DYNAMICS AND PERTURBATIONS
eds. E. Athanassoula, A. Bosma, and R. Mujica
ISBN: 1-58381-117-6

Vol. CS-276 SEEING THROUGH THE DUST:
THE DETECTION OF HI AND THE EXPLORATION OF THE ISM IN GALAXIES
eds. A. R. Taylor, T. L. Landecker, and A. G. Willis
ISBN: 1-58381-118-4

Vol. CS 277 STELLAR CORONAE IN THE CHANDRA AND XMM-NEWTON ERA
eds. Fabio Favata and Jeremy J. Drake
ISBN: 1-58381-119-2

Vol. CS 278 NAIC–NRAO SCHOOL ON SINGLE-DISH ASTRONOMY:
TECHNIQUES AND APPLICATIONS
eds. Snezana Stanimirovic, Daniel Altschuler, Paul Goldsmith, and Chris Salter
ISBN: 1-58381-120-6

Vol. CS 279 EXOTIC STARS AS CHALLENGES TO EVOLUTION, IAU Colloquium 187
eds. Christopher A. Tout and Walter Van Hamme
ISBN: 1-58381-122-2

Vol. CS 280 NEXT GENERATION WIDE-FIELD MULTI-OBJECT SPECTROSCOPY
eds. Michael J. I. Brown and Arjun Dey
ISBN: 1-58381-123-0

Vol. CS 281 ASTRONOMICAL DATA ANALYSIS SOFTWARE AND SYSTEM XI
eds. David A. Bohlender, Daniel Durand, and Thomas H. Handley
ISBN: 1-58381-124-9 ISSN: 1080-7926

Vol. CS 282 GALAXIES: THE THIRD DIMENSION
eds. Margarita Rosado, Luc Binette, and Lorena Arias
ISBN: 1-58381-125-7

Vol. CS 283 A NEW ERA IN COSMOLOGY
eds. Nigel Metcalfe and Tom Shanks
ISBN: 1-58381-126-5

ASP CONFERENCE SERIES VOLUMES

Published by the Astronomical Society of the Pacific

PUBLISHED: 2002 (* asterisk means OUT OF STOCK)

Vol. CS 284 AGN SURVEYS
eds. R. F. Green, E. Ye. Khachikian, and D. B. Sanders
ISBN: 1-58381-127-3

Vol. CS 285 MODES OF STAR FORMATION AND THE ORIGIN OF FIELD POPULATIONS
eds. Eva K. Grebel and Walfgang Brandner
ISBN: 1-58381-128-1

PUBLISHED: 2003

Vol. CS 286 CURRENT THEORETICAL MODESL AND HIGH RESOLUTION SOLAR OBSERVATIONS: PREPARING FOR ATST
eds. Alexei A. Pevtsov and Han Uitenbroek
ISBN: 1-58381-129-X

Vol. CS 287 GALACTIC STAR FORMATION ACROSS THE STELLAR MASS SPECTRUM
eds. J.M. De Buizer and N.S. van der Bliek
ISBN:1-58381-130-3

Vol. CS 288 STELLAR ATMOSPHERE MODELING
eds. I. Hubeny, D. Mihalas and K. Werner
ISBN: 1-58381-131-1

Vol. CS 289 THE PROCEEDINGS OF THE IAU 8TH ASIAN-PACIFIC REGIONAL MEETING, VOLUME 1
eds. Satoru Ikeuchi, John Hearnshaw and Tomoyuki Hanawa
ISBN: 1-58381-134-6

Vol. CS 290 ACTIVE GALACTIC NUCLEI: FROM CENTRAL ENGINE TO HOST GALAXY
eds. S. Collin, F. Combes and I. Shlosman
ISBN: 1-58381-135-4

Vol. CS-291 HUBBLE'S SCIENCE LEGACY:
FUTURE OPTICAL/ULTRAVIOLET ASTRONOMY FROM SPACE
eds. Kenneth R. Sembach, J. Chris Blades, Garth D. Illingworth and Robert C. Kennicutt, Jr.
ISBN: 1-58381-136-2

A LISTING OF IAU VOLUMES MAY BE FOUND ON THE NEXT PAGE

INTERNATIONAL ASTRONOMICAL UNION (IAU) VOLUMES

Published by the Astronomical Society of the Pacific

PUBLISHED: 1999 (* asterisk means OUT OF STOCK)

Vol. No. 190 NEW VIEWS OF THE MAGELLANIC CLOUDS
eds. You-Hua Chu, Nicholas B. Suntzeff, James E. Hesser, and David A. Bohlender
ISBN: 1-58381-021-8

Vol. No. 191 ASYMPTOTIC GIANT BRANCH STARS
eds. T. Le Bertre, A. Lèbre, and C. Waelkens
ISBN: 1-886733-90-2

Vol. No. 192 THE STELLAR CONTENT OF LOCAL GROUP GALAXIES
eds. Patricia Whitelock and Russell Cannon
ISBN: 1-886733-82-1

Vol. No. 193 WOLF-RAYET PHENOMENA IN MASSIVE STARS AND STARBURST GALAXIES
eds. Karel A. van der Hucht, Gloria Koenigsberger, and Philippe R. J. Eenens
ISBN: 1-58381-004-8

Vol. No. 194 ACTIVE GALACTIC NUCLEI AND RELATED PHENOMENA
eds. Yervant Terzian, Daniel Weedman, and Edward Khachikian
ISBN: 1-58381-008-0

PUBLISHED: 2000

Vol. XXIVA TRANSACTIONS OF THE INTERNATIONAL ASTRONOMICAL UNION REPORTS ON ASTRONOMY 1996-1999
ed. Johannes Andersen
ISBN: 1-58381-035-8

Vol. No. 195 HIGHLY ENERGETIC PHYSICAL PROCESSES AND MECHANISMS FOR EMISSION FROM ASTROPHYSICAL PLASMAS
eds. P. C. H. Martens, S. Tsuruta, and M. A. Weber
ISBN: 1-58381-038-2

Vol. No. 197 ASTROCHEMISTRY: FROM MOLECULAR CLOUDS TO PLANETARY SYSTEMS
eds. Y. C. Minh and E. F. van Dishoeck
ISBN: 1-58381-034-X

Vol. No. 198 THE LIGHT ELEMENTS AND THEIR EVOLUTION
eds. L. da Silva, M. Spite, and J. R. de Medeiros
ISBN: 1-58381-048-X

PUBLISHED: 2001

IAU SPS ASTRONOMY FOR DEVELOPING COUNTRIES
Special Session of the XXIV General Assembly of the IAU
ed. Alan H. Batten
ISBN: 1-58381-067-6

Vol. No. 196 PRESERVING THE ASTRONOMICAL SKY
eds. R. J. Cohen and W. T. Sullivan, III
ISBN: 1-58381-078-1

Vol. No. 200 THE FORMATION OF BINARY STARS
eds. Hans Zinnecker and Robert D. Mathieu
ISBN: 1-58381-068-4

Vol. No. 203 RECENT INSIGHTS INTO THE PHYSICS OF THE SUN AND HELIOSPHERE: HIGHLIGHTS FROM SOHO AND OTHER SPACE MISSIONS
eds. Pål Brekke, Bernhard Fleck, and Joseph B. Gurman
ISBN: 1-58381-069-2

Vol. No. 204 THE EXTRAGALACTIC INFRARED BACKGROUND AND ITS COSMOLOGICAL IMPLICATIONS
eds. Martin Harwit and Michael G. Hauser
ISBN: 1-58381-062-5

INTERNATIONAL ASTRONOMICAL UNION (IAU) VOLUMES

Published by the Astronomical Society of the Pacific

PUBLISHED: 2001 (* asterisk means OUT OF STOCK)

Vol. No. 205 — GALAXIES AND THEIR CONSTITUENTS AT THE HIGHEST ANGULAR RESOLUTIONS
eds. Richard T. Schilizzi, Stuart N. Vogel, Francesco Paresce, and Martin S. Elvis
ISBN: 1-58381-066-8

Vol. XXIVB — TRANSACTIONS OF THE INTERNATIONAL ASTRONOMICAL UNION REPORTS ON ASTRONOMY
ed. Hans Rickman
ISBN: 1-58381-087-0

PUBLISHED: 2002

Vol. No. 12 — HIGHLIGHTS OF ASTRONOMY
ed. Hans Rickman
ISBN: 1-58381-086-2

Vol. No. 199 — THE UNIVERSE AT LOW RADIO FREQUENCIES
eds. A. Pramesh Rao, G. Swarup, and Gopal-Krishna
ISBN: 58381-121-4

Vol. No. 206 — COSMIC MASERS: FROM PROTOSTARS TO BLACKHOLES
eds. Victor Migenes and Mark J. Reid
ISBN: 1-58381-112-5

Vol. No. 207 — EXTRAGALACTIC STAR CLUSTERS
eds. Doug Geisler, Eva K. Grebel, and Dante Minniti
ISBN: 1-58381-115-X

PUBLISHED: 2003

Vol. XXVA — TRANSACTIONS OF THE INTERNATIONAL ASTRONOMICAL UNION REPORTS ON ASTRONOMY 1999-2002
ed. Hans Rickman
ISBN: 1-58381-137-0

Vol. No. 208 — ASTROPHYSICAL SUPERCOMPUTING USING PARTICLE SIMULATIONS
eds. Junichiro Makino and Piet Hut
ISBN: 1-58381-139-7

Vol. No. 211 — BROWN DWARFS
ed. Eduardo Martín
ISBN: 1-58381-132-X

Vol. No. 212 — A MASSIVE STAR ODYSSEY: FROM MAIN SEQUENCE TO SUPERNOVA
ed. Karel A. van der Hucht, Artemio Herrero and César Esteban
ISBN: 1-58381-133-8

Ordering information is available at the beginning of the listing